# THE TREES OF SONORA, MEXICO

# THE TREES OF SONORA, MEXICO

Richard Stephen Felger

Matthew Brian Johnson

Michael Francis Wilson

OXFORD
UNIVERSITY PRESS
2001

OXFORD
UNIVERSITY PRESS

Oxford New York
Athens Auckland Bangkok Bogatá Buenos Aires Calcutta
Cape Town Chennai Dar es Salaam Delhi Florence Hong Kong Istanbul
Karachi Kuala Lumpar Madrid Melbourne Mexico City Mumbai
Nairobi Paris São Paulo Shanghai Singapore Taipei Tokyo Toronto Warsaw

and associated companies in
Berlin Ibadan



Published by Oxford University Press, Inc.
198 Madison Avenue, New York, New York 10016

Oxford is a registered trademark of Oxford University Press



Library of Congress Cataloging-in-Publication Data
Felger, Richard Stephen.
The trees of Sonora, Mexico / by Richard Stephen Felger, Matthew Brian Johnson, Michael Francis Wilson.
p. cm.
Includes bibliographical references.
ISBN 9780195128918
1. Trees—Mexico—Sonora (State)—Identification. I. Johnson, Matthew Brian II. Wilson, Michael Francis. III. Title.

QK211. F388 2000
582.16'0972'17—dc21 99-053338

# Acknowledgments

This project was made possible by grants and donations to Drylands Institute of Arizona. Primary support came from the Wallace Research Foundation. Additional funds were provided by the National Geographic Society grant no. #6063-97, the Sunny Knickerbocker Foundation, and James Aronson, Gil Gillenwater, Jim Hills, and an anonymous donor.

We thank all who have helped with the project, especially the following: Rita Jo Anthony (field work); James Aronson (encouragment and botanical information); Dan Austin (Convolvulaceae); Marc Baker (Cactaceae); Tom Bowen (field work); Bill Broyles (field work, geography, history, and sustained encouragement); Alberto Búrquez (field work, ecology, geography, and biological, historical, scientific, and technical information as well as sustained encouragement); Fernando Chiang-Cabrera (botanical data); Jane Cole (bibliography); Charles Conner (field work and technical information); Kevin Dahl (field work and manuscript assistance); Mark A. Dimmitt (field work and botanical and ecological information); Amy Evans (field work); Exequiel Ezcurra (field work, ecological and geographical information, and inspiration); George M. Ferguson (field work and botanical information); Lloyd Findley (field work and geographic information); Mark Fishbein (field work, Apocynaceae, and other botanical information); Bernard Fontana (history); Paul Fryxell (Bombacaceae); Judy Gibson (botanical and archival data); Gil Gillenwater, Todd Gillenwater, and Troy Gillenwater (field work); Edward Glenn (field work, ecology of the Río Colorado delta); Julie Hawkins (*Parkinsonia*, Fabaceae); Colin Hughes (Fabaceae); Phil Jenkins (Solanaceae, review of the manuscript, and a great deal of botanical and technical expertise); Gene Joseph (field work); Mark Kaib (field work); Les Landrum (Fagaceae and Myrtaceae); Mervin Larson (information on palms, and hospitality); Linda Leigh (field work and extensive help with the manuscript); Angelina Martínez Yrízar (ecology, geography, and technical assistance); Michael McClure (field work); Lucinda McDade (extensive manuscript review and botanical information); Margaret McClelland (field work); Stephanie Meyer (field work and botanical information); John Palting (botanical information); Bruce Parfitt (Cactaceae and botanical information); William D. Peachey (geology and field observations); Donald Pinkava (Cactaceae); Jon Rebman (Cactaceae); Charlotte Reeder (bibliographic and botanical information); John Reeder (bibliographic and botanical information, and manuscript review); James Riser (conifers); Patricia Rorabaugh (encouragement and support); Sue Rutman (field work and technical information); Andrew Salywon (field work); Dean Saxton (O'odham names); Silke Schneider (field work, encouragement, and tolerance); Richard Spellenberg (Fagaceae and botanical information); Victor Steinmann (field work, Euphorbiaceae, and botanical information); Laurence Toolin (manuscript review); Dale S. Turner (field work and manuscript assistance); Raymond M. Turner (ecological and geographical information, and sharing illustrations); Tom Van Devender (manuscript review, and extensive botanical, ecological and paleo-botanical information); John Wiens (plant material); George Yatskievych (extensive botanical information and manuscript review); David Yetman (field work and geography); Allan Zimmerman (Cactaceae and other taxa).

We especially thank the curators and staff of the various herbaria who provided facilities, hospitality, information, and loans of specimens. In this regard we thank researchers and staff of the herbaria at the Arizona State University, California Academy of Sciences, Desert Botanical Garden, Jepson Herbarium, Instituto de Biología de la Universidad Nacional Autónoma de México, Missouri Botanical Garden, Rancho Santa Ana Botanic Garden, San Diego Museum of Natural History, University of Texas, and University of California, Berkeley.

Institutions and presses that have granted permission for use of illustrations include the Hunt Institute for Botanical Documentation, Stanford University Press, and The University of Arizona Press. Stanford University Press gave permission to use illustrations from the *Cacti of the United States and Canada*, by Lyman Benson (1982), *Flora of Baja California*, by Ira L. Wiggins (1980) and the *Illustrated Flora of the Pacific States*, vols. 3, by LeRoy Abrams (1951), copyright by the Board of Trustees of the Leland Stanford Junior University.

James Henrickson has given unwavering encouragement and an enormous wealth of botanical information, shared numerous illustrations by Bobbi Angell and Felicia Bond, and lent his monumental and extremely useful unpublished *Flora of the Chihuahuan Desert Region* (Henrickson and M. C. Johnston, unpublished ms). A major part of our work was carried out at The University of Arizona herbarium, and we sincerely thank Phil Jenkins, Lucinda McDade, Charlotte Reeder, and John Reeder, and other staff members, volunteers, and herbarium scholars for their generous help.

## Illustrators

Nancy Bartels
Victor Rudolph Beneska
Felicia Bond
Kathie Stackpole Bunnell
David Burckhalter
Alberto Búrquez
M. Ornelas Carillo
Eleanor Catherine
Rosa Maria Diez
Mark A. Dimmitt
Jack W. Dykinga
Elvia Esparza
Matilda Essig
Charles Edward Faxon
Richard Stephen Felger
George M. Ferguson
Joel Floyd
Lucretia Brezeale Hamilton
William Hartman
Rodney Hastings
Leta Hughey
A. Kristin Jakob
Matthew B. Johnson
Paul Mirocha
C. Allan Morgan
Cathy Moser Marlett
E. Tad Nichols
Nancy L. Nicholson
Manuel Ornelas Carillo
Francis Runyan
George Russell Shaw
Homer Shantz
Dale S. Turner
Raymond M. Turner
Merlin D. Tuttle
Nancy Evans Weaver
Ira L. Wiggins
Rosemary Wise
David Yetman

The Illustrator of each figure is noted in its caption.

# Contents

# THE TREES OF SONORA, MEXICO

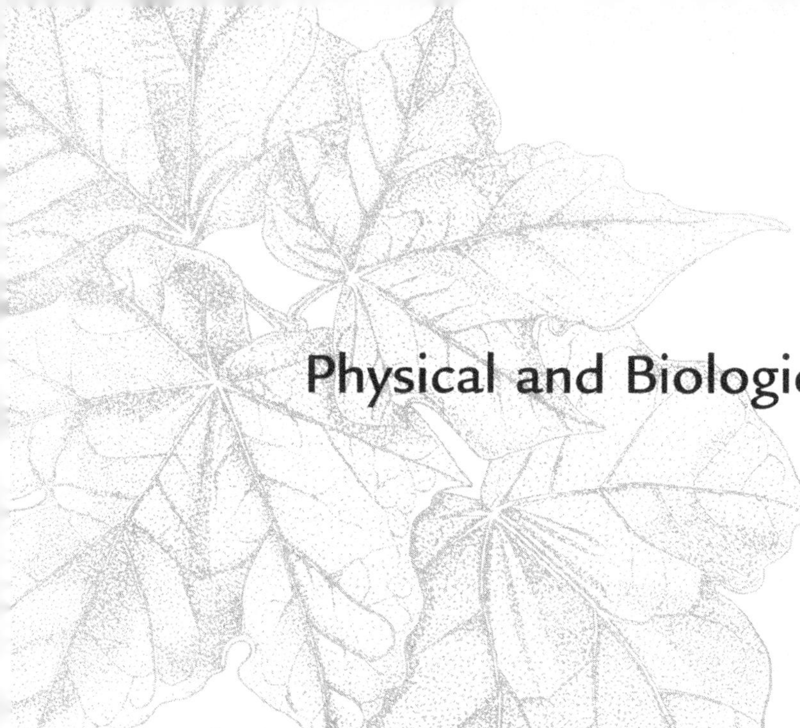

# Physical and Biological Setting

Trees are the cornerstone life form in the vegetation of Sonora. They are the species that affect and influence the survival of other species in the ecosystem, whether plants, animals, or people. Trees grow virtually everywhere in the state except in the extremely arid sandy desert bordering the Río Colorado. Across the deserts and the grasslands, trees cover a minority of the land surface, but perhaps more than half of the state is (or was) forested with trees, including large evergreen trees in the more tropical canyons. This book covers all of the tree species known to us that grow naturally in Sonora.

An estimated 5000 species of vascular plants are present in Sonora (e.g., Felger and Wilson 1995). Among this rich flora about 285 species, nearly 6% of the total estimated flora, are trees or treelike species (see the Appendix). These tree species are distributed among 160 genera in 69 families. Some barely enter Sonora, or are usually shrubs and only rarely develop into trees. A few are known to grow so close to the state boundary that they doubtless occur in the state. Excluding species that are geographically marginal, usually shrubby, or non-native, there are about 200 common or widespread Sonoran tree species.

Sonora encompasses 185,430 $km^2$ and is the second largest state in Mexico after Chihuahua (figure 1). It is topographically and biologically diverse. Four major river systems, the Río Colorado, the Río Yaqui, the Río Mayo, and tributaries of the Río Fuerte, course through the state and empty into the Gulf of California. In addition, several other minor rivers originate in the state, such as the Río Sonoyta and the Río Sonora. The eastern part of the state is composed of numerous north-south–trending mountain ranges, forming the western cordillera of the Sierra Madre Occidental. The highest elevation is 2625 m in the Sierra de los Ajos in northeastern Sonora. The Sierra San Luis in the extreme northeastern corner is nearly as high, and another high peak, farther south on the Chihuahua border near Mesa Tres Rios, may be equally high. Numerous other ranges, generally decreasing in elevation westward, spread across the rest of the state, interspersed with broad valleys and expansive plains.

A land of extremes, the state of Sonora includes the northern limits of tropical deciduous forest and thornscrub as well as some of the most arid desert regions of North America. The political borders of Sonora are, to varying degrees, biologically significant. The western boundary is the sea. The eastern boundary is close to the continental divide. The northern boundary marks a division between the better known and more continental flora of Arizona and the poorly known flora of northern Sonora. The Sonora–Sinaloa border separates a continuous flora but is far enough south to include the northern limits of tropical Mexican flora in southern Sonora.

Sonora's location makes it a transition zone for faunas and floras, especially those of tropical affinity. Many trees in the state represent extensions of wide-ranging species that occur far to the south in Latin America. Sixteen families, 78 genera, and 183 species of trees are at their northern range limits in the Americas or on the Pacific slope of the continent (table 1). Fully 64% of the tree species of Sonora range no farther north in western North America. In contrast, only 2 genera and 30 species and infraspecific taxa of trees reach their southern limits in Sonora (table 2). Very few tree species are endemic to Sonora (e.g, *Caesalpinia caladenia*, *Schoepfia shreveana*, and *Yucca declinata*). *Acacia willardiana*, *Leucaena involucrata*, and *Sabal uresana* scarcely extend outside of the state. Several Sonoran trees species are shared with the Baja California peninsula: *Fouquieria columnaris*, *F. diguetii*, *Lysiloma candidum*, *Parkinsonia ×sonorae*, *Sideroxylon occidentale*, *Washingtonia robusta*, and *Populus brandegeei*.

Mexican governmental agencies list various conservation status categories for certain Sonoran tree species (e.g.,

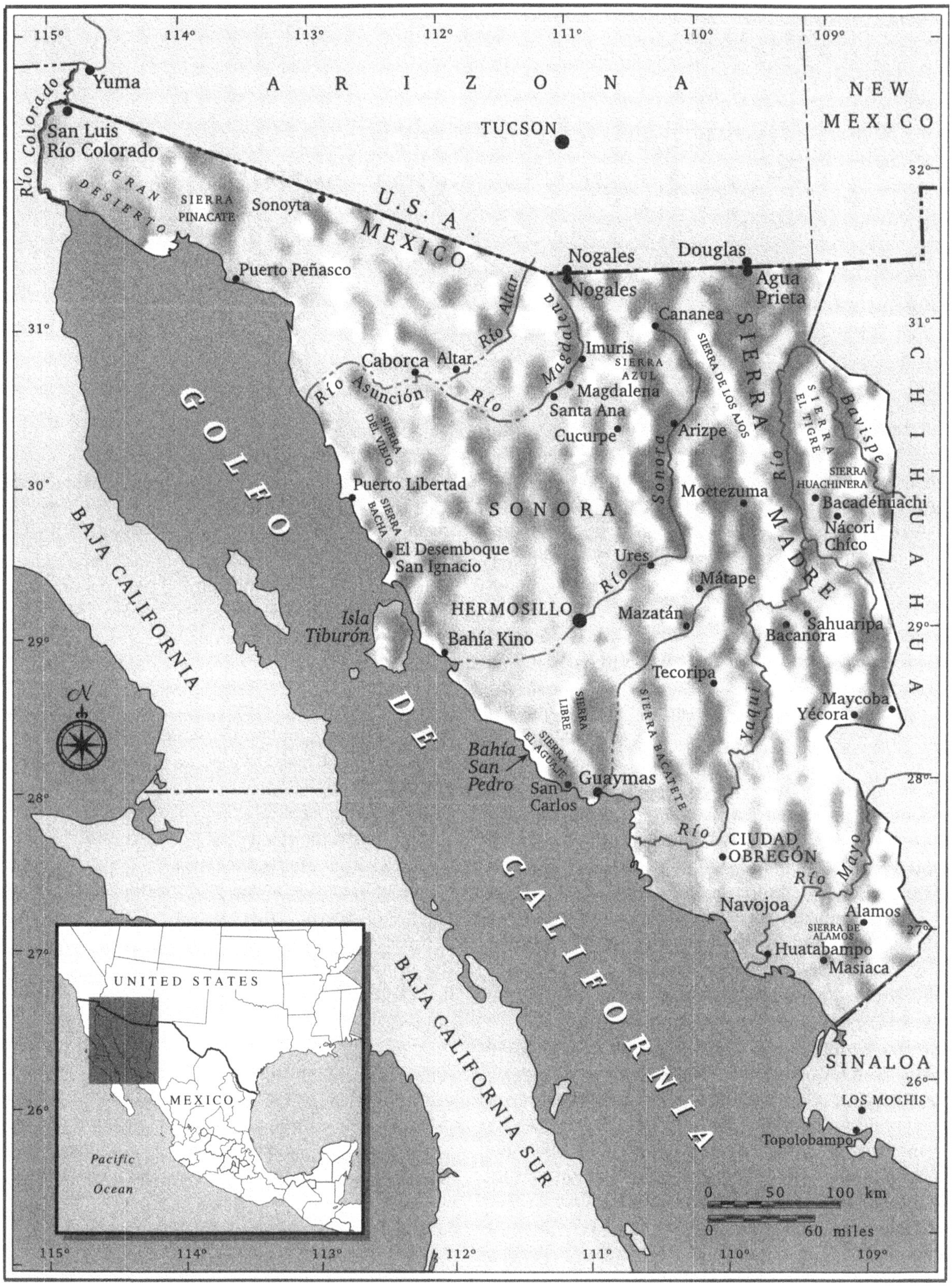


Figure 1. Map of Sonora, Mexico. Drawn by P. Mirocha.

Table 1. Trees reaching their northern limits on the Pacific slope of North America in Sonora. Genera and families at their northern limits indicated in bold. *Non-native species.

**CONIFERS**

Cupressaceae

*Cupressus lusitanica*
*Juniperus durangensis*
***Taxodium** distichum*

Pinaceae

*Abies durangensis*
*Pinus herrerae*
*P. maximinoi*
*P. oocarpa*
*P. yecorensis*

**DICOTS**

Anacardiaceae

*****Spondias** purpurea*

Apocynaceae

***Plumeria** rubra*
***Stemmadenia** tomentosa*
***Vallesia** glabra*
*V. laciniata*

**Aquifoliaceae**

***Ilex** rubra*
*I. tolucana*

Araliaceae

***Oreopanax** peltatum*

Asteraceae

***Montanoa** tomentosa*
*Parthenium tomentosum*

**Avicenniaceae**

***Avicennia** germinans*

Berberidaceae

*Berberis longipes*

Bignoniaceae

*****Crescentia** alata*
***Tabebuia** chrysantha*
*T. impetiginosa*

**Bombacaceae**

***Ceiba** acuminata*
***Pseudobombax** palmeri*

Boraginaceae

***Cordia** sonorae*

Buddlejaceae

*Buddleja cordata*
*B. parviflora*

Burseraceae

*Bursera grandifolia*
*B. hindsiana*
*B. lancifolia*
*B. laxiflora*
*B. penicillata*
*B. simaruba*
*B. stenophylla*

Cactaceae

*Opuntia thurberi*
*O. wilcoxii*
*Pachycereus pecten-aboriginum*
*P. pringlei*
***Pilosocereus** alensis*
*Stenocereus montanus*

Capparaceae

***Capparis** flexuosa*
***Crateva** palmeri*
***Forchhammeria** watsonii*

Celastraceae

***Maytenus** phyllanthoides*
***Wimmeria** mexicana*

**Clethraceae**

***Clethra** mexicana*

Cochlospermaceae

***Cochlospermum** vitifolium*

**Combretaceae**

***Laguncularia** racemosa*

Convolvulaceae

*Ipomoea arborescens*

Cornaceae

*Cornus disciflora*

**Ebenaceae**

***Diospyros** sonorae*

**Erythroxylaceae**

***Erythroxylum** mexicanum*

Euphorbiaceae

***Adelia** virgata*
*Croton fantzianus*
***Drypetes** gentryi*
*****Hura** polyandra*
*Jatropha cordata*
*J. malacophylla*
*Manihot aesculifolia*
*Manihot* sp.
*Sebastiania cornuta*
*S. pavoniana*

Fabaceae

*Acacia cochliacantha*
*A. coulteri*
*A. occidentalis*
*Acacia* cf. *pacensis*
*A. pennatula*
*A. pringlei*
*A. willardiana*
***Albizia** sinaloënsis*
***Bauhinia** pringlei*
***Brongniartia** alamosana*
*Caesalpinia caladenia*
*C. palmeri*
*C. platyloba*
*C. sclerocarpa*
***Chloroleucon** mangense*
***Conzattia** multiflora*
***Diphysa** occidentalis*
*D. suberosa*
***Haematoxylum** brasiletto*
***Havardia** mexicana*
*H. sonorae*
***Leucaena** involucrata*
*L. lanceolata*
***Lonchocarpus** hermannii*
*Lysiloma candidum*
*L. divaricatum*
*Mimosa palmeri*
*Parkinsonia praecox*
*P.* x *sonorae*
***Piscidia** mollis*
***Pithecellobium** dulce*
***Platymiscium** trifoliolatum*
*Prosopis articulata*
*Senna atomaria*

(*continued*)

Table 1. (*continued*)

Fagaceae

*Quercus albocincta*
*Q. chihuahuensis*
*Q. coccolobifolia*
*Q. durifolia*
*Q. mcvaughii*
*Q. perpallida*
*Q. scytophylla*
*Q. sideroxyla*
*Q. subspathulata*
*Q. tarahumara*
*Q. tuberculata*
*Q. viminea*

**Flacourtiaceae**

***Xylosma** flexuosum*
*Xylosma* sp.

Fouquieriaceae

*Fouquieria macdougalii*

Lauraceae

***Cinnamomum** hartmanii*
***Persea** podadenia*

**Magnoliaceae**

***Magnolia** pacifica*

Malpighiaceae

***Bunchosia** sonorensis*
***Malpighia** emarginata*

**Meliaceae**

***Cedrela** odorata*
***Trichilia** americana*
*T. hirta*

Moraceae

***Ficus** cotinifolia*
*F. insipida*
*F. maxima*
*F. pertusa*
*F. petiolaris*
*F. trigonata*
***Maclura** tinctoria*
***Trophis** racemosa*

**Myrsinaceae**

***Ardisia** revoluta*
***Myrsine** coriacea*

**Myrtaceae**

***Psidium** guajava*
*P. sartorianum*

Nyctaginaceae

*Pisonia capitata*

**Olacaceae**

***Schoepfia** schreberi*
*S. shreveana*

**Opiliaceae**

***Agonandra** racemosa*

Polygonaceae

***Coccoloba** goldmanii*
***Ruprechtia** fusca*

Rhamnaceae

***Colubrina** triflora*
***Karwinskia** humboldtiana*
*Rhamnus scopulorum*
*Ziziphus amole*

**Rhizophoraceae**

***Rhizophora** mangle*

Rosaceae

*Prunus gentryi*
*P. zinggii*

Rubiaceae

***Hamelia** xorullensis*
***Hintonia** latiflora*
***Randia** echinocarpa*
*R. laevigata*
*R. obcordata*

Rutaceae

***Amyris** balsamifera*
***Casimiroa** edulis*
***Esenbeckia** hartmanii*
***Zanthoxylum** fagara*

Salicaceae

*Populus brandegeei*
*P. mexicana*

Sapindaceae

*Sapindus saponaria*
***Thouinia** acuminata*
*T. villosa*

Sapotaceae

*Sideroxylon occidentale*
*S. persimile*
*S. tepicense*

Simaroubaceae

***Alvaradoa** amorphoides*

Solanaceae

***Cestrum** tomentosum*
*Solanum erianthum*

Sterculiaceae

***Guazuma** ulmifolia*

**Theophrastaceae**

***Jacquinia** macrocarpa*

Tiliaceae

***Heliocarpus** attenuatus*
*H. palmeri*

Ulmaceae

***Aphananthe** monoica*
*Celtis iguanea*

Urticaceae

***Urera** cf. caracasana*

Verbenaceae

***Citharexylum** scabrum*
*Lippia umbellata*
***Vitex** mollis*
*V. pyramidata*

Zygophyllaceae

***Guajacum** coulteri*

**MONOCOTS**

Agavaceae

*Yucca declinata*
*Y. grandiflora*

Arecaceae

*Brahea aculeata*
*B. brandegeei*
*B. dulcis*
*B. nitida*
***Sabal** uresana*
*Washingtonia robusta*

Nolinaceae

*Nolina matapensis*

Table 2. Tree species and infraspecific taxa reaching their southern limits on the Pacific slope of North America in Sonora. Genera at their southern limits indicated in bold.

**CONIFERS**

Cupressaceae

*Juniperus scopulorum*

Pinaceae

*Abies concolor*

**DICOTS**

Aceraceae

*Acer grandidentatum*

Betulaceae

*Alnus oblongifolia*

Bignoniaceae

*Chilopsis linearis* subsp. *arcuata*

Cactaceae

***Carnegiea*** *gigantea*
*Opuntia fulgida* var. *fulgida*
*O. versicolor*

Celastraceae

***Canotia*** *holacantha*

Fabaceae

*Parkinsonia microphylla*
*Prosopis pubescens*
*P. velutina*
*Psorothamnus spinosus*
*Robinia neomexicana*

Fagaceae

*Quercus chrysolepis*
*Q. emoryi*
*Q. gambelii*

Fouquieriaceae

*Fouquieria splendens*

Oleaceae

*Fraxinus gooddingii*

Rosaceae

*Vauquelinia californica* subsp. *pauciflora*
*V. californica* subsp. *sonorensis*

Salicaceae

*Populus angustifolia*
*P. fremontii*

Sapindaceae

*Sapindus drummondii*

Sapotaceae

*Sideroxylon lanuginosum* subsp. *rigidum*

Simaroubaceae

*Castela emoryi*

**MONOCOTS**

Agavaceae

*Yucca arizonica*
*Y. elata*
*Y. schottii*

Nolinaceae

*Nolina bigelovii*

SEDESOL 1994). As additional information becomes available and the threats of habitat loss and harvesting escalate, the conservation status for many tree species will undoubtedly change rapidly.

Sonoran climate grades from nearly frost-free, dry tropical to temperate and from moist canyons and mountain tops to extremely arid desert. The eastern and higher elevations of Sonora are semiarid, and the land becomes increasingly arid northward and westward. In general, the total annual precipitation generally decreases from higher to lower elevation, from south to north, and from east to west. Toward the drier regions the rainfall becomes increasingly sporadic and unpredictable. Climatic data for 10 localities in Sonora are summarized in figure 2.

In pine-oak forest regions such as at Yécora, at 1550 m, average annual precipitation is more than 900 mm. However, annual precipitation statewide can be highly variable; at Yécora it has varied from 0.5 m to nearly 1.5 m. Tropical deciduous forest at Alamos, at 400 m, receives about 645 mm annually. Coastal thornscrub of less than 25 m elevation in the southwestern part of the state may have an annual rainfall of about 280 mm (Friedman 1996). Average rainfall in the Pinacate region in the northwestern part of the state may be about 250 mm per year, and the extremely arid Gran Desierto near the Río Colorado may have less than 40 mm per year. Especially in the Gran Desierto there is great annual variation, and in the Sierra del Rosario one two-and-a-half year period passed with no recorded rainfall (May 1973; see *Nolina bigelovii*). The driest areas of the Gran Desierto are treeless but abut the Río Colorado, which once supported extensive riverine gallery forest.

Summertime weather of the higher sky-island forests is generally pleasant. The higher mountains, especially in northeastern Sonora, may experience several snowstorms annually, and snow may remain for several weeks. Farther south, such as in the Yécora region, there likewise may be several snowstorms during the winter, but the snow melts quickly. Snow is virtually unknown across the lowlands. Along the coast, maritime dew and occasional fog can be an important moisture source during the cooler months (see *Fouquieria columnaris*).

At mid- to lower elevations, the summers are long and very hot, but for the rest of the year the weather is gener-

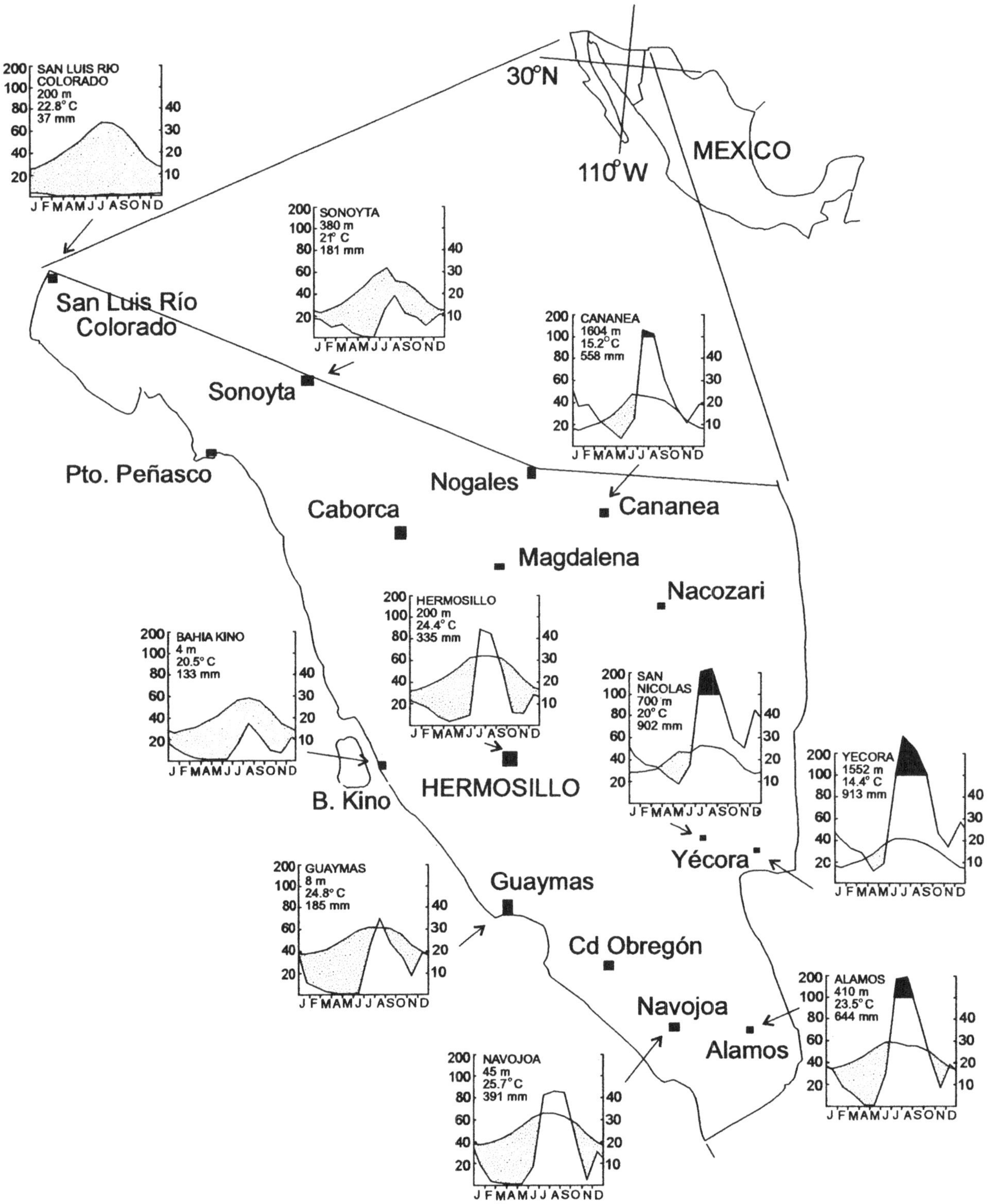


Figure 2. Ombrothermal climograms for selected localities in Sonora. Solid lines = mean monthly temperature and mean monthly precipitation. Precipitation values on left vertical scale, temperature on right. Locality, elevation, mean annual temperature, and mean annual precipitation are at the upper left corner of each diagram. The gray areas indicate periods of water deficit and the black areas indicate water surplus. After Búrquez et al. (1999). Drawn by A. Búrquez.

ally pleasantly warm and mild during the daytime. The dry season of late spring and early summer brings stifling heat and intense sunlight, and many animals and plants are inactive. During May, June and early July, before the summer rains, temperatures are oppressive, often exceeding 40°C. In the lowlands the onset of summer rain does little to ameliorate the heat, and increased humidity can be hard on people but beneficial for the plant life: "The nights fail utterly to refresh after the stultifying closeness of the days. In Navojoa (coastal thornscrub), for example, the July and August average 24-hour temperature is 35°C" (Martin et al. 1998: 23).

Similar climate prevails in the lowlands farther north, although the extremes can be more pronounced. In the Gran Desierto in the northwestern corner of the state, average maximum daily temperature away from the coast exceeds 38°C during June, July, and August, and temperatures exceeding 38°C are common from late April to early October. The highest temperature reported for anywhere in northwestern Mexico was 56.7°C in the Sierra Blanca in the Pinacate region (May 1973) in late June 1971.

The short, monsoonlike, summer rainy season is most pronounced and dependable toward the south and east. Summer rains, called *las aguas*, are characterized by thunderstorms, often local in extent, and may bring large quantities of rain in a few hours or less. The summer rains are especially crucial for most species of tropical origin. In the wettest montane regions in the southeastern part of the state, the summer rains may begin around the summer solstice, or San Juan's day (June 24). In the drier regions the summer rains might not begin until mid-July, or they may fail to occur in some years. The spring drought often ends abruptly with summer rainfall, and there is a rapid transformation of singed brown plains, valleys, and hills to verdant green. For the native Sonoran people, the new year or calendar commences with the summer rains, the time of greatest renewal of life in this part of the world (Felger and Moser 1985). In some years hurricane fringe-storms in late summer and early fall may bring significant quantities of rain in a short time.

Winter rains, called *equipatas*, are derived from Pacific frontal storms. They are extensive in area and tend to deliver gentle rains of relatively long duration. Winter-spring rains become increasingly important toward the northwestern part of the state and generally decrease to the southeast. There is a corresponding shift from plants and animals dependent on hot season rains in the subtropical southern part of the state, to a biota adapted to cool season rains in the northwestern portion of the desert. In the northwestern deserts the majority of the meager rainfall occurs during winter and spring.

While soil moisture is the usual principal limiting factor in this dry region, freezing temperatures limit the northern distributions of many of the more tropical or subtropical species. Freezing weather is infrequent toward the southern part of the state at lower elevations where many habitats are essentially frost-free. In general, frost is more severe to the north and at higher elevations, but even areas such as Alamos and environs in the southeast have endured winter frosts.

Catastrophic freezes with widespread damage to vegetation occur about every decade or so (Bowers 1980). Winter penetration of arctic air masses can be sudden, and damage can be especially severe when freezing weather follows relatively warm periods. The catastrophic freeze of December 5–7, 1978, resulted in severe damage to many usually hardy trees in Sonora. For example, many mesquite trees (*Prosopis velutina*) and Mexican blue oaks (*Quercus oblongifolia*) in northeastern Sonora and southeastern Arizona froze to the ground, and some were killed. A minimum temperature of −18°C was recorded at Sierra Vista in southeastern Arizona, and similar temperatures occurred in adjacent Sonora. Cliff and rock faces, canyons, and other shelter including the closed canopy of tropical deciduous forest ameliorate the effects of frost, but plants in exposed places may be mortally damaged.

Convoluted topography contributes to microclimate effects that lead to patchy distributions for many trees. Furthermore, steep elevational gradients, particularly along the Sierra Madre Occidental, create unusual assemblages of temperate plants intermixed with more tropical species (e.g., see *Acer grandidentatum*). In the north, evidence of repeated frosts is common, and consequently the stature and growth pattern of many species may be profoundly affected (e.g., see *Erythrina flabelliformis*, *Lysiloma watsonii*, and *Stenocereus thurberi*). Many of the plants of tropical origin in northwestern Mexico are highly frost sensitive, while their close relatives from comparable habitats in northeastern Mexico are significantly more frost hardy.

We have arbitrarily selected general heights to delimit small trees (5–8 m tall), medium-sized trees (9–18 m), and large trees (19 m or more). The demarcation between large shrubs and smaller trees is highly subjective, especially in an arid region like Sonora. We have chosen the 5 m height as the artificial limit between shrubs and trees, and when in doubt have favored including woody plants and excluding the more herbaceous ones with seasonal die-back. By our definition, a tree should have a single trunk at least 10 cm in diameter at about 1 m or more above ground level. Furthermore, we have decided that to qualify as a tree, it should be climbable without collapsing, though in practice thorns and spines often prevented field tests of this characteristic. Some species included in this listing have multiple trunks arising at or near the ground and could be classified as shrubs rather than trees. In some cases a particular species may be a shrub across most of its distribution, but in favorable habitats (such as moist canyons) may develop into a sizable tree. Other species are shrublike in the northern part of the state and treelike in the more tropical southern parts of the state.

Very large trees, 20–30 m tall, occur (or formerly occurred) scattered through much of the pine forest and mixed conifer forest. Such large trees also are (or were) represented by a number of nonconifer species at lower elevations but mostly in riparian habitats. Many of the tree species found in Sonora, especially those of tropical affinity, reach larger sizes southward in moister and more tropical regions.

Tree cover has been rapidly receding worldwide as well as in Sonora at the destructive hand of humans. The human population in Sonora remained low, and major roads were few until the latter part of the twentieth century. Much diversity and richness of near natural habitat remains, but assaults on natural habitats are escalating by way of river dams, logging, firewood cutting, charcoal production, urbanization, and clearing for agriculture, especially for buffelgrass (*Pennisetum ciliare*) pasture.

In some ways the plant life of Sonora is better known than that of many other parts of Latin America, but there are enormous gaps. There is no overall work on the flora of the state of Sonora, and much of the nondesert area lacks regional floristic coverage. Principal works on the flora and vegetation of Sonora include Brown (1982), Búrquez et al. (1992), Búrquez et al. (1999), Ezcurra et al. (1987), Felger (1980, 1992, 1999, 2000), Felger and Lowe (1976), Felger et al. (1992, 1997), Felger and Johnson (1995), Felger and Wilson (1995), Friedman (1996), Gentry (1942, 1949), Hastings et al. (1972), Johnston (1924), Laferrière (1994), Marshall (1957), Martin et al. (1998), Mearns (1907), Robichaux (1999), Shreve and Wiggins 1964, Turner et al. (1995), Van Devender et al. (2000), White (1948). Other selected floristic works, relevant to the Sonoran region, include Benson and Darrow (1981), Gentry (1946a, 1946b, 1948), Henrickson and Johnston (*Flora of the Chihuahuan Desert Region*, unpublished ms.), Hickman (1993), Johnston (1990), Kearney and Peebles (1960), Little (1950, 1953, 1976, 1979), McVaugh (1974, 1987, 1992), Munz (1968, 1974), Pennington and Sarukhán (1968, 1998), Pesman (1962), Powell (1988), Rzedowski (1978), Spellenberg et al. (1996), Standley (1920–1926), and Wiggins (1980).

The major vegetation types in Sonora are mangrove, tropical deciduous forest, thornscrub, Sonoran desertscrub, Chihuahuan desertscrub, grassland, oak forest and woodland, pine-oak forest and woodland, and mixed conifer forest. These are briefly discussed below (also see figure 3 and the Appendix).

## Mangroves

Mangroves are colonizing species with large geographic ranges fringing tropical coasts in protected waters around the world. Along the shores of Sonora and Baja California mangroves seldom exceed 5 m in height. Their size gradually increases southward, and in the humid tropics these trees can reach 25 m or more in height. In western North America mangroves reach their northern limits along the shores of the Gulf of California. South of the Guaymas Region, south of the desert, mangroves extend in a broad and perhaps once nearly unbroken coastal band southward into the tropics; these are the most extensive mangroves in Sonora and occur primarily in the tidal inlets and mudflats on the inland side of low, sandy barrier islands.

North of Guaymas mangroves occur in disjunct pockets where protected bays and inlets (*esteros*) provide a suitable habitat. The northernmost mangroves in Sonora are at Estero Sargento at the north end of the Canal del Infiernillo (figure 4), and there is a small northern outlier population at Puerto Lobos (see *Avicennia germinans*). Occasional winter freezing is the obvious factor limiting the northern distribution. Mangroves at Estero Sargento periodically suffer from severe freeze damage.

In southwestern Sonora the alluvial plain gradually dips into the sea and is bordered by many sandbar islands. These low-lying, shifting habitats provide extensive and protected tidal waters supporting mangroves. The greatest stands fringe the coastal thornscrub in areas once profoundly influenced by the Río Yaqui, Río Mayo, and the Río Fuerte with its delta in northernmost Sinaloa and other lesser drainages. The destruction of the deltas of these rivers due to upriver dams, has undoubtedly affected the mangroves.

From at least Guaymas northward in Sonora the mangroves grow in hypersaline (saltier than the sea) water. In contrast, in the tropics with increased rainfall and freshwater runoff from rivers and the land, the mangrove waters become increasingly brackish. The Gulf of California mangroves grow in muddy-sandy substrate in shallow waters which drain and fill daily with the tides. They do not tolerate stagnant water and soon perish if cut off from tidal circulation.

Partly because they harbor the unpleasant biting midge fly or "no-see 'um" called *jejen* (Ceratopogonidae: *Leptoconops* spp.), mangroves have been largely eliminated from places where people live such as Guaymas harbor and Bahía San Carlos. Mangroves provide crucial habitat and nutrients for fish and other marine organisms and for many species of birds. These wetland habitats deserve vigorous protection, and there is an urgent need for additional studies of the Gulf of California mangroves.

The Sonoran mangrove vegetation is made up of three arborescent species in three different families: red mangrove (*Rhizophora mangle*), black mangrove (*Avicennia germinans*), and white mangrove (*Laguncularia racemosa*). Within the mangroves there is overlapping zonation, with red mangrove extending into the deepest water, black mangrove in the shallowest water, and white mangrove reaching its highest density between the peak zones of the other two.

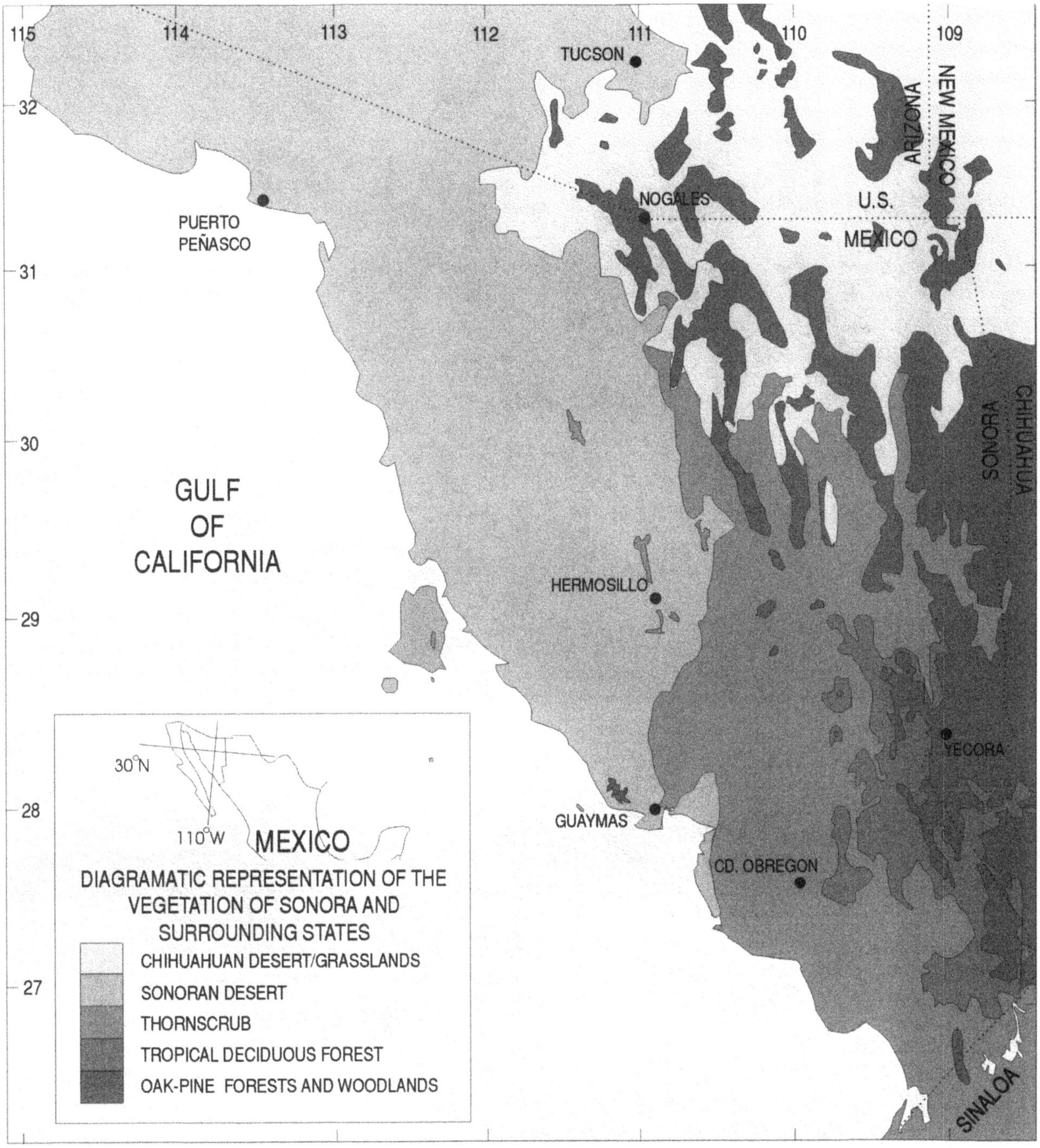


Figure 3. Major vegetation types in Sonora. Drawn by A. Búrquez.

*Rhizophora*, *Laguncularia*, and *Avicennia* are among the most important mangroves in the world and share many characteristics of ecology, growth form, and reproductive biology. The sizes of the plants and their leaves, inflorescences, and other organs of the Sonoran populations seem generally smaller than those of mangroves from more tropical regions.

Various succulent halophytes and salt grasses and other halophytes border the mangroves at their inland margins. These include *Allenrolfea occidentalis*, *Atriplex barclayana*, *Batis maritima*, *Cressa truxillensis*, *Monanthochloë littoralis*, *Salicornia bigelovii*, *S. subterminalis*, *S. virginica*, *Sesuvium portulacastrum*, *Suaeda esteroa*, and *S. moquinii*.

Figure 4. Mangroves at low tide, Infiernillo Coast. Note stilt roots of red mangrove (*Rhizophora mangle*). (Felger and Moser 1985). Photographs by N. L. Nicholson.

*Maytenus phyllanthoides*, a large woody shrub or rarely a small tree, is predictable along the landward margin of the mangroves. During the cooler months, and especially in spring, floating strands of eelgrass (*Zostera marina*) often festoon the lower branches of the mangroves.

### Riparian and Wetland Habitats

Everywhere that there is extra moisture, there is a concentration and increase in diversity and size of trees. Many kinds of trees occur only along watercourses, while others are more common in wetland habitats. The greater and lesser Sonoran rivers run through all variety of communities. Canyon bottoms and arroyos act as corridors for intrusions into adjacent habitats. Oaks extend into grassland and thornscrub trees into desert. Galleries of tropical deciduous forest follow drainageways into coastal thornscrub. Ribbons of evergreen forests course through deciduous vegetation. Deep riparian canyons in the mountains of southeastern Sonora are the greenest and most tropical places in the state. Lowland rivers used to support great gallery forests of salicaceous trees (figure 5; see Salicaceae). The mighty willow and cottonwood forests were limited to the big flood-prone rivers, and every one of them has been choked by upriver dams.

Trees generally restricted to riparian and wetland habitats include:

*Albizia sinaloënsis*
*Alnus oblongifolia*
*Aphananthe monoica*
*Cephalanthus salicifolius*
*Cinnamomum hartmanii*
*Drypetes gentryi*
*Fraxinus velutina*
*Ilex* spp.
*Juglans major*
*Magnolia pacifica*
*Pithecellobium dulce*
*Platanus wrightii*
*Populus* spp.
*Salix* spp.
*Sambucus nigra*
*Taxodium distichum*
*Trophis racemosa*
*Washingtonia robusta*

Figure 5. Remnant cottonwood-willow forest (*Populus mexicana* subsp. *dimorpha* and *Phoenix dactylifera*) along the lower Río Mayo at Rosales, 1994. Photograph by D. Burckhalter.

## Tropical Deciduous Forest

Tropical deciduous forest is characteristic of the dry tropics of a major part of the world. The dry tropics are also experiencing alarming human population increases and rapid environmental devastation. Much of the original tropical deciduous forest (TDF) of the world has undergone extreme desertification. Tropical deciduous forest is perhaps the least studied of the earth's major vegetation types. Because it is hardly fashionable to save poorly known, scrawny and often thorny trees in hot, uncomfortable climates, world conservation efforts have largely passed over tropical deciduous forest. Although deforestation and desertification of TDF in Sonora is accelerating, it remains remarkably intact in comparison with TDF in most parts of the world.

Tropical deciduous forest in Sonora represents a dry end of the spectrum of this broad vegetation type. In Sonora and adjacent southwestern Chihuahua, the northern arm of TDF sweeps northward from the tropics, and is the world's northernmost TDF (Búrquez et al. 1999, Martínez-Yrízar 1995, Martínez-Yrízar et al. 2000, Robichaux and Yetman 2000). In its natural condition in Sonora there is 100% ground cover of forest, often 10–15 m tall, made up of species of tropical origin or affinity (figures 6–8). Martin referred to TDF in Sonora as the "Secret Forest" (Bowden et al. 1993). Gentry (1942) first called it the "short-tree forest," and later Gentry (1982) and Brown (1982) called it "Sinaloan deciduous forest." Because this vegetation type is widespread in the New World, the term "Sinaloan" seems rather provincial. In contrast to open desert and thornscrub, there is sufficient forest cover to produce a substantial litter of leaves, bark, flowers, and other detritus. In the Alamos region of southeastern Sonora, there is an average production of 3–4 tons/ha/year of fine litter, which compares favorably with TDF farther south in coastal Jalisco (Martínez-Yrízar 1995).

Locally, in Sonora, the dry season landscape, or appearance of the forest, is called *monte mojino*, which refers to the brownish gray, drab color and absence of green from November to the end of June. The weather gradually becomes hotter and drier from March and April to June. Even the tree chollas and prickly pears become flaccid and droopy from water loss. Cicadas call loudly, and writers who visit the Sonoran TDF during the height of the spring drought write about skeleton forests and the lack of greenery:

> In the long spring dry season the forest is a dreary scene: a naked infinite host of trunks and branches, spreading interminably over the volcanic hills and mesas, bared to the fiery sun, under which the last leaf seems to have withered and died. He who walks this land in the month of May walks with a parched throat. The plants are waiting for the rains and their union with the soil, when like a piece of magic they turn the look of the dead into a fiesta of voluptuous growth, all in a few brief days. The sudden starting of new leaves with the first summer rain is aided greatly in many plants by the advanced development they have attained in the bud. (Gentry 1942: 33).

Summers are long and hot and winters short and mild. Freezing weather is rare, and most of the TDF species are highly frost sensitive. The monsoon rains begin soon after summer solstice and El Día de San Juan (June 24). The days grow hotter and hotter, approaching the long-awaited day every year when the summer rains begin. The clouds build up day by day, and finally when the violent thunderstorms begin, the trees and vines and undergrowth burst forth in a blaze of green (Gentry 1942, Van Devender et al. 2000). The rains typically continue through August and into early September. Defoliation of the forest begins with fall and winter drought, spring drought completes the defoliation of most of the TDF trees. Some trees, such as most of the *Bursera* species, *Cochlospermum*, *Erythrina*, *Jatropha*, *Ipomoea* and *Pseudobombax* defoliate soon after the rains cease. However, most of the trees defoliate more gradually. Midwinter rains may stay leaf fall of certain species.

Flowering trees can be found at virtually any time of the year, but there are some significant peaks of color display. Mid-winter brings the *amapas* (*Tabebuia* spp.), then the *palo santo* (*Ipomoea arborescens*) and *etcho* (*Pachycereus pecten-aboriginum*), and later in spring the *cuajilote* (*Pseudobombax palmeri*) and *palo barril* (*Cochlospermum vitifolium*). When the leaves are finally gone and dry and crackling under foot in the searing heat of the premonsoon drought, the hillsides blaze with the dark blue of *guayacán* (*Guajacum coulteri*) and rose-purple of *nesco* (*Lonchocarpus hermannii*). These are just a few of the more conspicuous flowering trees. The timing of fruit ripening largely coincides with the beginning of the summer rains.

Legumes rule the tropical-derived TDF as well as the thornscrub and Sonoran desertscrub. The fast-growing *mauto* (*Lysiloma divaricatum*) and many other legumes account for the majority of the vegetative cover. No single species or small number of species dominates—the forest is shared by a horde of species. Biological diversity is high. Fully 51 percent of the tree flora, 144 species, occurs in TDF (Appendix). Among the most prominent arborescent members of TDF in Sonora are *Bursera* spp., *Caesalpinia platyloba*, *Ceiba acuminata*, *Chloroleucon mangense*, *Conzattia multiflora*, *Haematoxylum brasiletto*, *Ipomoea arborescens*, *Lonchocarpus hermannii*, *Lysiloma divaricatum*, *L. watsonii*, *Pachycereus pecten-aboriginum*, *Pilosocereus alensis*, *Piscidia mollis*, *Senna atomaria*, *Stenocereus montanus*, *S. thurberi*, *Tabebuia chrysantha*, and *T. impetiginosa*.

Sonoran TDF is sandwiched between thornscrub to the west at lower elevations and the oak zone eastward at

A

B

Figure 6. Tropical deciduous forest vicinity of Alamos. (A) Wet season, (B) Dry season, *Lonchocarpus hermannii* in flower and *Pachycereus pecten-aboriginum* (Bowden 1993). Photograph by J. W. Dykinga.

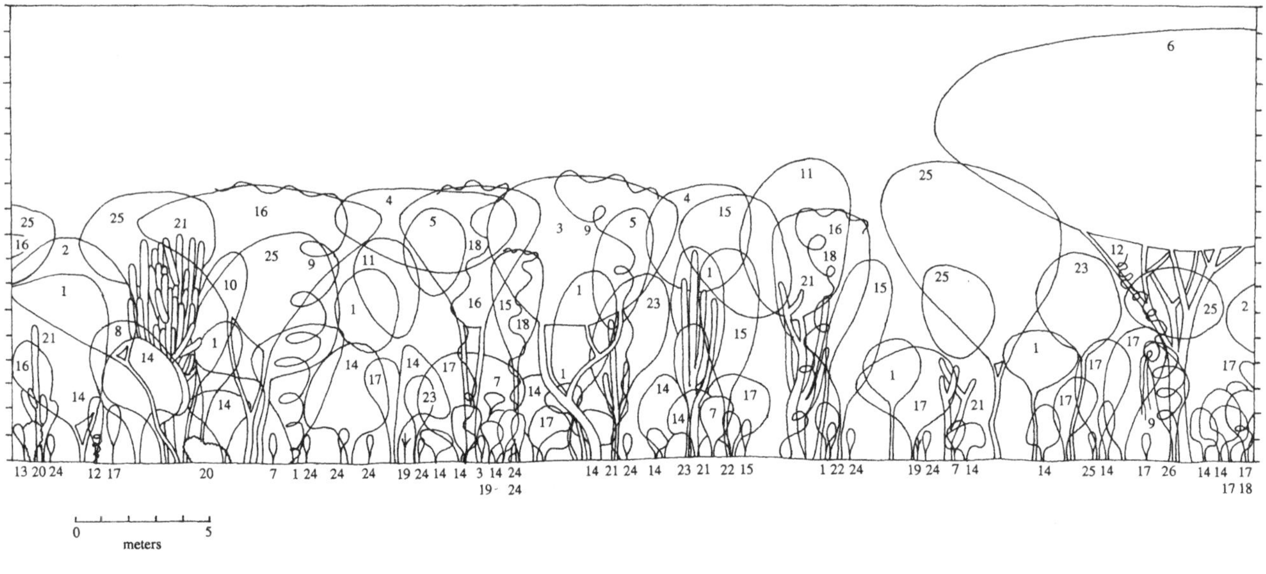


Figure 7. (*top*) Tropical deciduous forest, dry season aspect, north of Alamos. Photograph by D. Burckhalter.

Figure 8. (*bottom*) Diagram of tropical deciduous forest, vicinity of Minas Nuevas, west of Alamos. Modified from Krizman (1972). 1, *Brongniartia alamosana*; 2, *Bursera grandifolia*; 3, *B. laxiflora*; 4, *B. stenophylla*; 5, *Ceiba acuminata*; 6, *Conzattia multiflora*; 7, *Croton flavescens*; 8, *Erythroxylum mexicanum*; 9, *Heteropteris palmeri*; 10, *Hintonia latiflora*; 11, *Ipomoea arborescens*; 12, *I. bracteata*; 13, *Jatropha cordata*; 14, *J. malacophylla*; 15, *Lonchocarpus hermannii*; 16, *Lysiloma divaricatum*; 17, *Montanoa tomentosa*; 18, *Nissolia microptera*; 19, *Opuntia thurberi*; 20, *O. wilcoxii*; 21, *Pachycereus pecten-aboriginum*; 22, *Randia echinocarpa*; 23, *Senna atomaria*; 24, *Senna pallida*; 25, *Tabebuia impetiginosa*; 26, *Wimmeria mexicana*. Drawn by M. B. Johnson.

higher elevations. Northward, along the east side of the Sonoran Desert, the TDF merges into an inland thornscrub. In TDF the trees generally overtop the columnar cacti, while in thornscrub the columnar cacti overtop the small trees and shrubs (Gentry 1942). Epiphytic bromeliads and orchids extend from more tropical regions into TDF in Sonora.

### Thornscrub

Thornscrub in Mexico is of tropical origin. It is essentially a drier version of TDF and intermediate between TDF and desertscrub, their boundaries are fuzzy. Like TDF, the plants are adapted to a summer monsoon climate and for the most part are highly frost sensitive. Wholly natural tropical thornscrub generally forms 100% coverage. The stature of the nonriparian trees is lower, and the leaves are generally smaller than those of the TDF. Both TDF and thornscrub show the strong seasonality linked with hot-weather rains.

About 114 tree species occur in the thornscrub of Sonora (Appendix). The dominant trees and shrubs are legumes, together with arborescent burseras, jatrophas, and many cacti. There are spines everywhere, and there is enough leaf production to form a substantial leaf litter.

Thornscrub in Sonora was called "Sinaloan thornscrub" by Brown (1982). Two thornscrub formations can be discerned in Sonora (Búrquez et al. 1999): coastal thornscrub of southwestern Sonora (Freidman 1996) (figure 9), and the interior, foothills thornscrub along the east side of the Sonoran Desert and at higher elevations within the Sonoran Desert (figure 10). Coastal thornscrub of southwestern Sonora extends southward on the coastal plain from the southern margin of the Sonoran Desert in the vicinity of Guaymas into coastal, extreme northwestern Sinaloa. Southward and eastward it merges with TDF.

Coastal thornscrub in Sonora was called "thorn forest" by Gentry (1942). He emphasized the predominance of the boat-spine acacia (*Acacia cochliacantha*). Summer ephemerals can produce nearly 100% cover beneath the low green canopy. The diversity of winter ephemerals is low, and many of them are species that respond to rainfall at any season. Thornscrub in Sonora is devoid of orchids and supports only two species of epiphytic bromeliads (*Tillandsia*) and only along coastal regions experiencing maritime dew. This maritime dew, often heavy during the cooler months, probably contributes significant moisture for the vegetation. The soil is deep and alluvial, and most of the coastal thornscrub has been converted to large-scale irrigated agriculture. The coastal thornscrub is one of the least studied and most endangered habitats in the world.

Characteristic trees or treelike species of the coastal thornscrub include *Acacia cochliacantha*, *Bursera fagaroides*, *B. laxiflora*, *Forchhammeria watsonii*, *Fouquieria macdougalii*, *Guajacum coulteri*, *Haematoxylum brasiletto*, *Havardia sonorae*, *Jacquinia macrocarpa*, *Jatropha cordata*, *Pachycereus pecten-aboriginum*, *Parkinsonia praecox*, *Prosopis glandulosa*, and *Stenocereus thurberi*. The lower Río Yaqui and Río Mayo course through the coastal thornscrub and once supported great forests of *Populus mexicana* and *Salix gooddingii*. Elsewhere tall trees occur as gallery formations along the larger arroyos such as near Camahuiroa. Here one finds such trees as *Caesalpinia sclerocarpa*, *Ceiba acuminata*, *Diospyros sonorae*, *Pisonia capitata*, and *Tabebuia impetiginosa*.

Foothills thornscrub is essentially synonymous with Shreve's (1951) foothills of Sonora subdivision of the Sonoran Desert (Búrquez et al. 1999, Felger and Lowe 1976). This is a shrubby or semiarborescent scrub with a nearly closed canopy of small trees and large shrubs. Foothills thornscrub extends northward along the eastern side of the Sonoran Desert like the tail of TDF narrowing in geographic width and elevation. Northward it gives out at about the vicinity of Arizpe, where elevations rise and winter freezing as well as drier conditions become pronounced.

Foothills thornscrub is a highly diverse vegetation reflecting the diversity of the habitat and substrate. Among the more prominent, common species are *Acacia coulteri*, *Brahea brandegeei*, *Bursera fagaroides*, *B. lancifolia*, *B. laxiflora*, *Ceiba acuminata*, *Cordia sonorae*, *Coursetia glandulosa*, *Erythrina flabelliformis*, *Fouquieria macdougalii*, *Ipomoea arborescens*, *Lysiloma divaricatum*, *L. watsonii*, *Parkinsonia praecox*, *Plumeria rubra*, *Sabal uresana*, and *Stenocereus thurberi*.

Three isolated sky-island areas of thornscrub are enclosed in the Sonoran Desert in Sonora: (1) the higher elevations of the Sierra el Aguaje north of Bahía San Carlos, especially on northward-facing slopes. Gentry (1949) called this mountain mass the Guaymas Monadnock; (2) higher elevations of Sierra Kunkaak on Isla Tiburón, also mostly on north-facing slopes (Felger and Moser 1985); (3) peak elevations of Sierra Libre between Hermosillo and Guaymas. A dense cover of *Lysiloma divaricatum* is among the most conspicuous feature of these thornscrub islands. Although there are large expanses of barren, mostly rhyolitic rock, wherever there is a buildup of soil, or even cracks in the rock, there is a dense growth of thornscrub. On Sierra Libre and the Sierra el Aguaje, large populations of hesper palms, *Brahea brandegeei*, crowd drainageways and march up otherwise barren rock faces, as long as there are crevices into which the roots can penetrate.

### Sonoran Desertscrub

The Sonoran Desert, defined and elegantly described by Shreve in 1951, covers more than 300,000 km$^2$. Shreve divided the Sonoran Desert into seven geographic vegetation zones, five of which occur in Sonora. His "foothills of

Figure 9. Coastal thornscrub near Presa Mocuzari. The columnar cacti are mostly etcho (*Pachycereus pecten-aboriginum*). Photograph by D. Burckhalter.

Sonora" zone is essentially thornscrub. The remaining four zones are included here in the Sonoran portion of the Sonoran Desert. These are the Sonoran portion of the central Gulf Coast (figures 11–14), the plains of Sonora (figure 15), the lower Colorado Valley, and the Arizona upland (figure 16). The single unifying factor is the presence of creosotebush, *Larrea divaricata* subsp. *tridentata*. This plant more or less defines the Sonoran, Chihuahuan, and Mojave deserts in North America, and subspecies *divaricata* defines the deserts in South America.

Within the span of environments in the Sonora portion of the Sonoran Desert, there is great variation in vegetation cover and makeup and in the number and species present. The driest, harshest part of the desert, near the Río Colorado, is devoid of trees. Elsewhere this desert is characterized by a predominance of shrubs such as creosotebush, bursages (e.g., *Ambrosia deltoidea* and *A. dumosa*), brittlebush (*Encelia farinosa*), little-leaf cordia (*Cordia parvifolia*), and many others. Cacti are common. Unlike thornscrub, there is much open ground (you can run through the desert) and there is generally little or no buildup of leaf litter except along major drainageways. Across most of the Sonoran Desert the majority of species making up the flora are annuals that complete their life cycle in a single season. The numbers and percentage of annual species diminishes in the other vegetation zones.

The closest relatives of most perennial Sonoran desert plants are in tropical thornscrub and TDF. Many of the Sonoran Desert perennials, especially the trees, are sensitive to winter freezing, and their northern limits are fixed by freezing temperatures. The desert is essentially frost free at its southern limits, such as near Guaymas, and severity and duration of freezing increases northward. Similarly, drought is more severe northward and westward.

About 97 trees species are recorded from Sonoran desertscrub within the state (Appendix). The most common and ubiquitous trees include the desert legume trinity of ironwood (*Olneya*), palo verde (*Parkinsonia*), and mesquite (*Prosopis*). Other widespread, common arborescent species include large cacti such as saguaro (*Carnegiea gigantea*), cardón (*Pachycereus pringlei*), and organpipe (*Stenocereus thurberi*) and many small trees such as *Bursera* spp.

The only major river cutting across the Sonoran portion of the Sonoran Desert is the once mighty Río Colorado. The lower reaches and delta of the river supported vast riparian forests of cottonwood (*Populus fremontii*) and willow (*Salix gooddingii*) and a virtually extirpated flora of perhaps

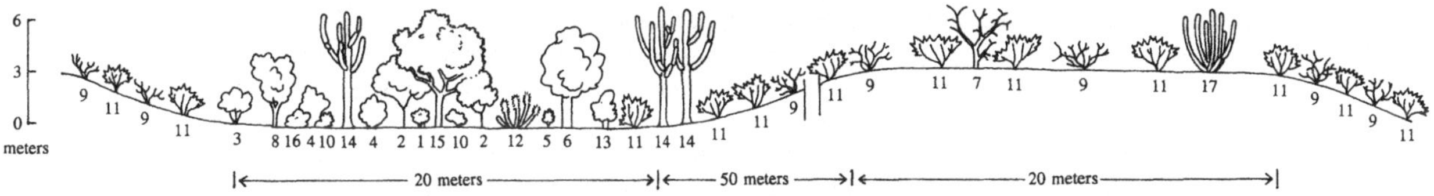


Figure 10. (*top*) Diagram of a thornscrub community. Southeast-facing slope in mountains 10.2 km by road west of Rayón (on road to Carbo), 750 m, 29°42'80" N, 110°37'50" W. 1, *Ambrosia cordifolia*; 2, *Bursera laxiflora*, 3, *Ceiba acuminata*; 4, *Coursetia glandulosa*; 5, *Eysenhardtia orthocarpa*; 6, *Jatropha cordata*; 7, *Lycium* sp., 8, *Pachycereus pecten-aboriginum*. Drawn by M. B. Johnson.

Figure 11. (*bottom*) Diagram of a community of Sonoran desert scrub, Gulf Coast of Sonora with interface of coastal thornscrub, 0.5 km inland and 5 km east of Bahía San Carlos. 1, *Holographis virgata*; 2, *Caesalpinia palmeri*; 3, *Citharexylum flabellifolium*; 4, *Cordia parvifolia*; 5, *Croton sonorae*; 6, *Forchhammeria watsonii*; 7, *Fouquieria diguetii*; 8, *Guajacum coulteri*; 9, *Jatropha cuneata*; 10, *Justicia candicans*; 11, *Larrea divaricata*; 12, *Lophocereus schotti* var. *schottii*; 13, *Maytenus phyllanthoides*; 14, *Pachycereus pecten-oboriginum*; 15, *Prosopis glandulosa* var. *torreyana*; 16, *Ruellia californica*; 17, *Stenocereus thurberi*. Drawn by M. B. Johnson.

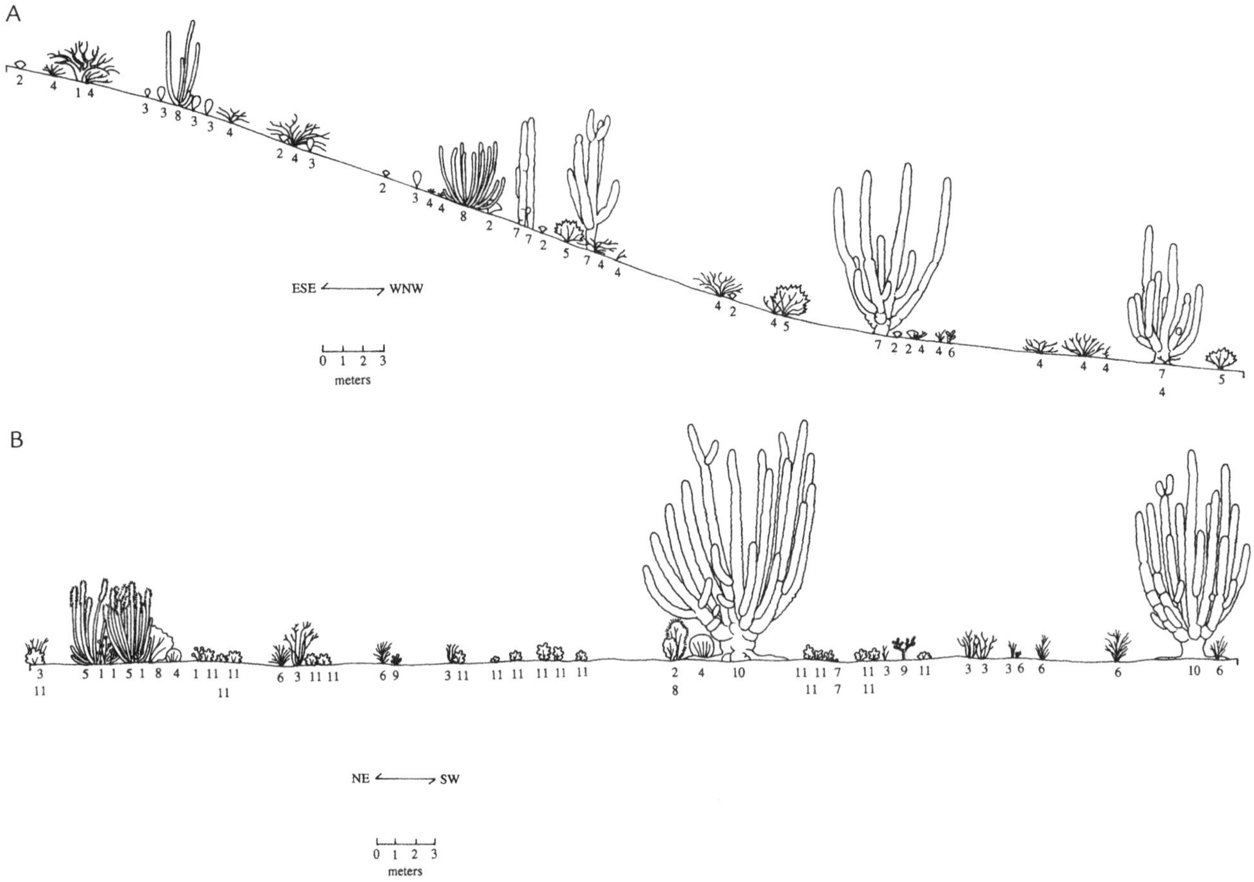


Figure 12. Diagrams of Gulf Coast of Sonora, Sonoran desertscrub community. (A) West-facing slope near base of a hill to the east of Kino Nuevo (Bahía Kino). Rhyolitic rocky soil; 10–20 m, 28°51'30" N, 111°59'45" W. 1, *Bursera microphylla*; 2, *Encelia farinosa*, 3, *Hibiscus denudatus*; 4, *Jatropha cuneata*; 5, *Larrea divaricata*; 6, *Opuntia fulgida* var. *fulgida*; 7, *Pachycereus pringlei*; 8, *Stenocereus thurberi*. (B) Coastal plain north of Sonora Highway 16, about 1.5 km inland from Playa Kino, north-northwest of Kino Viejo (Bahía Kino). Sandy soil. Near sea level; 28°50'55" N, 111°57'30" W. 1, *Abutilon californicum*; 2, *Capparis atamisquea*; 3, *Jatropha cinerea*; 4, *Justicia californica*; 5, *Lophocereus schottii* var. *schottii*; 6, *Lycium brevipes*; 7, *Mammillaria swinglei*; 8, *Maytenus phyllanthoides*; 9, *Opuntia fulgida* var. *mamillata*; 10, *Pachycereus pringlei*; 11, *Suaeda moquinii*. Drawn by M. B. Johnson.

200–400 species (Ezcurra et al. 1988). Elsewhere larger canyons and larger washes support the major diversity of Sonoran Desert trees. Countless desert washes course across the desert floor in an endless array of writhing and anastomosing watercourses. Dry most of the year, they occasionally carry flash floods. *Olneya*, *Parkinsonia*, and *Prosopis* are among the most common trees along xeroriparian drainageways.

## Chihuahuan Desertscrub

The northwestern corner of the Chihuahuan Desert extends into limited areas in the northeastern corner of Sonora and adjacent southeastern Arizona at elevations below about 1430 m (4700 ft). It is an inland desert, covering much of north-central Mexico between the Sierra Madre Occidental and Sierra Madre Oriental and extending into adjacent inland areas of the southwestern United States. At its northern limits freezing weather is much more severe than in the Sonoran Desert, which probably accounts for the absence of columnar cacti. Rainfall occurs mostly during the summer. Only 18 tree species are recorded from this habitat within the state (Appendix).

The Sonoran portion of the Chihuahuan Desert is bordered by grassland and oak woodland. As with the Sonoran Desert, there is open ground, and shrubs predominate. Characteristic shrub species include Chihuahuan white-

Figure 13. (*top*) Sonoran desertscrub, Gulf Coast of Sonora, near Puerto Libertad. *Bursera microphylla*, *Fouquieria columnaris*, *F. splendens*, *Lophocereus schottii*, *Pachycereus pringlei* (Turner et al. 1995). Photograph by R. M. Turner.

Figure 14. (*bottom*) Sonoran desertscrub, Gulf Coast of Sonora, near Punta Baja (between Bahía Kino and Tastiota). The columnar cacti are *Lophocerus schottii* var. *schottii*, *Pachycereus pringlei*, and *Stenocereus thurberi* (Felger and Moser 1985). Photograph by R. S. Felger.

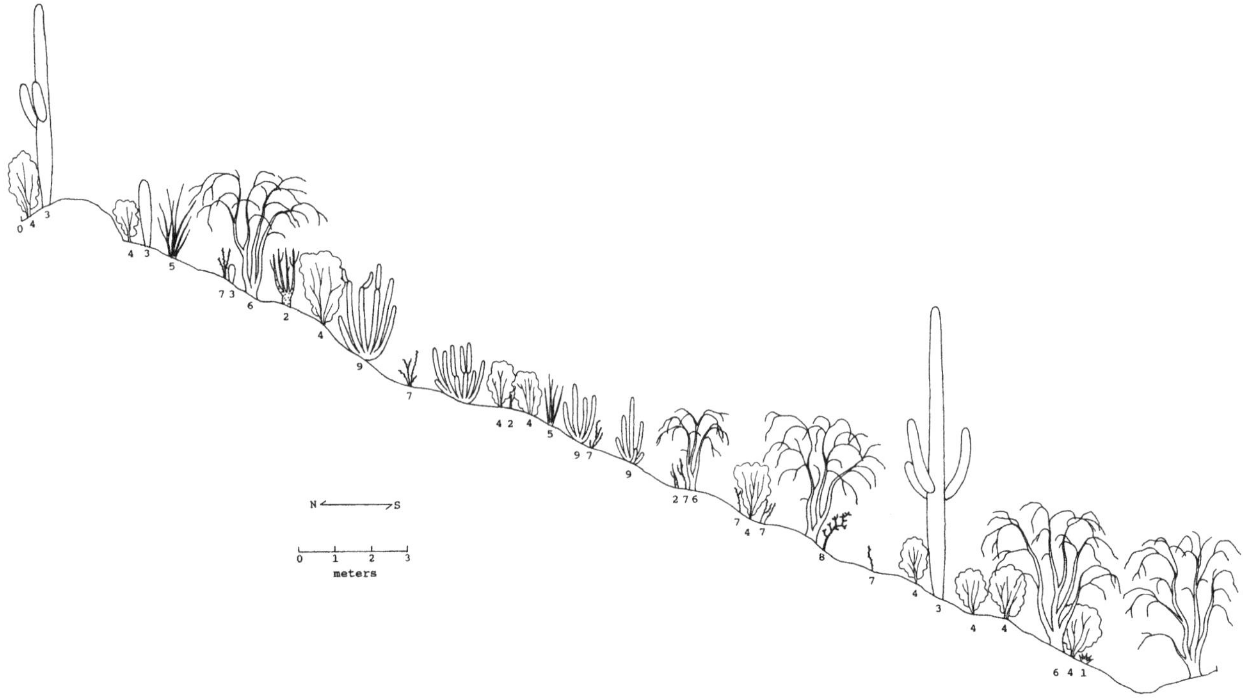


Figure 15. (*top*) Diagram of a Plains of Sonora, Sonoran desertscrub community. East-facing slope of hill along the west side of Mexico Highway 15 between kilometer marker 46 and 47, north of Hermosillo. Rocky granitic soil, 525 m. 29°29'55" N, 111°00'40" W. 1, *Bursera fagaroides* var. *elongata*; 2, *B. laxiflora*; 3, *Cardiospermum corindum*; 4, *Parkinsonia microphylla*; 5, *Coursetia glandulosa*; 6, *Croton sonorae*; 7, *Ipomoea arborescens*; 8, *Jatropha cordata*. Drawn by M. B. Johnson.

Figure 16. (*bottom*) Diagram of an Arizona Upland, Sonoran desertscrub community. Hill on south- to southeast-facing slope of Cerro Huequechi, 9 km by road east of Cucurpe, 1050 m, 30°19'50" N, 110°39'30" W. 1, *Agave palmeri*; 2, *Bursera fagaroides* var. *elongata*; 3, *Carnegiea gigantea*; 4, *Coursetia glandulosa*; 5, *Erythrina flabelliformis*; 6, *Fouquieria macdougalii*; 7, *Jatropha cardiophylla*; 8, *Opuntia thurberi*; 9, *Stenocereus thurberi*. Drawn by M. B. Johnson.

thorn (*Acacia neovernicosa*), creosotebush (*Larrea divaricata*), tarbush (*Flourensia cernua*), and shrubby honey mesquite (*Prosopis glandulosa* var. *torreyana*). Much of the substrate is limestone and alkaline soils. Apart from riparian or semiriparian habitats, there are few tree species in Chihuahuan Desert in Sonora. Much of the Chihuahuan Desert region in Sonora was probably grassland before the advent of intensive cattle grazing.

## Grassland

Grassland habitats are primarily found in northeastern Sonora at elevations above desert or thornscrub and below oak woodland (see McClaran and Van Devender 1995). The most extensive area of grassland of the state stretches across the valleys and low hills of northeastern Sonora (figure 17). Prime grassland occurs from the international border west of Agua Prieta to near Cananea, and also east of Agua Prieta to the foothills of the Sierra San Luis near the Chihuahua border. This grassland has been called "plains grassland" or "short-grass prairie" (Brown 1982). The tree flora includes 42 species (Appendix), mostly restricted to washes and arroyos. Freezing weather is more severe than in semidesert grassland. The winter freezing apparently excludes the mesquites and other subtropical woody plants that commonly occur in the warmer semidesert grassland. Common perennial grasses in grassland areas of northeastern Sonora include species of *Aristida, Bothriochloa, Bouteloua, Eragrostis, Leptochloa, Lycuris, Panicum, Schizachyrium, Setaria*, and *Sporobolus*.

Another major grassland formation in Sonora has been termed "desert grassland" or "semidesert grassland" (Brown 1982). These are the habitats with extensive grass cover among a minority of shrub or tree cover, often resulting in a woodland landscape. Woody plants were probably much less common before intensive cattle grazing beginning in the late nineteenth century. Common grass genera include species of *Aristida, Bouteloua, Eragrostis, Hilaria, Muhlenbergia, Panicum, Sporobolus*, and *Tridens*. Conspicuous woody plants include species of *Acacia, Atriplex, Dodonaea, Fouquieria, Mimosa, Opuntia, Prosopis*, and *Yucca*, and at higher elevations, *Juniperus* and *Quercus*. Other smaller, specialized grassland areas are not distinguished here because of their limited extent.

## Oak Woodland

Oak woodland vegetation is widely distributed at elevations above desert, grassland, thornscrub, and TDF, but below pine-oak woodland. Across Sonora and northern Mexico,

Figure 17. Grassland with Emory oak (*Quercus emoryi*) on hillsides; north side of Sierra de los Ajos, about 1550 m. Photograph by D. S. Turner.

the species composition and density of oak woodland changes both with elevation and from south to north. A total of 87 species of trees are known from this habitat in Sonora (Appendix).

Although the Sonoran oak zones have been called "Madrean evergreen woodland" (Brown 1982), many Sonoran oaks and associated species are deciduous during the late spring drought. In these regions "autumn colors" occur in late spring as the air and ground desiccate, temperatures soar, and oak leaves senesce and fall. The biggest flush of new foliage occurs before the summer rains, or with the summer rains especially during a severe drought year, and it is then that hills and mountainsides of mixed oak species become mosaics of subtle hues—pink, green, orange, tan and yellow—as the new buds burst forth. Within about two weeks the forest turns bright green.

Extensive areas in northeast and north-central Sonora are dominated by open woodland of Emory oak (*Quercus emoryi*). Emory oak, Mexican blue oak (*Q. oblongifolia*), and Arizona oak (*Q. arizonica*) are the most common low-elevation oaks in the northern part of the state. At lower elevations the oaks border grassland or desertscrub (figure 18). There is sometimes a broad ecotone between oak woodland and grassland where the oak trees become widely spaced and grasses predominate. Such areas have been called oak grassland or oak savannah.

In east-central and southeastern Sonora, the oak woodland sometimes forms a narrow belt between tropical deciduous forest and pine-oak woodland, or occasionally it may be locally absent with pine-oak woodland occurring immediately above TDF. In these mountains oak woodland sometimes occurs on acidic, hydrothermally altered soils within TDF (Goldberg 1982). The transition between the two plant communities is often only a few meters in these areas.

Oak woodland in southeastern Sonora (figure 19), called "oak forest" by Gentry (1942), shows considerable tropical affinity. At its lower limits it borders TDF. Their boundaries are often remarkably well defined, apparently maintained by fire. Across the Río Mayo and Río Fuerte mountain drainages, low fires are often seen creeping almost harmlessly through dry grasses and forbs among the leafless oaks in May and June. During visits to the region, our horses and mules would sometimes weave their way between smoldering ash and blackened grasses. These fires destroy small TDF trees and shrubs but not the perennial grasses, forbs, and oaks. Many of the oaks in southeastern Sonora and nearby southwestern Chihuahua are large-leaved compared to the oaks of northern Sonora. The southern oaks often support epi-

Figure 18. Lower oak woodland; near mouth of Cañón Evans, north side of Sierra de los Ajos, about 1690 m. *Agave palmeri, Quercus emoryi, Q. oblongifolia, Juniperus coahuilensis, Yucca arizonica.* Photograph by D. S. Turner.

Figure 19. Oak woodland, Sierra de Alamos. Chihuahua oak (*Quercus chihuahuensis*); 1100 m (Martin et al. 1998). Photograph by G. M. Ferguson.

phytic bromeliads (*Tillandsia* spp.) and orchids (e.g., *Encyclia* spp., *Laelia eyermaniana*, and *Oncidium cebolleta*) (Felger and Dimmitt 1998a, 1998b).

The oak zone in Sonora is highly variable, both in species composition, stature, and geographic affinity. The drier, generally more northern and lower elevation oak zones are clearly woodland, but especially at higher elevations toward the south the oak zone should be termed forest rather than woodland. However, for the sake of simplicity we refer to the collective oak zones as oak woodland.

## Pine-Oak Woodland and Forest

Extensive areas of pine-oak woodland and forest are found along the east side of the Sierra Madre Occidental, east of our region in Chihuahua. Along the western slope of the Sierra Madre Occidental the climate is generally somewhat wetter, with milder winter temperatures than the interior. Although pine-oak woodland and forest is structurally and ecologically similar from the north to the south, there is a gradual introduction of more southerly and tropical species and then a replacement of northern temperate species. The southern pine zones have a diverse flora with subtropical elements including egg-cone, Herrera, Maximino, and Yécora pines and tropical-montane oaks. Seventy-five tree species are found in pine-oak habitats in Sonora (Appendix).

Pine-oak woodland is continuous with oak woodland at lower elevations. In pine-oak woodland, the pines form the overstory, and the oaks generally form an understory. There are extensive areas of pine-oak woodland in the mountains of eastern Sonora. Pine-oak woodland has been included within the concept of Madrean evergreen woodland (Brown 1982).

At higher elevations within the pine-oak zones, the pines become increasingly conspicuous, and the vegetation is distinctive enough to be termed pine-oak forest or pine forest (figures 20, 21). Pine forest is characteristically dominated by Arizona pine (*Pinus arizonica*), Apache pine (*P. engelmannii*), or white pine (*P. strobiformis*), with scattered individuals or small groups of other trees such as oaks, especially *Q. gambelii* and *Q. rugosa*. Pine forest is more widespread in Chihuahua and Durango than in Sonora. This pine forest has been called "Madrean montane conifer forest" (Brown 1982). As with the oak zones, for the purposes of this book it is generally not practical to distinguish pine-oak woodland from pine forest in discussing plant distributions.

Figure 20. Diagram of pine-oak community. Southeast-facing slope in Willow Canyon, Santa Catalina Mountains, Arizona, 2205 m, 32°23'20" N, 110°42'20" W. Similar communities in northeastern Sonora would have *Pinus arizonica* rather than *P. ponderosa*. 1, *Arbutus arizonica*; 2, *Muhlenbergia emersleyi*; 3, *Pinus ponderosa* var. *scopulorum*; 4, *Quercus hypoleucoides*; 5, *Yucca schottii*. Drawn by M. B. Johnson.

## Mixed Conifer Forest

Mixed conifer forest occurs on several high mountain tops in northeastern Sonora: Sierras de los Ajos (figure 22), San José, San Luis, and el Tigre. Limited areas, especially on north-facing slopes, support mixed conifer forests at elevations mostly above 2135 m (7000 ft). Extensive mixed conifer forest at high elevations in Chihuahua barely extends into east-central Sonora near Yécora (see *Abies durangensis*). Mixed conifer forest in Sonora supports about 13 tree species (Appendix).

Three coniferous genera, *Abies*, *Pinus*, and *Pseudotsuga*, define this vegetation. These trees are commercially valuable for lumber. Most of the old growth forest has been logged, but in some places it is recovering from extensive logging in the mid-twentieth century. Most of the broadleaf (dicot) trees are winter deciduous (e.g., *Quercus gambelii*, *Fraxinus papillosa*, and *Populus tremuloides*). The most common pine is *Pinus strobiformis*. Riparian canyons are shaded with tall forests that include *Acer grandidentatum* and *Alnus oblongifolia*. Winters are cool or cold, and summers relatively cool and moist.

## How to Use This Book: Species Accounts and Keys

Information included here is based on our field experience, herbarium records (primarily the herbaria of The University of Arizona, Tucson; California Academy of Sciences, San Francisco; Jepson Herbarium, Berkeley; Instituto de Biología, Universidad Nacional Autónoma de México, Mexico City; San Diego Society of Natural History, San Diego; University of California, Berkeley; University of Texas, Austin), the published record, and information shared by colleagues. The trees are listed alphabetically by family, genus, and species, except that we have grouped the families of conifers (gymnosperms), dicots, and monocots and treat them in that order. For the purposes of better communication, we are following a conservative taxonomic approach. Family designations mostly follow Cronquist (1981). Taxonomy of the genera and species is based on our own studies and interpretation of the literature.

The descriptions and measurements for the various species and infraspecific taxa, and some genera (when more than one tree species is present in Sonora), pertain only to

Figure 21. (*top*) Pine-oak forest, west of Yécora, about 2000 m. Photograph by D. Burckhalter.

Figure 22. (*bottom*) Mixed conifer forest, east side of Sierra de los Ajos, about 2200 m. Photograph by D. S. Turner.

plants and populations from Sonora and immediately adjacent areas unless otherwise stated. Often there is little agreement on the number of taxa in a genus or family, and for this reason we have not repeated the qualification "about" or "approximately" every time numbers of species or other taxa are cited, but this imprecision is implied. The identification keys for the various taxa (e.g., families, genera, and species) are designed only for plants from Sonora. The keys are artificial and not phylogenetic. In some cases large shrubs that might approach tree size are included in the keys but are not otherwise treated.

In the descriptions we emphasize characters that seem important to understanding the variation and adaptations of plants in this generally arid region. For this reason there is sometimes more emphasis on vegetative characters and less on other features that are emphasized in other floras. More detailed description, discussion, and interpretation is offered for taxa not covered well elsewhere or for those of special regional interest, uniqueness, or significance.

The accepted names are in **boldface**, both in the species accounts and in the index. Throughout the text, nonnative plants are marked with an asterisk (*). Selected pertinent synonyms are listed, especially those not appearing in the standard regional references. Synonyms are in brackets [ ] following the accepted scientific name. Authors of scientific names are in the index and are given in the text only when necessary to avoid confusion. Common names, when available, are listed first in English (not italicized), then in Spanish (italicized), and then in the indigenous native language(s) such as Guarijío or Yaqui (which are indicated). Sources of local and indigenous names are many, e.g., Felger (1999, 2000), Felger et al. (1992, 1997), Felger and Moser (1985), Martin et al. (1998), Rea (1997), Pennington (1980), Reina Guerrero (1993), Sobarzo (1966), Yetman (1998, in press) and Yetman and Van Devender (in press); our own field studies as well as unpublished notes and works by Alberto Búrquez, Felipe Molina, Amadeo M. Rea, Ana Lilia Reina Guerrero, and Thomas Van Devender; and information on herbarium specimen labels.

We are not totally consistent with the presentation and format of the infraspecific taxa (subspecies or variety). After all, variation and evolution are not consistent. When an infraspecific taxon seems worthy of recognition, then it is given in the heading and may be included in the keys. When the biological significance is more questionable, then we often submerge the infraspecific taxon or taxa in the discussion for the species.

Nearly all tree heights are rough estimates and should be taken only as relative values. When feasible we measured reasonable samples of specimens for each taxon (often 10–20+ measurements depending upon variation encountered). In some cases information for certain species was available from only a small number of specimens, and a greater range of variation in nature can be expected. In addition to the usual, or commonly encountered range of variation, the extreme or uncommon variation is often provided in parenthesis [e.g., (5) 10–15 (20)]. Measurements for length or height precede those for width, and the terms "length," "long," or "height" are omitted unless needed to avoid confusion. Measurements, especially of leaves, are for the usual or "normal" range of variation. The small, first leaves of the season are generally not included in the measurements. Large shade leaves or "water shoot" leaves of trees such as oaks and ashes are likewise generally excluded. Unless otherwise stated, measurements for leaf lengths include the combined length of the petiole and the blade.

Unless otherwise noted, it is assumed that the flowers are bisexual (perfect) and the ovary superior (hypogynous flowers). The term "radial" refers to flowers that are radially symmetrical, and "bilateral" refers to those that are bilaterally symmetric. Flower color refers to the dominant, or most conspicuous color, and not necessarily just the perianth or corolla, unless individual floral parts are specified. Season or months of flowering are based largely on observations and records on herbarium specimens. In an arid region such as Sonora, season of flowering, fruiting, leafout, leaf fall, and other phenological attributes can be highly variable. For the most part the phenological information is necessarily imprecise.

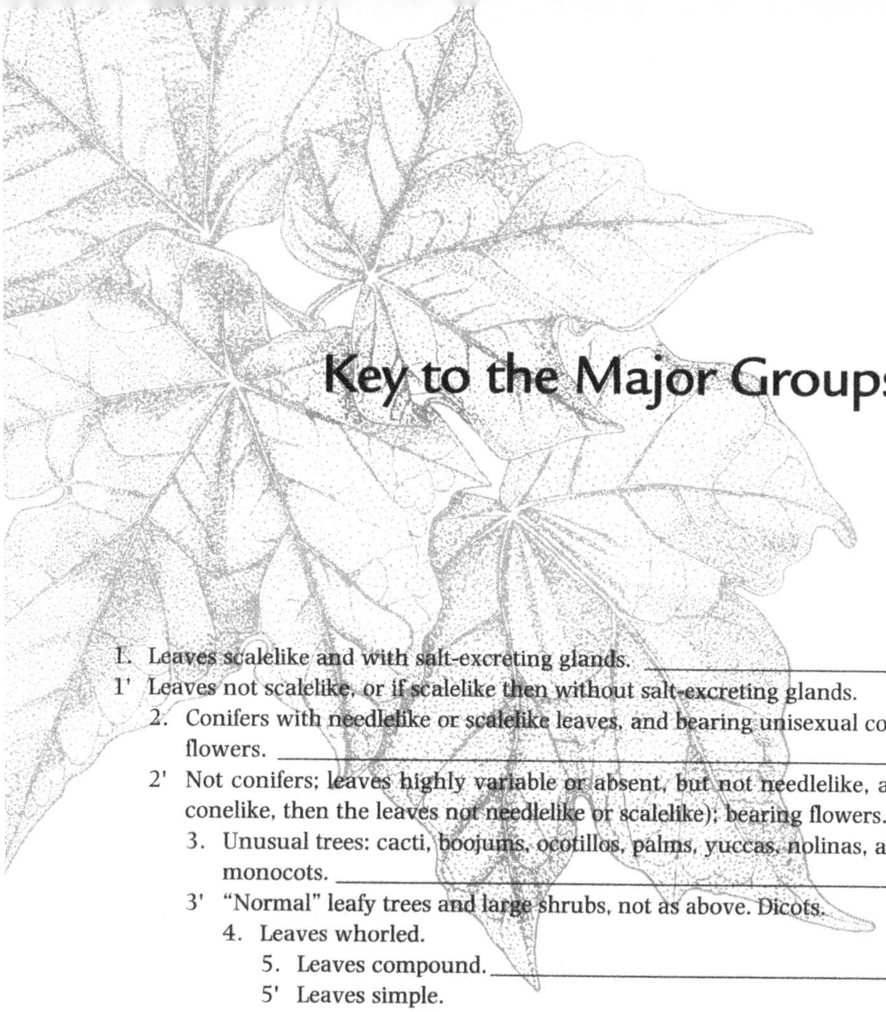

# Key to the Major Groups and Families

1. Leaves scalelike and with salt-excreting glands. ______ **Tamaricaceae**
1' Leaves not scalelike, or if scalelike then without salt-excreting glands.
2. Conifers with needlelike or scalelike leaves, and bearing unisexual cones or conelike structures; without flowers. ______ **Gymnosperms: Conifers**
2' Not conifers; leaves highly variable or absent, but not needlelike, and without unisexual cones (if inflorescences conelike, then the leaves not needlelike or scalelike); bearing flowers. Angiosperms:
3. Unusual trees: cacti, boojums, ocotillos, palms, yuccas, nolinas, and "crucifixion thorns." Dicots and monocots. ______ **Key I**
3' "Normal" leafy trees and large shrubs, not as above. Dicots.
4. Leaves whorled.
5. Leaves compound. ______ **Fabaceae (*Platymiscium*)**
5' Leaves simple.
6. Leaves linear; stipules absent; fruits of slender capsules at least 15 cm. _ **Bignoniaceae (*Chilopsis*)**
6' Leaves elliptic to lanceolate, oval or ovate; stipules present but small; fruits nutlike or fleshy, ≤ 1 cm in length. ______ **Rubiaceae (*Cephalanthus, Hamelia*)**
4' Leaves alternate (sometimes fascicled on short shoots) or opposite.
7. Leaves simple (includes unifoliolate leaves), entire to deeply divided or variously dissected.
8. Often with sharp, curved spines; leaves 5–14 cm, opposite, one in the pair larger, or sometimes alternate; blades rounded, entire to shallow toothed; stipules absent; fruits club-shaped, 1-seeded, 7–13 mm, 5-angled with glandular hairs. ______ **Nyctaginaceae**
8' Not as above, not with all of the above characteristics.
9. Composites; flowers in heads < 10 mm with disk florets ringed by white ray florets or rays absent (female heads burlike in *Hymenoclea*); fruits dry, < 3 mm. ______ **Asteraceae**
9' Not composites; not with all of the above characteristics.
10. Leaves alternate (sometimes crowded on short shoots and appearing whorled). __ **Key II**
10' Leaves opposite. ______ **Key III**
7' Leaves compound.
11. Leaves alternate. ______ **Key IV**
11' Leaves opposite. ______ **Key V**

## Key I: Unusual Trees

1. Trees leafy and/or succulents: cacti, boojums and ocotillos, palms, yuccas and yuccalike plants.
2. Cacti; stems succulent, leafless or leaves small and soon deciduous; areoles present. Dicots: ______ **Cactaceae**
2' Not cacti; without areoles.
3. Dicots; branches spiny; leaves, when present, < 6 cm. ______ **Fouquieriaceae**
3' Monocots; stems not thorny; leaves > 30 cm.
4. Palms; leaves with a broad, expanded blade. ______ **Arecaceae**
4' Not palms; leaves swordlike or straplike, the blade not expanded.

5. Leaves concave, swordlike, semisucculent, the tip ending in a sharp spine; fruits > 3 cm. ____________ **Agavaceae**
5' Leaves flat and straplike, not succulent, the tip not spine tipped; fruit ≤ 1 cm. ____________ **Nolinaceae**
1' Dicots; not as above, not succulent; leafless or sometimes with few, reduced leaves; twigs or branchlets spinescent.
6. Twigs silvery or grayish; twigs and calyx dotted with orange (purplish when dry) glands; flowers dark blue. ____________ **Fabaceae (*Psorothamnus*)**
6' Twigs and branchlets green or yellowish, not gland-dotted; flowers not dark blue; calyx lacking orange glands.
7. Twigs pubescent, the hairs small (use magnification); stems becoming glabrous with age.
8. Fruits falling soon after ripening; fruit a 1-carpelled pod more than three times as long as wide; (leafy during favorable seasons). ____________ **Fabaceae (*Parkinsonia microphylla*)**
8' Fruits hard and woody, tenaciously persistent for > 1 year, in dense clusters; each fruit as wide or wider than long with several to 10 carpels in a starlike pattern. ____________ **Simaroubaceae (*Castela*)**
7' Twigs glabrous.
9. Twigs mostly ascending at acute angles; fruits longer than wide, 10–20 mm including beaks. ____________ **Celastraceae (*Canotia*)**
9' Twigs spreading at right angles; fruits rounded, 5–6.5 mm in diameter. ____________ **Koeberliniaceae**

## Key II: Leaves Simple and Alternate

1. Leaves toothed, lobed, or parted; not entire.
2. Leaves palmately veined, lobed, or parted.
3. Leaf margins with small teeth, not lobed or parted; fruits disk shaped with a double row of feathery bristles. ____________ **Tiliaceae (*Heliocarpus*)**
3' Leaves lobed or parted; fruits not as above.
4. Plants with stellate hairs; petioles expanded at base, clasping and partially covering the twig.
5. Bark not flaking; twigs thick; leaves clustered at branch tips; fruits fleshy. ____________ **Araliaceae (*Oreopanax*)**
5' Bark flaking in puzzlelike pieces; twigs slender; leaves not clustered; fruits dry. ____________ **Platanaceae**
4' Glabrous or with simple hairs; petiole bases not expanded as above.
6. Flowers bisexual, 8–12 cm wide, bright yellow and showy; fruits about 10 cm; seeds with long, white hairs. ____________ **Cochlospermaceae**
6' Flowers unisexual, < 1 cm wide, dull; fruits < 4 cm; seeds glabrous. ____________ **Euphorbiaceae (*Manihot, Ricinus*)**
2' Leaves pinnately veined and/or lobed or parted.
7. Herbage with stellate hairs.
8. Leaves usually rather firm, oaklike.
9. Not oaks; flowers bisexual, in firm, spreading racemes; fruits rounded capsules 3–5 mm wide, not nested in a cup, not acorns; seeds many and about 1 mm. ____________ **Clethraceae**
9' Oaks; flowers unisexual; male flowers in slender, pendent catkins; fruit an acorn nested in a cuplike structure, 1-seeded, mostly at least 1 cm. ____________ **Fagaceae**
8' Leaves thin or relatively so, not firm, not oaklike.
10. Fruits 2.3–4.5 cm, conelike, with short, blunt prickles. ____________ **Sterculiaceae**
10' Fruits 0.7–1.3 cm, disklike, edged with a double row of feathery bristles. ____________ **Tiliaceae (*Heliocarpus*)**
7' Plants glabrous or with simple hairs.
11. Male flowers in catkins or dense unisexual spikes.
12. Leaf margins double serrated. ____________ **Betulaceae**
12' Leaf margins not double serrated.
13. New growth and floral buds enclosed by brownish scales; seeds with cottony hairs. ____________ **Salicaceae**
13' New growth buds not enclosed by brownish scales; seeds without cottony hairs.
14. Stipules small or quickly deciduous, scalelike, or represented by glands, or absent; fruits not fleshy. ____________ **Euphorbiaceae**
14' Stipules conspicuous, enclosing the leaf bud and leaving a scar on the twig; fruits fleshy. ____________ **Moraceae (*Maclura, Morus*)**

11' Flowers not in catkins or unisexual spikes.

15. Fruits dry, woody capsules or winged samaras.

16. Fruits capsules, not winged. ____ **Rosaceae (*Vauquelinia*)**

16' Fruits of winged samaras.

17. Leaves narrowly elliptic to obovate; fruits 3-winged; southeastern Sonora. ____ **Celastraceae (*Wimmeria*)**

17' Leaves broadly elliptic to lanceolate; fruits 1-winged; northeastern Sonora. ____ **Ulmaceae (*Ulmus*)**

15' Fruits fleshy or at least with an aril, not woody and not winged.

18. Flowers inconspicuous; petals none.

19. Herbage with stinging hairs; leaves 18–40+ cm; fruits (1.5) 2–3 mm in diameter. ____ **Urticaceae**

19' Hairs not stinging; leaves 3–11.5 cm; fruits 7–11 mm in diameter.

20. Leaves symmetrical; pedicels jointed near base; fruits several-seeded. ____ **Flacourtiaceae**

20' Leaves asymmetric at base; pedicels not jointed; fruits 1-seeded. ____ **Ulmaceae (*Aphananthe*, *Celtis*)**

18' Flowers conspicuous or not; petals present.

21. Wood soft; coastal desert to 305 m elevation. ____ **Burseraceae (*Bursera hindsiana*)**

21' Wood hard; mountains in eastern or north-central Sonora above 600 m elevation.

22. Twigs with thin, flaking, reddish bark; flowers bell-shaped; corolla baglike. ____ **Ericaceae**

22' Twigs not exfoliating; flowers not bell-shaped; corolla not baglike.

23. Leaves without conspicuous glands; stipules none; dioecious; flowers 4-merous with 4 stamens; fruits several-seeded. ____ **Aquifoliaceae**

23' Petioles and leaf blade bases often with conspicuous glands; stipules usually present; flowers bisexual, 5-merous with many stamens; fruits 1-seeded. ____ **Rosaceae (*Prunus*)**

1' Leaves entire (or margins sometimes slightly wavy).

24. Leaf buds enclosed by stipules, these leaving a circular scar on the stem.

25. Stipules tightly clasping and encircling the stem but not leafy; fruits dry, 1-seeded. ____ **Polygonaceae.**

25' Stipules paired, bractlike, or leafy; fruits or fruiting structure fleshy or if dry then multiple seeded.

26. Sap not milky; flowers large, bisexual; fruits woody, conelike; seeds red, > 10 mm. ____ **Magnoliaceae**

26' Sap milky; flowers minute, inconspicuous, unisexual; fruits berrylike and 1-seeded or aggregated into a fig; seeds minute or 10 mm, not red. ____ **Moraceae (*Ficus*, *Trophis*)**

24' Stipules absent or, if present then not enclosing the leaf buds as above.

27. Leaves glabrous or essentially so, moderately thickened to semisucculent, the midrib and veins inconspicuous.

28. Leaves obovate, broadest above middle, the tips rounded or notched; fruit a capsule, 3-angled and splitting open to reveal a red aril; coastal. ____ **Celastraceae (*Maytenus*)**

28' Leaves elliptic to ovate or rounded, broadest at or below the middle, the tips acute to rounded with an apiculate tip; fruit a drupe, rounded, and indehiscent; mostly inland, not coastal. ____ **Opiliaceae**

27' Leaves glabrous or pubescent, not especially thickened, not semisucculent or if so then pubescent, the midrib and often the lateral veins conspicuous.

29. Flowers unisexual (sometimes also with some bisexual flowers in *Celtis*).

30. Leaf scabrous; the stipules conspicuous; flowers small, petals none; stamens spring loaded.

31. Leaves mostly ≥ 8 cm, blades longer than wide, symmetric, net veined. ____ **Moraceae (*Trophis*)**

31' Leaves mostly < 6 cm; blades about as long as wide, usually asymmetric at base, palmately 3-veined at base, pinnately veined above with 4–6 or fewer pairs of lateral veins. ____ **Ulmaceae (*Celtis reticulata*)**

30' Leaves various but not scabrous; stipules present and mostly small and inconspicuous or absent; flowers minute to rather large, petals present or absent; stamens not spring loaded.

32. Leaves sessile, narrow (elliptic to oblanceolate).

33. Plants unarmed; herbage resinous-viscous. ____ **Sapindaceae (*Dodonaea*)**

33' Branches spiny; herbage not viscous. ____ **Flacourtiaceae**

32' Leaves with petioles, broadly lanceolate to ovate or oval.

34. Sap not milky; leaf margins entire, petioles short; stipules none; flowers > 2 cm wide; fruits very fleshy, indehiscent, about 2.5+ cm. ____ **Ebenaceae**

34' Sap milky or not; leaf margin entire or not; petioles short to long; stipules usually present, often small or quickly deciduous; flowers ≤ 1 cm in width; fruits dehiscent (capsules) mostly ≤ 1 cm in length or indehiscent and drupes (*Drypetes*) but not fleshy. ____ **Euphorbiaceae**

29' Flowers bisexual or generally bisexual.
35. Leaves spine tipped; flowers bright orange. ______ **Theophrastaceae**
35' Leaves not spine tipped; flowers not bright orange.
36. Fruits dehiscent.
37. Herbage and inflorescences densely pubescent with orange-brown stellate hairs. ______ **Clethraceae**
37' Herbage and inflorescences glabrous or pubescent, the hairs not stellate and orange.
38. Herbage gland dotted; capsules woody, 2–2.5 cm wide. ______ **Rutaceae**
38' Herbage not gland dotted; capsules not woody or if slightly woody then < 2 cm wide.
39. Plants with conspicuous spines; flowers tubular, red or reddish. ______ **Fouquieriaceae**
39' Plants unarmed; flowers not red or reddish.
40. Sepals and petals usually 4 each; fruits with red arils or red pulp. ______ **Capparaceae (*Capparis*)**
40' Flowers 5-merous; fruits without arils or red pulp.
41. Fruits (pods or follicles) much longer than wide.
42. Sap milky; stems thick and succulent; fruits with a pair of horizontally spreading tubular follicles, the seeds thin and papery winged. ___ **Apocynaceae (*Plumeria*)**
42' Sap not milky; stems not thick, not succulent; fruits of flattened pods; seeds lens-shaped, not winged and not papery. ______ **Fabaceae (*Bauhinia*, *Acacia willardiana*)**
41' Capsules, about as long to less than twice as long as wide.
43. Trees often > 6 m; trunk thick, semisucculent, and pithy; sap milky; plants pubescent, especially the young shoots. ______ **Convolvulaceae (*Ipomoea*)**
43' Shrubs or small trees to about 6 m; trunks slender, neither succulent nor pithy; sap not milky; glabrous. ______ **Solanaceae (*Nicotiana*)**
36' Fruits indehiscent.
44. Perianth 4- or 6-merous, or of 2 whorls of 3 tepals; leaves aromatic or not.
45. Fertile stamens 6 or numerous; flowers small to large; base of fruit (ovary) elongated as a stipe (stalk). ______ **Capparaceae (*Capparis*, *Forchhammeria*)**
45' Fertile stamens 4 or 9 in 3 whorls; flowers small; fruits not on a stipe.
46. Leaves not aromatic; perianth with a 4-lobed calyx and 4-lobed corolla; stamens 4 in one whorl, staminodes absent; fruits several seeded; sepals not persistent. ______ **Aquifoliaceae**
46' Leaves aromatic; perianth of 2 whorls of 3 tepals; fertile stamens 9 in 3 whorls plus 3 staminodes in an inner whorl; fruits 1-seeded; sepals or sepal bases persistent. ______ **Lauraceae**
44' Perianth 5-merous; leaves not aromatic.
47. Leaves relatively thin, 3–8 cm; petioles 2 mm; stipules scalelike, fringed below. ___ **Erythroxylaceae.**
47' Leaves often thickish, often > 8 cm; petioles mostly > 2 mm; stipules not fringed or absent.
48. Fruits with 2 or more seeds.
49. Stipules to 3 mm; leaf margins minutely serrated; inflorescences of few-flowered sessile clusters; fruits 3-seeded; rare in Sonora as trees. ______ **Rhamnaceae (*Rhamnus*)**
49' Stipules none; leaf margins entire to serrated or lobed; inflorescences few- to many-flowered; fruits 2- to many-seeded.
50. Leaf margins entire to serrated; inflorescences paniculate, often longer than the petioles; corollas baglike with minute lobes. ______ **Ericaceae**
50' Leaf margins entire to lobed; corollas open or with an elongated tube, not baglike. ______ **Solanacaceae (*Cestrum*, *Solanum*)**
48' Fruits 1-seeded, or a pair of 1-seeded segments.
51. Major leaf veins 3 from base of blade; stipules usually small or modified as spines. ______ **Rhamnaceae (*Condalia*, *Ziziphus*)**
51' Leaf veins pinnate; stipules minute and not spinescent (*Vallesia*) or none.
52. Herbage with sunken or buried dotlike or streaked glands. ______ **Myrsinaceae.**
52' Herbage without sunken or buried glands as above.
53. Sap not milky or colored; fruits dry, with nutlets 6–7 mm enclosed by the calyx and corolla base, the dry corollas persistent and falling with the fruit. ______ **Boraginaceae (*Cordia*)**
53' Sap milky or colored; fruits fleshy or leathery, at least 10 mm, not enclosed in calyx and corollas bases; corollas not persistent.

54. Shrubs or small trees; leaves not fascicled; flowers fragrant; sepals minute, inconspicuous; fruits mostly paired. ___ **Apocynaceae (*Plumeria, Vallesia*)**
54' Shrubs to large trees; leaves sometimes fascicled on short shoots; flowers not especially fragrant; sepals conspicuous; fruits not paired. ___ **Sapotaceae (*Sideroxylon*)**

## Key III: Leaves Simple and Opposite

1. Plants with stellate and branched hairs. ___ **Buddlejaceae**
1' Plants glabrous or with simple hairs.
  2. Leaf margins toothed, incised, lobed, or parted.
    3. Leaves palmately veined and 5- or 7-lobed, not rough-surfaced; fruits of 2-winged samaras; eastern Sonora. ___ **Aceraceae (*Acer grandidentatum*)**
    3' Leaves pinnately veined, the margins with small teeth but not lobed; fruits of small, capsules or fleshy; southern Sonora. ___ **Verbenaceae (*Citharexylum, Lippia*)**
  2' Leaf margins entire or nearly so.
    4. Mangroves, the roots in tidally inundated sea water.
      5. Leaves dull gray-green, the upper surface darker; twigs more or less square in cross section. ___ **Avicenniaceae**
      5' Leaves shiny green, same color on both surfaces; twigs rounded in cross section.
        6. Plants not forming stilt roots; petioles with a conspicuous pair of glands; stipules none. ___ **Combretaceae**
        6' Plants with stilt roots; petioles without glands; stipules large and sheathing the leaf buds. ___ **Rhizophoraceae**
    4' Not mangroves, not in sea water.
      7. Leaves linear, more than several times longer than wide. ___ **Bignoniaceae (*Chilopsis*)**
      7' Leaves mostly elliptic to lanceolate or ovate, to about twice as long as wide.
        8. Herbage densely pubescent with 2-armed hairs or glabrous with fruits 3-lobed. ___ **Malpighiaceae**
        8' Herbage glabrous or pubescent with simple hairs; fruits not 3-lobed.
          9. Leaf veins dotted and dashed with blackish glands. ___ **Rhamnaceae (*Karwinskia*)**
          9' Veins without blackish glands as above.
            10. Leaf tips blunt or rounded; plants not sharp spined; fruits not hard shelled..
              11. Herbage pubescent when young; flowers > 2.5 cm; fruits 2.5+ cm, many seeded; southeastern Sonora and also cultivated. ___ **Myrtaceae (*Psidium guaiava*)**
              11' Glabrous or pubescent; flowers > 1 cm; fruits to 1 cm, 1- or 2-seeded.
                12. Glabrous; fruits 1-seeded. ___ **Olacaceae**
                12' Pubescent with short, thick hairs; fruits 2-seeded. ___ **Verbenaceae (*Citharexylum*)**
            10' Leaf tips pointed, or if blunt, then plants sharp-spined and/or fruits hard shelled.
              13. Flowers yellow and large; fruits paired. ___ **Apocynaceae (*Stemmadenia*)**
              13' Flowers white to greenish white or red-orange; fruits not paired.
                14. Sepals united in bud, without lobes; stamens numerous. ___ **Myrtaceae (*Psidium sartorianum*)**
                14' Sepal lobes or teeth evident; stamens 4 or 5.
                  15. Stipules none; flowers 6 mm, 4-merous; Sierra Saguaribo, 1065+ m. ___ **Cornaceae**
                  15' Stipules present but small (if deciduous then leaving a scar); flowers at least (6) 10 mm, 4- to 6-merous; mostly below 1065 m, widespread. ___ **Rubiaceae**

## Key IV: Leaves Compound and Alternate

1. Leaves palmately or digitately compound.
  2. Leaflets not conspicuously stalked, not gland-dotted; flowers > 10 cm. ___ **Bombacaceae**
  2' Leaflets stalked, dotted with small glands flush with the surface; flowers < 1 cm. ___ **Rutaceae (*Casimiroa, Ptelea*)**
1' Leaves pinnately or bipinnately compound.
  3. Inner bark bright yellow; leaflets rather firm; flowers yellow, 3-merous, perianth segments 12. ___ **Berberidaceae**
  3' Inner bark not bright yellow; flowers not 3-merous or perianth segments 6 and not bright yellow.

4. Leaflet margins entire.
   5. Plants armed or unarmed; stipules often well developed and persistent. ________ **Fabaceae (*Eysenhardtia, Haematoxylum, Olneya, Parkinsonia*)**
   5' Plants unarmed; stipules absent or small and deciduous (*Crateva*).
      6. Leaflets 3 (southern Sonora; not common)
         7. Petioles broadly winged; flowers and fruits borne on trunk and lower limbs. ________ **Bignoniaceae (*Crescentia*)**
         7' Petioles slender, not winged; flowers and fruits on upper branches. ________ **Capparaceae (*Crateva*)**
      6' Leaflets >3.
         8. Leaves odd-pinnate; fruits glabrous, ovoid to globose, 1-seeded or with 1–3 one-seeded lobes.
            9. Foliage often aromatic; fruits < 10 mm in diameter, exocarp splitting into 2 or 3 leathery valves. ________ **Burseraceae**
            9' Foliage not aromatic; fruits (or lobes) > 12 mm in diameter, indehiscent. ________ **Sapindaceae (*Sapindus*)**
         8' Leaves odd- or even-pinnate; fruits pubescent, globose or winged, the seeds 2 or more, or if 1 then samaras.
            10. Leaflets 10–21 in number, 3–17 cm. ________ **Meliaceae**
            10' Leaflets 22–41, 1.1–2.5 cm. ________ **Simaroubaceae (*Alvaradoa*)**

4' Leaflet margins serrated, toothed, or lobed.
   11. Leaflets 3 per leaf.
      12. Lateral leaflets not stalked; fruits rounded, not winged. ________ **Burseraceae** (in part)
      12' Lateral leaflets on short stalks; fruits winged. ________ **Sapindaceae (*Thouinia*)**
   11' Leaflets > 3 per leaf.
      13. Leaves 2 (3) times pinnate, > 25 cm. ________ **Meliaceae (*Melia*)**
      13' Leaves pinnate or if 2 (3) times pinnate < 25 cm.
         14. Pubescent at least early in the season with branched or stellate hairs; fruits ribbed or pubescent.
            15. Stipules partially fused to the petiole base and partially clasping the stem; hairs branched; fruits 0.5 cm. ________ **Araliaceae (*Aralia*)**
            15' Stipules absent; hairs stellate and simple; fruits > 2 cm. ________ **Juglandaceae**
         14' Glabrous or with simple hairs; fruits smooth and glabrous.
            16. Fruits fleshy, edible, > 2 cm; near Alamos and cultivated. ________ **Anacardiaceae**
            16' Fruits scarcely fleshy, not edible, < 1 cm; widespread. ________ **Burseraceae**

## Key V: Leaves Compound and Opposite

1. Leaves palmately or digitately compound.
   2. Flowers at least 5 cm; fruits dry, much longer than broad. ________ **Bignoniaceae (*Tabebuia*)**
   2' Flowers 2 cm or less in length; fruits fleshy, about as wide as long. ________ **Verbenaceae (*Vitex*)**

1' Leaves pinnately compound.
   3. Leaves large, with 3 (5) leaflets.
      4. Leaves > 30 cm, with 3 leaflets, the margins entire. ________ **Fabaceae (*Platymiscium*)**
      4' Leaves 10–20 cm; leaflets 3 (5), the margins coarsely toothed. ________ **Aceraceae (*Acer negundo*)**
   3' Leaves large or small, with 5 or more leaflets.
      5. Leaves even-pinnate, 1.5–6 cm; stipules present; flowers dark blue. ________ **Zygophyllaceae**
      5' Leaves odd-pinnate, mostly more than 6 cm; stipules none; flowers not blue.
         6. Lenticels conspicuous; fruits of single samaras, laterally compressed, > twice as long as wide. ________ **Oleaceae**
         6' Lenticels not especially conspicuous; fruits rounded or ovoid, < twice as long as wide.
            7. Leaves 8.5–15 cm with 3–7 leaflets; flowers bisexual; fruits berrylike and juicy. ________ **Caprifoliaceae**
            7' Leaves 17–31.5 cm with 7–19 leaflets; flowers unisexual; fruit firm, not juicy. ________ **Juglandaceae**

# Species Accounts

## CONIFERS

*George McNeil Ferguson*
*Richard S. Felger*
*Matthew B. Johnson*

Trees and shrubs; resin-producing and aromatic with resin canals widespread in all organs; often with long and short shoots. Leaves scalelike, linear, or needlelike. Wind pollinated with unisexual cones; cone scales spirally arranged. Pollen (staminate or male) cones relatively small, herbaceous, soon deciduous (developing and falling within a single season); seed (pistillate or female) cones larger and more persistent (developing in 1–3 years), often woody, or berrylike and deciduous in *Juniperus*. Mostly in the temperate parts of the world and some tropical; 9 families, 68 genera. References: Flora of North America (1993), Martínez (1963), McVaugh (1992).

The two conifer families in Sonora occur in mountains and foothills mostly at elevations above desert, thornscrub, or tropical deciduous forest, except *Taxodium*, which grows in riparian habitats in thornscrub and tropical deciduous forest. Descriptions and measurements of cones are for mature, opened seed cones.

1. Leaves scalelike to needlelike, < 2.5 cm; cones < 2.1 cm. ________ ***Cupressaceae***
1' Leaves linear or needlelike, at least (1.5) 2.5 cm; cones at least 2.5 cm. ________ ***Pinaceae***

### Cupressaceae—Cypress Family

Trees and shrubs. Usually monoecious except most junipers. Worldwide; 25+ genera. Often treated as two families: Cupressaceae, *sensu stricto*, mostly in temperate regions; 20 genera, 125 species; includes *Cupressus* and *Juniperus*. Herbage aromatic. Branchlets of mature plants covered with scalelike leaves, opposite or in whorls of 3; juvenile plants with longer, needlelike leaves. Cones rounded to ovoid, with relatively few scales. Taxodiaceae, southeastern and western United States, Mexico, eastern Asia, and Tasmania; 10 genera, 16 species; includes *Taxodium* and some of the rarest, largest, and longest-lived trees in the world such as the California redwood, giant sequoia, and dawn redwood. References: Bartel (1993); Felger (1998).

1' Leaves alternate, distichous (2-ranked) on pinnate branchlets, linear and flattened, 1–2.5 cm; riparian thornscrub or tropical deciduous forest. __ ***Taxodium***
1. Leaves opposite and decussate (4-ranked) or in whorls of 3, scalelike (at least on branchlets of mature plants), < 1 cm; mostly grasslands and oak woodland or higher elevations.
    2. Monoecious; cones 15–27 mm, dry and woody at maturity, separating (opening) between scales, persistent for several years; seeds winged. ______ ***Cupressus***
    2' Dioecious (in Sonora); cones 5–14 mm, berrylike, fleshy or semifleshy, not woody, not separating between scales, maturing and deciduous in first or second year; seeds not winged. ________ ***Juniperus***

#### *Cupressus*—Cypress

Bark smooth and often red-brown on younger branches, becoming checkered on older limbs and longitudinally furrowed on trunks. Leaves scalelike. Cones rounded, woody, persistent for several years; scales peltate, broad and flattened at the apex with a short central projection (the umbo). Seeds several to many per scale, each with a small papery wing.

Temperate and montane subtropical–tropical regions in the Northern Hemisphere; 15 species. References: Farjon (1993), Little (1970), Martinez (1947), Wolf and Wagener (1948), Zanoni (1980).

The lumber of these trees is much valued, but in Sonora the trees are too few and too inaccessible to be of commercial significance. These large, handsome trees are widely cultivated in temperate regions worldwide and are recommended for higher elevations in Sonora.

1. Herbage glaucous gray-green; northeastern and east-central Sonora. ______ ***C. arizonica***
1' Herbage dark green; east-central and southeastern Sonora. ______ ***C. lusitanica***

***Cupressus arizonica.*** Arizona cypress; *sabino, táscate* (figure 23)

Conical trees 12–20+ m, usually with a single, well-formed trunk. (Sometimes reaching 25+ m with trunks to more than 1–1.5 m in diameter in favorable situations in the Sierra San Luis. Such trees may have been fairly common but have been logged.) Bark soon becoming scaly, on older limbs and trunks gray to dark brown, peeling in rectangular plates separated by low furrows, the outer surface somewhat fibrous. Leaves on ultimate, smallest twigs 1.2–2.3 mm and to 4.5 mm on larger twigs, glaucous, pale bluish green, overlapping, with or without a resin gland; leaf tips closely appressed to the twigs to form 4-sided branchlets. Cones 17–20 mm, single or paired and in small clusters, appearing sessile to short-stalked on stems with scale leaves, maturing during second year, persistent long after seeds fall. Seeds 4.5–5 mm, red-brown, flattened, with angular ridges edged with a narrow wing. Pollen shed in March.

Slopes, canyons, and streams; oak woodland and pine-oak woodland; northeastern and east-central Sonora, 1500–2250+ m (4920–7380+ ft). Distribution in Sonora spotty, but sometimes forming forests in mountain canyons. The southernmost Sonoran population occurs along a riparian canyon bottom in Municipio de Sahuaripa southeast of Tarachi on the road to Matarachi (vicinity of 28°44' N, 108°53' W; see *C. lusitanica*). Southern California (rare) to west Texas, northern Baja California, Sonora, Chihuahua, Coahuila, and Durango. Sonoran populations are subsp. *arizonica*, the most wide-ranging of the three subspecies, although the subspecies appear to be only weakly differentiated.

This species has been widely planted as an ornamental and lumber tree in various semiarid temperate regions of the world. Most horticultural stock has been selected from unusually bluish glaucous-leaved trees. Cypress trees are recommended for elevations above the desert.

***Cupressus lusitanica*** [*C. benthamii; C. lindleyi; C. lusitanica* [var.] *lindleyi*]. Mexican cypress; *sabino, táscate*

Conical trees to 20+ m with green foliage. Mature (opened) cones 13.3–21.5 mm. Pollen shed in March. Resembling Arizona cypress, except the trees tend to become more conical (more slender), the foliage conspicuously dull, dark green, perhaps less often with glandular pits on the scale leaves (specimens show both conditions), and the cones are smaller in average size. Leaf sizes are similar.

Slopes, canyons, and streams; oak woodland and pine-oak woodland; southeastern and east-cental Sonora, 1200–1980 m (3935–6500 ft). In Sonora documented for the Sierra Tecuarahui (ca. 10 km north of Las Chinacas), Sierra Saguaribo, the vicinity of Yécora, and northward to the vicinity of Tarachi. More common in nearby southwestern Chihuahua. Southward in the Sierra Madre Occidental and other montane regions of Mexico to Honduras.

Perhaps the green-foliage cypress trees in southeastern Sonora are conspecific with *C. arizonica*, and not the same trees McVaugh (1992) treated as *C. lusitanica*. The cypress trees with green foliage of the Sierra Madre in Sonora and nearby Chihuahua belong to a species complex with many variable forms. *Cupressus lusitanica* was described from trees cultivated by Carmelite friars in Portugal, as indicated by the specific epithet (Farjon 1993, McVaugh 1992). Other authors consider *C. benthamii* to be the correct name for this common Mexican cypress tree. Zanoni (1980) recognizes two varieties, calling the northern one from Chihuahua and Sonora to Coahuila and Tamaulipas *C. benthamii* var. *lindleyi*. It is replaced southward by var. *benthamii*, which extends to Central America.

This handsome tree is widely cultivated in the highlands of Mexico and many other parts of the world. At least eight named horticultural varieties are grown in Europe and elsewhere. Apparently no selections have been made from the Sonora-Chihuahua populations, which should be relatively drought tolerant.

The relationship of the northernmost populations of *C. lusitanica* and the southernmost *C. arizonica* in Sonora are unresolved. A population in the vicinity of Tarachi (see *C. arizonica*) consists of individual trees resembling either species in foliage color with a range of intermediates. It is not resolved whether this population represents a hybrid zone or natural variation within a single taxon.

### *Juniperus*—Juniper

Shrubs and trees. Rapid-growing shoots (whip branches) and juvenile plants with slender, relatively elongated or needlelike leaves (whip leaves); adult, lateral branchlets or twigs with scalelike leaves (scale leaves) 1–3 mm, commonly with a resin gland near the center of the leaf. Mostly dioecious. Cones fleshy or fibrous, berrylike with 1–3 pairs

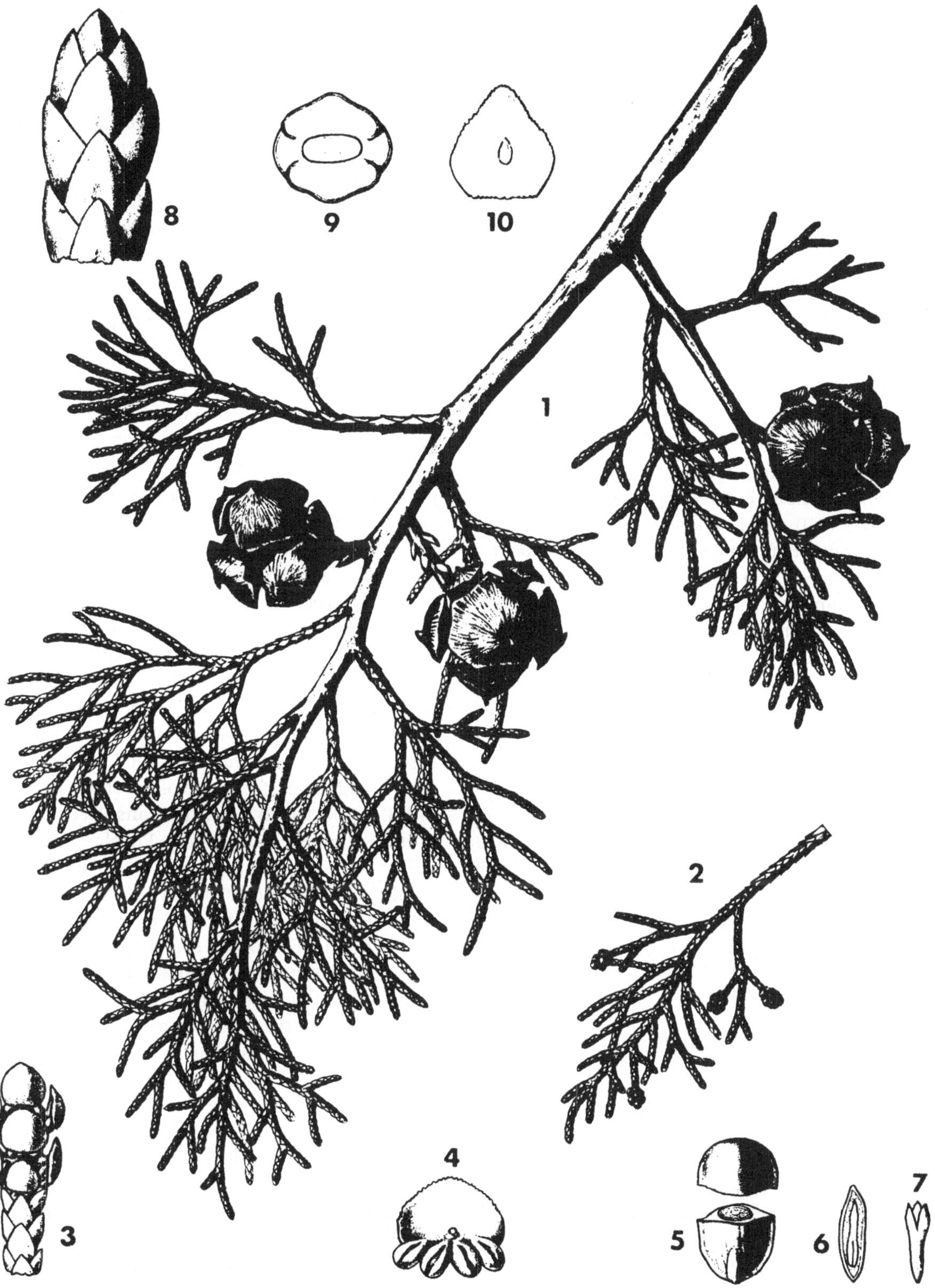


Figure 23. *Cupressus arizonica.* (1) Branch with seed cones, (2) branch with pollen cones, (3) twig with pollen cones, (4) pollen cone scale with pollen sacs, (5) seed, cross-section, (6) seed, longitudinal section, (7) embryo, (8) twig, (9) twig, cross-section, (10) leaf. (Sargent 1890). Drawn by C. E. Faxon.

of coalesced fleshy scales; ovules 1 or 2 per scale. Seeds wingless, 1–7 per cone (in Sonora), germinating within the fallen cone except when separated by animals (presumably birds and mammals). All in the Northern Hemisphere; 50 species. References: Adams (1993), Adams and Zanoni (1979), Martínez (1946), Zanoni and Adams (1975, 1979).

Juniper wood is used for firewood and fence posts in northern Mexico and the southwestern United States. Several ornamental junipers are cultivated in Sonora, and worldwide more than 250 horticultural varieties are known. In Europe "berries" of the common juniper (*J. communis*) are used to flavor gin, and most wooden pencils are made from juniper wood.

1. Bark checkered; mature cones (8) 10–20 mm in diameter, light to dark red-brown (beneath a glaucous waxy bloom); pulp hard, somewhat fibrous; cones not shrinking and not wrinkling on drying. _____ ***J. deppeana***

1' Bark longitudinally fissured, not checkered; mature cones to 9 (10) mm in diameter, orange, red, red-brown, bronze, dark blue, or bluish black (beneath a glaucous waxy bloom); pulp soft and fleshy; cones often wrinkling and shrinking on drying.

  2. Leaves with entire margins (using 30× magnification); mature cones dark bluish-black. ***J. scopulorum***

  2' Leaves with minutely toothed or denticulate margins (seen with 30× magnification; a subtle character that works best if you are familiar with the variation); mature cones lighter colored.

    3. Scale leaves not beadlike, the surface low and often angled, the apex conspicuously pointed; seeds 1 (2) per cone. _______ ***J. coahuilensis***

    3' Scale leaves at ends of branchlets usually beadlike, the surface arched and rounded, the apex blunt; seeds 1–4 (6) per cone. ***J. durangensis***

***Juniperus coahuilensis*** [*J. erythrocarpa* var. *coahuilensis*]. Rose-fruited juniper; *huata, táscale, táscate* (figure 24)

Usually large shrubs and occasionally small trees to 8 m, often branching from near the base but also with trunks to 0.5 m or more in diameter. Bark thin, grayish, shredding longitudinally in long, ragged strips. Scale leaves pointed, often angled when dry, the margins minutely toothed. Cones 6–7 mm, reddish orange beneath the rosy-glaucous surface, usually 1-seeded, maturing in 1 year, juicy (soft), sweet pulp. Pollen shed December–February.

Slopes, ridges, mesas, canyons, and arroyos; grassland, oak woodland, pine-oak woodland, and occasional in upper Chihuahuan desertscrub and Sonoran desertscrub (sometimes intermixed with the uppermost palo verdes, organpipe cacti, and saguaros). Common and widespread in northeastern and north-central Sonora at least as far south as the vicinities of Aconchi and Nacozari, 915–1700 m (3000–5500 ft). Arizona to Trans-Pecos Texas and southward to Durango, Zacatecas, and Nuevo León.

The Sonora and Arizona populations were known as *J. monosperma* and later as *J. erythrocarpa. Juniperus coahuilensis, J. pinchotii* (*J. erythrocarpa*), and *J. monosperma sensu stricto* form a complex of closely related one-seeded junipers in the southwestern United States and northern and central Mexico.

***Juniperus deppeana.*** Alligator juniper; *huata, táscale, táscate* (figure 25)

Large shrubs or trees to 10–13 (18) m; trunks often massive with distinctive gray bark checkered like that of an alligator hide with squarish or quadrangular plates. Terminal whip branches ascending or erect. Scale leaves opposite and decussate, forming twigs appearing squarish in cross section, the margins minutely toothed. Cones at first very bluish-glaucous, maturing in two years, reaching 10–14 mm in diameter, rounded, with the dry cone scale tips protruding from the body of the cone (a unique character among the junipers of northwestern Mexico and Arizona). Seeds 1–4 (7) per cone. Pollen shed February–March.

Slopes, ridges, canyons and arroyos; oak and pine-oak woodland, sometimes in mixed conifer forest, and occasionally in grassland. Widespread and common in northeastern and north-central Sonora and southward in mountains along the eastern edge of the state, 1290–2500 m (4230–8200 ft). Arizona to Trans-Pecos Texas in the United States and southward to Puebla.

This is the largest juniper in Mexico and southwestern North America. The often large size, checkered bark, squarish twigs, and large fruits set it apart from other junipers in the region. Trees of this species appear to be long lived but are difficult to date. Populations in the northern part of Sonora often include large trees with massive straight trunks, whereas the trees in the southern part of the state tend to be much smaller in stature, often shrubby, the checkering of the bark is generally not as pronounced, and the berries tend to be larger.

Several geographic varieties are described, but they are poorly understood and in need of study (Adams 1993). The Sonoran populations are reported to be var. *robusta*. A rare and unusual form, *J. deppeana* forma *sperryi*, is reported from Sonora based on a single specimen in the northeastern part of the state (14 miles north of El Coyote on the road to Huachinera; Zanoni and Adams 1979). It is otherwise known from the mountains in western Texas and distinguished by drooping, or "weeping," branches and by the stem bark, which is longitudinally furrowed into interconnected strips.

Figure 24. *Juniperus coahuilensis* (Sargent 1890). Drawn by C. E. Faxon.

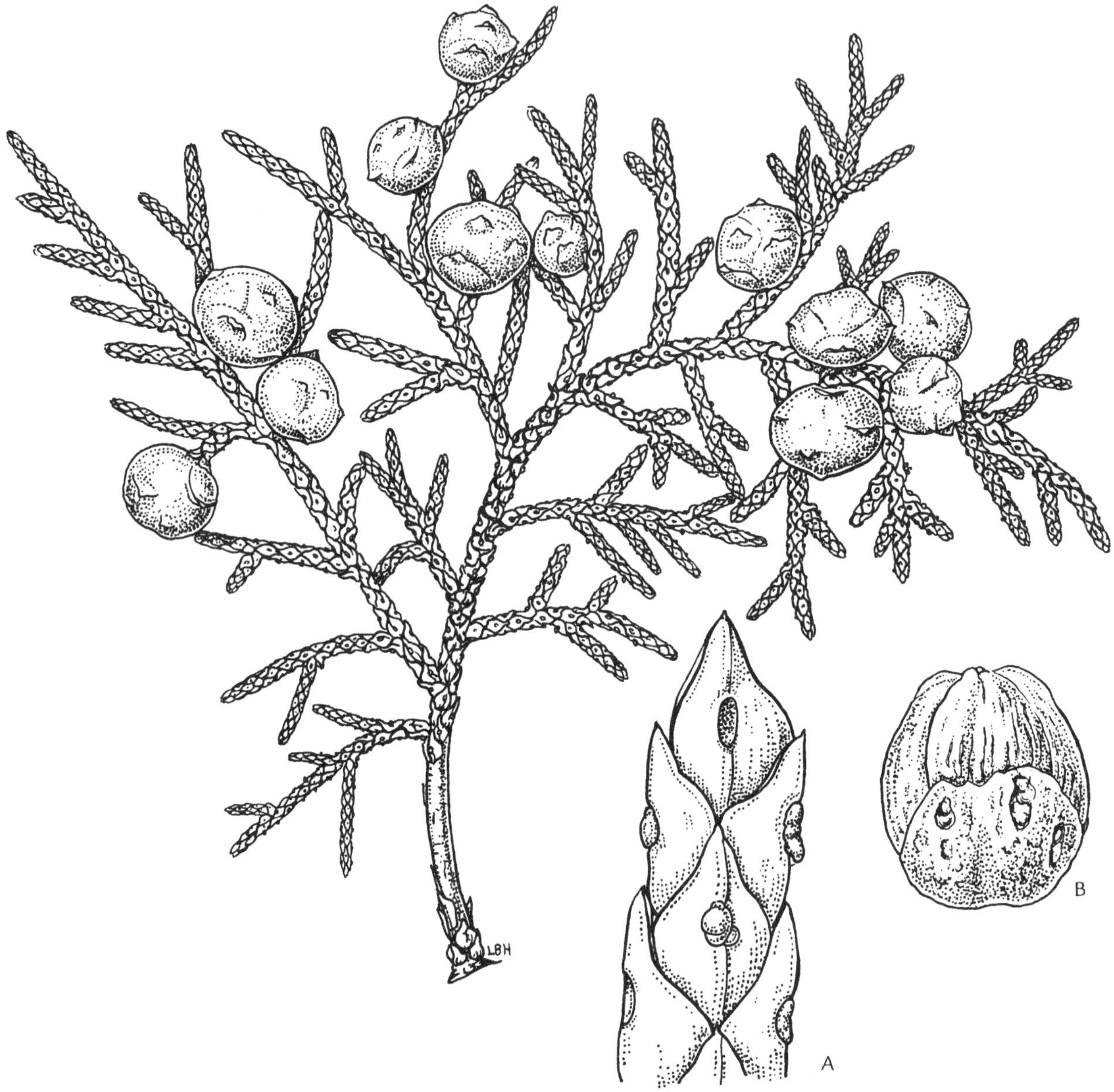


Figure 25. *Juniperus deppeana* (A) twig, (B) seed. (Parker 1958). Drawn by L. B. Hamilton.

***Juniperus durangensis.*** Durango juniper; *táscate espinosa* (in Maycoba)

Shrubs or shrubby trees to 5 m, branching from near the base; trunk occasionally to 50 cm in diameter; juvenile growth (with spinescent leaves) often occurs even on branches of mature plants. Bark light brown, shredding off in long, fibrous strips. Scale leaves opposite, forming twigs appearing squarish in cross section, often beadlike, high and arched with broad, blunt tips, the margins minutely toothed. Apparently dioecious. Cones 6–9 mm in diameter, globose, glaucous, mauve; pulp soft and juicy; probably maturing in 1 year. Seeds 1–4 (6) per cone. Pollen shed in summer, or at least after March. Cones smaller (as small as those of *J. coahuilensis*) and with fewer seeds (sometimes mostly 1-seeded) during drier years or in more xeric habitats.

Slopes in pine-oak woodland; east-central, southeastern, and northeastern (Sierra el Tigre) Sonora, 1525–2100 m (5000–6890 ft). Southwestern Chihuahua near the Sonora border in the upper Río Mayo drainage on volcanic or igneous soils. Reported for the Río Bavispe region as far north as Fronteras in northeastern Sonora (White 1948; Zanoni and Adams 1979). Sonora and Chihuahua to Durango, Zacatecas, Jalisco, and Aguascalientes.

This juniper resembles *J. coahuilensis* but is distinguished by the distinctive beadlike scale leaves and by having generally more than one seed per cone. The northern specimens, from the Sierra el Tigre, seem closer to *J. coahuilensis* but with somewhat blunter, more rounded scale leaves than those of the usual *J. coahuilensis*.

***Juniperus flaccida*** var. ***flaccida***. Weeping juniper

Reported from oak woodland in northeastern Sonora (White 1948; Zanoni and Adams 1979). Primarily a species of the Sierra Madre Oriental and mountains within the Chihuahuan Desert extending to the Chisos Mountains in Texas.

We have not seen unequivocal specimens with cones from Sonora, and based on our field work we doubt that it is present in Sonora. It occurs in far eastern Chihuahua.

***Juniperus scopulorum*** [*J. mucronata*]. Rocky Mountain juniper; *sabino, táscate; ga'a* (Pima Bajo)

Large shrubs or trees to 10 m; trunk single, often 30–50 cm in diameter. Bark brown, shallowly furrowed, thin, longitudinally shredding into straplike plates. Twigs drooping, markedly slender; leaves shiny green. Scale leaves with entire margins; resin gland especially large and conspicuous, translucent when fresh. Cones 6–8 mm, dark bluish-purple beneath a light blue glaucous surface; pulp soft and juicy, reportedly maturing in 2 years. Seeds usually 2, filling the cone; cone bilobed, or seeds occasionally 1 or 3 and cone rounded. Pollen shed February–March.

Canyon bottoms along streams and sometimes extending onto generally north-facing slopes; pine-oak woodland; northeastern and east-central Sonora, 1200–1735 m (3940–5700 ft). Locally common in Cañón Evans on the northern side of Sierra de los Ajos and the Sierra el Tigre in northeastern Sonora and mountains in east-central Sonora. Also adjacent western Chihuahua, locally in Coahuila, and widespread in western United States and southwestern Canada.

This species is closely related to and not always readily distinguished from the eastern red cedar, *J. virginiana*, of the eastern United States and adjacent Canada to Texas. The reports of *J. blancoi* from the Río Bavispe Region and the Sierra de los Ajos (Martínez 1963; Zanoni and Adams 1979) are based on misidentified specimens of *J. scopulorum*. White (1948) originally reported this juniper as *J. flaccida*. Adams (2000) segregated the east-central Sonora population of *J. scopulorum* as a new taxon, *J. mucronata*, based on leaf essential oils and RAPD DNA banding. Both analyses only weakly distinguished this population from *J. scopulorum* of the western United States.

## *Taxodium*—Bald-cypress

Variously treated as having 1 to 3 species or 1 species with 2 or 3 geographic varieties. Guatemala, Mexico, and southeastern United States. Reference: Watson (1993).

***Taxodium distichum*** var. ***mexicanum*** [*T. mucronatum*]. Montezuma bald-cypress, Mexican bald-cypress; *sabino, ahuehuete* (southern Mexico); *gagüorí* (Guarijío) (figures 26 and 27)

Trees to 25+ m with a massive trunk enlarged near the base often reaching 1–2 m in diameter; bark red-brown to brown with deep longitudinal furrows. Roots spreading and often partly exposed among streambed boulders; occasionally forming irregularly conic "knees" (e.g., in a wet canyon at La Estrella, east of Tesopaco). Essentially evergreen, although often becoming nearly leafless as branchlets are shed when the pollen cones ripen in late December and early January, or during late spring drought in seasonally dry watercourses; new leaves (on branchlets) appearing shortly after maturation of the pollen cones or with summer rains. Leaves often 1–2.5 cm (at middle portion of branchlet), relatively soft, distichous (2-ranked and lying flat in a plane) on secondary branchlets or twigs, these leafy branchlets 4–22 cm resembling a once-pinnate leaf.

Pollen cones small, produced in prodigious quantities on slender, drooping, cone-bearing branchlets to approximately 30 cm. Seed cones one to several at the distal end of the branches, maturing and scales breaking apart in one season, often emerging in late December above the staminate cone branches, these young female cones green with glistening resin droplets in scale axils; mature cones rounded, 11–20 mm wide, sessile or nearly so. Seeds 6–10 mm, angled, and wingless.

Along rivers, streams, and canyons with permanent water or sometimes seasonally dry, in habitats often subject to high-energy floods; riparian tropical deciduous forest; southeastern and east-central Sonora, 180–950 m (600–3120 ft). Guatemala to northern Mexico. Two other varieties in southeastern United States.

There are a number of fine groves of these magnificent trees in Sonora; some of the well-known stands occur in the vicinity of Nuri, numerous places along the Río Cuchujaqui, at Guirocoba, and near Cedros. The northernmost locality (near 28°32' N) in Sonora is near San Javier, on a tributary of the Río Yaqui. At this locality the surrounding vegetation is intermediate between thornscrub and tropical deciduous forest.

The durable wood is highly prized for house beams and furniture and was used for coffins. These trees are protected by law, although some of the groves have been seriously depleted. *Sabino* has ornamental potential in well-watered or riparian situations. The *Arbol de la Noche Triste*, near Mexico City, under which Hernán Cortés wept on the night of July 1, 1520, after a disastrous battle, still stands. The *Arbol del Tule* in Oaxaca is considered to have the largest

Figure 26. *Taxodium distichum* var. *mexicanum*. Río Cuchujaqui, southeast of Alamos (Bowden et al. 1993). Photograph by J. W. Dykinga.

trunk circumference of any tree in the world (Dorado et al. 1996). There are several other famous living *ahuehuete* trees that were known nearly five centuries ago.

### Pinaceae—Pine Family

Monoecious (in Sonora, except *Pinus discolor*), evergreen (in Sonora) trees with needlelike or flattened, linear leaves. Winter buds distinct with dry, membranous scales. Pollen cones relatively small, herbaceous, deciduous. Seed cones larger, mostly woody, and persisting longer, with many to numerous spirally arranged scales; scales appressed and cone closed before maturity; end or apex of scale often (in Sonora) conspicuously thickened, the exposed part (apophysis) commonly wider and generally rhombic in outline, usually transversely keeled and usually with a prominent terminal or dorsal protuberance (umbo) ending in a spine or prickle (mucro). Each scale (except the few uppermost and lowermost ones) bearing two ovules near its base. Seeds winged or not; wings persistent or detachable (deciduous).

Almost all in the Northern Hemisphere; 10 genera, 200 species. Reference: Ferguson (1998).

The family includes some of the world's most important lumber and paper-pulp trees. There are four genera in Mexico; spruce (*Picea*) occurs in adjacent Chihuahua and Arizona but not in Sonora. The three genera in Sonora are restricted to mountains in the eastern part of the state.

1. Leaves 2–6 in short-shoot fascicles, the fascicle bases with membranous sheaths. Subfamily Pinoideae: ***Pinus***

1' Leaves single, not in fascicles, without basal sheaths. Subfamily Abietoideae:
   2. Branching of twigs mostly opposite; winter buds blunt and covered with gooey resin; leaf scars circular, not on a woody peglike base (nearly flush with the stem, or if slightly raised then evenly raised all around); cones erect, breaking up scale by scale; the cone axis persistent; the scale bracts much shorter than the mature scale and not conspicuously 3-pronged. ________ ***Abies***
   
   2' Branching of twigs mostly alternate; winter buds mostly pointed, not resinous or with thin, varnish-like resin; leaf scars usually oval, on a small woody peglike base, tilted, slightly higher on the lower

Figure 27. *Taxodium distichum* var. *mexicanum*. A medium-sized tree, southeast of Alamos. Photograph by R. S. Felger.

(proximal) side and nearly flush on the upper side; cones pendulous, falling intact, the scales persistent with a conspicuous 3-pronged bract longer than the scale. ______________________ ***Pseudotsuga***

### *Abies*—Fir, *Oyamel, Pinabete*

Temperate and cool regions of the Northern Hemisphere, mostly Eurasia and North America; 42 species. References: Ferguson (1998), Liu (1971), Martínez (1948a).

This genus includes important timber trees and many horticultural selections in cool temperate and subarctic regions.

1. Northeastern Sonora; twigs gray-brown; leaves mostly > 2 mm wide, relatively robust and convex dorsally. ______________________ ***A. concolor***

1' East-central Sonora; twigs dark brown; many leaves < 2 mm wide, generally somewhat flattened with a longitudinal furrow or groove on the upper surface. _______________ ***A. durangensis***

***Abies concolor.*** White fir; *pinabete* (figure 28)

Conical, spirelike trees to about 30 m; bark on large trees dark, rough, and thick with deep longitudinal fissures; primary branches repeatedly branched in one plane to form large, flat, horizontal sprays. Twigs mostly opposite, gray-brown, at least sometimes glabrous; younger twigs with resin blisters. Winter buds subglobose, covered with globs of resin. Leaves single (not in fascicles), 2.5–7.5 cm × (1.5) 2.0–2.8 mm, green or glaucous-bluish (both colors may occur on the same or different trees), slightly narrowed and often twisted near the base to turn the leaves into a single plane (two ranked), somewhat flattened but with a convex dorsal surface, often slightly to strongly curved upward, with a prominent midrib on the lower surface, the tips blunt or short-pointed. Leaf scars circular, nearly flush with the stem, or if slightly raised then evenly raised all around. Cones in upper part of tree; erect, cylindrical, rounded-blunt at the apex, longer than wide, probably 7.5–12.5 × 5 cm (based on Sonora and southern Arizona trees), smooth surfaced (the scales not bristly or spinescent), green, maturing to tan and scales falling apart within a single season, leaving a persistent cone axis. Cone scales deciduous, many, thin, crowded, and closely overlapping before falling away; resinous, broader than long, and fan-shaped, 2.6–4.5 cm wide; bracts oblanceolate, slightly more than half the scale length, and with a mucronate tip. Seeds pyramid-shaped with large, broad wings. Pollen shed in May.

Mountain slopes, canyons, and streams; mixed conifer forest; northeastern Sonora, 2150–2300 m (7050–7545 ft). In Sonora only known from one drainage in the Sierra de los Ajos, a single locality in the Sierra el Tigre at Rancho de Cruz Díaz (collected in 1940 and not seen during our recent fieldwork), and reported in the Sierra San Jose by Mearns (1907). Also northern Chihuahua, Baja California, and western United States. There are numerous horticultural selections of this species.

***Abies durangensis.*** Durango fir; *Pinabete*

Trees to 25 m, with a trunk to 70–96 cm in diameter (to 33 m with a trunk to 1 m in diameter in nearby Chihuahua). Resembling *A. concolor*, but bark not so deeply furrowed, darker twigs, generally shorter and narrower leaves with a median dorsal groove, and smaller cones. Twigs dark gray, the youngest ones sometimes with minute hairs. Leaves mostly 2.5–4.0 (6) cm × 1.5–2 mm. Cones 6.5–7.7 × up to 4.2 cm; scales 1.5–2.5 cm wide.

Watercourses in canyons; primarily within pine-oak forest; east-central Sonora, (ca. 1800) 1900–2060 m [(ca. 5900) 6230–6760 ft]. In Sonora known from one canyon below the northwestern side of Mesa el Campanero near the Chihuahua border. Widespread but localized in southwestern Chihuahua where there is more high-elevation habitat, such as moist canyons in the upper Río Mayo drainage at 1700–2700 m (5575–8850 ft). Also Durango and Jalisco.

Another fir, from Coahuila, in northeastern Mexico, described as *A. coahuilensis*, has been treated as *A. durangensis* var. *coahuilensis*. *Abies durangensis* is closely related to *A. concolor;* both are placed in Series *Grandes* by Liu (1971), which includes several other species. Other than their non-overlapping ranges, the distinction between the two Sonoran firs is poorly understood. Differences in leaf width and dorsal surfaces offer convenient characters for comparison. There is a lack of a good series of cones for study, a difficult situation to overcome because the cones are at the top of very tall trees and disintegrate before falling. McVaugh (1992: 28) suggests that *A. durangensis*, at least in Jalisco, "might better be treated as a local population of *A. guatemalensis*." Leafy branches of *A. durangensis* might be confused with those of *Pseudotsuga*.

## *Pinus*—Pine, *Pino*

Small to large evergreen trees. Long-shoot buds (actually a compound bud) covered with many overlapping, resinous scales; long-shoot leaves dry, bractlike, variably deciduous. Short shoots axillary from long-shoot scale leaves, the bases enclosed by persistent or deciduous sheaths of membranous scales, with 2–5 (6) green, slender leaves (needles) in bundles (fascicles). Pollen cones clustered near ends of twigs at the base of the current season's growth, shedding pollen in spring or early summer and soon deciduous. Seed cones produced near the apex of twigs (subterminal—actually axial), after fertilization remaining without appreciable growth until the second season, becoming woody, maturing in the second or sometimes the third year, falling intact (or sometimes a few basal scales remain on the peduncle attached to the branch), or persistent for a number of years, the cone scales often with a mucro (prickle or spine) on the umbo (tip of cone scale). Seeds normally 2 per cone scale, elliptic, with or without a thin, papery, persistent or detachable wing.

Northern Hemisphere; 110 species. References: Bailey and Hawksworth (1979), Carvajal and McVaugh (1992), Critchfield and Little (1966), Debreczy and Rácz (1995), Farjon (1984), Farjon and Styles (1997), Ferguson (1998), Klaus (1987), Little and Critchfield (1969), Loock (1950), Malusa (1992), Martínez (1948b), Mirov (1967), Perry (1991), Price et al. (1998), Richardson (1998), Shaw (1909), Stead and Styles (1984).

Many are important timber trees and yield rosin and edible seeds (pine nuts) and are the source of turpentine.

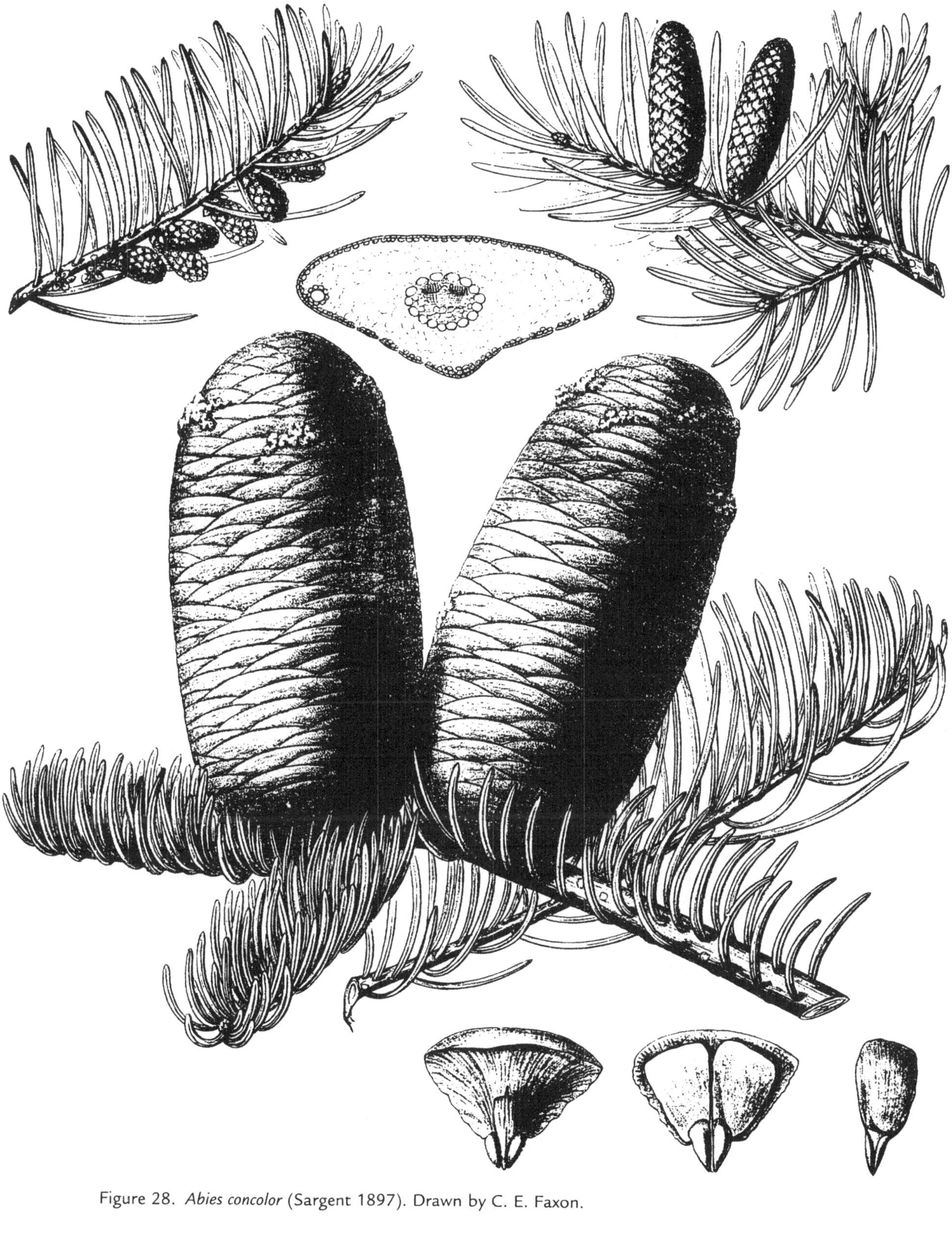

Figure 28. *Abies concolor* (Sargent 1897). Drawn by C. E. Faxon.

Pines and oaks are the most common trees at higher elevations in the eastern half of Sonora. In northern Sonora they generally occur above 1600 m, but in southeastern Sonora are as low as 1000 m or sometimes to 600 m on red hydrothermally altered soils. Pines are absent from the western half of Sonora. Five subtropical montane pines are at their northern limits in Sonora and adjacent western Chihuahua: *P. herrerae*, *P. lumholtzii*, *P. maximinoi*, *P. oocarpa*, and *P. yecorensis*.

Many of the great pine forests of the Sierra Madre Occidental and the sky islands of northeastern Sonora have been decimated by logging. Some forests, such as those in the Sierra de los Ajos, were cut during the first half of the twentieth century and are recovering. A few, such as those in parts of the Sierra San Luis and the Sierra el Tigre, remain in remarkably good condition. Recent catastrophic fires have destroyed large forest tracts in the Sierra San Luis and elsewhere. In earlier times loggers sought specific qualities and were selective about cutting certain pines and leaving others. For example, trees with harder wood were selected for mine props and brought higher prices. These were trees with closer, tighter growth rings and grew on xeric or harsher sites, such as higher slopes. Some species are preferred for lumber, largely because of their size (e.g., *P. engelmannii* and *P. yecorensis*).

Scotch pine (*Pinus sylvestris*), planted in Cañón Evans in the Sierra de los Ajos, is naturalized at an orchard; it has light gray bark, is reddish in the fissures, leaves 2–6.4 cm, two per bundle, and small, delicate cones 2.5–6.5 cm. The Aleppo pine (*P. halepensis*) is cultivated at lower elevations, especially in the northern part of the state.

Twig diameters are those of the previous year's growth, within the leafy portion of the twig. Number of leaves (needles) per bundle (fascicle) can be a useful character, although unusual leaf numbers are occasionally seen.

Worldwide there are two subgenera. The subgenera are often called "soft" and "hard" pines, but these terms are not very meaningful because some of the soft pines in Mexico have harder wood than some of the hard pines. The Sonoran pines may be classified as follows (see Price et al. 1998):

Subgenus **Strobus** (*Haploxylon*), white or soft (*blandos*) pines
- Section *Strobus*, subsection *Strobi*: *P. strobiformis*
- Section *Parrya*, subsection *Cembroides*: *P. discolor*

Subgenus **Pinus** (*Diploxylon*), hard (*duros*) pines
- Section *Pinus*
- Subsection *Leiophyllae*: *P. leiophylla*, *P. lumholtzii*
- Subsection *Oocarpae*: *P. herrerae*, *P. oocarpa*
- Subsection *Ponderosae*: *P. arizonica*, *P. engelmannii*, *P. maximinoi*, *P. ponderosa*, *P. yecorensis*

1. Leaf sheaths deciduous (second year or season leaf bundles without sheaths at bases) or reflexed (curled back).
   2. Leaves stiffly pendent from the base (drooping); southeastern Sonora. ______ ***P. lumholtzii***
   2' Leaves erect or spreading (not drooping, or if partially drooping then the leaves not pendent from the base).
      3. Leaves 5 per bundle; cones > 13 cm, more than twice as long as wide. ______ ***P. strobiformis***
      3' Leaves 3 (occasionally 2 or 4) per bundle; cones ≤ 6.5 cm, less than twice as long as wide.
         4. Leaves mostly < 5 cm; cones light brown, not persistent; seeds wingless. ___ ***P. discolor***
         4' Leaves mostly > 6 cm; cones dark gray or blackish, persistent; seeds winged. ***P. leiophylla***

1' Leaf sheaths persistent and not curled.
   5. Leaves usually 3 per bundle.
      6. Most twigs notably thick; leaves 20–40 cm; cones 8.5–22 cm. ______ ***P. engelmannii***
      6' Twigs not notably thick; leaves mostly < 18 cm; cones < 9 cm.
         7. Leaves relatively slender, bright green; cones mostly < 5 cm; southeastern Sonora. ***P. herrerae***
         7' Leaves coarse, dull green; cones 5–9 cm; higher elevations in northeastern Sonora. ***P. ponderosa***
   5' Leaves 5 (occasionally 4 or 6) per bundle.
      8. Most cone scales relatively flat, scarcely thickened to not thickened at tip.
         9. Leaves slender, partially drooping; cones not serotinous, mostly 6–11.5 cm; mature cones about 1.5 times longer than wide, the basal scales not conspicuously swollen, the inner (upper) scale surfaces with only a thin, T-shaped dark area. ______ ***P. maximinoi***
         9' Leaves coarse, more spreading than drooping; cones serotinous (see text), mostly 4.5–8 cm, the mature cones about as broad or broader than long, the basal scales somewhat swollen (convex-rounded), the middle and upper scales with sharply defined, broad, dark tips on inner (upper) surfaces. ____ ***P. oocarpa***
      8' Cone scales obviously thickened at tip.
         10. Leaves mostly 15–24 cm, spreading; cones relatively light weight, mostly < 8.5 cm, essentially sessile; higher elevations. ***P. arizonica***
         10' Leaves mostly 20–35 (38) cm, somewhat drooping; cones relatively heavy (woody), mostly > 8.5 cm, reflexed on a thick peduncle; lower elevations. __ ***P. yecorensis***

***Pinus arizonica*** [*P. ponderosa* var. *arizonica*; *P. durangensis* forma *quinquefoliata*]. Arizona pine, western yellow pine; *pino blanco*, *pino fino* (figure 29)

Handsome trees to > 20 m (to > 30 m in some areas of Chihuahua and Arizona), the larger ones often with tall boles, the trunk to 52 cm in diameter, the crown rounded. Bark grayish brown, relatively smooth with shallow furrows, often becoming reddish to orange-brown, flaking in large, irregular plates. Twigs generally < 10 mm in diameter, glaucous towards tip. Leaves (4) 5 (6) per bundle, (10.5) 15–24 cm, bright green, relatively straight, spreading (not drooping) and slender, approximately 1.0 mm wide; sheaths persistent, dark, and short. (In winter, older leaves turn a rusty golden color before falling.) Cones small to medium sized, essentially sessile, 4.5–8.5 (10) cm, about as long to longer than wide, asymmetrical, light brown to gray. Cone scales rigid, slightly reflexed, the basal scales usually remaining on the twig after the rest of the cone falls (so the lowermost scales of the cone are missing); scale apophyses pyramidal and moderately thickened at apex, transversely keeled; umbo gray, prominently downcurved, with a weak and sometimes inward-curved gray prickle. Seeds 6–8 mm, brown; wing detachable, extending 14–19 mm beyond the seed. Pollen shed April–May.

Slopes, ridges, canyons, and often on peaks; upper pine-oak woodland, mixed conifer forest; north-central, northeastern, and east-central Sonora, 1850–2450 m (6070–8035 ft). Also Chihuahua, Durango, southeastern Arizona north to the Pinaleño Mountains and southwestern New Mexico.

This is one of the most common pines in the higher mountains of northeastern Sonora southward to the Sierra Huachinera and often forms extensive forests which have been heavily exploited for lumber. A disjunct population in east-central Sonora occurs in the vicinity of Yécora on the north slope of Mesa el Campanero at 1920–2140 m; this pine is also common in nearby southwestern Chihuahua.

We classify *P. arizonica* as monotypic. Farjon and Styles (1997) treat the Sonoran populations as belonging to var. *arizonica*. They recognize *P. arizonica* var. *stormiae* from northeastern Mexico and western Texas and var. *cooperi* from southwestern Chihuahua and Durango. In our opinion the relationships of these taxa deserve further study, as well as the relationship of *P. arizonica* to *P. durangensis*. In southern Arizona *P. arizonica* may hybridize with *P. ponderosa*.

***Pinus discolor*** [*P. cembroides* var. *bicolor*; *P. c.* subsp. *cembroides* var. *bicolor*] Border pinyon; *piñón*, *pino piñonero* (figure 30)

Mostly large shrubs to small trees, or rarely to about 10 m, often branching near the base. Bark dark gray, rough, and flaking. Twigs relatively smooth, gray, and slender (ca. 5 mm in diameter). Leaves (rarely 2) 3 (4) per bundle, 2.5–5.2 cm, relatively stout, rigid, often curved upward, without serrated margins, conspicuously glaucous on inner faces, with stomata on the inner surfaces but absent from the outer surfaces; sheaths curled into rosettes and eventually deciduous. Usually dioecious (unique among the Sonoran pines). Cones small, 2.3–4.0 cm, usually wider than long, light brown, deciduous soon after the seeds ripen; scales relatively few, thick, the tips blunt, knobbed, and without spines; nearly sessile or the peduncles short. Seeds edible (piñones or pinyon nuts), obovoid, 10–13 mm, wingless; seeds falling mid-September–October. Pollen shed late May–June.

Slopes, ridges, and canyons; oak woodland and pine-oak woodland; north-central, northeastern, and east-central Sonora, 1525–2490 m (5000–8170 ft). Most major mountains in northeastern Sonora southward to the Sierra Huachinera and east-central Sonora east of Maycoba. Also southeastern Arizona and southwestern New Mexico to northern Mexico including Chihuahua and Durango. Readily distinguished from other Sonoran pines by its relatively small stature, short leaves, short and broad cones, and large, edible, and wingless seeds.

*Pinus discolor* was originally considered a variety (var. *bicolor*) of the Mexican pinyon *P. cembroides*, and some recent authors treat them as synonyms. *Pinus cembroides* is distinguished in part by its monoecious habit, leaves 2 or 3 per bundle, less distinctly bicolored leaves with stomata on both surfaces, larger cones, and generally larger trees. Passini (1994) and Farjon and Styles (1997) synonymized *P. johannis*, a dioecious pinyon of the Sierra Madre Oriental of Nuevo León and Zacatecas, with *P. discolor*, a decision that in our opinion merits further study.

*Pinus cembroides sensu stricto* occurs in western Chihuahua as a tree to 17 m with a trunk to 50 cm in diameter. It occurs immediately east of the Río Mayo drainage and extends southward to Puebla. *Pinus cembroides* may enter Sonora in areas adjacent to western Chihuahua. Farjon and Styles (1997) indicate that *P. cembroides* subsp. *cembroides* occurs in northeastern Sonora. We have seen numerous specimens from the two areas cited (Sierra San Luis and Sierra San José), and they are *P. discolor*. A single tree of *P. cembroides* was found along the roadside in east-central Sonora, just east of Santa Rosa on the old road to Yécora. Presumably it grew from seed transported from Chihuahua.

The Colorado pinyon, *P. edulis* of the pinyon-juniper regions of New Mexico, northern Arizona, Colorado, and Utah, produces the pinyon nuts of commerce. The raw or roasted, shelled seeds, or pinyon nuts, are the most important wild-harvested nut crop in the United States. Together with the acorns, or *bellotas*, from Emory oak (*Quercus emoryi*) in northern Sonora, pinyon nuts are one of the few remaining economically viable wild-harvested food crops

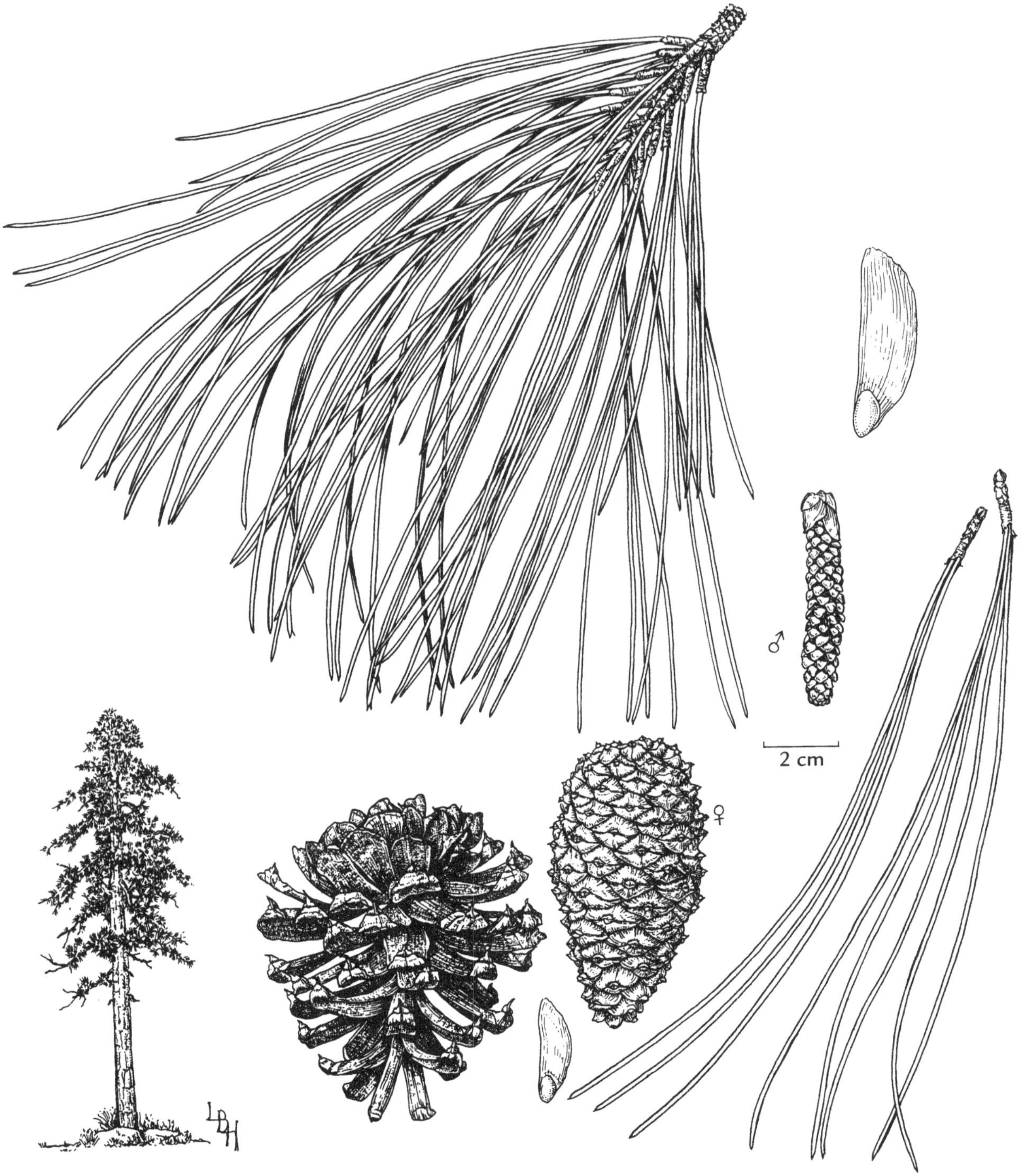


Figure 29. *Pinus ponderosa* var. *scopulorum*, tree, twig, leaf bundle with 3 leaves, pollen cone, immature seed cone, mature seed cone, and seeds; *P. arizonica* leaf bundle with 5 leaves. Drawn by L. B. Hamilton.

Figure 30. *Pinus discolor*. (Sargent 1897). Drawn by C. E. Faxon.

in southwestern North America. *Pinus discolor*, especially in Sonora, is not economically significant due to its relatively limited distribution and small, hard, thick-shelled seeds.

The Colorado pinyon (*P. edulis*) is used as a landscape subject at higher elevations in New Mexico. They are often transplanted from the wild as small trees rather than seed-grown because they are so slow growing. Our field observations indicate that *P. discolor* should be faster-growing than *P. edulis*, especially in cultivation. The border pinyon should be cultivated. It is recommended as a landscape subject for parks and public places at higher elevations in Arizona, Chihuahua, and Sonora in areas such as Nogales, Cananea, and Yécora.

***Pinus engelmannii*** [*P. apacheca; P. latifolia; P. macrophylla; P. mayriana*]. Apache pine, Arizona long-leaf pine; *pino real*.

Stately trees to > 25 m, forming a trunk to 70 cm in diameter, and a rather sparsely branched, open canopy. Bark reddish brown and rough with dark furrows. Twigs thick, 11.5–20 mm in diameter. Leaves in large pom-pom tufts at ends of branches, 3 per bundle (or sometimes a few with 4 or 5 leaves), 20–42.5 cm, coarse and very large (1.5–2.0 mm in diameter), the sheaths persistent, dark, and resinous. Cones single, or in twos or threes, medium-sized to large, 8.5–22 cm, longer than wide, ovoid and slightly asymmetric (curved), light brown to gray, sessile; lower scales usually remaining on the twig after the cone falls; scale apophysis much thickened and pyramidal; umbo spine usually sharp and prominent. Seeds (5.3) 6.2–8.1 mm, mottled, the wing broad, detachable, extending 17–24 mm beyond the seed. Pollen shed May–June.

Ridges, mesas, slopes, and canyons; oak woodland and pine-oak woodland; north-central and eastern Sonora, 1290–2600 m (4230–8530 ft). Southeastern Arizona and southwestern New Mexico southward in the Sierra Madre Occidental to Durango, Sinaloa, and Zacatecas.

The lowest elevation record for this species in Sonora is on the Río Maycoba east of Yécora. At higher elevations such as on the Mesa el Campanero (Mesa Enmedio), it is the most common pine, and it formed extensive forests before large-scale logging began in the mid-twentieth century. The wood is often hand-cut into shingles.

***Pinus herrerae***. Herrera pine, shag-bark pine; *pino chino* (figure 31)

Trees to > 15 m; trunk usually slender, to 41 cm in diameter; bark shaggy, longitudinally furrowed into large, rough plates, dark gray, exfoliating to cinnamon brown; crown dense, rounded; twigs slender. Leaves 3 per bundle, 7–16 cm × 1.0 mm or less, relatively flexible, bright green, the sheaths persistent. Cones small, 2.3–5.2 cm, symmetric or nearly so, broadly ovoid, dark gray; umbo spines minute or deciduous. Cones on reflexed peduncles < 10 mm, deciduous with the attached peduncle, although often persisting for several years after the seeds are shed. Seeds 4 mm, dark, the wing prominent, detachable, extending 8 mm beyond the seed. Pollen shed April–May.

Pine-oak woodland; southeastern and east-central Sonora, 1700–2000 m (5575–6560 ft). In Sonora known from Sierra Saguaribo (near Santa Bárbara) and the Chihuahua border near Maycoba and on Mesa el Campanero. Also in nearby Chihuahua to 2250 m, where it attains large sizes in mesic canyons in mountains near Ocampo and Basaseachic. Pacific slopes of the Sierra Madre southward to Michoacán, Guerrero, and the state of México.

This is one of the smallest-coned pines anywhere, and with *P. discolor* has the smallest cones in Sonora. *Pinus herrerae* somewhat resembles *P. leiophylla* var. *chihuahuana* but is distinguished by the plated gray bark, persistent fascicle sheaths, thinner leaves, and smaller gray and brown cones that mature in two rather than in three years. Critchfield and Little (1966) treated *P. herrerae* as a synonym of the related *P. teocote*.

***Pinus leiophylla*** var. ***chihuahuana*** [*P. chihuahuana*]. Chihuahua pine; *pino chino, pino chimonque* (figure 32)

Usually slender trees, often reaching 15–20 m with a trunk to 46 cm in diameter; bark rough and dark-colored. Crown open and often somewhat conical; branches peppered with persistent, dark cones. Twigs 2.5–5.5 mm

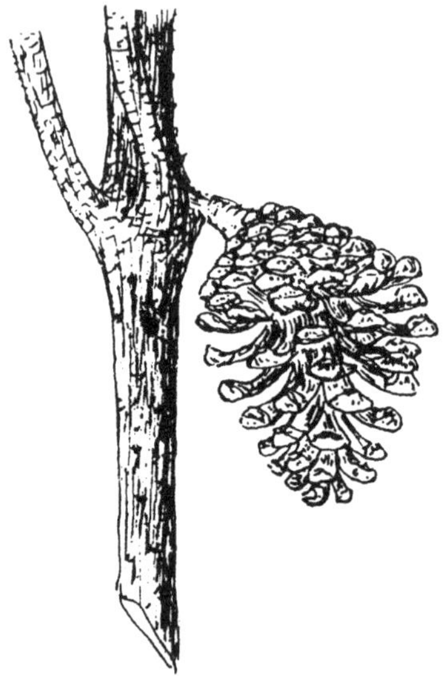

Figure 31. *Pinus herrerae*. (Martínez 1948b). Drawn by M. Ornelas Carillo.

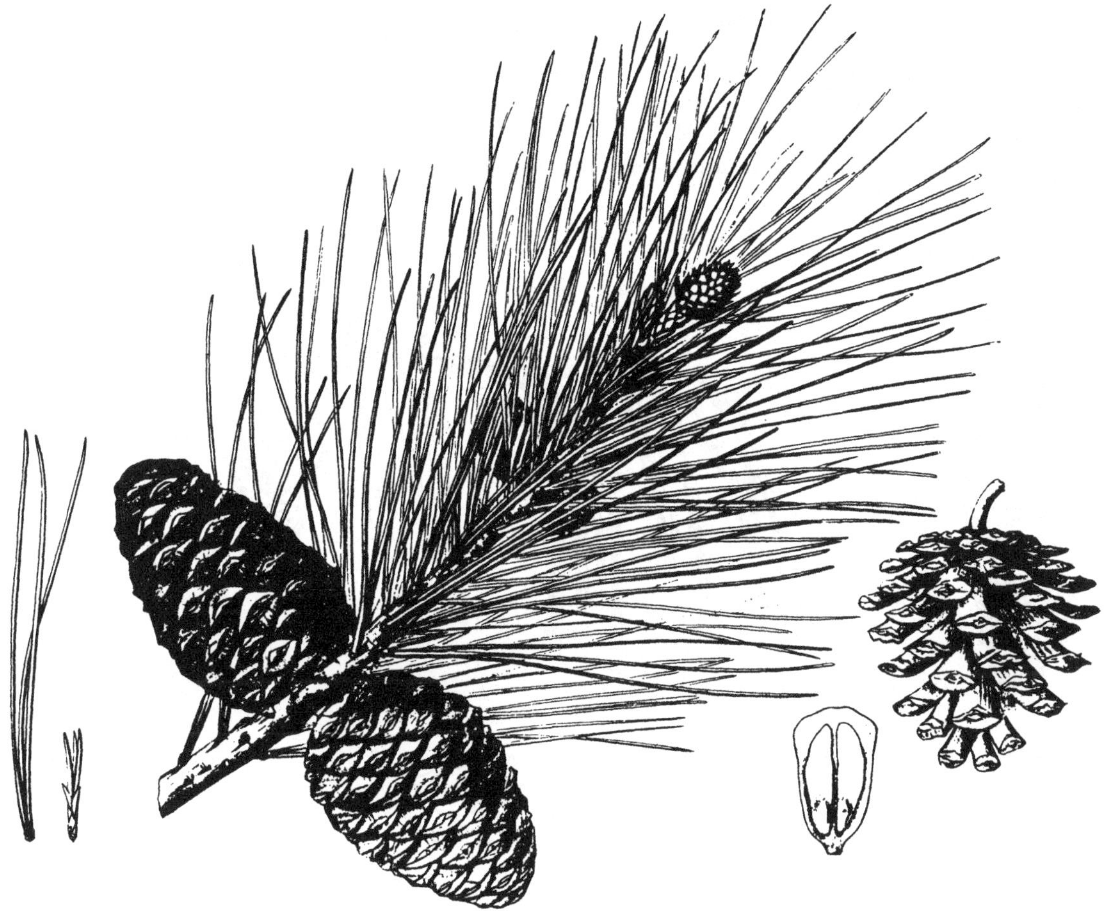

Figure 32. *Pinus leiophylla* var. *chihuahuana*. (Sargent 1897). Drawn by C. E. Faxon.

in diameter. Leaves (rarely 2) 3 (4) per bundle, (4.5) 6–12 (16) cm × 1.5 mm, dull green, relatively stout; sheaths quickly deciduous. Cones small, 3.2–6.6 (7.2) cm, often as wide or slightly wider than long, dark gray (often blackish), requiring three years to mature and persisting for years after opening; scales nearly flat; umbo prickle weak and inconspicuous; peduncles 10–13.7 mm, curved, relatively thick. (Cones infected with an orange-colored rust fungus, *Cronartium conigenum*, become abnormally swollen.) Seeds (4) 5.8–6.8 mm, mottled, the wing broad, detachable, extending 7–18.8 mm beyond the seed. Pollen shed June–July (later than for other pines in Sonora).

Slopes, ridges, and canyons; oak woodland and pine-oak woodland. Widespread and common; north-central and eastern Sonora, 1200–2060 m (3940–6760 ft). Southwestern New Mexico and southeastern and central Arizona to Jalisco, Aguascalientes, and Zacatecas.

Various authors (e.g., Carjaval and McVaugh 1992, Farjon and Styles 1997) treat *P. chihuahuana* as a variety of *P. leiophylla* and note that further study of the relationships of these two pines is desirable. Variety *chihuahuana* extends farthest north, while var. *leiophylla* ranges substantially farther southward (to Oaxaca and Veracruz), but they overlap at least from Chihuahua to Zacatecas. Carvajal and McVaugh (1992) indicate that var. *leiophylla* might extend into Sonora, and Farjon and Styles (1997) map it in northeastern Sonora adjacent to the populations in western Chihuahua but do not cite a locality.

Variety *leiophylla* is distinguished by having leaves markedly more slender (< 1.0 mm wide) and five per bundle. The two varieties retain their differences over large areas of Durango (Farjon and Styles 1997) and as our observations indicate, apparently in Chihuahua as well. In western Chihuahua, var. *leiophylla* occurs sporadically in relatively mesic, high-elevation sites, or along canyon bottoms, whereas var. *chihuahuana* is widespread in these sites and in more xeric areas. These are the only North American pines in which the cones require three years to mature. Another unusual character is the ability of the stump (or burned tree) to produce sprouts, although resprouted stumps generally do not develop into trees.

***Pinus lumholtzii.*** Lumholtz pine; *pino triste* (figure 33)

Trees usually < 15 m, sometimes to 18 m; trunk to 37 cm in diameter. Bark dark gray and rough. Twigs relatively slender, often ≤ 5 mm in diameter. Leaves (2) 3 (rarely 4) per bundle, 7.5–19 (23) cm, mostly in one plane and usually stiffly drooping, yellow green, somewhat stout (like those of *P. leiophylla* var. *chihuahuana*), the sheaths long, reddish brown and deciduous after the first season. Cones maturing in two years and often persisting for several more years, usually turned downward (peduncle curved), symmetrical, small, 3.3–5.2 (5.8) cm, gray, ovoid to ovoid-

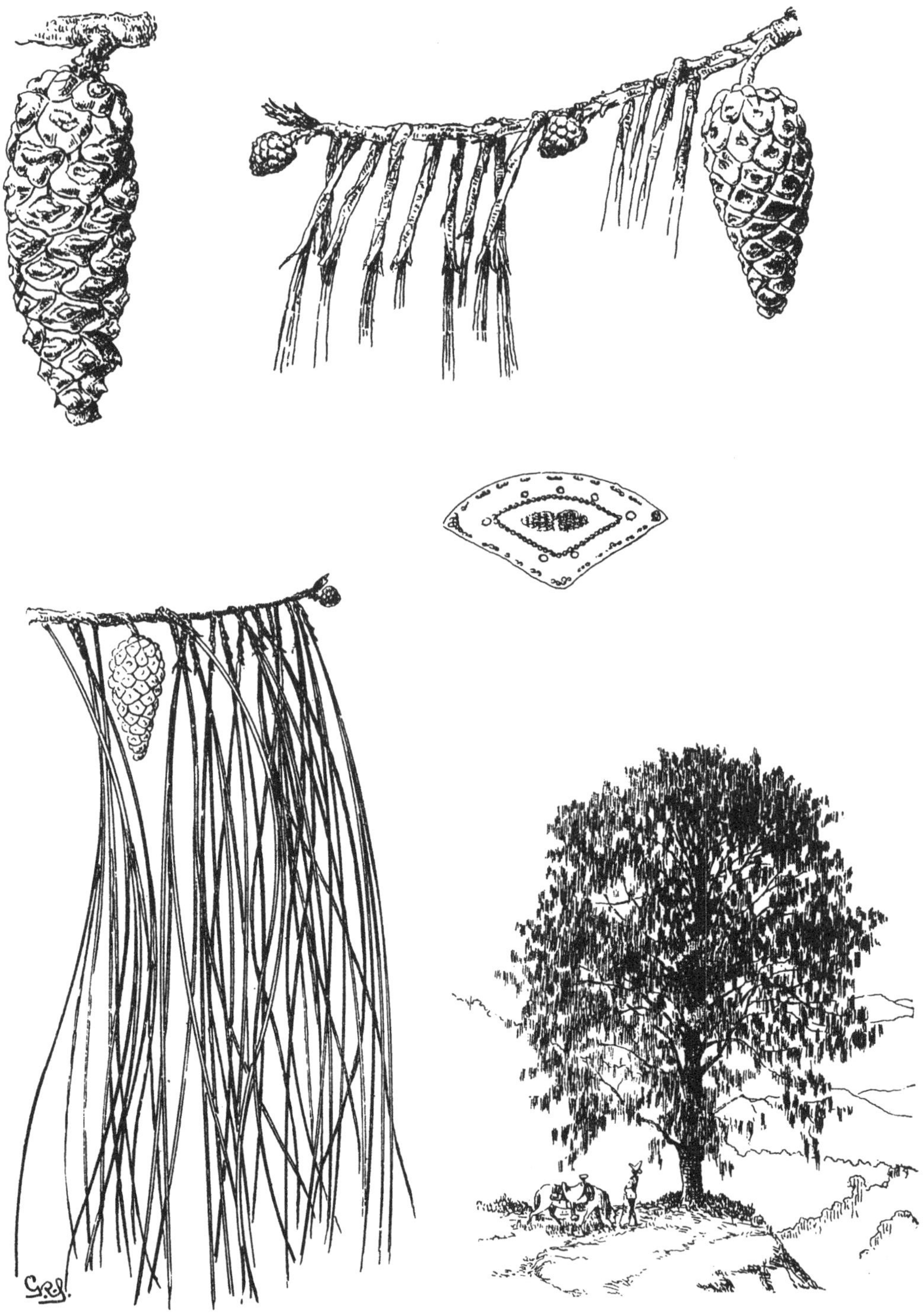

Figure 33. *Pinus lumholtzii.* (Shaw 1909). Drawn by G. R. Shaw.

conical; scales rather thin and often reflexed; umbo with a short, inconspicuous, or soon-deciduous spine; peduncle 8–16 mm, slender, usually falling with the cone. Seeds with a deciduous wing. Pollen shed before April.

Slopes and ridges; pine-oak woodland; east-central Sonora near Yécora at 1720 m and southeastern Sonora on Sierra Saguaribo above 1600 m. Sometimes common in the mountains of southwestern Chihuahua to 2300 m. Pacific slopes of the Sierra Madre Occidental southward to Jalisco and Guanajuato.

The stiffly pendent leaves are unique among pines in northwestern Mexico, giving the trees a wilted or drooping appearance, hence the name "pino triste." Farther south, such as in the Sierra Surotato, Sinaloa, the leaves tend to be longer (to 30 cm). Short-leaved specimens of *P. lumholtzii* might be difficult to distinguish from sterile specimens of *P. leiophylla* var. *chihuahuana*. The leaves of *P. lumholtzii* are characteristically rather glossy, yellow green, and pendent, whereas those of *P. leiophylla* are rather dull green and more spreading. There are indications of genetic introgression (Perry 1991; Spellenberg et al. 1996).

***Pinus maximinoi*** [*P. tenuifolia; P. pseudostrobus* var. *tenuifolia*]. Maximino pine (figure 34)

Trees to more than 20 m; trunk to 70 cm in diameter. Bark reddish brown to dark gray, with dark furrows, yet relatively smooth and forming thin, irregular plates. Twigs relatively slender, < 15 mm in diameter, glaucous toward the tip. Leaves 5 per bundle, 18–30 cm, slender (< 1 mm diameter), partially drooping, the inner surfaces glaucous (at least at first); sheaths persistent, light brown. Cones medium sized, 6–11.5 × 6–7.5 cm, ovate-oblong (longer than wide), often slightly curved, lustrous brown; upper scales often not opening or opening tardily; scales firm but rather thin throughout with the apophysis only scarcely raised, and the umbo flat and usually blunt, or the apophysis sometimes more prominent, pointed or with a minute prickle; peduncles relatively slender (< 10 mm in diameter), 13–20 mm, usually falling attached to the cone. Seeds 6–6.5 mm, with detachable wing extending 15–25 mm beyond the seed. Pollen shed in April.

Canyons with riparian habitats; pine-oak woodland; southeastern Sonora, 1250–1600 m (4100–5250 ft), and to 2040 m in nearby Chihuahua where the trees reach approximately 30 m. Sonora to Nicaragua. In Sonora in the Sierra Milpillas, and disjunct in scattered localities near Yécora and adjacent southwestern Chihuahua.

This species is a member of a species complex that includes the closely related and more southern *P. douglasiana* and *P. pseudostrobus*. The elongated, often curved cones with their thin, stiff scales, the peduncle falling with the cones, and the internal leaf anatomy are diagnostic. The cones are sometimes infected with the fungus *Cronartium*, resulting in deformed and enlarged cones, resembling a small pineapple and having fused cone scales.

***Pinus oocarpa***. Egg-cone pine; *pino; joj'có* (Guarijío) (figure 35)

Trees usually < 20 m; trunk to 50 cm in diameter, with a rounded crown. Bark rough, dark gray, somewhat checkered and thinly flaking to expose brown plates, and not deeply furrowed until the trunk becomes large. Twigs scaly, usually < 15 mm in diameter. Leaves (4) 5 per bundle, 15–27 (31.5) cm, relatively coarse (ca. ≥ 1 mm in diameter), light green, more spreading than drooping, the sheaths persistent. Cones small to medium sized, (3.5) 4.5–8.0 (8.5) cm, relatively heavy (woody), elegant and distinctive, often ovoid (egg-shaped) before opening, ochre-yellow or with a greenish tint, about as broad or broader than long after opening and conical or top-shaped, slightly asymmetrical, long-persistent, opening slowly after ripening (serotinous—the seeds released over a relatively long period of time; the only serotinous-coned pine in Sonora), the scales rigid, opening gradually from the apex toward the peduncle; lower scales spreading widely; upper scales remaining closed or partially spreading; basal scales small, numerous, thick, and rather bulbous, dome-shaped, closely overlapping, the scales otherwise flat; scale apex rounded in outline; umbo sometimes with a short, hard, prickle; peduncle 15–25 mm and scaly, falling with the cone. Color pattern unique; young cones somewhat greenish tan; open, older cone scales lustrous light brown (turning gray) with sharply defined, broad dark tips on the otherwise red-brown inner surfaces. Seeds 5.5–7.5 mm, dark and mottled, the wing detachable, thickened near the seed, extending 12–17 mm beyond the seed. Pollen shed March–May.

Slopes and ridges; pine-oak woodland; east-central and southeastern Sonora, 600–1700 m (1970–5575 ft). Sonora and southwestern Chihuahua to Nicaragua.

This is one of the most widespread Latin American pines. *Pinus oocarpa* and *P. caribaea* are the southernmost New World pines. *Pinus oocarpa* and *P. yecorensis* occur at lower elevations than any other pines in Sonora. The lowest elevation populations (600–1000 m) are on isolated pockets of altered soils within tropical deciduous forest and on some north-facing ridges; low-elevation stands among oaks begin at about 900 m. In southeast Sonora *P. oocarpa* is commonly the first pine species encountered upon climbing into the pine-oak woodland above 1100 m.

***Pinus ponderosa*** var. ***scopulorum*** [*P. scopulorum*]. Rocky Mountain ponderosa pine, western yellow pine; *pino real* (figure 29)

Large trees, similar to *P. arizonica* but distinguished by stouter, more rigid, and often shorter leaves (9–19.5 cm);

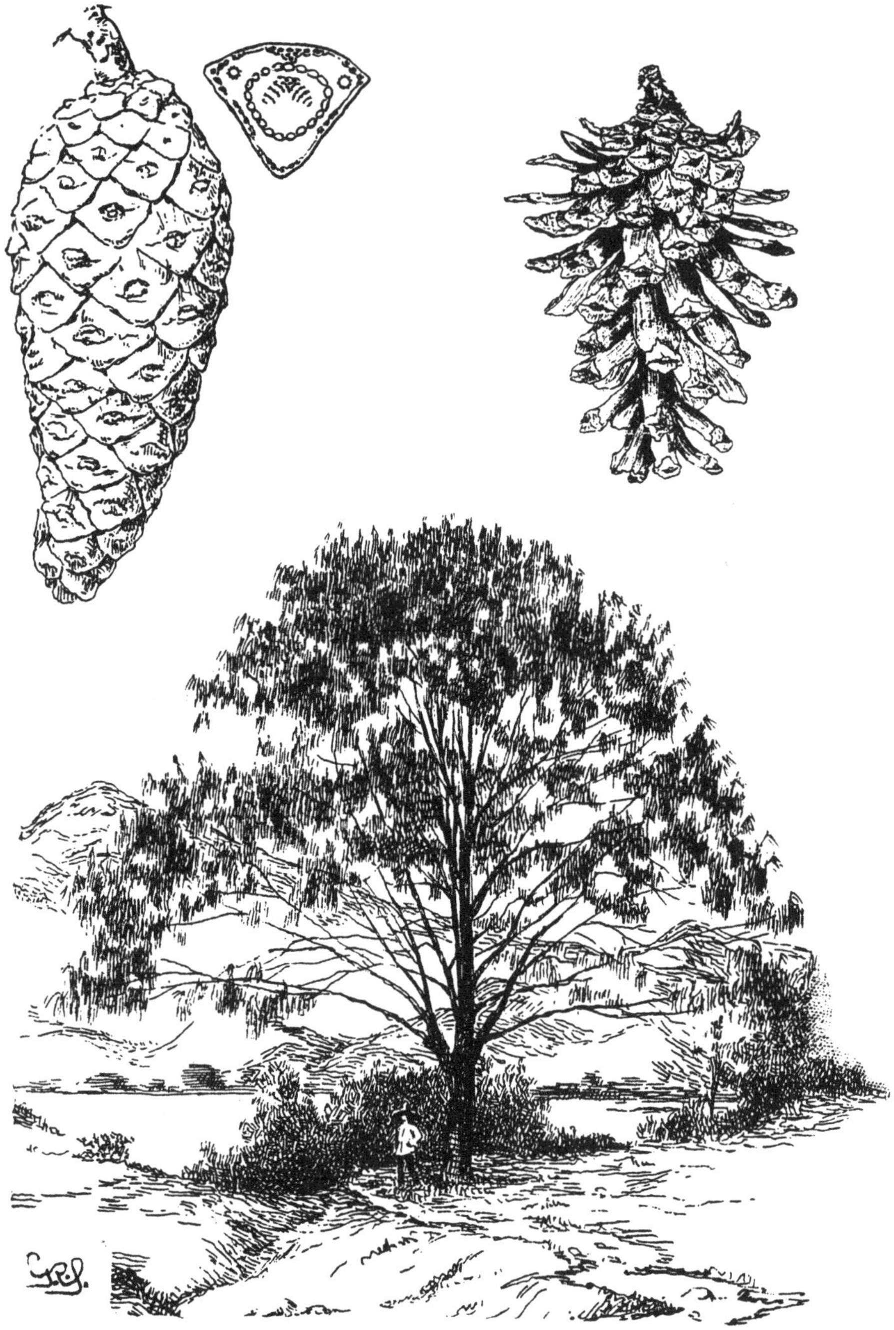

Figure 34. *Pinus maximinoi*. (Martínez 1948b, Shaw 1909). Drawn by M. Ornelas Carillo and G. R. Shaw.

leaves (2) 3 (4) per bundle, less glaucous twigs, and mostly larger cones with more prominent umbo prickles. Pollen shed in June (Arizona and Chihuahua).

Mixed conifer forest on north-facing slopes above 2200 m (7220 ft) in northeastern Sonora near the Arizona border. Mearns (1907) reported it to be abundant on the Sierra San Jose. Its occurrence there on the north-facing slope, on the highest bench below the peak, has been verified by Mark Kaib and James Riser (personal communication 1995). Perhaps also present in the Sierra San Luis in extreme northeastern Sonora. Common in nearby mountains in Arizona and at higher elevations in western Chihuahua. Western Chihuahua to inland western and midwestern United States and western Canada.

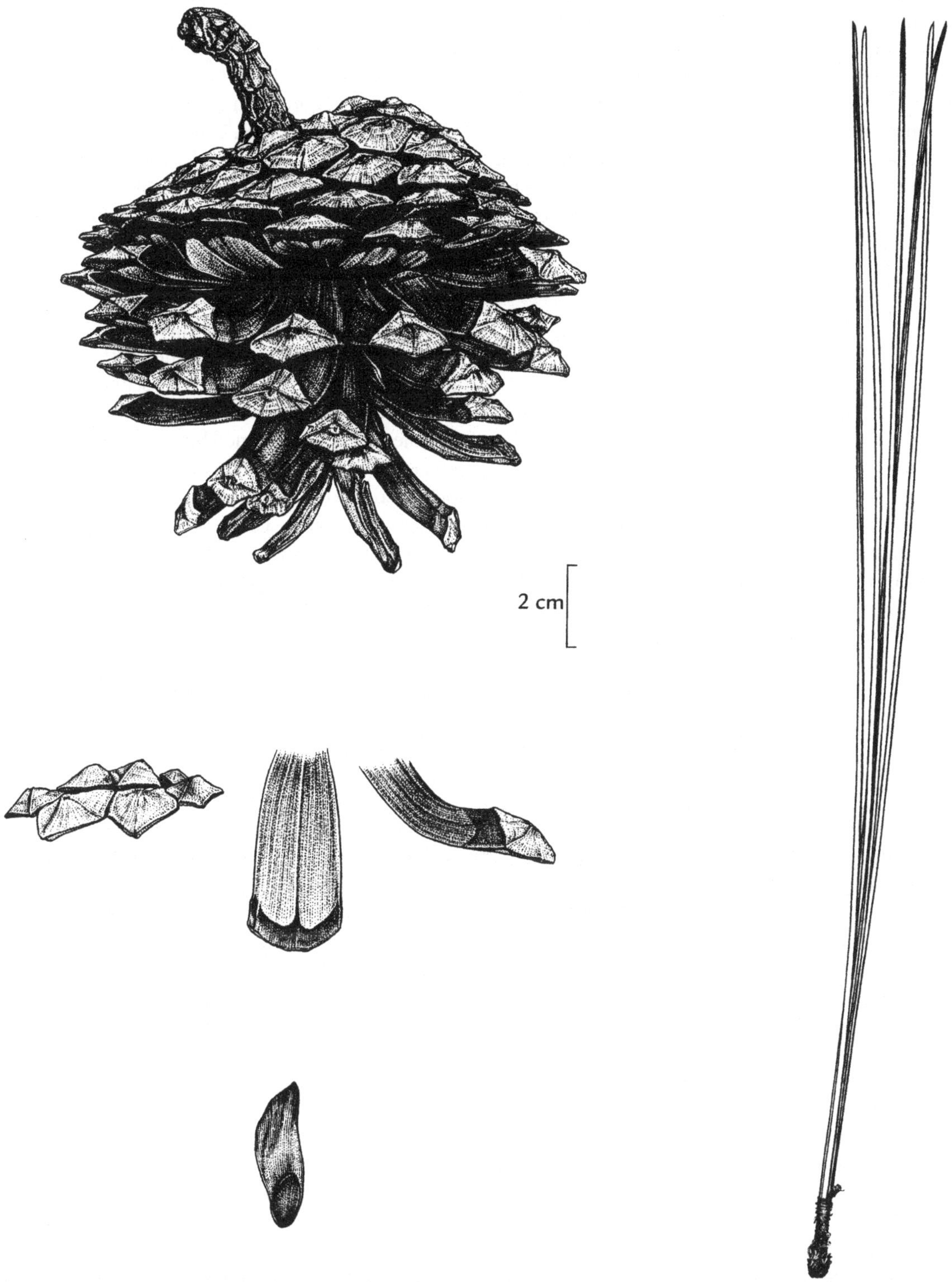


Figure 35. *Pinus oocarpa*. Leaf bundle, cone, cone scales, and seed. Drawn by A. K. Jakob.

Ponderosa and Arizona pines both occur in southeastern Arizona mountains including the Santa Catalina, Huachuca, Santa Rita, Pinaleño, and Chiricahua mountains. On these mountains *P. ponderosa* tends to occur at higher elevations on ridge tops, while *P. arizonica* tends to be more common at lower elevations in more mesic habitats such as canyon bottoms and north-facing slopes. The two taxa may hybridize (see *P. arizonica*). The relationship of *P. ponderosa* to members of the *P. arizonica* complex (including var. *cooperi*) deserves further study.

***Pinus strobiformis*** [*P. ayacahuite* var. *brachyptera*, *P. ayacahuite* var. *strobiformis*]. Southwestern white pine; *pino blanco, huiyoco, piñón* (near Yécora) (figure 36)

Large trees to 25 m with a tapering, conical crown; trunk to 70 cm in diameter. Bark light gray, becoming finely fissured with age. Twigs slender, < 10 mm in diameter, relatively smooth, light gray-brown. Leaves 5 per bundle, 4.7–11.5 cm × 1.0 mm wide or less, soft and flexible, conspicuously glaucous on the two inner surfaces of each needle, giving the foliage a bicolored appearance; leaf sheaths very soon deciduous. Cones (13) 15–33 cm, elongated (more than twice as long as wide), pendent, usually with gummy, sticky, whitish sap, at maturity soon falling from branches, with a relatively short peduncle (10–17 mm); cone scales large (to 35 mm wide), often strongly recurved at the tip; umbo terminal and unarmed. Seeds edible, 11–15 mm, wingless, the position of the wing marked by a scar, or with rudimentary, persistent wings to 5 mm. Pollen shed late May–June.

Usually north-facing slopes and canyons; pine-oak woodland and mixed conifer forest; north-central, northeastern, and east-central Sonora, 1950–2600 m (6400–8530 ft). Arizona, New Mexico, and western Texas to Jalisco, Zacatecas, and San Luis Potosí. Widespread and locally common in the higher mountains of eastern Sonora. Disjunct in east-central Sonora on the Mesa el Campanero on north-facing slopes and mesa tops from 1950 m to the summit at 2160 m. Often growing with Arizona pine (*P. arizonica*), firs (*Abies*), Douglas-fir (*Pseudotsuga*), and aspen (*Populus tremuloides*). The cones are the longest of any of the Sonoran pines, although the cones of *P. engelmannii* are sometimes more massive.

A complex of closely related and apparently intergrading white pines range from southwestern Canada to northern Central America. The northernmost member is limber pine (*P. flexilis*) and the southernmost is Mexican white pine (*P. ayacahuite*). Sandwiched between them is *P. strobiformis*, which at its southern limit is only weakly separable from *P. ayacahuite*. *Pinus ayacahuite* and *P. strobiformis* have been treated as varieties of a single species (Carvajal and McVaugh 1992, Shaw 1909).

Farjon and Styles (1997: 215) treat the Mexican white pines in a relatively restrictive sense (which we follow here), recognizing *P. strobiformis* primarily on the basis of seed characters: "The reduced and ineffective wing of the seed of *P. strobiformis* suggests a diversion from wind dispersal to animal vectors." *Pinus ayacahuite* is distinguished by slightly longer leaves, cones with thinner, more flexible scales, and invariably long and effective seed wings. The clinal nature of the complex deserves further study.

***Pinus yecorensis.*** Yécora pine; *pino colorado, pino rojo*

Trees to 30 m, often developing a spreading, dome-shaped crown; trunk to 95 cm in diameter, the bark rough, red-brown with dark furrows. Twigs scaly, relatively thick (> 10 mm in diameter), at first glaucous. Leaves (4) 5 (6) per bundle, 20–35 cm, somewhat coarse (ca. 1 mm in diameter) and firm but usually slightly drooping, the sheaths persistent, dark. Cones medium sized to large, relatively heavy (woody), 6–15.5 cm, slightly longer than wide, red-brown, symmetric to slightly asymmetric (curved), semipersistent, reflexed on the peduncle, subsessile or the peduncle thick, reaching 1–1.5 cm, falling with the cone, or sometimes remaining on the branch together with some basal scales. Cone scales relatively stout and firm, often broad; apophysis moderately to considerably thickened; umbo pointing down (dorsal), the prickle small, rather blunt, sometimes deciduous. Seeds 6.0–8.4 mm, thick and somewhat ovoid, the wing broad, detachable, extending 16–25 mm beyond the seed. Pollen shed March–April.

Slopes, canyons, and valleys; oak woodland and pine-oak woodland; eastern Sonora, 900–1975 m (2950–6480 ft). The northernmost population is on the Sierra Bacadéhuachi, where it grows with *P. leiophylla* var. *chihuahuana* and *P. engelmannii* at 1220–1620 m in the lowest elevation of the pine-oak zone.

This species occurs as a codominant tree with *P. oocarpa* on the Pacific slopes of the Sierra Madre Occidental in Sonora and southwestern Chihuahua to Sinaloa and Durango. The Sonoran population is var. *yecorensis*; another variety is described from Sinaloa. Our concept of this recently described species (Debreczy and Rácz 1995) includes specimens from Sonora previously identified as *P.* aff. *durangensis*. The relationship between *P. durangensis* and *P. yecorensis* deserves further study. Farjon and Styles (1997) indicate a close relationship of *P. yecorensis* to the Mexican false white pine *P. pseudostrobus*, possibly as a variety of the latter. It is our opinion that *P. yecorensis* is best allied with another member of the *pseudostrobus* complex, *P. douglasiana*, which occurs from Sinaloa to Oaxaca. In Sonora, *P. yecorensis* differs primarily in the lack of hypodermal intrusions, a character of the internal leaf anatomy.

Figure 36. *Pinus strobiformis* (Sargent 1897). Drawn by C. E. Faxon.

The lowest elevation records for this pine in Sonora are on hydrothermally altered (red) soils at 900–1000 m surrounded by tropical deciduous forest. On unaltered soils, the lowest elevation record is 1200 m.

*Pinus yecorensis* resembles *P. engelmannii* but differs in having a denser crown, usually more slender twigs, and brighter green, thinner, leaves with five or more per bundle. *Pinus yecorensis* can be distinguished from *P. arizonica* by its often darker, rough bark, thicker twigs, usually longer, more drooping leaves, larger, heavier cones, blunter and thicker cone scale tips, and usually blunter prickles. The leaves of *P. yecorensis* are longer and slightly thinner than those of *P. oocarpa* and moderately drooping because they are so long. Young, freshly opened cones may somewhat resemble those of *P. oocarpa*, but the cones are larger and the scale-tips are much thickened and lack a conspicuous blackish band on the inner surface.

### *Pseudotsuga*—Douglas-fir

Important timber trees in western North America and eastern Asia; 5 species.

***Pseudotsuga menziesii*** var. ***glauca*** [*P. macrolepis; P. mucronata; P. taxifolia*]. Douglas-fir; *pinabete, sabino monje* (Cajón Bonito) (figure 37)

Conical trees to about 30 m; trunk to 1.5 m in diameter but usually much smaller. Twigs often alternately branched, the youngest twigs usually with minute hairs. Winter buds somewhat ovate, mostly pointed, and not resinous or with thin, varnishlike resin. Leaves single, 1.5–4 cm, linear, flattened, narrowed at the base (short petioled), the upper surface with a median groove or channel; leaf scars usually horizontally oval, forming a small, "tilted" peglike woody base slightly higher on the proximal (lower) side and nearly flush on the distal (upper) side. Cones maturing in one year, (3) 3.5–7.2 cm (not including the protruding, terminal bracts), pendulous, red-brown, falling whole and not breaking apart, the scales overlapping at their bases, relatively thin, broad, and rounded at the apex, each with a 3-pronged bract projecting well beyond the scale. (These three-pronged bracts are unique among the regional flora.) Seeds often 6 mm, flattened on one side, with a broad wing often extending about 10 mm beyond the seed. Pollen shed April–May.

Mountain slopes and canyons to ridge tops; mixed conifer forest, and occasionally interdigitating into pine-oak forest; northeastern Sonora, (1675) 1880–2490 m [(5500–) 6165–8170 ft]. Locally common at higher elevations of the Sierras de los Ajos, Huachinera, San José, San Luis, and el Tigre. Also in the upper reaches of the Río Mayo drainage in nearby Chihuahua.

In Sonora, Douglas-fir is not of commercial significance because of its scarcity and inaccessibility, but it is being logged in Chihuahua. Large trees in Chihuahua and Sonora are scarce probably due to logging and fires.

This species occurs in western North America from Canada to northern Mexico southward to Zacatecas and Oaxaca. There are two varieties: The Rocky Mountain Douglas-fir, var. *glauca*, includes populations of the inland United States and Mexico. The Pacific Douglas-fir, var. *menziesii*, in the northwestern part of the range, is a magnificent tree reaching 70–100 m in height. It is one of the tallest trees in the world and the most important lumber tree in North America. Douglas-fir is cultivated in temperate climates, where it is prized for its stately form and fast growth. Young trees are often grown as Christmas trees. There are about 40 named horticultural selections, mostly derived from var. *menziesii*. Sterile specimens might be confused with *Abies* and would be most readily distinguished by differences in the winter buds.

# DICOTS

## Aceraceae—Maple Family

Two genera including *Dipteronia* in China with two species.

### *Acer*—Maple

Trees and shrubs. Leaves winter deciduous, opposite, petioled, without stipules; winter buds with deciduous scales leaving ringlike scars on the twigs. Flowers inconspicuous, mostly unisexual; sepals small, 4 or 5; petals absent (in Sonora). Fruits 2-winged, with two 1-seeded samaras that eventually separate from each other. Northern Hemisphere, temperate regions and mountains in the tropics; 110 species. References: Cronquist et al. (1997), Landrum (1995).

Many temperate maples have moderately hard, close grained, light-colored wood highly prized for furniture, house interiors, and other uses. The sap of *A. saccharum* yields maple syrup. Various maples are extensively cultivated in temperate regions but in Sonora would be suitable only for higher elevations.

1. Leaves simple. ______________ ***A. grandidentatum***
1' Leaves compound with 3 or 5 large leaflets. ***A. negundo***

***Acer grandidentatum*** [*A. grandidentatum* var. *brachypterum; A. saccharum* var. *grandidentatum*]. Big-tooth maple; *palo azucar* (figure 38)

Hardwood trees, often 6–15 m, reaching 20–24 m; trunks 75 cm in diameter in the Sierra de los Ajos and similar sizes

Figure 37. *Pseudotsuga menziesii*. Tree, branch, cone, scale, and seed. Drawn by L. B. Hamilton.

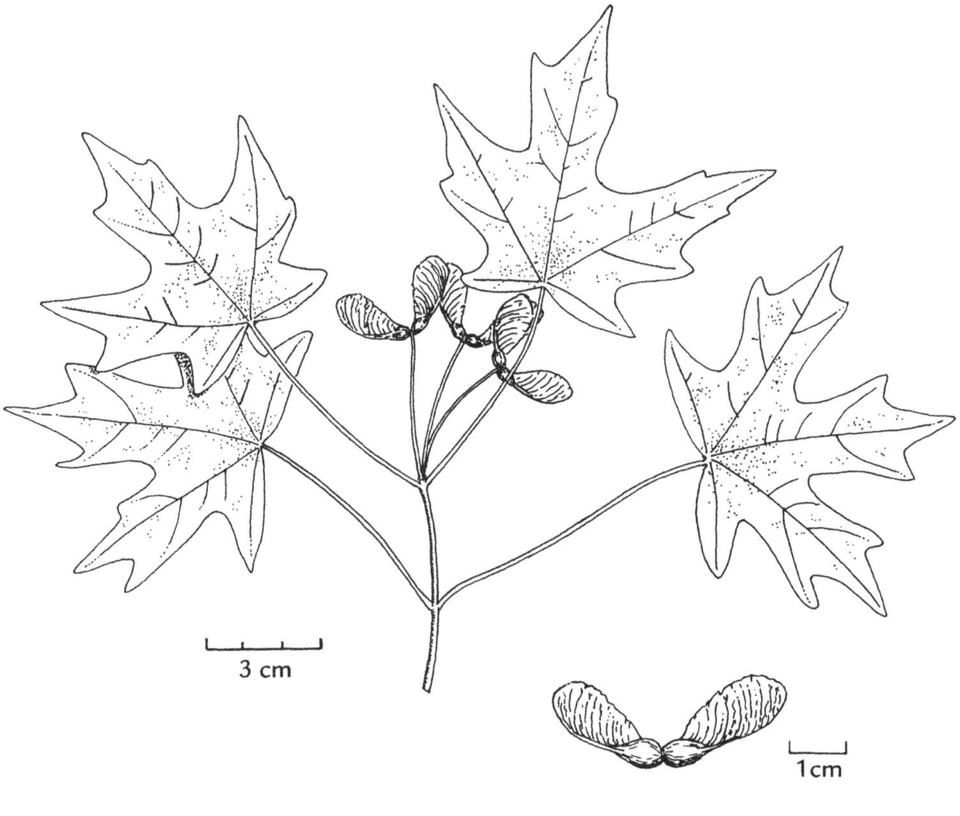


Figure 38. *Acer grandidentatum*. Drawn by M. B. Johnson.

reported for the Sierra San Luis by Mearns (1907). Mostly glabrate, often sparsely to densely pubescent with slender spreading whitish hairs when young, especially the petioles and veins on lower leaf surfaces. Leaves turning bright yellow to red in fall, simple, the blades (4) 5–9 cm, palmately lobed with 3 or 5 shallow to deep lobes, these usually with few large, blunt teeth; petioles 2.3–8 cm.

Unisexual and some bisexual flowers. Flowers in small terminal clusters, the pedicels long, slender, and drooping; sepals about 5 mm, cuplike and yellow-green. Samaras 2–3.5 cm, often reddish. Flowering April–July, before and with leaves; samaras ripening May–July.

Canyons, mostly along streams; oak and pine-oak woodland, mixed conifer forest, and rarely in riparian thornscrub-oak woodland ecotone; northeastern and east-central Sonora, (895) 1000–2135 m [(2930) 3300–7000 ft]. Northeastern Sonora including the Sierras de los Ajos, Nacori, San José, San Luis, and el Tigre. A remarkable low elevation population of small maple trees occurs at about 895 m in riparian canyon bottoms of the Río Sátachi near Nacori Chico, while adjacent southfacing slopes support organpipe cacti and palms. Also extreme east-central Sonora near Maycoba and more common in adjacent Chihuahua. Montana and Idaho to Sonora, Chihuahua, and Coahuila.

Big-tooth maple has been treated as a western subspecies or variety of the sugar maple, *A. saccharum*. However, big-tooth maple is also closely related to other maples of the eastern United States. Syrup can be made from the sap of big-tooth maple but there is no record of syrup production in Sonora.

**Acer negundo** [*A. negundo* var. *arizonicum; A. negundo* var. *interius*]. Box-elder (figure 39)

Small to medium-sized, fast-growing trees with soft, weak, light-colored wood; apparently not long lived. Herbage sparsely to moderately hairy when young and essentially glabrous at maturity, the young twigs often glaucous. Leaves turning bright yellow in fall, often 10–20 cm, pinnately compound, the leaflets stalked, 3 or sometimes 5, often 4.5–11 cm, ovate, coarsely toothed, the lateral leaflets asymmetric.

Usually dioecious or essentially so. Flowers green, about 5 mm, in terminal, drooping clusters, the pedicels long and slender; female flowers in racemes; male flowers in fascicled

Figure 39. *Acer negundo*. Drawn by L. B. Hamilton.

umbel-like clusters or some in few-flowered racemes. Fruiting racemes (including peduncles) 9–16 cm; samaras 2.7–3.5 cm, often reddish. Flowering and fruiting before and with new leaves, April–May. (Description based on trees from southeastern Arizona.)

Probably in canyon streams; pine-oak woodland and perhaps mixed conifer forest; northeastern Sonora, about 2000 m.

Mearns (1907) reported *A. negundo* in a canyon on the north side of Sierra San José near Naco. There is no other record for this tree in Sonora. It is common in nearby southern Arizona in the Huachuca and Chiricahua Mountains. Southern Canada to Central America.

## Anacardiaceae—Sumac Family

Woody trees, shrubs, and vines, often having toxic substances. Worldwide, especially tropical and subtropical; 70 genera, 850 species.

This family includes cashew, mango, pistachio, the Baja California elephant tree (*Pachycormus discolor*), *ciruela* (*Cyrtocarpa edulis*), and poison ivy and poison oak (*Toxicodendron* spp.). Poison ivy (*T. radicans*) is widespread in mostly montane regions of Sonora, often festooning tree trunks and cliffs. Other members of the family are important sources of dyes, lacquers, tannins, waxes, and varnishes. The Peruvian pepper tree or *pirul* (*Schinus molle*) is cultivated as a fast-growing shade tree in the Sonoran Desert region. The Brazilian pepper tree (*Schinus terebinthifolius*) is widely planted in low-elevation regions of Sonora.

There are about eight native species of *Rhus* in Sonora, all shrubs although some approach tree size. The African sumac (*Rhus lancea*) is a commonly cultivated tree in the desert regions of Arizona and Sonora, and occasionally persists or becomes locally established near habitations.

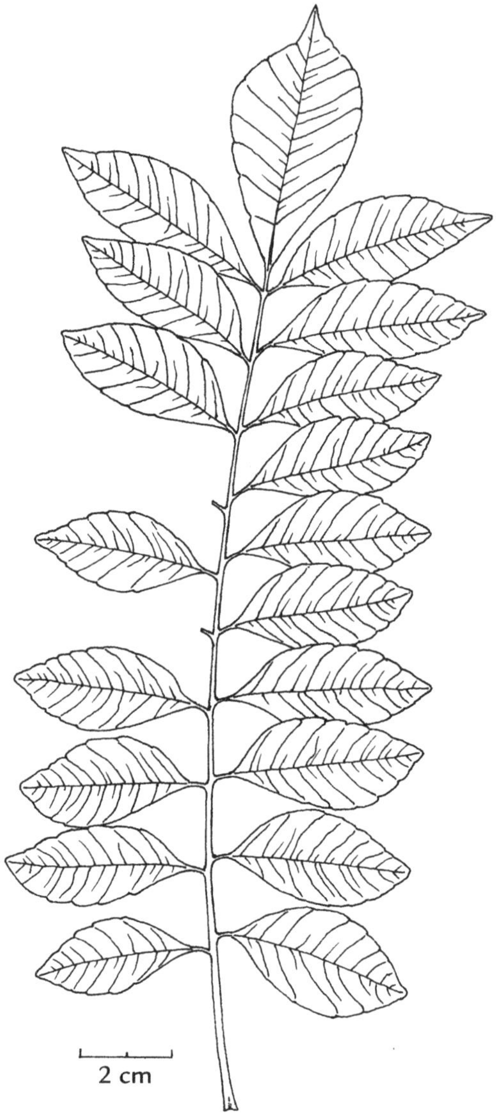


Figure 40. *Spondias purpurea*. Drawn by M. B. Johnson.

### *Spondias*

Trees and shrubs. Tropical America and Asia; 10 species.

***Spondias purpurea*.** Red mombin, Spanish plum; *ciruelo, yoyoma* (figure 40)

Shrubs or trees to 10+ m, with soft wood and thick trunks and limbs. Bark at first brown and gray, and shallowly furrowed, with age checkered in high relief (like a tortoise shell); lenticels conspicuous. Young leaves sparsely pubescent, especially along the veins, with short, often curved hairs. Leaves drought deciduous, and appearing with summer rains, alternate, often 20–28 cm, odd-pinnate; leaflets often 11–21, ovate to obovate, short stalked, margins shallowly serrated, the tip pointed to rounded or sometimes shallowly notched and with a small bristle, the terminal leaflet slightly to moderately larger or sometimes reduced or not developing, the lateral leaflets asymmetric, often 3.5–7.5 cm; stipules absent.

Inflorescences of small axillary panicles; flowering and fruiting when leafless. Flowers red-purple to green, unisexual, small, with a ring-shaped disk between the stamens and ovary; calyx 5-lobed; petals probably 5. Ovary superior. Fruits 3–3.5+ cm, fleshy, edible, plumlike drupes; different trees bear yellow or red-purple fruits. Seeds large. Flowering March–April; fruits May–June.

Tropical deciduous forest; southeastern Sonora, 260–600 m (855–1970 ft). Solitary trees rarely encountered in canyon bottoms or persisting at abandoned settlements. A grove in the canyon above La Aduana was probably planted. Native to tropical America and widely cultivated.

The fruit is highly esteemed; the fleshy pericarp is slightly acidic and eaten fresh or sun dried. The dried fruits, rolled in salt, are sold as *ciruelas*. The fruit is made into preserves, and the almondlike seeds are toasted and eaten like peanuts.

This tree is readily propagated by breaking off a branch and sticking it in the ground. In the Alamos region some people use the term *ciruelo* for the red-fruited form and *yoyoma* for the yellow-fruited form. Both are cultivated in southern Sonora and sometimes in Hermosillo. There are numerous cultivars in the American tropics.

## Apocynaceae—Dogbane Family

Trees, shrubs, and herbs; many vines, and some succulents, especially in the Old World. Many have milky sap (in Sonora) and toxic compounds including alkaloids. Leaves alternate or opposite, simple, entire; mostly without stipules; commonly with colleters (see *Rhizophora*) in the position of stipules or at the base of the blade and/or inside the calyx.

Flowers pedicelled; calyx 5-lobed, small and green, the segments united basally; corollas 5-lobed, often showy, the lobes imbricate (overlapping); stamens 5, attached to the corolla wall by short filaments or sessile; ovaries superior except in *Plumeria*, two with a single style and an often enlarged stigma. (The stigma and style-tip often form a complex structure, the style-stigma head.) Fruits of paired (or one sometimes not developing), dehiscent follicles or indehiscent fleshy drupes.

Worldwide, mostly tropical and subtropical, a few in temperate regions and some in arid regions; 200 genera, 2000 species.

Several shrub and perennial herbaceous species are widespread in Sonora. The arborescent Sonoran species are frost sensitive. Oleander or *laurel* (*Nerium oleander*) is extensively cultivated in the Sonoran Desert and occasionally becomes weakly adventive. Less commonly grown are the frost-tender yellow oleander (*Thevetia peruviana*) and stem succulents such as *Adenium* and *Pachypodium*. Many members of the family are highly poisonous (e.g., oleander) or well-known for their medicinal properties (e.g., *Rauvolfia*, tropical trees providing important cardiac medicines). Two herbaceous periwinkles, *Vinca major* and *Catharanthus roseus* (Madagascar periwinkle), are widely cultivated in Sonora. References: Rosatti (1989), Williams (1996), Woodson (1938).

1. Leaves opposite; flowers yellow. ____ ***Stemmadenia***

1' Leaves alternate; flowers white.

  2. Stems fleshy, thick; leaves > 20 cm; flowers > 5 cm wide; fruits dehiscent, of 2 divergent, cylindrical segments, each > 10 cm. ____________ ***Plumeria***

  2' Stems slender, not fleshy; leaves < 13 cm; flowers ≤ 1 cm in width; fruits indehiscent, opaque, succulent drupes about 1 cm. ____________ ***Vallesia***

### *Plumeria*

Seasonally dry tropical America but not native south of Colombia; seven species including two in Mexico.

Many forms of *Plumeria* are cultivated in the tropics for their magnificent, fragrant flowers. In Sonora they are grown from Hermosillo southward. The sap is poisonous.

***Plumeria rubra*** [*P. acutifolia*]. Frangipani; *cacalosúchil, súchil; turoco* (Guarijío) (figure 41)

Shrubs to small trees to 7+ m, with copious milky white sap throughout. Young stems and leaves and veins on lower leaf surfaces with short, soft hairs. Stems thick and succulent. Leaves quickly deciduous at the end of the summer–fall rainy season, plants remaining leafless until summer rains of the following year; leaves alternate, often 20–40 × 9–11 cm, elliptic to oblanceolate; the apex mostly acute to acuminate; petioles 1–2 cm.

Flowers 6–6.5 cm wide, showy and fragrant, crowded in 2- or 3-branched terminal cymes on thick peduncles 4.5–8 cm. Calyx scarcely differentiated from the pedicel; the lobes 1.5–2.0 mm, overlapping, thick, and broad. Corollas white with a yellow center, the tube 2–3 cm; tube and throat slender throughout; throat and inner surface of tube with sparse, upward pointing small white hairs; corolla lobes 2.5–6.0 cm. Stamens at base of tube; filaments shorter than anthers; anthers 2.3–2.5 mm, cream colored, held close together, and surrounding the approximately 2-mm green style–stigma head. Ovary inferior. Fruits of 2 spreading, thick, linear-cylindrical, leathery, dehiscent follicles, each one often 11–20 cm. Seeds many, wind dispersed, 28–35 mm, papery winged basally, the body 11–13 mm and flattened. Flowering preceding leaves or with new leaves at the end of the dry season or beginning of summer rains, May–July; fruits ripening the following spring (e.g., March) or earlier.

Cliffs, rocky slopes, and canyons; Sonoran desertscrub-thornscrub ecotone, thornscrub, tropical deciduous forest, and lower margins of oak woodland; east-central to west-central and southeastern Sonora, 240–1065 m (785–3500 ft). Also Panama to Chihuahua and Baja California Sur; widely cultivated and naturalized in tropical regions around the world.

The northern populations are many stemmed, trunkless shrubs as a result of repeated freeze damage. Farther south in Sonora *Plumeria* occurs as both a shrub and as a small tree. The trees are usually candelabralike, with a high, small crown, the stems becoming more slender with each successive branching. The white flowers dot the dry,

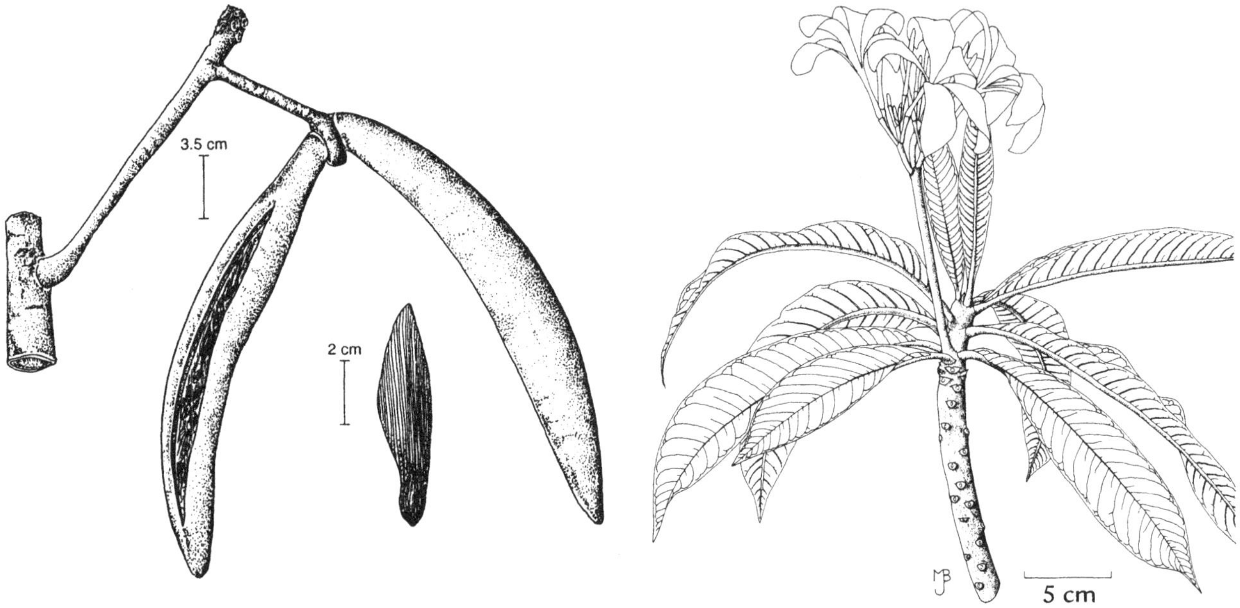


Figure 41. *Plumeria rubra.* Drawn by M. B. Johnson (leafy stem with flowers); (Pennington & Sarukhán 1998) drawn by R. M. Diez (fruit and seed).

parched hills in the last days of the premonsoon drought and in the early part of the summer rainy season. During the night the flowers are visited by sphingid moths.

The larvae of a common sphinx moth, *Isognathus rimosus*, feed on plumeria leaves. Before starting to feed, the caterpillars chew a notch of considerable size through the midrib on the lower surface of the leaf blade near its base, which cuts off the flow of the milky sap. By late summer many plumerias are defoliated by these caterpillars. During the day the cryptically colored larvae, especially the larger ones, rest near the base of the plant, often in grasses and forbs. Younger caterpillars are less secretive and even feed in the daytime (also see *Bursera*).

Plants from Sonora and Chihuahua have white flowers with a yellow center, and at least one population in Baja California Sur has pure white flowers. The native plumeria is not as fragrant as the commonly cultivated forms, and there may be differences in the intensity of fragrance between populations in Sonora and those in Baja California Sur. There are many horticultural varieties including ones with red, pink, or yellow flowers. Unlike the native Sonoran populations, the cultivated varieties flower with the leaves and leaf out earlier in the year, and some are evergreen. *Plumeria* is readily propagated from cuttings.

### *Stemmadenia*

Trees and shrubs; tropical Mexico to northern South America; 10 species. References: Leeuwenberg (1994), Woodson (1928).

***Stemmadenia tomentosa*** [*S. palmeri; S. tomentosa* var. *palmeri*]. *Berraco, huevos del toro* "bull's testes"; *peychí* (Guarijío) (figure 42)

Mostly shrubs, sometimes trees 5–12 m, with copious milky sap throughout. Bark gray, smooth except for horizontally ellipsoid lenticels. Leaves drought deciduous, opposite, 10–24.5 × 4–10.5 cm, glabrous except quickly deciduous white hairs along the midrib; petioles 5–18 mm; blades thin, elliptic to ovate or obovate, gradually tapering at the base; the tip acute to acuminate; midrib and lateral veins conspicuous on the lower surfaces, the lateral veins nearly perpendicular to the midrib; opposite leaves of a pair equal or mostly unequal in size.

Inflorescences of few-flowered cymes shorter than the larger leaves; flowers fragrant, open in the daytime, the likely pollinators being hawk moths. Pedicels approximately 1 cm, rather thick, with two or three small scales (bracts). Sepals green, united at very base, broad, thick, closely appressed and scalelike, moderately unequal in size, the inner one largest, (3.5) 5–6 mm; calyx with a continuous row of about 15 colleters 0.5–0.8 mm near the base on the inner surface. Corollas (5) 6–7 long × 6–8 cm wide (when fully open); base of corolla tube yellow-green, swollen, and nectar filled, the rest of the corolla bright yellow; mid-section of tube, in the region of the anthers, twisted and moderately constricted; corolla tube just above the anthers constricted by five fleshy wings or infoldings of the inner surface of the corolla tube—when seen from above, the five wings form a star pattern and block most of

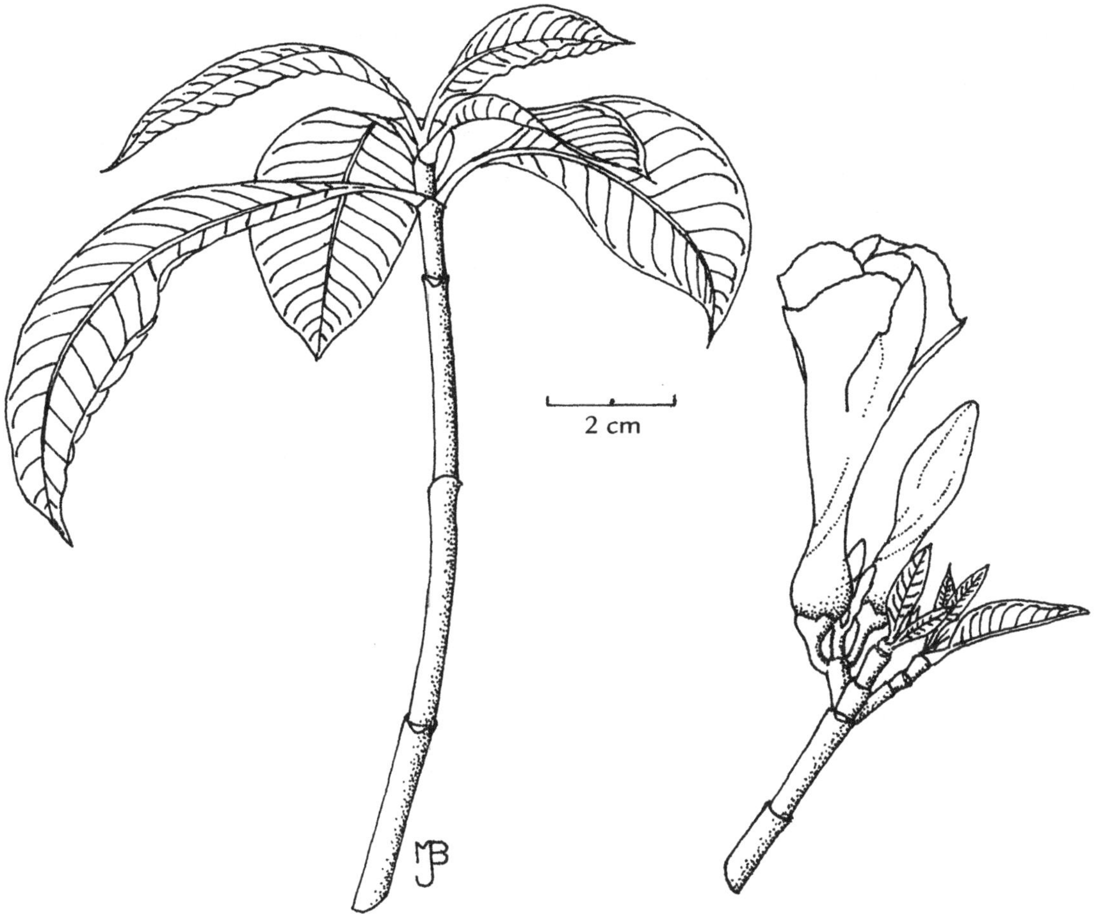


Figure 42. *Stemmadenia tomentosa*. Young leaves and partially closed flower. Drawn by M. B. Johnson.

the entrance to the anthers, style-stigma head, and the nectar-filled lower portion of the tube; throat 1.5–1.8 cm wide below apex. Anthers sessile. Ovary base surrounded by a ring-shaped yellow nectary. Style white and slender; upper style with a collar or veil just below the cylindrical or barrel-shaped style-stigma head; stigma white and wet when fresh.

Fruits with a pair of very thick, spreading, mericarps, each drying to 4–5 cm long, curved, hard shelled, and green, at maturity splitting longitudinally to reveal numerous seeds embedded in brilliant orange-red, fleshy, and oily arils (suggesting bird dispersal). Seeds 8–10 mm, ellipsoid, the surfaces brown and minutely pitted. Flowering May–August (November), often when leafless during the dry, hot premonsoon season or with the first new leaves.

Slopes, canyons, and arroyos; mostly tropical deciduous forest; southeastern Sonora, 270–1000 m (885–3280 ft). [A report of this species in northeastern Sonora (White 1948) was based on a misidentified specimen of *Randia laevigata*]. Especially large, arborescent plants grow in riparian canyons near Choquincahui and along the Río Guajaráy above the village of Guajaráy. Also in adjacent southwestern Chihuahua and southward on the Pacific slopes of Mexico to Oaxaca and adjacent areas of Morelos.

All parts of the plant are reported as poisonous, although Gentry (1942) reported that the milky sap is used medicinally for "bad eyes" and "all illnesses." This handsome plant is worthy of cultivation in hot, nearly frost-free regions such as from Hermosillo southward. The brilliant yellow flowers are spectacular during the end of the premonsoon dry season.

### *Vallesia*

Shrubs or occasionally small trees. Leaves evergreen or tardily deciduous in extreme drought, alternate, with small stipulelike structures (here called stipules) having colleters. Inflorescences shorter than the leaves. Flowers subtended by small, scalelike colleter-bearing bracts similar to the stipules. Flowers small, nectaries absent; corollas white, fading yellow when dried, the tube swollen above to accommodate the style–stigma and anthers, constricted at the top of the throat, the lobes spreading, their margins inrolled above. Style–stigma head cylindrical, with only a minute collar, the apex broadly 2-lobed. Fruits smooth and oblong-ellipsoid, opalescent (opaque) white to pinkish succulent drupes (1-seeded) in pairs, or single if one fails to develop. Tropical America; 6 species, 4 of them in Mexico.

1. Leaves 3.3–8 cm; stipules 1.3–1.5 mm, the margins nearly entire; corolla tube 3.6–4.3 mm; widespread. ______ ***V. glabra***

1' Leaves 5–12 cm; stipules 1.5–3.0 mm, the margins laciniate; corolla tube 9.5–10.5 mm; locally in canyons. ______ ***V. laciniata***

***Vallesia glabra*** var. ***glabra***. *Citavaro, huevito; sitavaro* (Mayo); *tonóopa* (Seri); *sita'avao* (Yaqui)

Multiple-stem shrubs or occasionally small trees to 6 m in southern Sonora, especially near the base of Sierra de Alamos. Petioles usually puberulent and midrib of lower leaf-surfaces often sparsely puberulent; herbage otherwise glabrous or sparsely pubescent. Leaf blades 3–7.5 × 0.7–2.0 cm, lanceolate to elliptic; petioles 3–5 mm; stipules 1.3–1.5 mm, triangular, apparently glandular, the margins nearly entire.

Inflorescences of small several-flowered racemose or paniculate cymes. Flowers not noticeably fragrant. Calyx lobes 0.5–0.8 mm, broadly triangular. Corolla tube 3.6–4.3 mm, the lobes 1.5–2.0 mm, puberulent above and surrounding the throat. Fruits 10–12 mm. Flowering at various seasons.

Canyon bottoms, washes, bajadas, plains; Sonoran desertscrub, thornscrub, and tropical deciduous forest. Often especially common in low-lying heavy soils and in riparian and semiriparian habitats. Widespread across the southern two-thirds of Sonora, near sea level to about 500 m (1640 ft). Also the Baja California peninsula, Chihuahua, and southern Florida to Argentina. Another variety occurs on the Galapagos Islands.

*Vallesia glabra* is not palatable to cattle and is often common in overgrazed habitats. The Seri eat the fleshy pericarp (Felger and Moser 1985) and the Mayo of Sonora use a few drops of the milky sap or fruit pulp, placed in a corner of the eye, to treat *mal de ojo*, which usually is pinkeye (probably bacterial conjunctivitis; Gentry 1942). This species is distinguished from *V. laciniata* by its smaller and shinier leaves, smaller stipules, and smaller, nonfragrant flowers.

***Vallesia laciniata*** [*V. baileyana; V. conzattii*] (figure 43)

Shrubs or occasionally small trees, 3–4 (5+) m in height, with multiple, woody stems. Herbage glabrous or glabrate, or the inflorescences, petioles, and lower leaf-surface midribs puberulent, especially when drought stressed. Leaf blades 4.5–11.5 × 1.3–2.8 cm, lanceolate to elliptic; petioles 0.5–1.0 cm, relatively thick; stipules 1.5–3.0 mm, triangular with laciniate margins of colleters, soon becoming red-brown. Seedling leaves opposite (figure 44).

Inflorescences of small umbellate clusters, often of two short, dichotomous branches, each with 10–25 crowded flowers. Flowers with a strong gardenialike fragrance day and night. Calyx lobes 0.8–1.2 mm, narrowly triangular. Corolla tube 9.5–10.5 mm; lobes 6 mm; throat surrounded by a star-shaped corona or "pad" (slightly thickened as a callous) of glistening, white, erect hairs. Style–stigma head approximately 1 mm. Fruits 10–13 mm. Flowering nonseasonally depending largely on soil moisture, with mass flowering often March–May.

Riparian canyon bottoms in desertscrub and thornscrub; west-central Sonora, 5–130 m (15–425 ft) in the Sierra el Aguaje including Cañón de Nacapule and Sierra Libre; also farther south in mountains southeast of Ciudad Obregón.

At Nacapule a locally dense stand grows along the winter-shaded portion of the canyon floor with *Celtis reticulata, Coccoloba goldmanii, Sapindus saponaria, Zanthoxylum mazatlanum*, and many other riparian species (Felger 1999). An indistinguishable *Vallesia* occurs in northern Sinaloa. Williams (1996) suggests that *V. conzattii* of southern Mexico and *V. baileyana* are synonyms of *V. laciniata*, the latter formerly considered to be endemic to Baja California Sur. We are following this suggestion.

This *Vallesia* is readily grown from seed and can be propagated by cuttings if supplied with bottom heat and high humidity. Although frost sensitive, it is gaining popularity as an ornamental garden plant in southern Arizona for its attractive foliage and powerfully sweet-scented flowers. It is worthy of wide cultivation in arid-tropical regions.

## Aquifoliaceae—Holly Family

Trees, mostly small, and shrubs. Worldwide; 4 genera, 420 species.

### *Ilex*—Holly

Handsome evergreen hardwood trees (in Sonora) or shrubs. Leaves alternate; short petioled; blades often thickish and dark green (in Sonora); without stipules (in Sonora). Generally dioecious but with some bisexual flowers. Flowers small, solitary, or in cymes in leaf axils;

Figure 43. *Vallesia laciniata*. Drawn by A. K. Jakob.

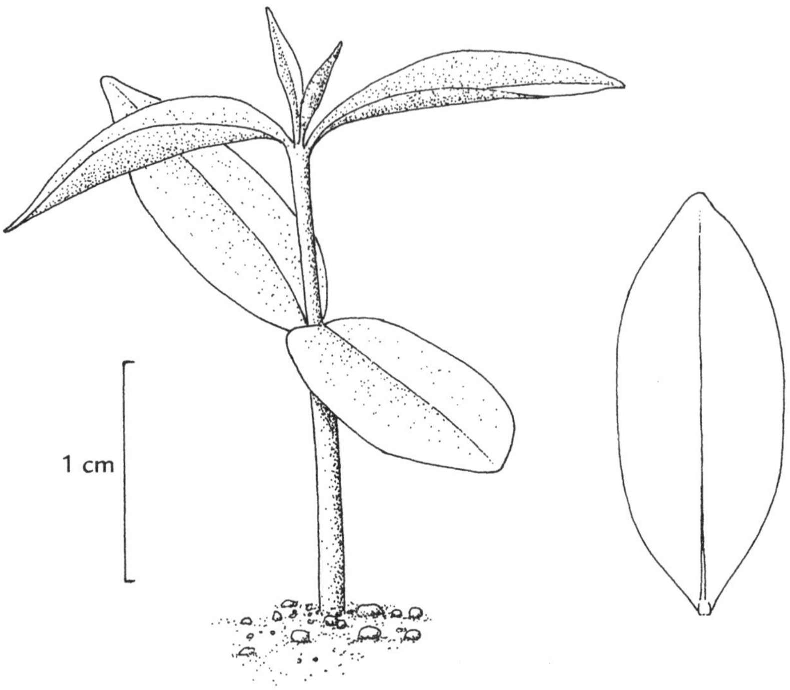


Figure 44. *Vallesia laciniata*. Seedling with opposite leaves. Drawn by M. B. Johnson.

calyx and corolla each 4-lobed, the stamens 4; ovary superior. Fruits red or blackish, several-seeded drupes.

Temperate and tropical regions mostly in eastern Asia and the Americas; 400 species. Some are important timber trees. Reference: Brizicky (1964a).

Maté, the national drink of Argentina, is made from the leaves of *I. paraguariensis*. In Europe and North America Christmas wreaths are made from holly.

1. Flowers and fruits mostly solitary in leaf axils. **_I. rubra_**

1' Flowers and fruits in small umbellate clusters in leaf axils. ________ **_I. tolucana_**

### *Ilex rubra*. *Chilillo* (figure 45)

Trees 6–20 m; bark gray, somewhat irregularly checkered and fissured. Herbage and pedicels glabrous or sparsely to moderately pubescent. Leaves 4–11 × 3–5 cm, usually ovate, sometimes broadly so, the margins sometimes entire but usually some leaves on each branch with spinescent teeth. Flowers mostly solitary or sometimes paired, approximately 1 cm wide, greenish white. Fruits approximately 1 cm in diameter, globose, on slender stalks (peduncles and pedicels) 4–10 mm, at least sometimes five seeded; seeds 5–6 mm. Flowering May–June; fruits ripening in the same summer.

Riparian mountain canyon bottoms, often along streams; oak and pine-oak woodland; southeastern Sonora; 1080–1555 m (3540–5100 ft). Sierra Saguaribo including Tepopa and near Milpillas, Rancho Santa Bárbara and mountains from near Yécora eastward. Also adjacent southwest Chihuahua to at least 2075 m, and probably farther south.

*Ilex rubra* is distinguished from *I. tolucana* by its usually larger, broader, thicker, and firmer leaves, sharper marginal spines, and often solitary and slightly larger fruits.

### *Ilex tolucana* (figure 46)

Trees to 15 m; bark light gray mottled with green. Glabrous except minute white hairs on pedicels and new shoots. Leaves 5–8 × 1.5–3 cm, lanceolate or oblanceolate to narrowly ovate or obovate, the margins entire or with small teeth mostly toward apex. Flowers in small, dense umbellate clusters, white or pink tinged, 7–8 mm wide; pedicels 5–10 mm. Fruits approximately 0.7 cm in diameter, globose and red, 4-seeded; seeds 2.5–3.0 mm. Flowering April–June; fruits in August.

Mountain streams and slopes of moist canyons; pine-oak woodland and pine forest; southeastern Sonora; 1200–1600+ m (3935–5250+ ft). In Sonora generally more widespread than *I. rubra* and extending into higher elevations; Sierras from the vicinity of Yécora eastward and southward. Also adjacent southwestern Chihuahua and southward in the mountains of Mexico to El Salvador. Appears to be closely related to *I. paraguariensis*.

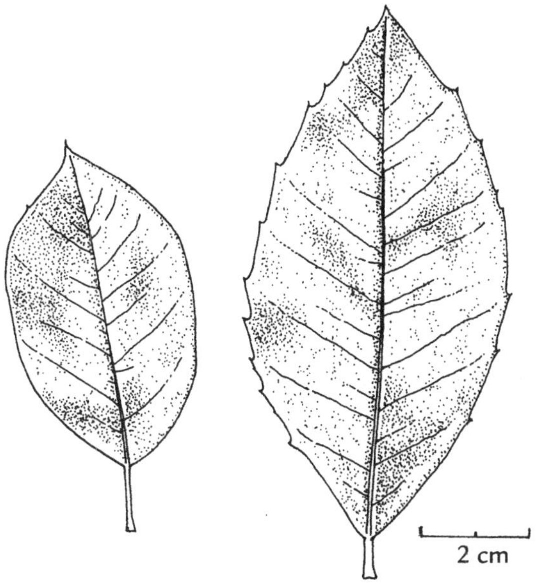

Figure 45. *Ilex rubra*. Drawn by M. B. Johnson.

## Araliaceae—Ginseng Family

Perennial herbs to shrubs, trees, or vines with branched or stellate hairs. Leaves alternate and large (in Sonora); stipules partially fused to the petiole base, the stipules and petiole base partially clasping the stem. Flowers unisexual or sometimes unisexual and bisexual on the same plant. Calyx tube fused to the ovary, the lobes minute, the petals larger than the calyx and soon deciduous. Ovary inferior. Fruits rounded, fleshy berries.

Worldwide, mostly tropical and subtropical; 57 genera, 800 species. References: Graham (1966), Smith (1944).

English ivy or *hiedra* (*Hedera helix*), is sometimes grown on walls in northernmost Sonora at elevations above the desert, but fares poorly at lower elevations in hot climates.

1. Leaves pinnately compound; fruits red with 5 grooves. ***Aralia***

1' Leaves simple; fruits white or blackish, not grooved. ***Oreopanax***

### *Aralia*

Large herbs to trees. North America and Asia; 36 species, 5 in Mexico.

A second species from Sonora, *A. racemosa*, occurs in high-elevation, moist montane habitats. It is herbaceous and has enormous, tripinnate leaves. Some aralias are medicinal and others in Asia serve as leaf vegetables.

**Aralia humilis** [*A. pubescens*]. *Tepetate* (figure 47)

Shrubs or small trees often 2.5–5 m, or occasionally trees 6–10 m with a broad, open crown, the bark corky and deeply furrowed. Herbage glabrate, at first with minute, often branched white hairs. Leaves winter deciduous and perhaps drought deciduous, aromatic, odd pinnate, highly variable, 15–38 cm, long petioled; leaflets (5) 7–9 (11) per leaf, on slender stalks; blades thin, (5) 7–13 × (3) 4.5–8 cm, ovate to ovate-triangular to sometimes oval, often with acuminate tips, the margins often coarsely toothed.

Monoecious or perhaps also with some bisexual flowers. Flowers greenish white, on slender pedicels in umbellate clusters often on paniculate inflorescences. Berries 5 mm in diameter, red, of 5 carpels with 5 conspicuous grooves, 5-seeded. Flowering spring to fall; fruits mostly ripening fall–winter.

Mountain canyons and slopes; oak and pine-oak woodland; occasionally in upper elevations of thornscrub and tropical deciduous forest; eastern Sonora, 800–1880 m (2625–6165 ft). Mostly in riparian habitats with rocky soils, sometimes on rocky outcrops and slopes. Tree-sized plants, apparently not common in Sonora, have been

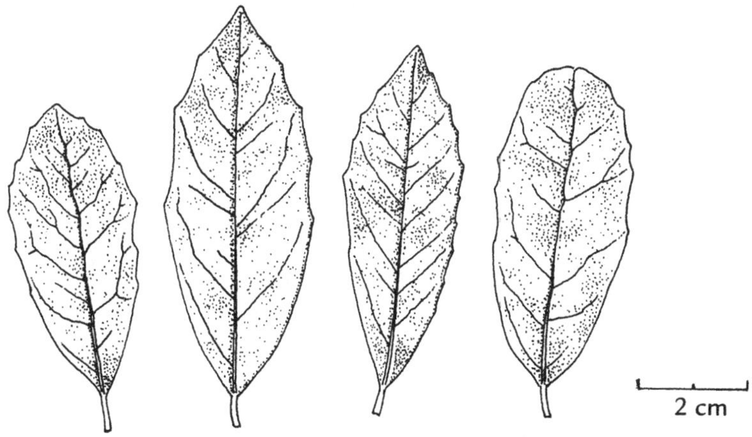

Figure 46. *Ilex tolucana*. Drawn by M. B. Johnson.

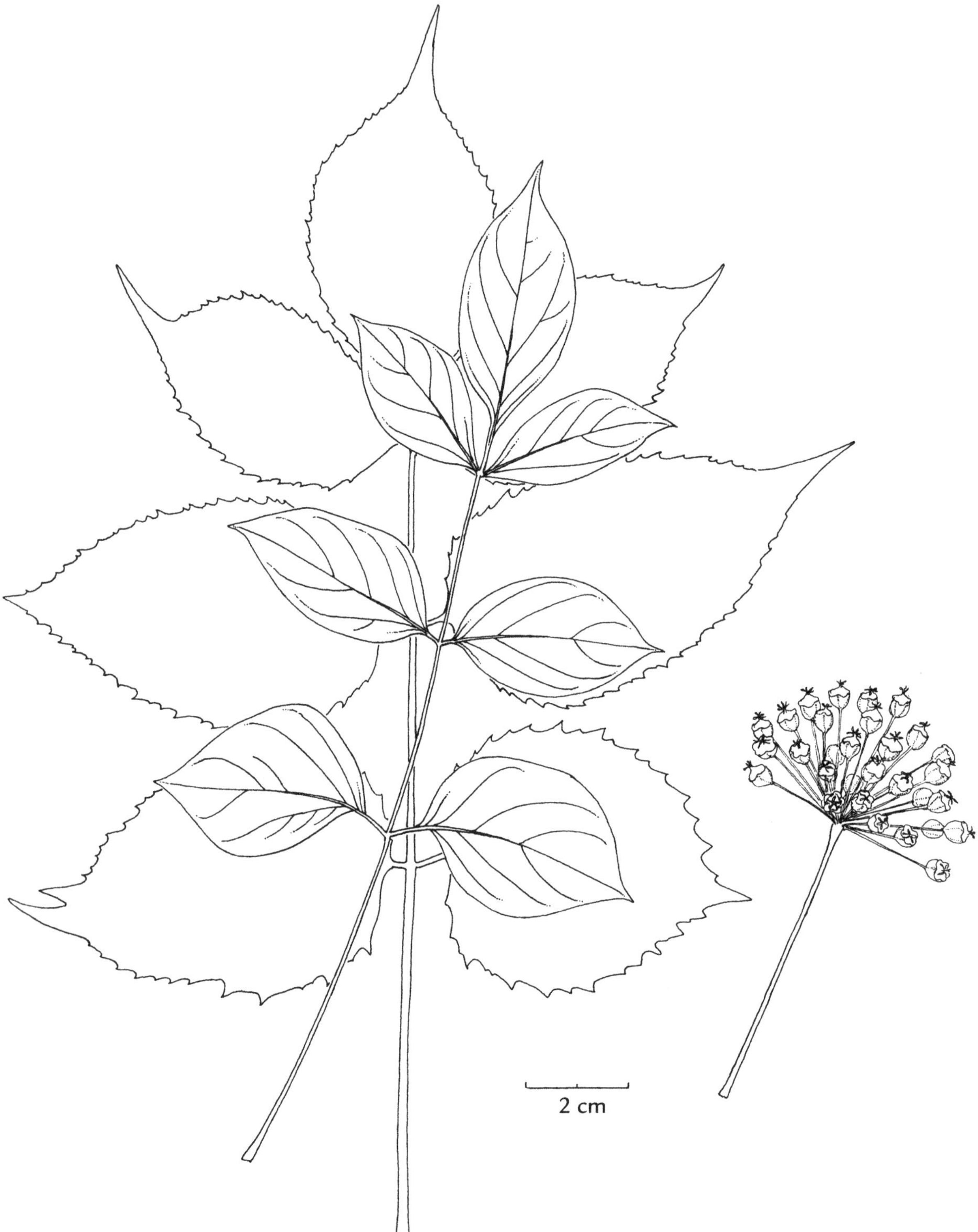


Figure 47. *Aralia humilis*. Drawn by M. B. Johnson.

found in scattered mountain localities including the vicinities of Moctezuma and San Javier in east-central Sonora, Sierra de Alamos, and Arroyo Santa Bárbara in southeastern Sonora. Oaxaca to Chihuahua and southern Arizona.

*Aralia scopulorum*, endemic to mountains in Baja California Sur, seems to be closely related to *A. humilis*.

### *Oreopanax*

Trees and shrubs. Tropical America; 80 species.

**Oreopanax peltatum** [*O. salvinii*]. *Papaya cimarrón, mano de león* (figure 48)

Usually large, spindly shrubs to small trees with high, spreading crowns, soft wood, and relatively thick twigs; rarely trees 12–15+ m with a tall, knobby trunk 30–50 cm in diameter in deep, tropical riparian canyons. Stems and leaves densely stellate-woolly pubescent; leaves often sparsely pubescent with age, especially the upper surfaces. Evergreen to nearly leafless in extreme drought. Leaves clustered at branch tips, often 45–75 cm; blades mostly 20–50 cm and about as wide, palmately and deeply lobed, with 7 major veins; petioles 13–40 cm.

Dioecious. Flowers small, in small umbellate clusters on large terminal panicles; female flowers mostly green, the male flowers white. Berries approximately 8 mm in diameter, rounded, white, ultimately turning blackish, one seeded. Sometimes propagating by root sprouts. Flowering November–December; fruits in March.

Mountain springs and streams in riparian canyons; oak and pine-oak woodland and occasionally in upper margins of tropical deciduous forest; southeastern Sonora, 800–1770 m (2625–5800 ft). Honduras to Sonora.

In Sonora and nearby southwestern Chihuahua *O. peltatum* occurs in some of the most mesic habitats in the region, often in the understory of large riparian trees such as sycamore (*Platanus wrightii*) and magnolia (*Magnolia pacifica*). This plant is suitable for cultivation as an ornamental in well-watered situations. It is probably frost sensitive and requires semishade in desert regions. It has been successfully grown in southern California.

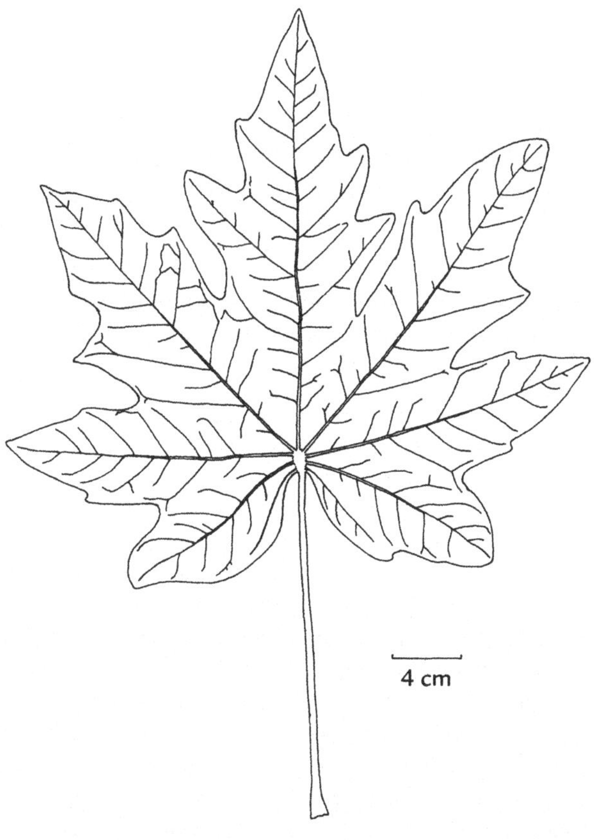


Figure 48. *Oreopanax peltatum*. Drawn by M. B. Johnson.

## Asteraceae—Daisy or Composite Family

Mostly herbaceous, some shrubs and few trees. Flowers in headlike clusters (heads or capitula) surrounded by bracts (phyllaries), highly variable, with disk and/or ray flowers (florets). Florets bisexual, unisexual, or sterile; fruit 1-seeded (generally called an achene, but technically a cypsela); calyx modified into diverse bristles (pappus) or absent. Worldwide; 1500 genera, 25,000 species.

This is the largest vascular plant family in the world. In Sonora the family comprises several hundred genera and perhaps about 1000 species, mostly herbs. Several of the larger shrubs in Sonora reach 4 or rarely 5 m in height, but have multiple and rather slender stems and are hardly trees. Among the larger *Baccharis* shrubs are the widespread seep willow or *batamote* (*B. salicifolia*, = *B. glutinosa*), desert broom or *romerillo* (*B. sarothroides*), and *B. emoryi* in the delta region of the Río Colorado. Burro brush or *jécota* (*Hymenoclea monogyra*), also widespread in Sonora, sometimes reaches 5 m in height in north-central and northeastern Sonora. The tree daisy (*Tithonia fruiticosa*) of southeastern and east-central Sonora is cultivated in southern Arizona.

1. Flower heads relatively inconspicuous; ray florets absent.
   2. Dioecious; male and female flower heads rather similar, the female heads not burlike. _____ ***Baccharis*** (not treated in text)
   2' Monoecious; male and female flower heads markedly dissimilar, the female heads burlike. ____ ***Hymenoclea monogyra*** (not treated in text)

1' Flower heads relatively conspicuous; ray and disk florets present.
   3. Heads sunflowerlike, > 5 cm in diameter; ray and disk florets yellow. ________ ***Tithonia fruticosa*** (not treated in text)
   3' Heads not sunflowerlike, < 2 cm in diameter; rays white.
      4. Leaves opposite, ovate, green on both surfaces; rays white; disk florets yellow. ___ ***Montanoa***
      4' Leaves alternate, triangular-lanceolate, green above, whitish below; ray and disk florets white. ________________ ***Parthenium***

### *Montanoa*

Mostly shrubs with a few small trees. Tropical America; 25 species. Reference: Funk (1982).

One other species in Sonora, *M. leucantha* subsp. *arborescens*, is a shrub in the southeastern and east-central part of the state.

#### ***Montanoa tomentosa*** subsp. ***rosei*** [*M. rosei*]. Tree daisy; *batayaqui* or *matayaqui* (Spanish, from the Mayo); *batayaqui* "vestido de Yaqui" or "dressed as a Yaqui" (Mayo); *talacáo* (Guarijío) (figure 49)

Mostly large, and slender-stem understory shrubs and sometimes small trees to 8 m with one to mostly several slender trunks. Wood weak and soft; bark light brown, checkered, soft, and corky; twigs brittle. Herbage, inflorescences, and flower heads densely white-hairy, the young foliage aromatic. Leaves opposite, mostly 7–20+ × 2.5–8+ cm; petioles 0.5–2.5 cm; blades ovate, with 3 main veins, the hairs of upper surface somewhat scabrous with expanded bases, the margins irregularly serrated.

Inflorescences of dense, much-branched and many headed, more or less rounded terminal panicles, 8–10 cm across; dry, brown flower heads conspicuous and persistent in the crown of the trees or shrubs. Flower heads 5–8 mm; phyllaries in a single series and ciliate. Flowers fragrant with a citruslike odor; ray florets sterile, several per head, the rays white, 5–7 mm; disk florets yellow, mostly 8–10 per head. Achenes often 2.6–3.0 mm, 4-angled, blackish, without a pappus, often only 2 maturing per head, enveloped by a firm, chaffy bract 6.5–7.0 mm. Flowering December–February.

Mountains, hills, canyons, and arroyos; tropical deciduous forest; southeastern Sonora; 260–540 m (850–1770 ft). Subspecies *rosei* also in southwestern Chihuahua and Sinaloa. This species, with three other subspecies, ranges southward to Costa Rica.

Many traditional houses in the mountains have walls woven with slats made from batayaqui. The herbage is used together with that of *Karwinskia* as an abortive.

Figure 49. *Montanoa tomentosa* subsp. *rosei*. Drawn by M. B. Johnson.

### *Parthenium*

Mostly herbaceous perennials or small shrubs. North and Central America with 1 or 2 species in South America; 18 species. Reference: Rollins (1950).

Guayule (*P. argentatum*) is a source of commercial rubber.

#### ***Parthenium tomentosum*** var. ***stramonium*** [*P. stramonium*]. *Ocotillo; guasaraco* (Guarijío and mestizo)

Large shrubs or small trees to 6 m. Branches ascending; trunk short but well formed, to sometimes about 2 m, with few major branches supporting a high, spreading crown with leaves clustered at the ends of the ultimate branches. Bark corky, brown to gray, finely to deeply furrowed, the furrows anastomosing. Herbage gland dotted and resinous; young herbage densely white-wooly; older twigs and smaller branches with conspicuous, corky lenticels, the twigs often relatively thick. Leaves alternate, 10–30 × 3–13 cm; blades triangular-lanceolate, densely white-wooly below, with age green and sparsely hairy or glabrate above, conspicuously and pinnately veined, mostly asymmetric basally, the margins mostly somewhat unevenly and bluntly toothed or serrated; petioles 2–8 cm; larger leaves shed in drought, often becoming nearly leafless before the summer rains.

Inflorescences much branched, often 10–18 cm across, with many small, crowded flower heads. Flower heads 3–4 mm high; the phyllaries in two series, membranous and deciduous. Flowers dull white, approximately 2 mm; ray florets 5, fertile; disk florets producing pollen. Achenes 1.9–2.2 mm, flattened, dark brown to blackish; pappus absent. Flowering January–March and July–October.

Hills and mountains; tropical deciduous forest and lower oak woodland; eastern Sonora as far north as the Río Bavispe region (ca. 20 km south of Arizpe) and the mountains west of Sinoquipe; 300–1070 m (985–3500 ft). Northernmost populations are shrubs often in thornscrub. Also southwestern Chihuahua and southward to northern Jalisco.

The gum is used to treat wounds and infections. There are two varieties with widely separated geographic ranges; var. *tomentosum* occurs in Oaxaca and Puebla.

## Avicenniaceae—Black Mangrove Family

This family has a single genus; sometimes placed in the Verbenaceae. Reference: Tomlinson (1986).

### *Avicennia*—Black Mangrove

Trees and shrubs. Tropical shores worldwide; 11 species. *Avicennia* is one of eight major genera of mangroves of the world and one of three in the Gulf of California (see *Laguncularia*, Combretaceae, and *Rhizophora*, Rhizophoraceae).

***Avicennia germinans*** [*A. nitida*; *Bontia germinans*]. Black mangrove; *mangle blanco*, *mangle negro*; *ciali* (Mayo); *pnaacoj-iscl* (Seri); *paseo* (Yaqui) (figures 50 and 51)

Hardwood shrubs or small trees forming impenetrable thickets in shallow tidal waters; roots and lower stems generally inundated daily. Root systems with cable roots, anchoring roots, and pneumatophores. Pneumatophores form a miniature "forest" emerging from the mud beneath the branches, erect, often 20–30 cm, straight and pencil-like, usually unbranched or sometimes few-branched, corky, and tapering to a blunt tip; they facilitate uptake of atmospheric oxygen into a root system growing in anoxic soils. Cable roots slender and extending outward horizontally beneath the mud often many meters from the base of the trunk. Both pneumatophores and anchoring roots arise alternately at rather regular intervals from the cable roots.

Bark gray, at first smooth. Inflorescences, flowers, fruits, youngest herbage, and lower leaf surfaces with short, coarse, appressed, golden brown to silvery hairs. Stems moderately 4-angled, slightly enlarged at the nodes. Stems and leaves opposite and decussate. Stipules absent. Leaves evergreen, often 6–12 cm, 1.7–4.0 cm wide; blades rather leathery, mostly narrowly to broadly elliptic or ovate with blunt tips, dull gray-green, the upper surface darker, and glabrous with age, the midrib prominent below; petioles 5–20 mm with a deep basal groove, this groove blackish inside with a dense, brushlike margin of tawny yellowish white hairs. (The leaf buds as well as the axillary meristems are protected by the closely appressed margins and its brush of dense hairs on the surrounding basal petiole groove of the ultimate leaf pair.) Xylem sap with high salt concentrations; excess salt exuded through microscopic salt glands in both leaf surfaces, but especially the upper surface encrusted with salt crystals.

Inflorescences of crowded, terminal and axillary simple or compound spikes shorter than leaves, each flower subtended by one bract and 2 bractlets. Flowers white, bisexual, 13–15 mm wide, bilateral, producing copious nectar, fragrant especially at night, remaining open several days. Corollas thick, with a short tube and 4 lobes; lobes overlapping and spreading, densely hairy on both sides, the upper one largest and opposite a bilobed, weakly differentiated lip. Stamens 4, attached to the corolla base and bent to orient the anthers to one side above the larger corolla lobe. Ovary superior. Stigmas with 2 spreading lobes, developing before the stamens, the lobes closing apparently after pollination and before the anthers open. Fruits of leathery capsules, 1–2 cm, asymmetric, laterally compressed, pointed at the tip (a short lateral beak), 1-seeded.

No resting terminal buds; growth slows or ceases during colder weather, especially among the northern populations. Rapid growth occurs during the long, hot summer. Flowering most heavily in late spring and early summer, with heavy seedling production in summer and early fall. Fruits ripen rapidly during the summer and at maturity consist of the thin fruit wall enclosing a single embryo. (No true seed in the usual sense since the embryo germinates while the fruit is still on the tree, said to be viviparous). Fruit wall splits almost immediately when the fruit falls, freeing the compact seedling, which consists of folded green cotyledons encasing a thick radicle densely covered by root hairs.

Intertidal margins of saltwater bays, inlets, and estuaries. Reaching maximum density on the landward side of the *Laguncularia* and *Rhizophora* zones and rarely extending onto sand soils slightly above the tidal zone. Invariably present at all mangrove *esteros* and sometimes the only mangrove present in the smaller mangrove areas including Estero Lobos (30°15' N), the northernmost mangrove area in western Mexico.

Both shores of the Gulf of California and the Pacific coast of southern Baja California Sur; Sonora northward to Estero Sargento and a small population of stunted plants further north at Puerto Lobos. Southward to Peru; Bahamas,

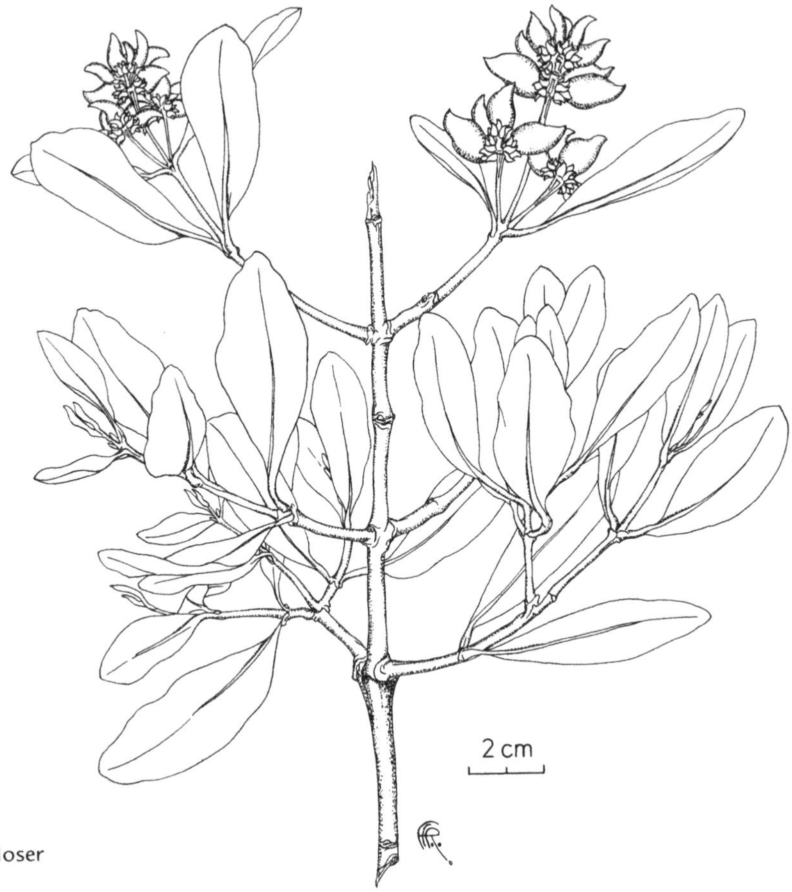


Figure 50. *Avicennia germinans.* (Felger and Moser 1985). Drawn by F. Runyan.

Florida, the Caribbean, and Texas to Brazil, and West Africa.

In the tropics *A. germinans* develops into a tree sometimes 25 m in height with a massive bole. Like other mangroves, black mangroves are severely damaged by freezing weather. This is the most widely distributed of the four New World species of *Avicennia* and has the largest flowers in the genus. There is some question whether *A. africana* is distinct from *A. germinans.*

### Berberidaceae—Barberry Family

Herbs, shrubs, or trees; 15 genera, 650 species. Members of this family are rich in alkaloids and have medicinal as well as poisonous properties.

#### *Berberis*—Barberry

Mostly shrubs, some trees. Leaves simple, actually unifoliolate (*Berberis sensu stricto*) or pinnately compound (*Mahonia*, often recognized in horticultural works, but seldom recognized by botanists).

Nearly worldwide in north temperate regions of both hemispheres and Andean South America; 500 species. References: Laferrière (1992), Marroquín (1972), Whittemore (1997).

The inner bark is a well-known source of yellow dye. The fruits are generally edible and can be made into jam and jelly. Some species harbor the black stem-rust of wheat (*Puccinia graminis*), and in the United States and elsewhere are illegal to sell or transport. Most *Mahonia* species are not susceptible to the black stem-rust.

Several shrub species are widely scattered in mostly mountainous regions in eastern and northern Sonora. *Berberis haematocarpa* is a large shrub common in some areas of north-central Sonora at the upper elevational limits of the Sonoran Desert and lower oak woodland; it sometimes reaches 4 m in height and develops a thick, woody base.

***Berberis longipes*** [*Odostemon longipes*]. *Palo amarillo; lolesí* (Guarijío) (figure 52)

Glabrous shrubs or sometimes small trees to 6 m; bark dark brown, rugose or fissured; twigs (spur branches) rough with prominent leaf scars; cambium (inner bark) bright yellow. Leaves alternate, clustered at the stem tips, 15–29 cm, odd-pinnate; leaflets 9–15, 3–8 × 1–2.6 cm, highly

Figure 51. *Avicennia germinans*. Pneumatophores at low tide. Photograph by R. S. Felger.

variable in shape, ovate to lanceolate or oblong or nearly oval, rather tough or leathery, bicolored and moderately glaucous, especially below, the margins entire or undulate, or with few blunt or spine-tipped teeth; stipules absent.

Flowers on slender, many-flowered racemes 12–25 cm including the peduncle. Pedicels (technically pseudo-pedicels) slender, 12–18 mm, arising from axil of a slender bract and bearing a minute bractlet well below the sepals and another slender bract subtending each flower. Flowers yellow, radial; 3-merous, with 3 sepal-like bracts; bracts, sepals, and petals cupped and soon deciduous. Sepals 6 in two series (outer ones may be bracts); petals 6 in two series, about 1 cm (presumably bearing nectaries, a generic character). Stamens 6, tactile (bending or closing inward when touched); anthers dehiscing by 2 valves hinged at the anther apex. Fruit a berry, the body broadly elliptic to obovoid, 9.5–12 mm, plus a discoid stigma on a persistent style 1.5–2.0 mm; at maturity the berry dark purplish brown beneath a bluish glaucous bloom. Seeds 7.5–8.0 mm, (usually?) two per fruit, obovoid and laterally compressed, smooth and shiny.

Highly localized in southeastern Sonora; tropical deciduous forest ecotone with lower oaks and oak forest, and pine-oak forest in adjacent southwestern Chihuahua 900–1450 m (2950–4760 ft). Also Durango and Sinaloa.

The yellow sapwood has been used for making yellow dye, and in mountains in northern Sinaloa it was used for dying wool (H.S. Gentry, specimen labels, University of Arizona herbarium). The fruits are edible (Standley 1923). This magnificent tree is worthy of cultivation.

## Betulaceae—Birch Family

Hardwood trees. Leaves winter deciduous, alternate, simple, with conspicuous pinnate veins; lateral veins alternate and straight; margins toothed; stipules deciduous, covering the leaves in bud. Monoecious. Flowers reduced, inconspicuous, and wind pollinated, the perianth of scalelike segments or absent. Male flowers in drooping catkins, female flowers in herbaceous or woody catkins. Fruits small, 1-seeded nutlets.

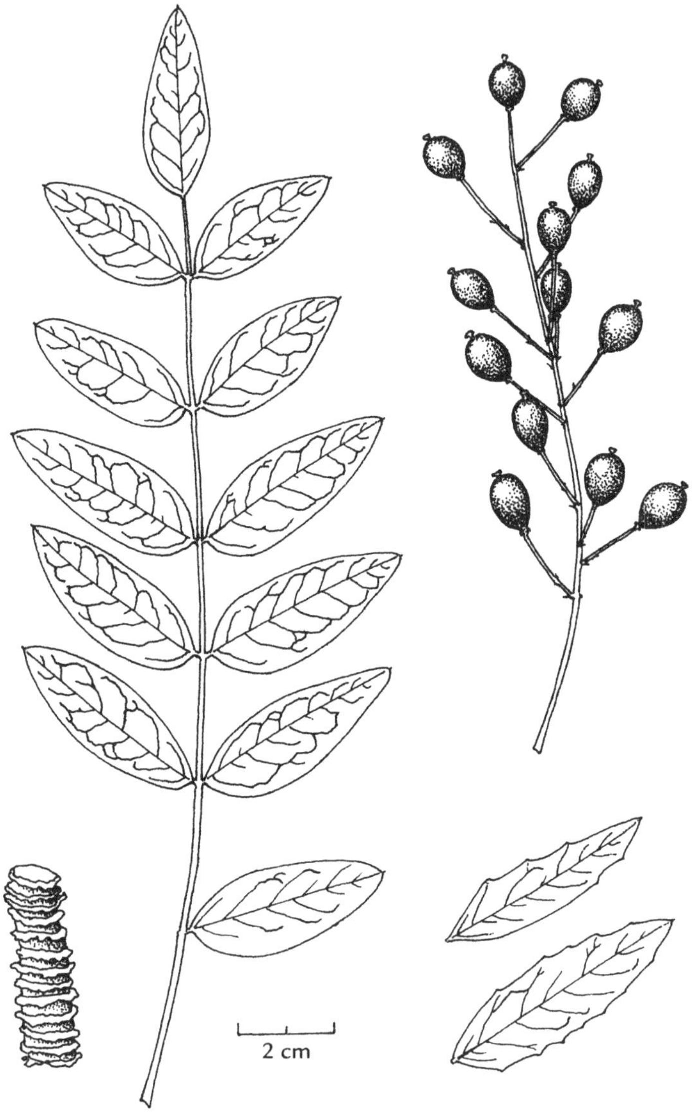


Figure 52. *Berberis longipes*. Drawn by M. B. Johnson.

Mostly temperate and cool to cold regions of Eurasia and North America, and mountains of Mexico, Central America, and the Andes; 6 genera, 125 species. Reference: Furlow (1997).

This family includes birch (*Betula*), hornbeam (*Carpinus*), and hazel (*Corylus*).

1. Young twigs and buds resinous; bud scales not striated; perianth present on male flowers, absent on female flowers; female catkins 8–25 × 6.5–10.5 mm, upright, woody, semipersistent, and not breaking into pieces. ________ ***Alnus***

1' Twigs and buds not resinous; bud scales distinctly striated; perianth absent on male flowers, present on female flowers; female catkins 25–55 × 20–25 mm, drooping, papery, herbaceous, and breaking into pieces. ________ ***Ostrya***

### *Alnus*—Alder, *Ailes*

Trees and shrubs. Northern Hemisphere, with one species extending to South America; 25 species, 9 in North America and 4 in Mexico. Reference: Furlow (1979).

The roots harbor symbiotic mycorrhizae containing nitrogen-fixing microorganisms.

**Alnus oblongifolia.** Arizona alder; *alamillo, júcaro* (figures 53 and 54A)

Slender, erect trees mostly 10–15 m, often 20–30 m in the Sierra de los Ajos. (The label on a herbarium specimen collected by Joe T. Marshall in 1952 reads: "Huge trees over 100 feet tall, some with lobed trunks 4 feet in diameter, forming dense groves along streams. Molina Canyon, 5400 feet, Sierra de los Ajos.") Apparently capable of sprouting

Figure 53. *Alnus oblongifolia*. Drawn by L. B. Hamilton.

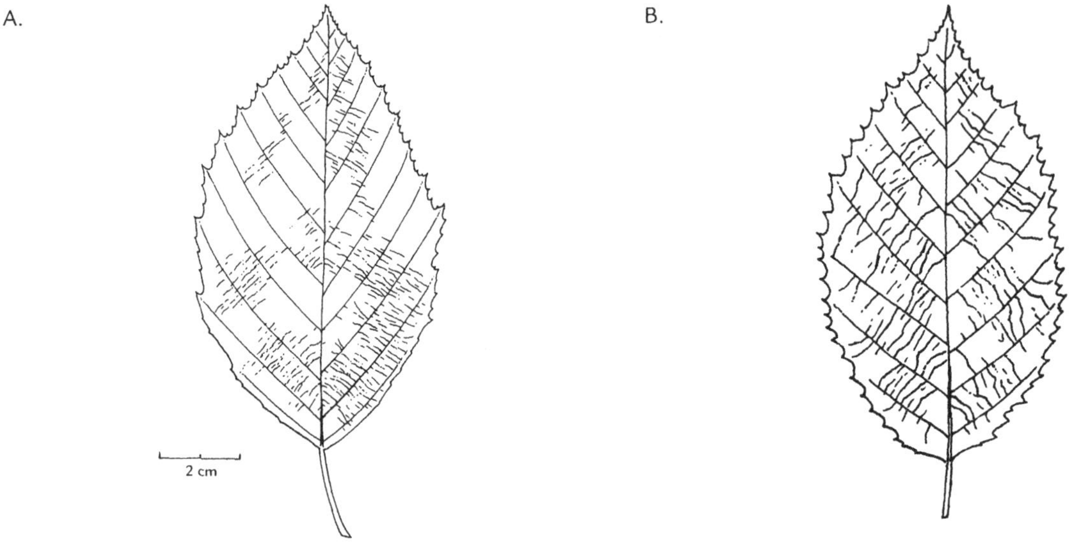


Figure 54. *Alnus* and *Ostrya*. (A) *A. oblongifolia*, (B) *O. virginiana*. Drawn by M. B. Johnson.

from exposed roots in streams. Bark mostly smooth and grayish; lenticels on twigs relatively inconspicuous. Winter buds protected by two slightly modified stipules. Young twigs, buds, and leaves reddish and conspicuously resinous; herbage moderately to densely hairy and glandular, especially the youngest growth, petioles, and lower leaf surfaces. Leaves 5–15 × 3–7.5 cm; petioles 1.5–2.5 cm; blades elliptic to ovate, the lateral veins 8–15 on each side, alternate to subopposite, the margins double-serrated with small, sharply pointed (but not spinose) teeth; stipules 10–15 mm, brown, longer than wide.

Male flowers 2 or 3 in axils of catkin scales, the perianth present. Female flowers in distinctive, woody, semipersistent, upright, and conelike catkins (8) 14–25 × 6.5–10.5 mm, the perianth absent. Fruit a minute, winged nutlet. Inflorescences emerge in fall; flowers appear at the beginning of the spring growing season in March and April.

Stream beds in moist, mountain canyons, usually with running water; oak woodland, pine-oak woodland, and mixed conifer forest; eastern Sonora, from the upper Río Mayo drainage northward; (895) 1220–2200 m [(2930) 4000–7220 ft]. Also Chihuahua, New Mexico, and Arizona south of the Mogollon Rim. The surprising low-elevation record of 895 m is from the Río Satachi near Nacori Chico (see *Acer grandidentatum* and *Brahea nitida*).

Furlow (1979: 88) reports that "In Sonora this species intergrades with *A. acuminata* subsp. *arguta*, and a case might be made to combine it with that taxon." However, the one specimen of *A. acuminata* he cites from Sonora is actually *A. oblongifolia* and is so annotated by him. *Alnus acuminata* ranges from Mexico to Argentina; subsp. *arguta* extends from Chihuahua and Durango to Panama.

## *Ostrya*—Hop Hornbeam

Trees, sometimes shrubs, usually with shaggy bark. Northern Hemisphere; five species. The name "hop hornbeam" refers to the hoplike fruiting clusters and the trees resemblance to hornbeam (*Carpinus*).

***Ostrya virginiana*.** American hop-hornbeam; *júcaro* (figure 54B)

Medium-sized, slender-trunked trees with an open crown; bark brown and furrowed but not shaggy. Scales covering the winter buds striated (the veins are raised and conspicuous), broad, and not resinous. Herbage, especially when young, with longer white hairs and often with small, stalked glandular hairs, both types of hairs enveloping youngest vegetative buds. Winter buds protected by relatively short, broad stipules. Leaves 3–9 cm; petioles 3–10 mm; blades nearly oval to lanceolate-elliptic, the lateral veins (9) 10–12 on each side, the margins serrated with often double-pointed teeth, the teeth of young, expanding leaves often gland tipped; stipules 6–8 mm, oblong to linear, thin, red-brown, and deciduous.

Male flowers single in axils of catkin scales, the perianth absent. Female flowers with a perianth; in herbaceous, drooping catkins 2.5–5.5 × 2–2.5 cm, with deciduous, papery, baglike pale green bracts 13–22 × 7–12 mm. Fruit a nutlet, not winged, enclosed in the saclike involucre. Flowering in spring; fruits in early summer.

In Sonora documented only for the Sierra el Tigre in oak and pine-oak woodland at approximately 1370 m, and Arroyo Hondo east of Yécora near the Chihuahua border; along streams in riparian canyons with ash, maple, oak, sycamore, and walnut. Canada to Central America including southwestern Chihuahua near the Sonora border and mountains in northeastern Sinaloa. Essentially a species of moist temperate and montane tropical regions; widespread across the eastern two-thirds of North America.

In temperate North America hop hornbeam has shaggy bark, and the female catkin bracts bear stinging hairs. These characters are not confirmed for northwestern Mexico. This tree is sometimes called ironwood because of its very hard wood and is noted for slow growth.

## Bignoniaceae—Bignonia Family

Trees, shrubs, and vines. Leaves usually opposite, sometimes whorled or alternate, simple or mostly compound with mostly entire leaflets; without stipules. Flowers bisexual, bilateral, large and showy; calyx with 5 usually unequal lobes; corollas 5-lobed and 2-lipped; often pollinated by large bees and hummingbirds. Fertile stamens 4, borne on the corolla (epipetalous) and arched beneath the roof of the tube with the anthers positioned for a pollinator, the pairs of different lengths, plus a staminode, or the staminode reduced or absent. (The corollas with their attached stamens usually fall free from the rest of the flower, often littering the ground, especially in the early morning and after rains.) Ovary superior; style 1; stigma 2-lobed. Fruits slender capsules opening lengthwise with many, flattened, papery winged seeds (except *Crescentia*).

Worldwide, mainly tropical, some in warm temperate regions, most diverse in South America; 112 genera (17 in Mexico), 725 species. References: Gentry (1980, 1992a, 1992b), Standley and Williams (1974).

This family includes a number of trees and vines grown for their spectacular flowers, mostly in tropical and subtropical regions. The African tulip tree (*Spathodea campanulata*) is often seen in gardens and parks from Hermosillo southward. It has large, brilliant orange-red flowers. Jacaranda (*Jacaranda mimosifolia*), cultivated for its blue-violet flowers, grows easily in lowland Sonora and is more frost tolerant than the African tulip tree. Western catalpa (*Catalpa speciosa*) is sometimes planted in northern Sonora, mostly at elevations above the desert. Catclaw vine (*Macfadyena unguis-cati*) is often seen covering walls and buildings in gardens throughout the Sonoran Desert; it has large, bright yellow flowers in spring.

1. Leaves simple and linear, sessile or the petioles < 0.5 cm. ______ ***Chilopsis***
1' Leaves compound (occasionally simple with a broadly winged petiole); petioles mostly > 3 cm.
    2. Leaves usually with 3 leaflets; petioles broadly winged. ______ ***Crescentia***
    2' Leaves with 5 or more leaflets (occasionally leaves with 3 leaflets); petioles not winged.
        3. Leaves palmately compound with 5 leaflets. ***Tabebuia***
        3' Leaves pinnately compound with mostly 7 or 9 leaflets. ______ ***Tecoma***

### *Chilopsis*

One species, 2 subspecies and 2 varieties. Reference: Henrickson (1985).

*Chilopsis* and *Catalpa* are closely related. ×*Chitalpa tashkentensis*, a sterile hybrid between the two genera, is occasionally grown in Arizona.

***Chilopsis linearis*** subsp. ***arcuata***. Desert willow; *mimbre; 'a:ñ, 'a:ñgam* (O'odham) (figure 55)

Large shrubs to small trees; trunks seldom straight; wood very hard; larger trunks irregular. Bark on trunks and larger limbs dark gray and fissured; twigs pale gray to yellow-brown. Twigs and leaves glandular viscid, sparsely hairy or mostly glabrous. (Large, sessile glands present on leaves and internodes of young twigs eventually rupture, leaving glandular, dotlike areas and coating the surfaces with viscid exudate. However, these glands do not break down at the nodes, where they seem to function as extrafloral nectaries.) Leaves winter deciduous, partially opposite to whorled or sometimes alternate, 10–20 cm × 3–8 mm, linear to linear-lanceolate, narrowed at the base to a short, indistinct petiole.

Stems terminating in inflorescences of short racemes or panicles; inflorescences and calyces densely hairy. Flowers 4.5–5.0 cm, sweet scented. Corollas 2-lipped with 5 crumpled lobes, pale to deep pink or rose-pink streaked and mottled with purple and yellow. Staminode filamentlike. Capsules linear, 15–35 cm, dark brown, pendent, persisting through the winter. Seeds with tufts of white hairs at each end. Flowering April–August.

Arroyos and washes; Chihuahuan desertscrub, grassland, and lower oak woodland; northeastern and north-central Sonora, 1220–1830 m (4000–6000 ft). Also Baja California and southeastern California to southwestern Utah and western New Mexico. Replaced by subsp. *linearis* farther east in New Mexico, western Texas, and north-central Mexico south to San Luis Potosí.

This is an attractive ornamental tree for desert gardens. There is considerable variation in flower size and corolla color—from pale pink or nearly white to purple with darker markings. Several darker-flowered selections are offered in the nursery trade. Like most members of the family, desert willow is readily propagated from cuttings as well as seeds. *Chilopsis* herbage produces a sweet and pleasant fragrance especially during and after rain. The tree is host to larvae

Figure 55. *Chilopsis linearis* var. *arcuata*. (Benson and Darrow 1944). Drawn by L. B. Hamilton.

of the geometrid moth, *Eucaterva variaria*, and the lacy, yellow cocoons of the moth are nearly always present among the foliage of trees in Sonora.

### *Crescentia*

Trees and shrubs in tropical America; 6 species.

****Crescentia alata*** [*Parmentiera alata*]. Gourd tree; *ayal; choca' ari* (Mayo) (figure 56)

Shrubs or trees 5–6 m; branches straight, relatively thick and rigid; herbage glabrous. Leaves evergreen, alternate on long shoots and clustered on short shoots, shiny green, 8.5–15 (27) cm, with 3 or rarely 5 leaflets; lateral leaflets usually much shorter or sometimes absent; leaflets oblanceolate; petioles broadly winged, often much longer than the leaflets; lower leaf surfaces minutely glandular-punctate.

Flowers and fruits borne directly on the trunks and larger limbs well below the leaves. Flowers bat pollinated, several in sessile clusters, with a musty odor, like garlic, cooked cabbage, or pond sludge, very strong at night. Calyx splitting to becoming spathelike, much shorter than the corolla. Corollas 5.5–7.0 cm and nearly as wide, somewhat bell shaped, the tube swollen near the base, the interior yellow-brown with purple-brown veins, the exterior shiny purple-brown. Stigma bilobed, the lobes spreading but closing when touched. Fruits hard shelled, rounded gourds, often to approximately 10+ cm in diameter; seeds numerous, flat, heart shaped, and not winged. Flowering March–July, sometimes in two or more flushes.

Tropical deciduous forest; southeastern Sonora. Rare, persisting from planted trees such as in the vicinity of La Aduana west of Alamos. Occasionally cultivated in Hermosillo and southward; frost sensitive. Northern Sinaloa to Costa Rica.

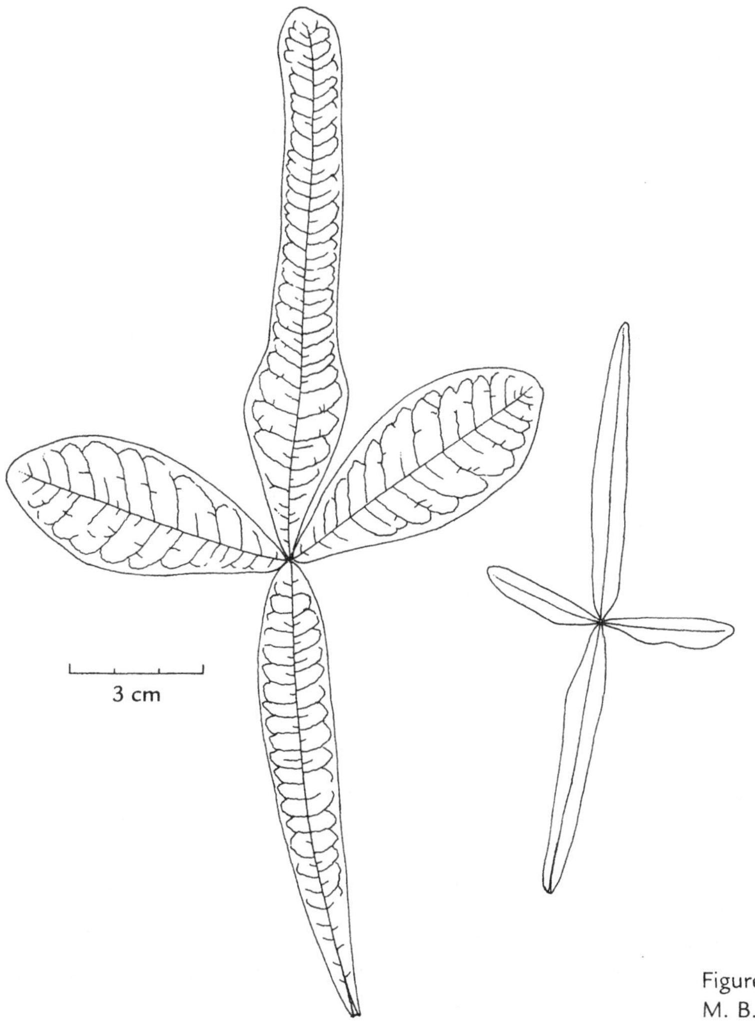


Figure 56. *Crescentia alata*. Drawn by M. B. Johnson.

The cauliflorous habit (flowers borne on trunks), a rather common occurrence in the tropics, is unique among the Sonoran flora. In the early sixteenth century the cross-shaped leaf was interpreted by Europeans as a significant Christian symbol (Oviedo y Valdes 1851–1855, Standley 1926). The gourds are used for utensils, although its more tropical and larger fruited relative *C. cujete* (calabash tree) is more widely utilized. In Tucson large numbers of honeybees have been observed to visit the blossoms. The unusual floral scent is similar to some species of *Manfreda* (Agavaceae).

### *Tabebuia*—Amapa

Hardwood trees and shrubs. Tropical to subtropical regions of the Americas; 100 species, greatest diversity in Cuba and Hispaniola.

Sonoran trees have a tall, slender trunk, bark of small, thick, longitudinal plates, and an open crown. Leaves appearing with the summer rains and drought deciduous; leafless or mostly so during the dry season, although during exceptionally wet fall or winter seasons the trees often do not entirely defoliate. Leaves large, opposite, palmately compound with 5 stalked leaflets, the margins entire to sometimes crenate-serrated, especially among young plants. Flowers in terminal clusters; mass flowering winter and early spring; bee pollinated. Staminode prominent but smaller than the fertile stamens. Capsules linear, pendent, produced in great quantity, mostly April–May.

Tabebuias are among thc world's most magnificent flowering trees. They are highly desirable as garden and avenue trees in hot climates with minimal frost. These trees are in full flower in the otherwise drab, dry-season landscape of tropical deciduous forests of the Sierra Madre Occidental and its foothills.

1. Pubescence of branched, and stalked and nonstalked stellate hairs, mostly golden brown, the leaf buds and calyces densely orange-brown woolly; corollas bright yellow; fruits rough surfaced with sparse, stellate hairs. ______ ***T. chrysantha***

1' Pubescence of simple or branched whitish hairs; leaf buds and calyces white-woolly (occasionally slightly yellowish); corollas rose-pink; fruits essentially smooth surfaced and glabrous. ______ ***T. impetiginosa***

***Tabebuia chrysantha*** subsp. ***chrysantha***. Yellow amapa; *amapa amarilla*; *tabuió yoresí* (Guarijío); *to'obo sahuali* (Mayo) (figure 57A)

Small to medium-sized trees to about 15 m. Bark dark brown. Herbage densely and minutely woolly with mostly branched and stalked, or nonstalked, stellate hairs, at first golden brown. Leaflets obovate to ovate or elliptic, with small, golden-brown stellate hairs, the upper surfaces often becoming glabrate, the terminal leaflet largest, 12–22 cm, often with a drip tip to 2.5 cm; petioles 5–14 cm, bases and distal portions minutely woolly.

Calyces 10–15 mm, very densely woolly with golden brown, often barbed or branched hairs. Corollas brilliant yellow, 6.6–8.5 cm. Capsules often 20–28 × approximately 1 cm, the surfaces rough with large papillae and moderately pubescent with stellate hairs. Flowering (January) February–March (June).

Slopes, mesas, and canyons; tropical deciduous forest; southeastern and east-central Sonora, 230–1035 m (750–3400 ft). Subspecies *chrysantha* in seasonally dry tropical forest from northern South America to Sonora and southwestern Chihuahua. Two other subspecies, in wetter regions, Panama and northern South America.

The flowering season is shorter and peak flowering usually later than that of *T. impetiginosa*. *Tabebuia chrysantha* occurs as scattered individuals or in small groves together with *T. impetiginosa* but is more restricted, both ecologically and geographically than the latter. The lumber of *T. chrysantha* is prized, as is that of *T. impetiginosa*.

***Tabebuia impetiginosa*** [*T. palmeri*]. Pink amapa; *amapa, amapa morada, amapa colorada; tabuió* (Guarijío); *to'bo* (Mayo); *pima hadunim* (Pima Bajo) (figure 57B)

Small to large trees sometimes reaching 20 (30) m. Bark dark brown and checkered. Herbage at first moderately pubescent with short, mostly white, simple or branched hairs. Leaflets elliptic to ovate or obovate, with slender hairs mostly on veins of lower surfaces or glabrate; larger or terminal leaflet 11–20 cm, often with a drip tip to 3 cm; petioles 5–11 cm, the proximal portion mostly glabrous, the distal portions often with minute, whitish hairs but not woolly.

Calyces 6–10 mm, with short, white or sometimes yellowish woolly hairs. Corollas bright rose-pink (trees very rarely white flowered), 4.4–8.5 cm, the throat white inside with 2 prominent yellow nectar guide lines plus 2 lesser lateral yellow lines. Filaments, anthers, style, and stigma at first white, turning rose-pink like the corolla but lighter. (As with many members of this family, the stamens, style, and stigma lie in a plane). Filament bases with stalked glandular hairs blocking entrance to the nectar chamber. Stigmas laterally flattened. Ovary base surrounded by a green to yellow ring-shaped nectary. Capsules often 25–45 × 1.5–2.2 cm, the surface more or less smooth and glabrous. Flowering (November) January–mid-March (April), the peak usually in late January with masses of flowers, and sometimes with a second flowering in March.

Mostly slopes and canyons in tropical deciduous forest; less common in thornscrub mostly along xeroriparian washes; southern and east-central Sonora, 5–1065 m (15–

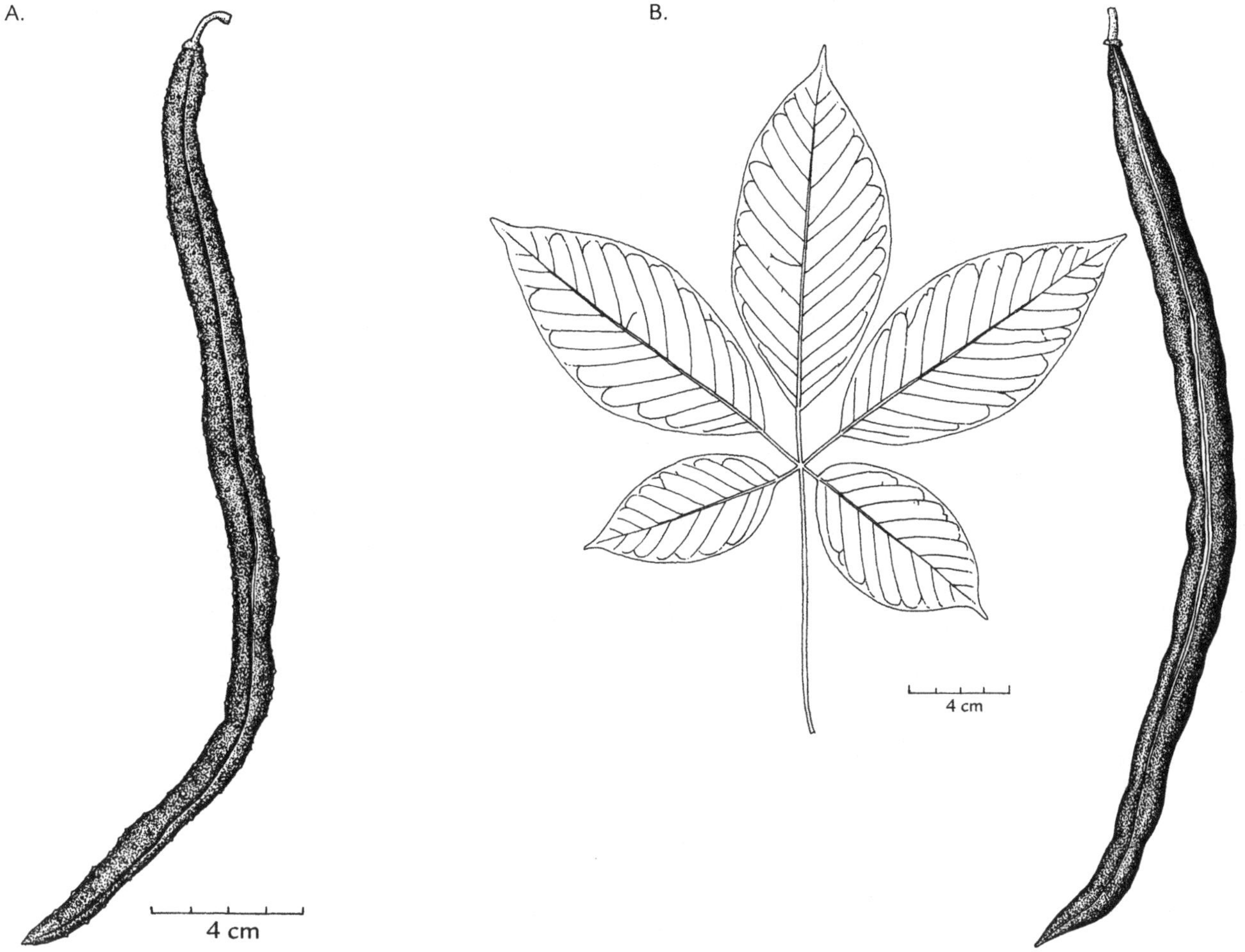


Figure 57. *Tabebuia*. (A) *T. chrysantha*, fruit; (B) *T. impetiginosa*, leaf and fruit. Drawn by M. B. Johnson.

3500 ft). The low elevation record is coastal thornscrub in southernmost Sonora. Guatemala to southwestern Chihuahua and Sonora. In Sonora extends into more xeric habitats and is much more common and widespread than *T. chrysantha*.

Flowering trees, visible from afar, dot the dry, parched hills and mountains with bright pink. They are sometimes planted along streets and near homes in southern Sonora. *Amapas* are prized for their virtually indestructible wood and were much used for *vigas* (roof supports) and posts. Cutting of *amapa* trees, generally *T. impetiginosa*, is prohibited by law for conservation purposes, and the law is generally respected. However, it is legal to clear the forest for planting buffel grass pasture. *Amapa* trees seem to be recolonizing the hills around Alamos, where the trees were cut out mostly in the nineteenth century for roof beams and supports for tunnels in the silver mines.

## *Tecoma*

Trees and shrubs; tropics and subtropics; 14 species, 2 in Africa and 12 in the Americas.

***Tecoma stans*** var. ***angustata*** [*Stenolobium incisum; Tecoma incisa*]. Trumpet bush; *gloria, lluvia de oro, palo de arco* (figure 58)

Shrubs or rarely trees to 6 m, glabrous or with minute, sparse hairs on newest growth. Leaves gradually drought deciduous (shrubs in northern Sonora and Arizona are winter deciduous), opposite, 10–24 cm, odd-pinnate; petioles 1.5–4.0 cm. Leaflets (3 or 5) 7 or 9; lateral leaflets 5–10 × 0.8–2.3 cm, thin, lanceolate, asymmetric at base, the margins coarsely toothed, the terminal leaflet longer than the lateral ones.

Figure 58. *Tecoma stans* var. *angustata*. (Benson and Darrow 1945). Drawn by L. B. Hamilton.

Inflorescences of terminal racemes or panicles 5–20 cm on peduncles to 5 cm. Calyx 7.5–8.0 mm, with 5 slender teeth. Corollas bright yellow, 4–5 cm, moderately bilateral, with a short basal tube and a swollen throat making up most of the corolla. Staminode minute. Capsules 10–15 cm, linear, pendent. Flowering March–December, but mostly during summer and fall.

Hills and mountains; Sonoran and Chihuahuan desertscrub, inland thornscrub, tropical deciduous forest, and oak woodland. Nearly statewide except most of western Sonora (occasional in mountains in the San Carlos-Sierra el Aguaje region); 50–1500 m (165–4920 ft). Tree-sized native tecomas are documented in only a few scattered localities, mostly in tropical deciduous forest in east-central and southeastern Sonora at 400–480 m (1310–1575 ft).

This species, with three broadly intergrading varieties, ranges from Florida, Texas, and Arizona to the Caribbean and northern Argentina. Variety *angustata* occupies the northern portion of the range of *T. stans* in southern Arizona, the Big Bend area of Texas, Baja California Sur, Chihuahua, Coahuila, Durango, Nuevo León, Sonora, and Zacatecas.

Variety *stans* occurs in Florida, the Caribbean, and Mexico generally south of the range of var. *angustata* to Argentina. It is extensively cultivated as an ornamental tree in tropical and subtropical regions and is widely grown in the lowlands virtually throughout Sonora. It occasionally persists or reproduces near habitations in southeastern Sonora. Variety *stans* is a large shrub or small tree with masses of brilliant yellow flowers and differs from var. *angustata* by its larger stature, larger leaves with leaflet margins not as coarsely toothed, larger flowers, and more vigorous habit of growth. Variety *angustata* is widely cultivated as an ornamental shrub in southern Arizona. In Sonora and elsewhere in Mexico the flowers of both varieties are brewed as tea to regulate blood pressure, and the leaves made into tea are taken for respiratory and other ailments.

## Bombacaceae—Kapok Family

Trees with stellate hairs. Leaves alternate and palmately compound; petioles prominent, with deciduous stipules. Flowers bisexual, large, and showy; sepals fused and cuplike; petals 5; ovary superior. Stamens numerous, or appearing as 5, the filaments fused at their bases. Fruits of capsules (in Sonora) with 5 woody valves; seeds many, enveloped in silky hairs (kapok) serving as parachutes for wind dispersal, the hairs not attached to the seeds.

Tropical and some subtropical, worldwide; 30 genera, 250 species.

This family includes spectacular tropical trees such as balsawood, the baobabs, kapok trees, silk-cotton trees, and the infamous durian. There are no native species north of Mexico. Bats and birds are among the most common pollinators. The Sonoran *Ceiba* and *Pseudobombax* are worthy of cultivation and easy to grow in hot climates but are frost sensitive.

1. Trunks and limbs with large thorns or ridges; leaflets 5–8, usually 7 per leaf, the margins serrated; stamens appearing as 5. ________ ***Ceiba***

1' Trunks unarmed; leaflets 5, the margins entire; stamens > 100. ________ ***Pseudobombax***

### *Ceiba*

Large, mostly deciduous trees. Tropical America and Africa; 10 species.

Commercial kapok is harvested from *Ceiba pentandra*, an enormous tree widely cultivated in tropical climates and grown for shade as far north as Hermosillo.

***Ceiba acuminata.*** Kapok; *pochote*; *guacapí* (Guarijío); *baogua* (Mayo); *vauwo* (Yaqui) (figure 59 and 60)

Medium-sized, drought-deciduous trees, with a tall, thick trunk and spreading canopy; bark gray-brown, densely armed with large conical "thorns" which become pyramidal and faceted, with age coalescing into intricate ridges (old trees occasionally unarmed). Even small plants form tuberous roots, the mature trees with many sometimes enormous, tough, fibrous but succulent tuberous roots. New herbage densely pubescent with golden brown stellate hairs; upper leaflet surfaces becoming sparsely pubescent with age. Leaflets (5) 7 (8), on short stalks, elliptic, lanceolate to oblanceolate or obovate, gradually tapered at both ends, the tips acute to acuminate, the margins serrated; terminal leaflet largest, 7–13 cm; petioles 6–18 cm; stipules slender, often 8–10 mm.

Flowers solitary on axillary peduncles. Calyx often 3.0–3.8 × 2.5–3 cm, thick and tough, shallowly lobed, glabrous outside, densely brown-velvety inside. Petals white, 10–16 × 1.8–2.5 cm, sparsely hairy inside, the outer surfaces densely pubescent basally with long, golden hairs. Staminal column 3.0–3.7 cm, thick and white, with cottony tufts of white hairs above (turning tawny brown when dry) covering 5 short, broad, hard spines, the column surmounted by 5 "stamens" with pink "filaments" 9.0–10.6 cm (each "stamen" actually a bundle of several stamens). Style often 12 cm; stigma approximately 2.5 mm, 5-lobed. Capsules 12–25 cm, ovoid-elliptic, pendulous, hard shelled, hanging from twigs in the canopy long after the leaves have fallen, dehiscing (November) late December through March, bursting open with a loud pop to release copious white kapok surrounding the seeds; fluffy, white kapok drifts through the forest. Seeds 6.8–

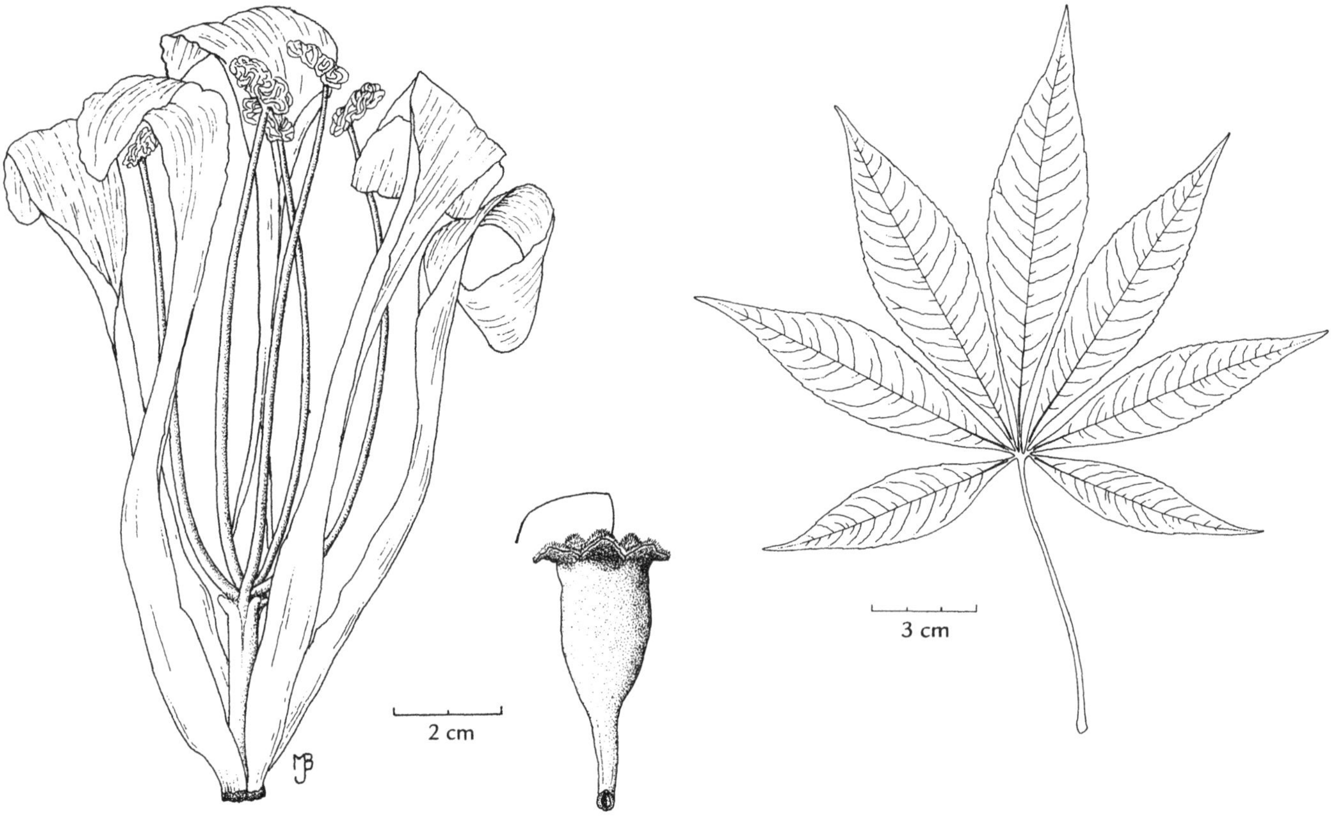


Figure 59. *Ceiba acuminata.* Drawn by M. B. Johnson.

7.7 × 6 mm, oval, dark brown, and smooth. Flowering late May–June (September).

Slopes, valleys and plains, canyons, and arroyos; common and widespread in tropical deciduous forest; less common in thornscrub and lower elevations in oak woodland; east, central, and southern Sonora, 60–1220 m (200–4000 ft). Best developed in tropical deciduous forest and in the thornscrub of southwestern Sonora in xeroriparian washes and arroyos. Northernmost populations grow on south-facing canyon slopes in the mountains west of Sinoquipe; these trees experience severe die-back during years of exceptional freezing weather. Also Baja California Sur and southwestern Chihuahua at least to Jalisco.

The Guarijío, Mayo, and Yaqui harvested the tender, tuberous roots from young saplings or young trees, generally as an emergency food; roots of larger trees are also edible but more fibrous and less desirable. The tuberous roots are peeled and eaten fresh or roasted. As emergency liquid one would just suck on the root. The Yaqui relied on this plant for food and liquid during the Yaqui–Mexican wars. The seeds are oily and edible, with a butterlike endosperm and a nutlike flavor; they are usually roasted and ground. The seeds have been used by the Yaqui for allergies. The fluffy kapok serves as stuffing for pillows. The wood is sometimes used for tomato crates and pallets, cots, and occasionally as house beams if protected from the rain. The trees are protected by law in Sonora. The trees are larval host of the giant, metallic buprestid beetle, *Euchroma gigantea*. During the spring, conspicuous feeding aggregations of bright scarlet nymphs of a true bug (*Dysdercus*) often form thick balls around *Ceiba* seeds on the ground.

### *Pseudobombax*

Small to medium trees. Tropical America; 20 species. Reference: Dugand (1943).

**_Pseudobombax palmeri_** [*Bombax palmeri*]. Shaving-brush tree; *cuajilote* (figures 61 and 62)

Striking trees, usually with a low, spreading crown, sometimes reaching 10 m, usually much shorter, unarmed, with massive and often irregular trunks and limbs. Bark peeling in large, very thick, rough pieces of several shades of red-brown. Twigs relatively thick; dormant, dry-season vegetative buds covered by broadly triangular to oblong, thick, green scales, the larger ones 1.0–3.2 cm. Leaves and twigs at first densely pubescent with soft, brownish hairs. Leaves large, present only during the summer rainy season; leaflets 5, sessile or nearly so, broadly obovate, the margins

Figure 60. *Ceiba acuminata*. Thorn on trunk, 2 cm in height. Photograph by M. A. Dimmitt.

entire, the terminal leaflet 20–37 cm, the lateral ones smaller; petioles often 8–14 cm, swollen at the base and moderately expanded at the apex; stipules often 8–22 mm, broadly oblong to linear.

Flowering when leafless from the upper branches; flowers large, upright, and bat pollinated. One to several flowers on very short, thick peduncles near the stem tips, the pedicels longer than the peduncles, both approximately 1 cm thick; each branch usually supporting only one open flower at a time. Flowering branchlets, pedicels, and calyx brown-pubescent like the young herbage, the buds held erect. Calyx 1–1.5 cm. Petals 12–20 × 1.8–2.2 cm, linear-oblong, thick, drooping like banana peels, white inside, the outside longitudinally white on one half, the other half (the portion exposed in the bud) pale pink, yellow, or brown. Stamens, style, and stigma white. Stamens > 250, held erect and spreading and some drooping, like a pom-pom, approximately 27 cm across, the filaments of unequal lengths, the longer ones often 14 cm. Style extending 1.8–2.2 cm beyond the longest stamens and hanging down below them, the stigma slightly thicker than the slender style. Capsules often 10–15 cm, elliptic, containing soft, golden yellow kapok. Seeds approximately 5.5 mm, ovoid, dark brown, and smooth. Flowering (February) March–April (May); fruits maturing June–July.

Cliffs, ridges, and steep rocky slopes; tropical deciduous forest and lower elevations in oak woodland; southeastern Sonora, 350–1100 m (1150–3600 ft). Also Jalisco to southwestern Chihuahua.

Two-year-old plants develop a swollen caudex, making them interesting subjects as container plants, and may produce flowers when only about 1 m tall. The trees are easily transplanted and readily grow from seeds or cuttings.

The flowers begin opening soon after sunset and are fully opened about 20 minutes later. A sweet but musty, pumpkinlike odor is produced at that time and through much of the following day. Upon opening, a small amount of non-sweet exudate is produced by glands just below the calyx. At this time no nectar is seen at the center of the flower, but later during the night sweet nectar accumulates at the center where the petal and filament bases form a cup.

Mark Fishbein (personal communication, 1995) made the following observations in hills east of Alamos:

> The flowers were visited at night by moths, which did not contact anthers or stigma, and bats which approached from below and contacted the drooping anthers and stigma. The bats struck the flowers forcefully, releasing a cloud of pollen, repeatedly, even

Figure 61. *Pseudobombax palmeri*. Drawn by M. B. Johnson.

Figure 62. *Pseudobombax palmeri*. Photograph R. S. Felger.

from the same flower. The flowers were being disassembled by leaf-cutter ants, which carried stamens, styles, and tepal pieces down the trunk. In the morning, the flowers were visited by several species of orioles, including the black-vented oriole.

Pretending to be bats, several curious botanists independently stuck their faces into a flower and promptly developed rather severe allergic reactions.

## Boraginaceae—Borage Family

Annuals to large trees. Worldwide, especially diverse in western North America and the Mediterranean Region east to Asia; 150 genera, 2500 species.

This family includes many herbs and some shrubs in Sonora such as *Cryptantha* and *Heliotropium*.

### *Cordia*

Mostly trees and shrubs. Old and New World tropics; 250 species. References: Borhidi et al. (1988), Miller (1985).

Two other species are native to Sonora: *C. curassavica*, a small-flowered, bushy plant mostly in thornscrub and tropical deciduous forest, and *C. parvifolia*, a common woody shrub in Sonoran desertscrub and thornscrub with masses of showy, white flowers. *Cordia boissieri*, with large, showy, white flowers and fleshy, somewhat edible fruits, is cultivated in southern Arizona and is suitable for most of Sonora; it is native to southern Texas and northeastern Mexico. The genus is sometimes segregated into as many as 10 different genera.

***Cordia sonorae.*** *Palo de asta; comahuí* (Guarijío); *pómajo* (Mayo); *pomahe* (Yaqui) (figure 63)

Large hardwood shrubs or trees to approximately 10 m; trunks single to multiple, usually erect to ascending with straight, slender stems. Twigs and petioles pubescent. Leaves alternate, simple, tardily drought deciduous (often nearly evergreen in cultivation), 4.5–17 (21) × 2–5.8 (10) cm, petioled; blades broadly elliptic or ovate to obovate or lanceolate to oblanceolate, dark green, rather firm, sparsely scabrous to glabrate above, the margins entire; stipules absent.

Flowers in short, dense, terminal panicles. Calyx 5-lobed, green, hairy, cylindrical with conspicuous longitudinal ribs and short lobes. Corollas, filaments, anthers, and style white; stigma yellow-green. Corollas 5-lobed, 3.0–3.7 cm wide. Stamens 5, attached to the corolla tube, the filaments with hairs covering the entrance to the nectar chamber. Ovary superior, white, narrowed at the base to a slender stipe surrounded by a yellow nectary disk (when the flower is opened and the nectary is cut, it has a sweet, rancid odor). Fruit a 1-seeded nutlet (the pericarp very thin and quickly drying), ripening in early summer, 6.0–7.0 × 3.9–4.2 mm, broadly ellipsoid, mottled tan and brown or red-brown, essentially smooth to faintly ribbed, minutely papillate, the dried style base forming an oblique, persistent knob at the fruit apex; fruit tightly held in the persistent calyx and corolla to form a parachutelike dissemination unit. Flowering (January) March–May.

Slopes, canyons, arroyos, and bajadas; Sonoran desertscrub, thornscrub, and tropical deciduous forest; central and southern Sonora, 30–800 m (100–2625 ft). Mostly a plant of the tropical deciduous forest, extending into thornscrub along the Río Sonora drainage, and the southern margins of the Sonoran Desert in riparian or semiriparian canyons and arroyos in the Sierra el Aguaje. Also Guerrero to southwestern Chihuahua.

The wood is used for making handles of implements such as axes, hoes, or shovels and is sometimes used in light construction. The profusion of showy, sweet-scented, white flowers in late spring and the dense foliage make it a suitable landscape subject for nearly frost-free regions. The corollas fade to a satiny brown color and keep their shape even after falling. The unusual parachutelike disseminule of the dried flower and the enclosed fruit is unique among Sonoran trees. Similar dispersal units also occur in several other Mexican cordias (e.g., *C. elaeagnoides*, a tree occurring from Sinaloa to Chiapas; also see *C. alliodora* in Janzen 1983).

## Buddlejaceae—Butterfly-bush Family

Trees, shrubs, and some herbs. Both hemispheres, mainly tropical and subtropical; 10 genera, 150 species.

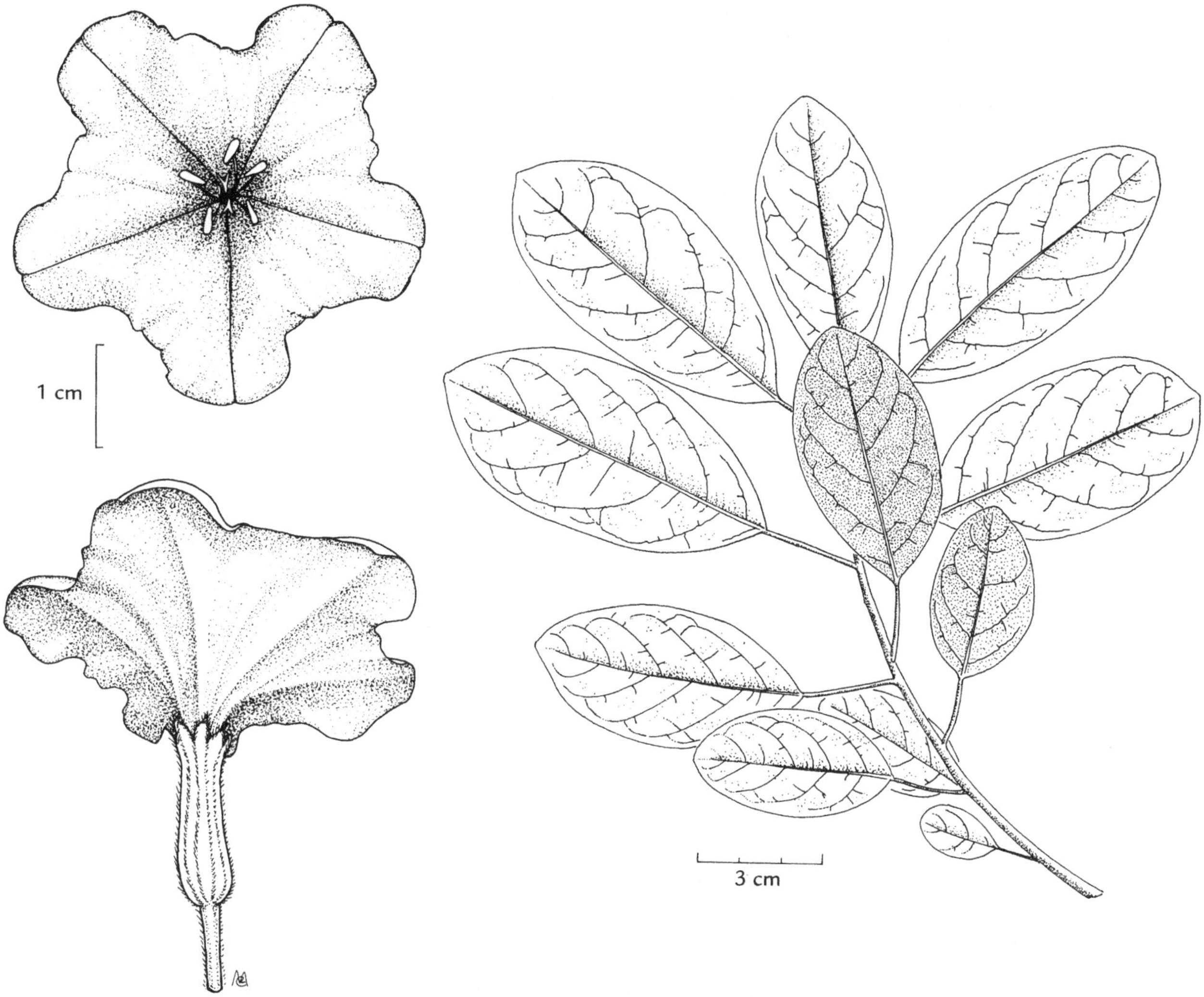


Figure 63. *Cordia sonorae.* Drawn by M. Essig (flower) and M. B. Johnson.

### *Buddleja*—Butterfly Bush

Trees and shrubs with stellate, branched, and often glandular hairs, especially the new herbage, inflorescences, calyces, and outer petal surfaces. Young stems square in cross section. Leaves opposite and decussate; stipules reduced to a thin but rather conspicuous line across the node connecting the opposite petioles.

Dioecious (in Sonora). Inflorescences of terminal panicles, 3- or 4-times branched, the branching decussate; branches subtended by leaves or bracts. Flowers numerous, small; male and female flowers of similar size and appearance. Calyx and corolla each united below into a tube and 4-lobed; stamens 4, attached to the corolla. Fruits of small, woody capsules, with 2 carpels partially separating to reveal a wrinkled, pitted central column (the placenta) and numerous, readily disseminating, minute, flat seeds.

Both hemispheres, mainly tropical and subtropical with greatest diversity in eastern Asia; 100 species. In North America concentrated in Mexico. References: Norman (1967, 1992).

There are two major species groups in North America: (1) shrubs with small inflorescences and usually wingless seeds, mostly in hot, arid regions. Two species in this group occur in Sonora; (2) large woody shrubs and trees with large inflorescences and winged seeds, largely in mesophytic conifer forests at higher elevations. This latter group includes the three arborescent Sonoran species, although *B. sessiliflora* extends into lowland, desert habitats.

1. Trunks and limbs thick, often massive; demarcation between leaf blade and petiole abrupt; flowers yellow. ______ ***B. cordata***
1' Trunks and limbs slender; leaf blades gradually narrowed at base; flowers white to greenish white.
   2. Leaves 2.5–15 cm; petioles not winged; inflorescences usually of much-branched pyramidal panicles; flowers pleasantly fragrant. ______ ***B. parviflora***
   2' Larger leaves (11) 15–30 cm; petioles conspicuously winged, the leaves thus often appearing sessile; inflorescences with an elongated main axis, sometimes branched below, and widely spaced pairs of headlike flower clusters; flowers foul smelling. ______ ***B. sessiliflora***

***Buddleja cordata*** var. ***cordata***. Rusty buddleja; *tepozana* (figure 64)

Large shrubs or small trees to approximately 6 m, often forming thick, massive trunks and lower limbs. Herbage sweet smelling when crushed; young twigs, petioles, and lower leaf surfaces yellow-brown to rusty colored with dense pubescence of stellate and branched hairs. Leaves mostly 10–16+ × 3–8.5 cm; petioles prominent; blades lanceolate to ovate, the upper surfaces dark green and sparsely pubescent or glabrate, the lower surfaces yellow-brown with very dense hairs consisting of a lower layer of appressed stellate hairs and taller branched-stellate hairs; leaf margins minutely toothed.

Panicles often 7–17 cm. Flowers with a sweet honeysuckle, citrus, or peppery fragrance, attracting bees. Corollas 5.5–6.0 mm wide, yellow, with hairs covering the entrance to the tube. Capsules 4.0–4.5 mm; seeds probably 1.2–2.0 mm. Flowering July–September.

Slopes and canyon bottoms; oak and pine-oak woodland; east-central Sonora, above about 1350 m. Barely reaching Sonora at the Chihuahua border east of Yécora and common in the mountains of adjacent southwestern Chihuahua to 2700 m. Southward in the highlands of Mexico to Guatemala.

In pine-oak and fir (*Abies*) forests in southwestern Chihuahua this species becomes a tree to 10 m; farther south it sometimes reaches 20 m. Another subspecies occurs in northeastern Mexico.

***Buddleja parviflora***. (figure 65)

Usually shrubs, occasionally small trees 5–7 m with several trunks or sometimes a single, well-formed, slender woody trunk 13–17 cm in diameter, often slightly irregular, gnarled, and twisted; bark brown and shredding; branches and twigs slender and brittle. Leaves highly variable, 2.5–15 × 0.8–7 cm; petioles 0.5–2.5 cm; blades lanceolate or ovate to nearly oblong, glandular and green above with branched hairs, silvery gray-green to yellow-brown below, the margins entire to coarsely toothed.

Panicles often 5–11 cm; flowers in small, short-stalked cymes, and with a sweet fragrance. Calyx (1.5) 2.2–3.0 mm. Corollas white, 5–6 mm wide, the tube approximately 1 mm longer than the calyx. Capsules 3.0–4.5 mm; seeds 0.4 mm. Flowering March–May.

Slopes, canyons, and arroyos. Oak woodland and pine-oak woodland; east-central and southeastern Sonora, 900–1500 m (2950–4920 ft). Also southwestern Chihuahua to Oaxaca.

The largest tree-sized plants in Sonora were seen in the region of Tarachi and Mulatos.

***Buddleja sessiliflora*** [*B. wrightii*]. *Lengua de buey, tepozana; guatohué* (Guarijío) (figure 66)

Usually shrubs, sometimes nearly herbaceous, and occasionally treelike to 5 m at the upper elevational margins of tropical deciduous forest in southeastern Sonora. Stems brittle. Leaves highly variable in size and texture, often reduced above, the larger leaves (11) 15–30 × (3.5) 5–11 cm; petioles to approximately 3 cm, winged; blades ovate to lanceolate, decurrent onto the petiole; margins nearly entire or minutely to coarsely toothed.

Panicles (or terminal floriferous branches) narrow with a well-defined main axis 6–32 cm and sometimes branched at the base, dense to open, leafy or not with the leaves often reduced above to bracts, the flowers in dense, sessile or peduncled headlike clusters, 1.2–1.5+ cm in diameter. Flowers yellow-green, the fragrance acridly sweet like carrion or stale urine and attracting flies, honeybees, butterflies, and ants which crawl all over the plants. Calyx 3.5–4.5 mm. Corollas 6.0–6.5 mm wide, the tube 1.0–1.5 mm longer than the calyx. Capsules 3.5–4.5 mm; seeds 0.7–1.1 mm. Flowering February–June.

Slopes, canyons, and arroyos; mostly in mountains in tropical deciduous forest and oak woodland, and occasionally in Sonoran desertscrub and thornscrub. Widely scattered but seldom common across most of the eastern two-thirds of the state, with a few records in western Sonora; 30–1170 m (100–3840 ft). Baja California, Arizona, and southern Texas to Oaxaca.

## Burseraceae—Torchwood Family

Trees and shrubs with resin ducts containing aromatic triterpenes and ethereal oils. Worldwide, tropical to subtropical, and especially numerous in the dry tropics of the Americas and northeast Africa; 20 genera, 600 species.

Old World members are the source of frankincense, made from the gum of *Boswellia carteri* and related species, and myrrh, made from *Commiphora myrrha* and related species;

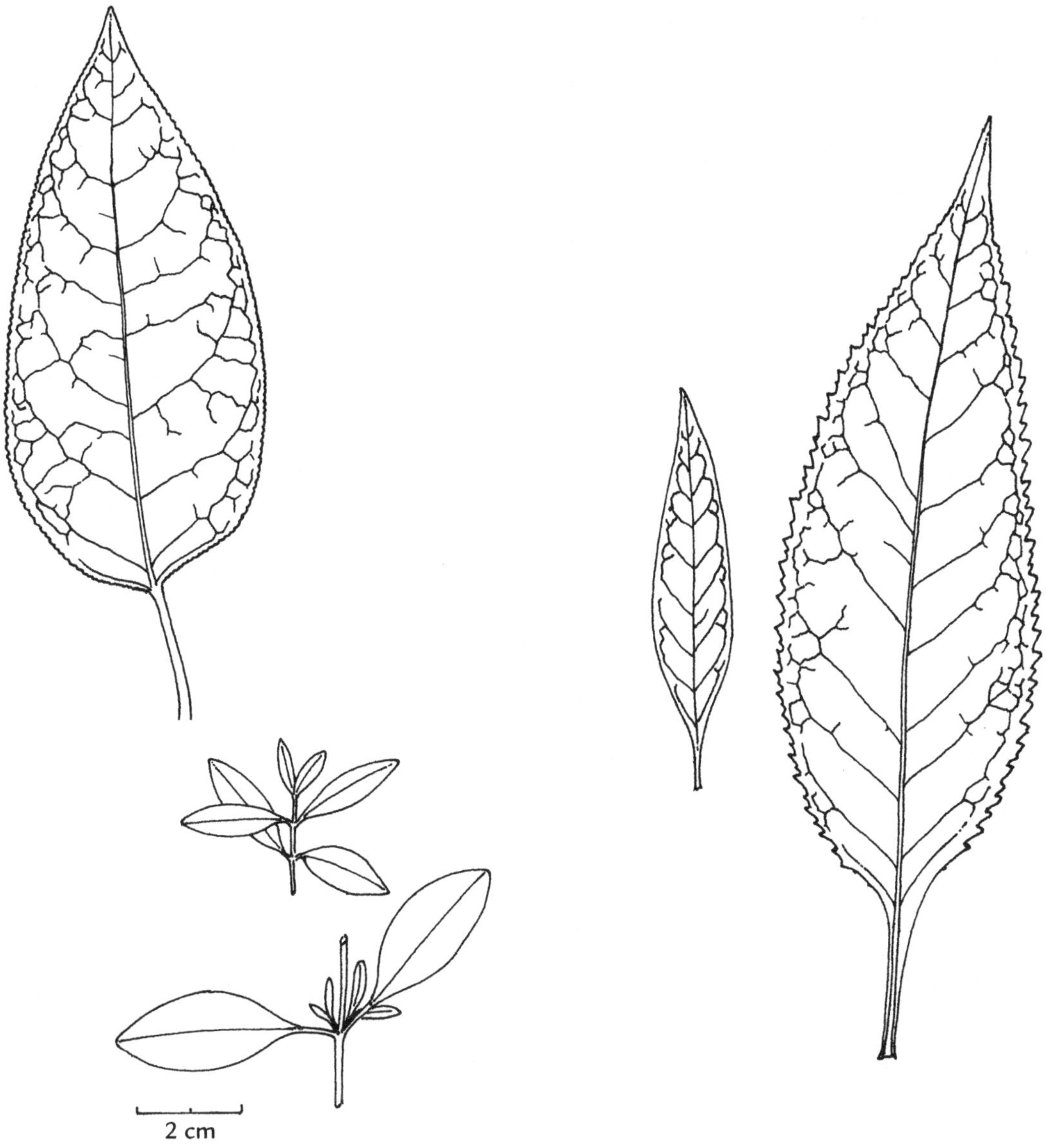


Figure 64. (*top*) *Buddleja cordata* var. *cordata*. Drawn by M. B. Johnson.

Figure 65. (*bottom*) *Buddleja parviflora*. Drawn by M. B. Johnson.

Figure 66. (*right*) *Buddleja sessiliflora*. Drawn by M. B. Johnson.

both genera, especially *Commiphora*, are closely related to *Bursera*. All members of the family are frost sensitive. The Burseraceae and Anacardiaceae are anatomically similar and closely related.

### *Bursera*

Trees and shrubs, rich in terpenoids and usually highly aromatic. Leaves alternate, pinnate or bipinnate, exceptionally tripinnate or rarely reduced to one leaflet (e.g., *B. hindsiana*); stipules absent; twigs often with the first few leaves of the season unifoliolate or with reduced numbers of leaflets (see *B. grandifolia*).

Mostly dioecious or flowers unisexual and bisexual on the same plant (polygamo-dioecious). Flowers small, solitary, or few- to sometimes many-flowered axillary racemes or panicles near stem tips (often on short shoots, especially the xerophytic species; measurements for inflorescences given here include the peduncle). Inflorescence axes and branches generally slender, often becoming thicker as fruits develop (especially among the trivalvate species). Calyx 3–5-lobed, cuplike, opening in the bud; petals 3–5, often cupped with the margin slightly to strongly inrolled above. Stamens in two whorls, twice as many as the petals (except sometimes reduced to 5 seemingly fertile stamens in some bisexual or pistil-bearing flowers); fertile anthers bright yellow, sterile anthers (on functionally female or some stamens on bisexual flowers) white. Fruits small, 1-seeded, glabrous at least when fully formed (in Sonora), drupelike, the exocarp splitting into 2 or 3 leathery valves that fall away to reveal a brightly colored aril-like mesocarp (here called the aril) partially to fully covering the seed.

Tropical and subtropical America, mostly in tropical deciduous forest and thornscrub, 100 species. Greatest diversity on the Pacific slope of Mexico, where 80 species occur; 70 of them endemic. Only genus of the family in northern Mexico and the United States. Eight species occur in tropical deciduous forest in southeastern Sonora, four in the Sonoran Desert, and one reaches the extreme desert in northwestern Sonora. References: Becerra and Venable (1999), Bullock (1936), Johnson (1992), Kohlmann and Sánchez-Colón (1984), McVaugh and Rzedowski (1965), Porter (1974), Rzedowski and Kruse (1979).

There are two distinctive, natural groups of *Bursera*, the bivalvate-fruited species (section *Bullockia*) and the trivalvate-fruited species (section *Bursera*; also see key to the species). Occasional, apparent hybrids occur among some Sonoran trivalvate burseras. Differences between these sections are summarized below:

| | Bivalvate species | Trivalvate species |
|---|---|---|
| Bark | Smooth, not peeling | Peeling in papery sheets |
| Wood | Harder, less succulent | Softer, semi-succulent |
| Leaves | Once pinnate or bipinnate (sometimes unifoliolate or tripinnate) | Once pinnate |
| Flowers | Mostly 4-merous | 3-, 4-, or 5-Merous (often even on the same plant; often (4) 5-merous in male flowers, and 3-merous in female flowers |
| Ovary (female flower) | 2-Chambered | 3-Chambered |
| Fruits | Globose, 2-valved, maturing the same season | Ovoid (narrowed at each end), 3-valved, maturing the following year |
| Seeds | 2-Angled, blackish | 3-Angled, light colored |
| Aril | Thin but fleshy, not papery | Thin and papery, not fleshy |
| Cotyledons | 3-Lobed | Highly dissected or 3-lobed |

The flowers and reproductive biology are poorly known for the Sonoran species. The flowers generally appear in early summer just before leaf emergence or with the first tender, new leaves (e.g., *B. fagaroides*, *B. grandifolia*, *B. lancifolia*, and *B. penicillata*), or among the leaves during hot, humid weather (e.g., *B. hindsiana* and *B. laxiflora*). In the tropical deciduous forest, the stronghold of the Sonoran burseras, flowering occurs during the first days of the summer rainy season, or with the buildup of clouds and humidity that heralds the first rains. The flowering pulse is massive and of short duration.

The following observations, reported for *B. simaruba* in Costa Rica by Janzen (1983:201), may apply to at least some of the Sonoran burseras:

> The flowers . . . grow in clusters at the ends of the branches. The male and female flowers look alike; the female has nonfunctional stamens, but the male produces much more nectar and at an earlier time each day than does the female. Nectar production starts before dawn, and the flowers last only one day. Male plants produce many more flowers (up to five thousand at one time) than do females (rarely more than one thousand at one time) and also flower for longer periods, starting before and finishing after all the females in the area. The flowers attract large numbers of *Trigona* and *Hypotrigona* (stingless bees) as well as flies, ants, and some small cerambycid beetles. Once the flowers have been pollinated the fruits expand to full size in less than a week (about 9 mm in diameter) and remain on the tree a full eight months before they ripen and are dispersed. The embryo of these fully expanded fruits is minute and remains that way until just before the fruit ripens. I assume this reduces seed predation by making the nutrient-rich embryo unavailable to seed-eating animals.

The fruits of the Sonoran burseras develop rapidly. Those of the bivalvate species appear to mature within one or two months and quickly disappear from the trees. Many of the fruits of the trivalvate species remain on the trees until just before flowering the following year. The embryos of the trivalvate species appear to fill out at the last moment, and the seeds are hollow if cut open earlier in the year. The arils do not develop their bright orange color until the seed is mature. Birds appear to be primarily responsible for seed dispersal in *Bursera*, the brightly colored arils serving as an advertisement. Rodents (see *B. microphylla*) and insects such as ants have also been observed to take the fruits or seeds.

The aromatic sap and resins seem to be effective in repelling herbivores. When leaves are broken off turgid stems of some species, a tiny jet of clear, aromatic sap may squirt from the stem (the "squirt-gun" response), and this terpene-rich liquid also quickly covers the stem surface around the broken leaf base (the "rapid-bath" response). Any hole chewed in a leaf or stem surface likewise produces the rapid-bath response, quickly covering an insect or other potential herbivore (Becerra 1997, Becerra and Venable 1990). Certain chrysomelid beetle larvae in the Tehuacán valley and elsewhere circumvent rapid-bath terpene dousing by very slowly biting a hole in the petiole to release turgor pressure before starting to feed on the leaf. These beetles also occur in Sonora (Judith X. Becerra, personal communication 1999).

The cotyledons of *B. fagaroides*, *B. lancifolia*, and *B. microphylla* have several thin, linear lobes and also often have secondary lobes. In contrast, the cotyledons of *B. hindsiana*, *B. laxiflora*, *B. grandifolia*, and probably *B. penicillata*, *B. simaruba*, and *B. stenophylla*, are palmately three lobed.

The foliage and gum, or copal, are aromatic and have a long history of religious and medicinal use. The wood is soft and light weight, easy to carve, and usually does not split during drying. The wood has been used for utensils, bowls, masks among the Guarijío, Mayo, Yaqui, and others, and for small boxes, fetishes, and other objects by the Seri. An attractive mistletoe, *Psitticanthus sonorae*, with tubular red flowers, is parasitic on *B. microphylla*, *B. laxiflora*, and *B. hindsiana* in coastal regions of Sonora and on *B. fagaroides* near Alamos.

Burseras in Sonora are generally called *torote*, a name applied to various semisucculent or sarcocaulescent shrubs and trees including certain species of *Jatropha* (Euphorbiaceae). The term *torote* is probably derived from the Piman or Uto-Aztecan name (see common names below, e.g., *toro*). The thick-trunked desert burseras are sometimes called elephant trees, a name usually applied to *Pachycormus discolor* (Anacardiaceae) of Baja California.

1. Outer bark not exfoliating or peeling in sheets; leaves once pinnate or bipinnate, sometimes reduced to one leaflet; fruits bivalvate with a thin, fleshy orange aril covering one-half to two-thirds of the blackish seed: bivalvate burseras.
   2. Leaves with (1) 3 (5) leaflets. ____ ***B. hindsiana***
   2' Leaves with (5) 7 or more leaflets, the leaves once pinnate, bipinnate, or tripinnate.
      3. Leaves once pinnate, hairy (fuzzy to the touch); twigs thick. ____ ***B. penicillata***
      3' Leaves usually bipinnate (occasionally pinnate or tripinnate), glabrous or glabrate (not fuzzy); twigs relatively slender.
         4. Individual leaflets fewer (see text), the margins usually several toothed. ____ ***B. laxiflora***
         4' Individual leaflets numerous, the margins entire. ____ ***B. stenophylla***

1' Outer bark exfoliating in papery flakes or sheets; leaves once pinnate; fruits trivalvate with a thin, papery aril covering the light-colored seed: trivalvate burseras.
   5. Leaves 16–30 cm, the leafstalk rounded or oval in cross section, neither winged, grooved, nor obviously flattened; leaflet margins entire.
      6. Leaflets densely pubescent; inflorescences pubescent, densely and many flowered, the pedicels shorter than the fruits, the fruits crowded. ____ ***B. grandifolia***
      6' Leaflets glabrous or sparsely pubescent; inflorescences glabrous, usually few-flowered or flowers solitary, the flowers and fruits not crowded, at least some pedicels longer than the fruits. ____ ***B. simaruba***
   5' Leaves 1.5–15 (17) cm, the leafstalk narrowly winged, flattened (especially when fresh), or grooved; leaflet margins entire, serrulate, or dentate.
      7. Bark of twigs reddish; leaflets 1–2.5 mm wide, the margins entire or occasionally with a few small lobes. ____ ***B. microphylla***
      7' Bark of twigs generally not reddish; leaflets 3–20 mm wide, the margins serrulate, irregularly dentate or rarely entire.
         8. Leaflet margins irregularly toothed, the teeth coarse to sometimes small, or some leaflets entire; leaflets 1.5–6 × 0.3–1.0 cm. ***B. fagaroides***
         8' Leaflet margins finely and uniformly serrulate; leaflets 3–7 × 0.7–2.0 cm. ***B. lancifolia***

***Bursera fagaroides*** var. ***elongata*** [*B. confusa* in part]. Fragrant elephant tree; *copal, torote, torote amarillo, torote de venado, torote de vaca; to'oro* (Guarijío); *to'ro, to'ro sahuali* (Mayo); *saa tooro* (Yaqui) (figure 67)

Glabrous trees to 8 m with a thick, semisucculent trunk and limbs, and a spreading crown, or multiple-trunked large shrubs at the northern limits of their distribution (resulting from freeze damage). Wood soft and light colored, the inner bark with blood-red sap. Bark on limbs and trunks greenish during summer rainy season, becoming pale yellowish or straw colored and exfoliating in papery sheets during dry seasons. Leaves once pinnate, 2.5–14 cm, the leafstalk narrowly winged; leaflets (5) 7–13 (17) per leaf, 1.5–6 × 0.3–1.0 cm, lanceolate to elliptic; margins entire or irregularly toothed.

Plants dioecious or not. Inflorescences 1–4 cm or flowers sometimes solitary. Sepals and petals each (3) 4 or 5 (even on the same plant), although male flowers mostly with (4) 5 petals, and female flowers generally with 3 petals and 3 sepals. Five-merous male flowers with 10 functional stamens; bisexual flowers (mostly?) with 5 functional stamens (anthers pollen bearing) and a few vestigial sterile stamens, fertile anthers touching the stigma. Calyx lobes or tips often reddish; petals 3.5–4.5 mm, pale yellow-green or greenish white. Fruits trivalvate; seeds approximately 6 mm, covered by a thin, papery, orange-red aril. Flowering June–July.

Slopes, ridges, canyons, cliffs, and plains; Sonoran desertscrub, thornscrub, tropical deciduous forest, and lower elevations in oak woodland. Nearly statewide except northern Sonora north of approximately 30°30' N and the easternmost margin of the state, near sea level–1350 m (4430 ft). Nayarit to southwestern Chihuahua and Sonora. Also disjunct in the Baboquivari Mountains at Fresnal Canyon in southern Arizona.

The leaves appear with the summer rains and briefly turn yellow to orange or reddish in September or early October and then quickly fall. In cultivation in southern Arizona, leaves do not appear until the onset of humid, summer monsoon weather in early July, even when amply watered during the very hot late spring and early summer. We have seen massive blooming of pale yellowish flowers on leafless trees in the thornscrub west of Alamos. This is one of the most common and widespread burseras in Sonora. Yaqui and Mayo men make pascola masks from the wood, which is easy to carve and does not split upon drying. The bark has been used in tanning (Standley 1923) and medicinally, including treatment of scorpion stings (Kearney and Peebles 1960).

The size, number, and margins of the leaflets show considerable variation among different individuals and populations. The variation seems to be genetic because these characteristics tend to be maintained under cultivation. The populations in Baja California Sur, distinguished by having relatively broader and blunter tipped or more oval leaflets and completely entire margins, are known as *B. odorata*. Several varieties of *B. fagaroides* are generally recognized for western Mexico.

*Bursera fagaroides* and *B. microphylla* appear closely related, although Becerra and Venable (1999) indicate they are in separate but related clades. They sometimes grow intermixed, but *B. fagaroides* generally reaches higher densities in habitats slightly less xeric than does *B. microphylla*. Their flowers are remarkably similar. Apparent hybrids, or plants with intermediate characteristics, are occasionally encountered in places where both species are present (e.g., in the Guaymas region, the Sierra Libre, and the Sierra Seri). Plants grown from seeds of the apparent hybrid found near the summit of Cerro el Vigía at Guaymas maintain characteristics of the parent and show remarkably little variation. These plants produce female flowers with three sepals and three petals 3–4 mm long. As with *B. microphylla*, the putative hybrids leaf out at any time during hot weather if given ample water. Plants of an enigmatic population in the Waterman Mountains northwest of Tucson at the easternmost limit of the genus in Arizona, have relatively larger leaflets like those of *B. fagaroides* but resemble *B. microphylla* in producing leaves almost any time of the year with sufficient water (see Johnson 1992).

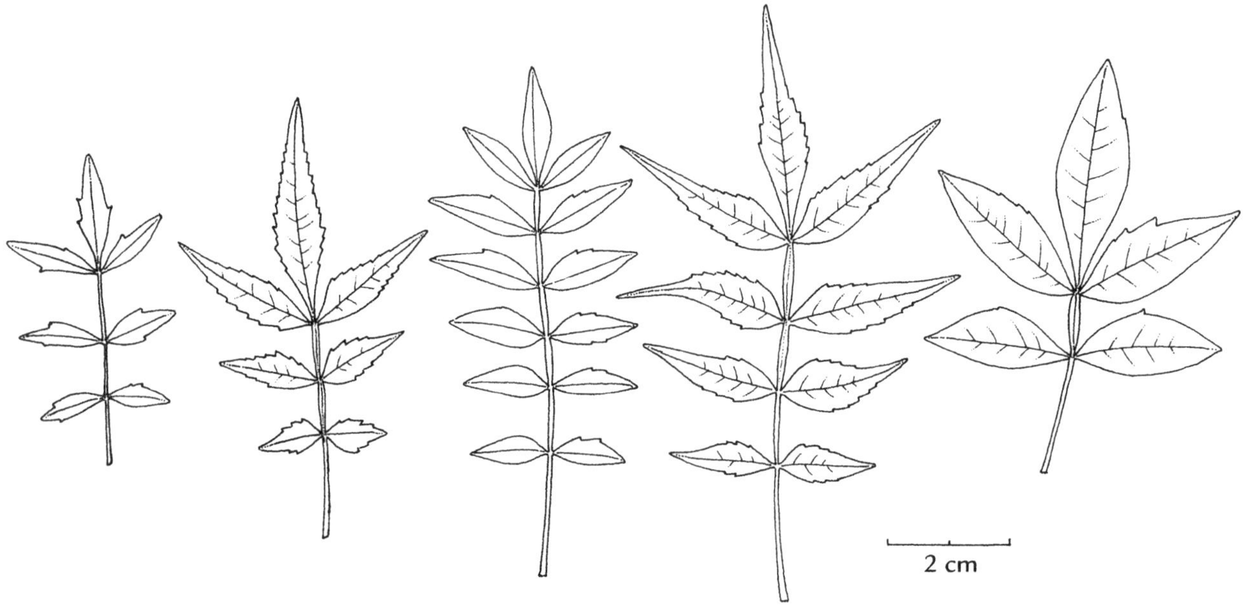


Figure 67. *Bursera fagaroides* var. *elongata*. Drawn by M. B. Johnson.

*Bursera fagaroides* is also related to *B. lancifolia*, although they occur in separate but related clades, and intermediates are not known. They sometimes grow intermixed. Their flowers are similar. *Bursera fagaroides* can be distinguished from *B. lancifolia* by its usually more numerous and smaller leaflets with entire to irregularly toothed margins. In the leafless state they are virtually impossible to distinguish except by finding dried, fallen leaves.

When leafless, *Jatropha cordata*, with its succulent trunk and papery, straw-colored bark, superficially resembles *B. fagaroides* as well as *B. lancifolia*. The branching habit of *J. cordata* is more upright, the twigs have persistent petiole bases at the nodes, and the sap lacks the aromatic properties of *Bursera*.

**Bursera grandifolia** [*B. occidentalis*]. *Palo mulato; napitoro* (Guarijío); *to'oro mulato* (Mayo) (figure 68)

Trees 6–15 m, trunk to 40–60 cm in diameter, often forming a very open, broad crown. Twigs at first finely pubescent, becoming glabrous. Bark of limbs and trunk dark green to bluish green during summer rainy season, in dry seasons exfoliating in large copper-colored papery strips or sheets rustling in the wind, the bark beneath smooth and often with a powdery bloom (glaucous) that can be rubbed off. Inner bark or wood blood red when cut; stems and inflorescences exuding aromatic, white sap when cut. Leafy only during summer rainy season, quickly drought deciduous in fall. Leaves once pinnate, 15–43 cm. Leaflets (3) 5 or 7 per leaf, 5–16 × 3–9 cm, oval to obovate with a caudate tip and entire margins, velvety pubescent, bright green with a velvety sheen above, the veins prominent. Twigs often with the first several (commonly 4 or 5) leaves of the season unifoliolate. Rachis not winged; petiole and rachis with short, white hairs.

Flowers pink and white, attractive, in compact to loose panicles as long as 6–20 cm; pedicels and calyces with short, white hairs; sepals and petals each 5, petals 2–3 mm; stamens 10. Fruits reddish, trivalvate; seeds 7–8 mm with a thin, cloudy pink aril. Seedling cotyledons 3-lobed (resembling those of the bivalvate species). Flowering June, before leaves emerge.

Slopes, and most common in canyons, often in small dispersed stands or as scattered individuals; tropical deciduous forest; southeastern and east-central Sonora south of the Río Yaqui, 275–1065 m (900–3500 ft). Also Guerrero to southwestern Chihuahua.

The thick trunks, unusual bark, and large leaves make these trees striking elements of the dry tropical forests. Flowering occurs a few weeks before that of the other common tropical deciduous forest burseras. Saplings only a few years old and 1.5 m may bear substantial quantities of flowers. This is the only Sonoran bursera that produces truly attractive masses of flowers. This species appears somewhat similar to the related *B. simaruba* but is readily distinguished by its unusual bark, generally fewer, larger, velvety-pubescent leaflets with prominent veins, and masses of pink to whitish flowers.

Figure 68. *Bursera grandifolia*. Drawn by M. B. Johnson.

This striking tree has potential as a landscape subject in essentially frost-free areas. Growth is more rapid than for many other members of the genus. Limbs of this tree are sometimes used for fence posts that often form roots and develop into a new tree. Cuttings made in summer may form roots and begin vegetative growth within eight weeks. Seeds germinate in 4–14 days, and seedlings develop rapidly. It is esteemed for medicinal tea and is said to "give strength." Guarijío men make their pascola masks from this bursera.

**Bursera hindsiana.** Red elephant tree; *copalquín, torote prieto; xoop inl* (Seri) (figures 69–71)

Shrubs or spreading trees to 5 (6) m; trunks and lower limbs generally thick and massive. Bark smooth, dark reddish brown on small branches, gray on older branches and trunk; young branches pubescent, soon becoming glabrous. Leaves highly variable and produced at various seasons depending on soil moisture, reaching 5 cm, the rachis sometimes winged. Leaflets (1) 3–5, 8–35 × 6–25 mm, ovate, obovate or broadly lanceolate, with soft, velvety hairs on both surfaces, the margins irregularly crenate.

Flowers white, in few-flowered clusters 1–5 (8) cm or solitary. Fruits bivalvate, globose, red and green, resembling a miniature McIntosh apple. Seed 7–8 mm, blackish, about two-thirds covered with a fleshy orange aril. Flowering August–October.

Slopes, bajadas, canyons, and washes; Sonoran desertscrub. Coastal Sonora from the vicinity of Bahía San Carlos northward to the vicinity of Puerto Lobos, near sea level –305 m (1000 ft). Attains its best development along sandy washes draining coastal mountain ranges. Also on many islands in the Gulf of California and through most of the Baja California peninsula.

The plant is ornamental when young. Seedlings quickly develop a swollen trunk above the soil line. The branches may be pruned and trained to encourage a bonsai effect, the green leaves contrasting with the red bark. Seed viability appears high, regardless of season of collection. Fully formed seeds are produced three to four weeks after flowering, and germination occurs in 5–12 days in warm weather. The wood is excellent for carving and was preferred by the Seri because it is relatively soft but firm and does not split upon drying. Tea made from chips of the wood together with desert lavender (*Hyptis emoryi*) was used as a remedy for asthma (Felger and Moser 1985). During the winter months, gray vireos (*Vireo vicinior*) feed heavily on the fruits (Bates 1987). This bursera is distinctive from other members of the genus in Sonora, but appears to be related to *B. epinnata* and *B. cerasifolia* from Baja California Sur. Becerra and Venable (1999) indicate that *B. hindsiana* is most closely related to *B. stenophylla* and in the same clade as *B. penicillata*. These species are, however, distinctive and readily distinguished from each other.

***Bursera lancifolia*** [*B. fragilis*]. *Torote papelío, torote amarillo, torote colorado; toró* (Guarijío); *to'oro chutama* (Mayo); *s-vüg kotpitkam* (Pima Bajo) (figure 72)

Glabrous shrubs or trees 6–10 m with one to several thick, semisucculent trunks. Bark on older branches and trunks smooth and greenish during summer rainy season, turning pale gray or straw colored and exfoliating in papery sheets during dry seasons. All parts of the plant aromatic including the gum; larger branches and trunk exuding milky white sap when cut. Leaves appearing with summer

Figure 69. *Bursera hindsiana*. East of Punta Sargento, north of Bahía Kino (Felger and Moser 1985). Photograph by R. S. Felger.

Figure 70. *Bursera hindsiana* (Felger and Moser 1985). Drawn by F. Runyan.

rains, turning yellow, orange, or red and making an attractive display for about 1–2 weeks before falling in September or early October; leaves once pinnate, 7–15 (17) cm; leaflets 3–7 (or 9) per leaf, 3–7 × 0.7–2.0 cm, lanceolate to elliptic, the rachis scarcely winged, the margins uniformly finely toothed.

Inflorescences slender, few flowered, 1–5 cm or flowers sometimes solitary. Flowers pale greenish white; petals 3.5–4.0+ mm. Fruits purplish, trivalvate. Seeds 5 mm, covered with a thin, papery orange aril. Flowering June–July, just before or during leaf emergence.

Slopes, ridges, cliffs, and canyons; thornscrub, tropical deciduous forest, and lower elevations in oak woodland; eastern Sonora, 150–1220 m (490–4000 ft). Sierra el Tigre westward to near Magdalena de Kino and southward to Sinaloa and southwestern Chihuahua, and disjunct in Guerrero, Morelos, Puebla, and Oaxaca.

In the northern part of the range in Sonora the plants remain as shrubs and show evidence of repeated freeze damage. *Bursera lancifolia* is widespread across much of eastern Sonora but not as abundant as *B. fagaroides*, with which it often grows. However, *B. lancifolia* scarcely enters the Sonoran Desert, whereas *B. fagaroides* is often common in the desert. The larger surface area of *B. lancifolia* leaves seem suited to a more moist environment. *Bursera lancifolia* is distinguished by its larger leaflets with uniformly serrulate margins, but when leafless they are virtually indistinguishable. According to Stephanie Meyer (personal communication 1996), the leaves of *B. lancifolia* are very fragrant, while those of *B. fagaroides* from the Alamos region are not fragrant. *Bursera lancifolia* can be propagated by cuttings. The gum is used as a poultice for backache, bruises, and broken bones (Gentry 1942). Tea made from this bursera is used by the Guarijío as a remedy for cough.

Figure 71. *Bursera hindsiana*. Drawn by M. B. Johnson.

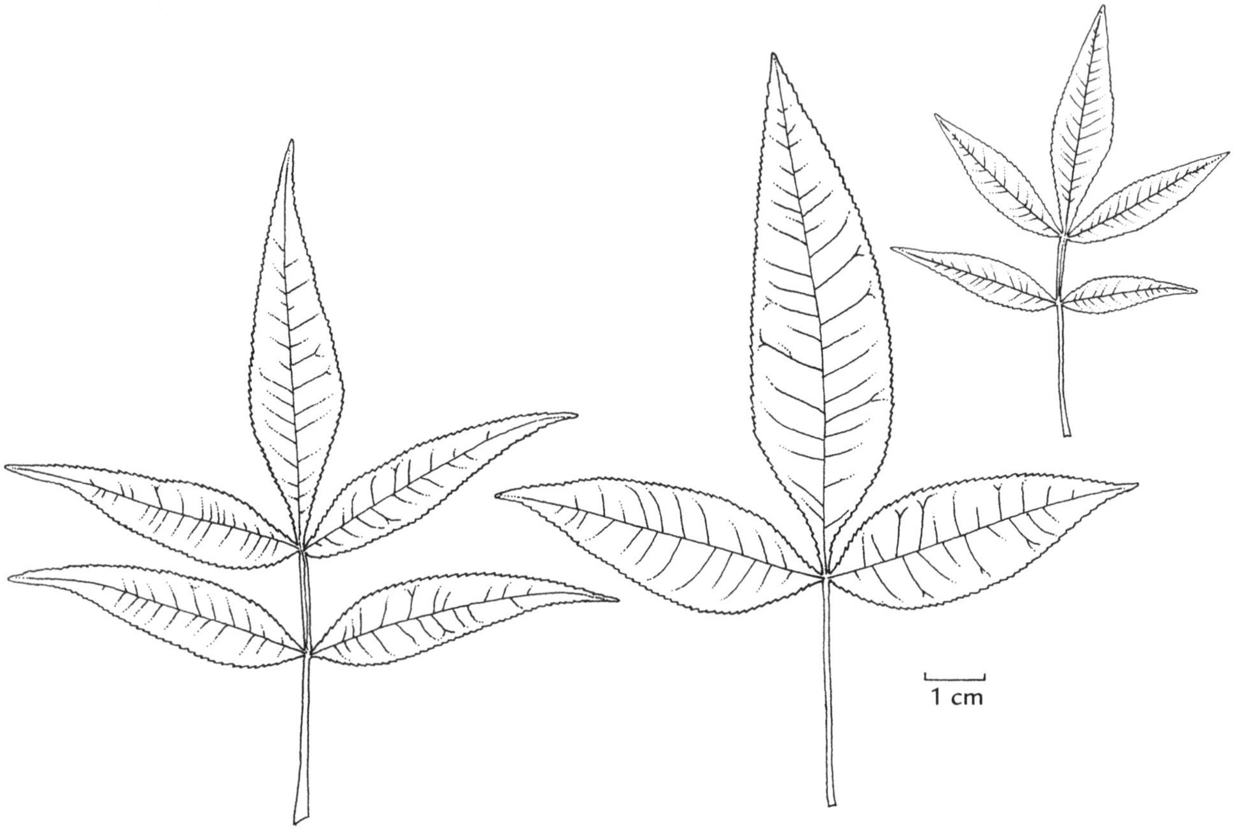


Figure 72. *Bursera lancifolia*. Drawn by M. B. Johnson.

The Guarijío also mix the bark with that of *guásima* (*Guazuma ulmifolia*) and tepeguaje (*Lysiloma watsonii*) to clean wounds.

***Bursera laxiflora.*** *Torote prieto; to'oró ochónome* (Guarijío); *to'oro chucuri* (Mayo); *kotpitkam* (Pima Bajo); *xoop caacöl* (Seri); *chukui tooro* (Yaqui) (figure 73)

Shrubs or trees to 6 (10) m. Trunks and branches neither succulent nor noticeably thickened. Bark reddish brown, smooth, and not peeling but somewhat scaly on older branches. Twigs slender, at first with short hairs, becoming glabrous. All parts of the plants aromatic when crushed. Leaves fernlike, produced at almost any season, pinnate or bipinnate (exceptionally tripinnate), 2.5–12 cm; rachis slightly winged; primary leaflets 5–15, secondary leaflets (when present) 3–7 per pinnae, obovate or rarely lanceolate, to 2 × 1 cm, usually with several uneven crenate teeth on the outer half of the margin.

Inflorescences few flowered, emerging with the new growth; branches and peduncles slender even in fruit, sometimes with small leafy bracts at the base of the pedicels. Flowers white to cream yellow; petals approximately 4 mm. Fruits bivalvate, approximately 8 mm. Seeds blackish, with a fleshy orange aril covering the upper one-half to two-thirds. Flowering June–August.

Slopes, bajadas, valleys and plains, and canyons; Sonoran desertscrub, thornscrub, and tropical deciduous forest. Nearly statewide except the northern and most of the easternmost portions, near sea level–1065 m (3500 ft). Also Sinaloa, southwestern Chihuahua, and Baja California Sur.

The bark has been used medicinally; the Seri made tea from the bark as a remedy for a cold, sore throat, or cough, or for scorpion stings and black widow bites (Felger and Moser 1985). In east-central Sonora it is used as a remedy for cough, and in southern Sonora the gum is used to relieve toothache (Gentry 1942). The Mayo put sticks of this bursera on top of agave "hearts" or "cabezas" in roasting pits to impart a better flavor. The seeds germinate in one to two weeks in hot weather, and the plants can grow rapidly.

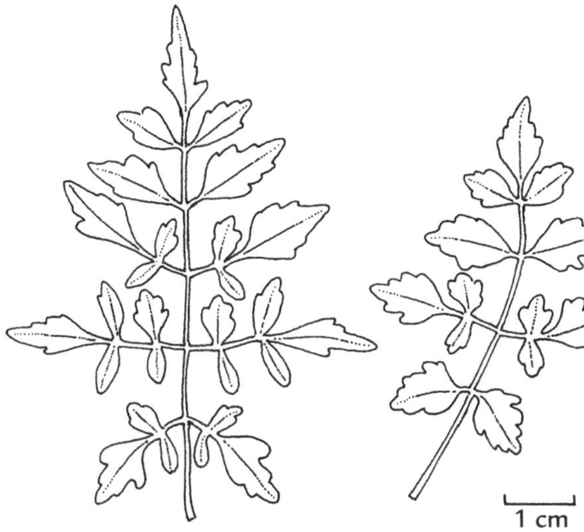


Figure 73. *Bursera laxiflora*. Drawn by M. B. Johnson.

***Bursera microphylla*.** Elephant tree; *copal, torote blanco, torote; to'oro* (Mayo); *'uṣabakam* (O'odham); *xoop* (Seri); *tooro, saa tooro* (Yaqui) (figures 74 and 75)

Glabrous, short-trunked large shrubs or small trees 2–6 (8) m, spreading and low branched. Trunk and lower limbs fat and semisucculent, often contorted and forming bizarre shapes, the wood pithy and soft. The branches or roots, when deeply cut, exude blood-red sap. Bark of twigs and smaller branches reddish brown; bark of trunk and larger limbs whitish and papery, peeling in large flakes or sheets, especially during spring dry season. Sap and leaves extremely aromatic. Long shoots with relatively long internodes; most leaves crowded at apices of short shoots. Leaves bright green, once pinnate, 1.5–8 cm, produced at almost any time of year with sufficient soil moisture; leafstalk flattened, especially when young and grooved; rachis often narrowly winged; leaflets 7–31 per leaf, 5–25 × 1–2 mm, linear to narrowly oblong, or rarely lanceolate or oval.

Plants dioecious or mostly so. Flowers 1 or 2 to several per cluster, cream white to pale greenish yellow, approximately 2 mm wide; male flowers usually with 5 sepals and 5 petals; female flowers generally with 3 sepals and 3 petals. Fruits fleshy but hard, dull reddish purplish and glaucous when fresh, 7–9 mm, trivalvate with a leathery exocarp. Seeds approximately 6 mm, 3-angled, enveloped in a thin, bright red-orange aril. Flowering mostly in July.

Slopes, bajadas, canyons, arroyos, and plains; Sonoran desertscrub; northwestern and west-central Sonora southward to the Guaymas region, near sea level–915 m (3000 ft). Common and conspicuous in western Sonora, especially in coastal regions and desert mountains. Also southeastern California to the Cape Region of Baja California Sur and southwestern Arizona; its distribution nearly coincides with the extent of the Sonoran Desert (Turner et al. 1995).

Mooney and Emboden (1968) demonstrated geographic variation in the terpene content. Populations in the Cape Region of Baja California have exceptionally tall, stout boles with a high, small crown. Gray vireos feed heavily on the seeds and arils during winter months (Bates 1987), and their winter range closely matches the distribution of *B. microphylla* in Sonora. Along the coast, south of Tastiota, Sonoran chipmunks (*Eutamias dorsalis*) open the fruits and eat both seeds and arils (Reyna Castillo, personal communication 1991).

This plant featured prominently in religious practices of diverse Sonoran Desert people, especially the Coahuilla and Seri, and has been utilized for many practical purposes including pitch for caulking boats, facepaint, fire-making drills, kindling, and especially for many medicinal purposes such as treatment of headache, cuts and scratches, lice, skin diseases, sores, stingray wounds, and venereal disease (e.g., Bean and Saubel 1972, Kearney and Peebles 1960,

Figure 74. *Bursera microphylla*. Vicinity of Campo Viboras, north of Bahía Kino (Felger and Moser 1985). Photograph by R. S. Felger.

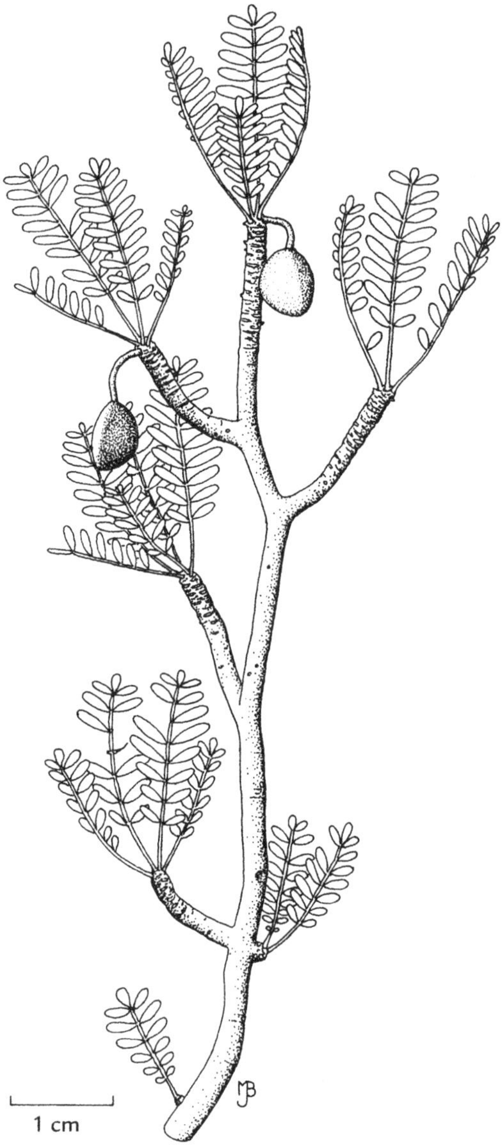


Figure 75. *Bursera microphylla*. Drawn by M. B. Johnson.

Felger and Moser 1985). The tannin-containing bark used to be harvested for export (Kearney and Peebles 1960). The Yaqui have carved pascola masks from the wood. A unique landscape subject, it is recommended for essentially frost-free desert regions and is suitable as a container or bonsai plant. The seeds germinate in 6–14 days in hot weather.

***Bursera penicillata*** [*B. inopinnata*]. *Torote copal, torote prieto, torote incienso, torote de puntagruesa; topocá* (Guarijío); *to'oro* (Mayo); *kopaar* (Pima Bajo) (figure 76)

Erect to ascending, sparingly branched trees reaching 10–17 m; sap and foliage highly aromatic with a licoricelike scent, the fragrance often wafting through the air. Twigs conspicuously thickened, the bark pubescent, red-brown, becoming slightly scaly and reddish gray on older limbs. Leaves once pinnate, 12–38 cm; rachis winged; leaflets 5–13 per leaf, 2.5–12 × 1–4 cm, broadly lanceolate to elliptic, finely pubescent on both surfaces, the margins coarsely toothed.

Flowers white, 3 mm, on delicate, few- to many-flowered panicles to 14 cm. Fruits purplish, bivalvate. Seeds 8 mm, blackish, about one-half covered with a fleshy orange aril. Flowering June–July.

Slopes and canyons; primarily tropical deciduous forest and sometimes extending into thornscrub and lower elevations in oak woodland; southeastern Sonora, 180–915 m (600–3000 ft). Ranging northward in eastern Sonora at least to the vicinity of Onavas. Also Michoacán northward along the west coast to southwest Chihuahua.

*Bursera penicillata* is conspicuous by its upright growth habit, thick twigs, fuzzy leaves with conspicuously toothed margins, and distinctive scent. It is often locally abundant. Although it enters the lower oak zone, at these elevations

Figure 76. *Bursera penicillata*. Drawn by M. B. Johnson.

it is usually comparatively stunted. The herbage has been used to treat catarrh and the resin or sap used to treat toothache, flu, and other ailments. This species is closely related to, or possibly conspecific with *B. graveolens* from southern Mexico to South America.

***Bursera simaruba*** [*B. arborea*]. Gumbo limbo; *torote colorado* (figures 77 and 78)

Trees reaching 20–30 m with a tall trunk to 1 m in diameter and a high, spreading crown; inner bark or wood whitish to pinkish when cut. Glabrous, except sometimes with brownish to whitish hairs on lower (adaxial) surface of petioles, especially the younger leaves, or the leaves sometimes with sparse, short, white hairs. Bark green, turning coppery red and exfoliating in large papery flakes or sheets. Evergreen to drought deciduous; leaves once pinnate, 17–30 cm; rachis not winged; leaflets (5) 7–11 (13) per leaf, (5) 6–10 (11.5) × 2–4 (6.3) cm, oval to obovate with a caudate tip, often asymmetric, shiny dark green, the lateral veins not prominent, the margins entire.

Inflorescences 1.5–12.5 cm, glabrous, and relatively stout, the pedicels 0.7–1.5 mm in diameter. Flowers white, approximately 1 mm. Fruits trivalvate, reddish. Seeds 7–8 mm, covered with a thin, pale pink aril. Flowering June–July.

Riparian canyons in tropical deciduous forest, mostly with tropical evergreen vegetation; southeastern Sonora, 305–915 m (1000–3000 ft). In Sonora known only from scattered, deep riparian canyons such as the narrows of Arroyo Gochico east of San Bernardo and Arroyo Verde of the upper Río Cuchujaqui, where it grows with a number of other isolated, tropical species. Northern South America to Sonora, Florida, and the Caribbean.

This is the most widely distributed member of the genus and is the largest bursera in western Mexico. We are unable to distinguish Sonoran specimens of *B. arborea* and *B. simaruba*, and follow Standley (1923) and others in treating them as a single species, although McVaugh and Rzedowski (1965) and Becerra and Venable (1999) regard them as distinct. The leaves of *B. arborea* are reported to have fewer leaflets, be pubescent when young and have sparse, white hairs when mature, whereas the leaves of *B. simaruba* are glabrous. Specimens called *B. arborea* seem to have thinner pedicels and slightly smaller fruits, but the number of Sonoran specimens is far too small to be conclusive.

***Bursera stenophylla.*** *Torote blanco, torote camforado, torote copal; to'ro, to'oro sajo* (Mayo); *stoa kotpitkam* (Pima Bajo) (figure 79)

Slender, delicately branched trees to 7 (12) m, glabrate. Branches slender; bark smooth, reddish gray becoming gray; twigs aromatic. Leafing out with summer rains, quickly drought deciduous after the summer rainy season. Leaves distinctly bipinnate, 4.5–24 cm, with (9) 13–15 (17) pinnae and primary leaflets per leaf; rachis scarcely winged with a few short, white hairs on the underside; secondary leaflets 3–10 (15) × 0.7–1 mm, oblanceolate to narrowly oblong, the margins usually entire.

Flowers in short clusters, 4 mm. Petals of male flowers greenish white and not spreading; petals of female flowers distinctly white and the petals spreading. Fruits bivalvate, approximately 1 cm. Seeds 8 mm, blackish, about two-thirds covered with a fleshy orange aril. Flowering June–July.

Slopes, and canyons; tropical deciduous forest and lower elevations in oak woodland; southeastern Sonora, 180–1065 m (600–3500 ft). Also Chihuahua and Sinaloa.

These trees usually grow as scattered individuals and sometimes fruit when only 1.5 m tall. The fernlike leaves are the most highly divided of the Sonoran burseras. The herbage, made into tea, is drunk as a remedy for cough.

## Cactaceae—Cactus Family

Stem succulents of diverse growth forms and size. Spines, leaves (if present), flowers, and branches arising from a specialized organ, the areole, derived from a modified short shoot. Leaves (Sonoran tree cacti) absent, or reduced and deciduous. Flowers (Sonoran tree cacti) large and showy, radial or nearly so, with many petal-like perianth segments (larger or inner tepals) intergrading with the outer, smaller, sepal-like segments (smaller or outer tepals); floral tube, especially among the columnar cacti, often filled with nectar. Stamens numerous. Ovary almost always inferior (in Sonoran cacti), deeply embedded in the receptacle, which often bears areoles and/or scalelike bracts. (The proximal part of the receptacle, covering the ovary, is called the pericarpel; the distal part of the receptacle is fused to the hypanthium and the stamens on its inner surface to form the floral tube.) Fruits generally succulent, sometimes dry at maturity; seeds numerous.

New World (except one group of *Rhipsalis*); 122 genera, 1200+ species; about 100 species in 17 genera in Sonora. References: Benson (1982), Bravo-Hollis (1978), Britton and Rose (1919–1923), Castetter and Bell (1937); Felger (2000), Felger and Moser (1985), Gibson and Horak (1978), Gibson and Nobel (1986), Paredes Aguilar et al. (2000), Pinkava (1995).

Although cacti are conspicuous in hot deserts, the family has its greatest development toward the margins of deserts and in semiarid subtropical regions; some are found in the humid tropics and some occur in dry temperate regions.

Most cacti have fleshy, edible fruits, although the fruits of some taste better than others. The sweet, succulent fruits

Figure 77. *Bursera simaruba*. Canyon near Rancho San Pedro, Sierra Sutucame, Río Mayo. Region. Photograph by R. S. Felger.

entice animals to consume the fruit and disperse the seeds. The seeds usually are passed undigested. Among the large Sonoran columnar cacti (e.g., *Carnegiea*, *Pachycereus*, and *Stenocereus*), the most important pollinators in Sonora are the lesser long-nosed bat (*Leptonycteris curasoae*) and the white-winged dove (*Zenaida asiatica*), although the flowers also are visited by insects (e.g., Fleming et al. 1996, Moran 1962). The long-nosed bat is one of the most important seed dispersers.

Several columnar cacti (e.g., *Pachycereus pecten-aboriginum*, *Stenocereus montanus*, *S. thurberi*) grow readily from stem cuttings, which were made into living fences, now usually replaced with barbed wire. The woody ribs of most of the columner cacti are used for light construction, fencing, and many other purposes.

In some cases the seeds, especially among the columnar cacti, rich in oil and protein, have been used by people for food. Sonoran people used the fleshy fruits of most of the larger cacti as food resources and prepared wine from the sweet, juicy fruits of columnar cacti such as the saguaro and organpipe.

1. Chollas and prickly pears; stems constricted into joints or pads, not ribbed; areoles with glochids (small spines deciduous at a touch) in addition to the larger persistent spines, or larger spines sometimes absent. ______________ ***Opuntia***

1' Columnar cacti; stems with conspicuous ribs, not constricted into joints or pads; areoles spiny or not, but without glochids.

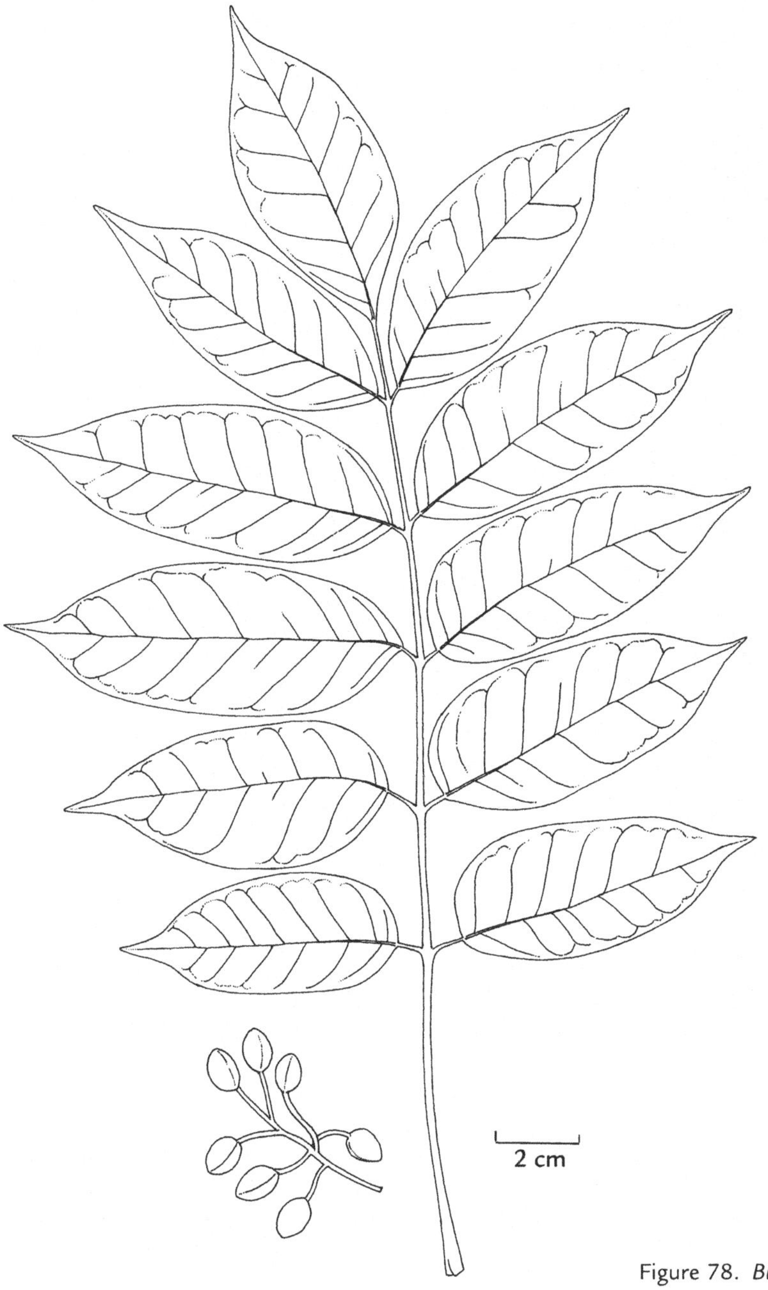


Figure 78. *Bursera simaruba*. Drawn by M. B. Johnson.

2. Stems and spines not conspicuously dimorphic; spines similar on juvenile (lower or sterile portion) and adult (upper or fertile portion) stems. ***Stenocereus***

2' Stems and spines markedly dimorphic; juvenile and adult stems markedly different (e.g., in stem diameter, distance between areoles, spine lengths and morphology, hair, and rib numbers).

3. Spines on adult stems absent or shorter than those of juvenile stems and not obscuring the stem surfaces; fruits spiny or not.

4. Areoles of adult stems close but not coalesced, with bristly spines; fruits spineless or nearly so and smooth. ___ ***Carnegiea***

4' Areoles of adult stems coalesced, spines short and stout or absent; fruits densely spiny or covered with dense feltlike hairs. _ ***Pachycereus***

3' Spines or hair on adult stems longer than those of juvenile stems and obscuring the stem surfaces; fruits spineless.

5. Stem ribs ≥ 12; upper stems with beardlike, white hairs on one side. ___ ***Pilosocereus***

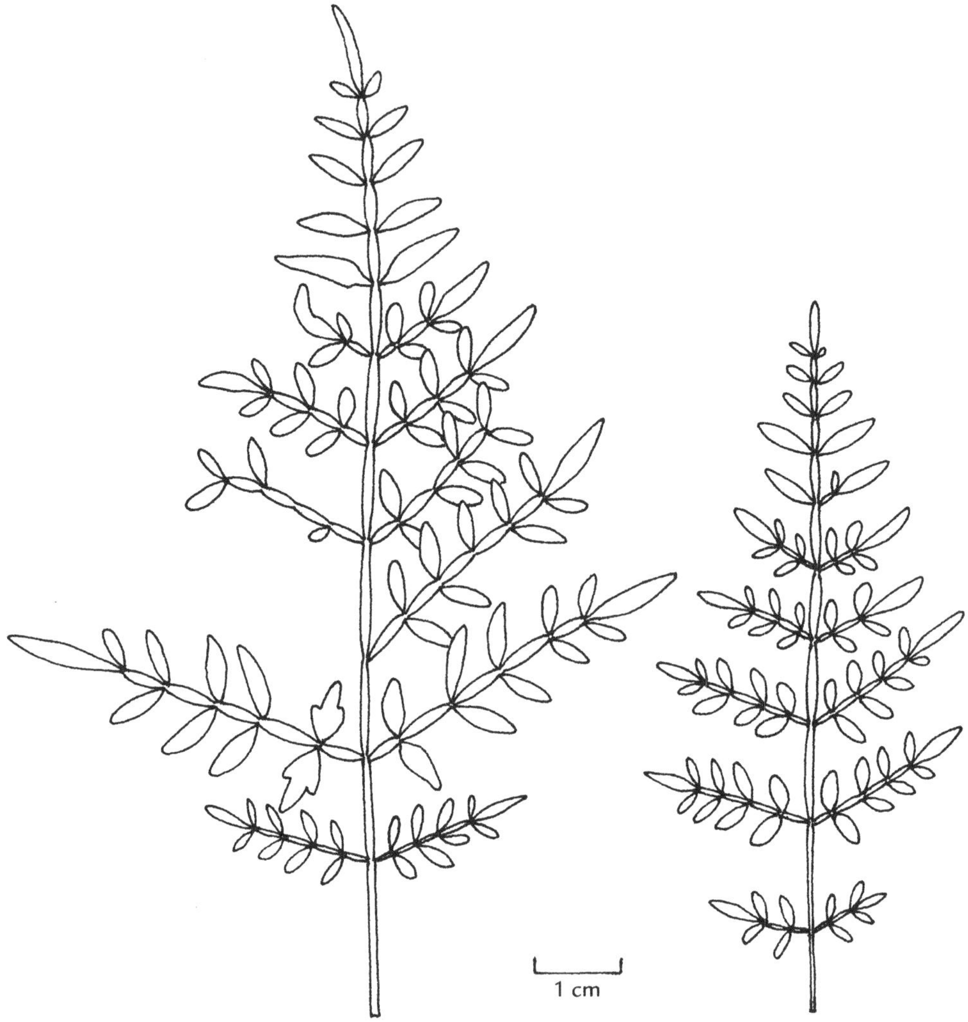


Figure 79. *Bursera stenophylla*. Drawn by M. B. Johnson.

5' Stem ribs 5–10; upper (fertile) part of stem with long, shaggy, twisted, and often flattened spines on all stem ribs. ______ ***Lophocereus***

## *Carnegiea*

This monotypic genus seems most closely related to *Neobuxbaumia*, which occurs much farther south in Mexico. Certain *Neobuxbaumia*, such as *N. mezcalaensis* and *N. tetetzo*, are vegetatively similar to saguaros and likewise form "forests" in the drier regions of Puebla and Oaxaca.

***Carnegiea gigantea*** [*Cereus giganteus*]. Saguaro; *sahuaro*; *sahu'o* (Mayo); *ha:shañ* (plant), *bahidaj* (fruit) (O'odham); *tuduhua* (Opata); *mojépe* (Seri); *sauwo* (Yaqui) (figures 80–82)

Giant columnar cacti; stems unbranched, erect, and straight to several-branched. Stem tissue turning black when cut or injured. Stems dimorphic: sterile (lower or juvenile) portion often extending to 2–3+ m in height, mostly with 11–15 stem ribs; areoles separate and bearing stout, inflexible spines, the longer central spines (3.7) 5.5–11 cm × (0.8) 1.1–2.3 mm, these areoles not producing flowers. Fertile (upper) portion of stems with shallow ribs often 19–25; areoles closely set with flexible, bristly spines 2.2–3 cm × 0.3–0.4 mm, producing flowers. First branches usually arising in the region of transition between the sterile and fertile area, the branches fertile.

Flowers white, large, and showy, opening in the evening and usually remaining open until about midday or later depending on temperature. First buds of each spring emerging on south or southwest side of the stems. Fruits spineless (or occasionally with a few bristly spines), green outside, the rind at maturity thick and splitting open to reveal the bright red inner surface; the pulp juicy, bright red, sweet, and edible. Seeds 1.8–2.1 mm in diameter, dark red-brown

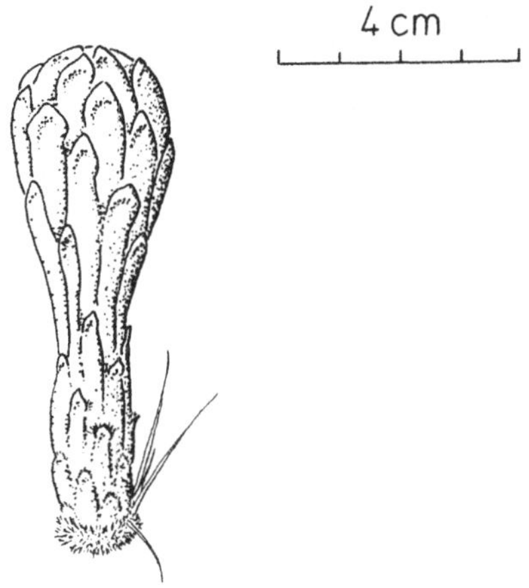


Figure 80. *Carnegiea gigantea*. Drawn by M. B. Johnson (plant) and N. L. Nicholson.

to blackish, shiny, nearly smooth, and edible. Flowering mostly late April–May; fruits June–early July.

Slopes, canyons, bajadas, plains, and arroyos; Sonoran desertscrub, thornscrub, and occasionally in lower elevations in grassland and oak woodland; western Sonora from near sea level–1370 m (4500 ft). Common across much of the western two-thirds of the state from near the Sinaloa border northward. Characteristic of the Sonoran Desert in Arizona and Sonora including Isla Tiburón, and very locally near the Colorado River in southeastern California; not in Baja California.

As with other large cacti, saguaros characteristically begin life beneath desert shrubs and trees, especially legumes such as *Olneya* and *Parkinsonia*, which serve as nurse plants. However, in some parts of the arid northwestern part of the state, the usual nurse plants are absent and saguaros begin life beneath white bursage (*Ambrosia dumosa*), galleta grass (*Pleuraphis rigida*, = *Hilaria rigida*), and even desert club-cholla (*Opuntia kunzei*). The fruits are delicious and have been harvested for food and wine-making by Sonoran Desert people.

Saguaro and *cardón* (*Pachycereus pringlei*), the two largest cacti in the Sonoran Desert, often occur together in western Sonora. Saguaros generally initiate branches higher off the ground than do *cardones*, and the mature stems are green rather than glaucous. The young plants, however, may be difficult to distinguish. Saguaro tolerates considerable freezing weather, whereas *cardón* is highly frost sensitive. Young saguaros, to about 2 m in height, can readily be transplanted. Larger saguaros seldom survive transplanting without special care, although it may take several years for them to succumb. Seed-grown small plants are readily available from nurseries.

Figure 81. *Carnegiea gigantea*. Lesser long-nosed bat feeding from flower. Photograph by M. D. Tuttle. (copyright Merlin D. Tuttle, Bat Conservation International).

### *Lophocereus*

Northwestern Sinaloa to southwestern Arizona and both states of Baja California; two closely related species. References: Felger and Lowe (1967), Lindsay (1963).

***Lophocereus schottii*** var. ***schottii*** [*Cereus schottii; Pachycereus schottii*]. Senita; *sinita, sina, músaro; musue* (Mayo); *ce:mĭ* (O'odham); *temi* (Opata); *hasahcápöj* (Seri); *museo* (Yaqui) (figures 83 and 84)

Columnar cacti; trunkless or with short, multiple trunks; stems mostly erect, 2–7 m, arising mostly at or near the base of the plant; cut stem surfaces quickly turning black. Stems markedly dimorphic: lower (sterile) portion of the stems 5- or 6-ribbed, the larger stems 8–16 cm in diameter; upper (fertile) portion 5–8-ribbed, 5–10 cm in diameter. Spines grayish. Spines of lower portion of stems as short as 5–7 mm, thick, rigid, not twisted, their bases enlarged. Spines of upper portion of stems often 4–10 cm, strongly twisted, slender, flexible, bristlelike.

Flowers white to pale pink, nocturnal, but on cool days (e.g., in March) sometimes remaining open part of the day, 3.8–4.5 cm wide. Fruits rounded, mostly 2.3–4.0 cm in diameter, red, and spineless or nearly so. Seeds blackish, 2.2–2.8 mm in diameter, shiny, and smooth. Seeds and pulp edible; pulp red and sweet but not as delicious as that of organpipe or sahuaro. Flowering (March) April–September (December). Fruiting April–December, the peak often in June when the upper stems may be loaded with fruits.

Plains and bajadas, rarely on rocky slopes; Sonoran desertscrub; western half of Sonora, near sea level–680+ m (2230+ ft).

Much like the mutualism between yuccas and yucca moths, there is an obligate relationship between sinita and

Figure 82. *Carnegiea gigantea*. White-wing dove feeding on ripe fruit. Photograph by C. A. Morgan.

Figure 83. *Lophocereus schottii* var. *schottii*. South of Sonoyta (left), photograph by R. S. Felger. Bahía Kino, photograph by E. T. Nichols.

a pyralid moth, *Upiga virescens*, the larvae of which feed on sinita ovules. Females of this moth collect pollen on specialized abdominal scales and actively deposit pollen on the stigmas of sinita flowers. As much as 90% of the fruit set may be due to these moths (Fleming and Holland 1998). Patrolling ants attracted to extrafloral nectaries in the areoles may provide protection from herbivorous insects (Turner et al. 1995). Birds and ants remove the pulp and seeds, leaving only the hollowed fruit case.

Variety *schottii* is characterized by relatively thick stems, fewer (five to eight) stem ribs, and an essentially trunkless habit. This is the northern variety ranging from near the international border in southwestern Arizona to the Guaymas region and inland to Hermosillo, also Baja California to northern Baja California Sur.

These cacti generally grows on finer textured soils of level or nearly level terrain. The tallest and most massive *Lophocereus* occur in cinder soils on the east side of the Pinacate volcanic field and between Caborca and the coast. In Sonora var. *schottii* is replaced to the south in thornscrub habitat by var. *tenuis*, which ranges to northwestern Sinaloa, and a similar situation occurs in Baja California Sur. Variety *tenuis* usually forms a short trunk, has more slender stems with 6–10 ribs, and is shrub sized.

There is a clinal shift in architecture and stem morphology to plants with lower surface-to-volume ratio in the north, which is interpreted as an adaptation to increasing aridity. Similar adaptations are seen among other columnar cacti in the region. The northern distribution is limited by freezing temperatures (Felger and Lowe 1967, Nobel 1982, Turner et al. 1995).

Considerable quantities of *músaro* stems are harvested in northern Sonora and sold in marketplaces and herbal shops from Tijuana to Mexico City. The five-ribbed stems are highly esteemed and much used as remedies for diabetes, stomach ulcers, cancer, poor blood circulation, and hair loss. (The five-ribbed stems are thicker and more mucilaginous than the fertile or upper stems.) Stem cross-sections are boiled and the liquid drunk throughout the day in place of drinking water.

## *Opuntia*—Chollas and Prickly Pears

Stems constricted into cladodes (joints or pads); areoles with glochids (tiny obnoxious spines, readily deciduous at a touch and difficult to remove from the skin) and often also with persistent larger spines. Two major kinds in North America: subgenus *Cylindropuntia*, the chollas, with stems

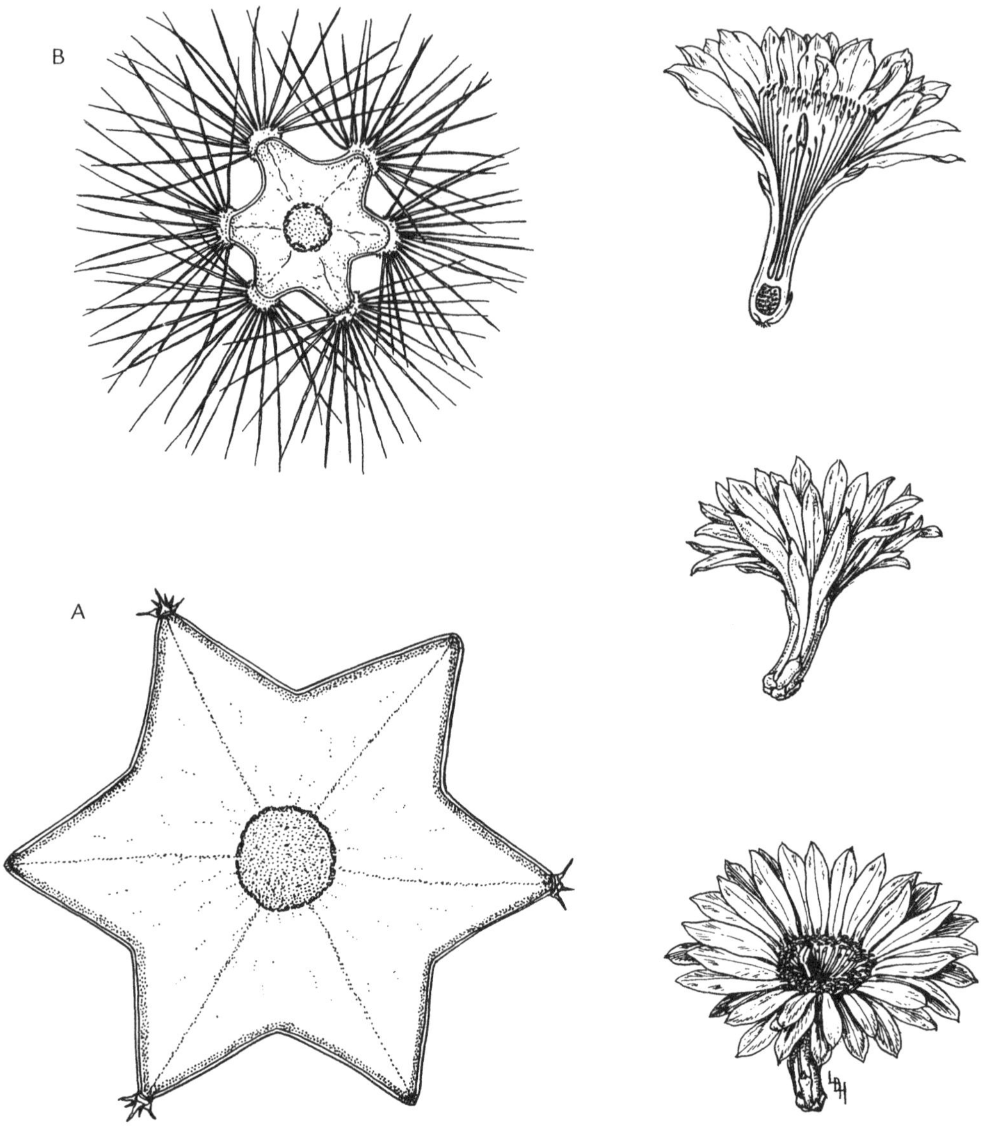


Figure 84. *Lophocereus schottii* var. *schottii*. (A) Cross section of young stem, (B) cross-section of reproductive stem. (Benson 1982). Drawn by L. B. Hamilton.

of joints, more or less rounded in cross-section and sheathed spines, and subgenus *Opuntia* (*Platyopuntia*), the prickly pears, with stems of flattened pads and spines without sheaths. The several subgenera are often treated as distinct genera.

Flowers usually diurnal; stamens tactile (filaments closing inward when touched). Fruits mostly edible. "Seeds" relatively large among the family, not edible, composed of a bony aril enclosing the seed.

Southern South America to Canada; 160+ species. Many species in Sonora, only a few are arborescent.

Unlike most other cacti, the young stems (pads or joints) have leaves that are small, succulent, and quickly deciduous. The stem tissue has the usual succulent plant CAM photosynthesis, while the leaves have a $C_3$ metabolic pathway. Mucilage from the stems of *O. ficus-indica* and *O. thurberi* is used in making plaster whitewash for walls.

1. Prickly pears; stem segments of laterally compressed cladodes (pads); spines without sheaths.
   2. Widely cultivated; pads glabrous; fruits > 5 cm. ***O. ficus-indica***
   2' Southern Sonora, not cultivated; pads pubescent with small hairs; fruits < 5 cm. _____ ***O. wilcoxii***

1' Chollas; stem segments of cylindrical cladodes (joints); spines with sheaths.
   3. Fruits and stems always green, the fruits proliferating in pendulous chains of 3 or more. _____ ***O. fulgida***

3' Fruits and stems green or often reddish purple in winter and dry seasons; fruits solitary, not proliferating in chains.
   4. Tubercles often 3–4 cm; longer spines 10–35 mm, often unequal in length within a cluster. ***O. thurberi***
   4' Tubercles often 1–2.5 cm; longer spines 9–25 mm, often equal in length within a cluster. ***O. versicolor***

***Opuntia ficus-indica.** Tree prickly pear, Indian fig, mission cactus; nopal

Large shrubs or small trees, forming a thick trunk. Pads green, glabrous, often with few or no spines. Flowers yellow-orange. Fruits green or pale yellow or orange-yellow when ripe, 5–10 × 3.5–7.3 cm; pulp often green or yellowish; fruiting areoles with short glochids. Flowering April–May; fruits ripening early to mid-summer.

Opuntia ficus-indica is widely cultivated across most of the state and occasionally persists near settlements, mostly below the oak zones; it does not persist in the drier desert regions. This is the common prickly pear of commerce and was domesticated in south-central Mexico. It is extensively cultivated in Mexico and elsewhere in the world. There are a number of cultivars, including spiny and nearly spineless forms. The young pads (nopales) serve as a green vegetable, and the fruits (tunas) are eaten for their sweet pulp.

***Opuntia fulgida*** var. ***fulgida*** [*Cylindropuntia fulgida*]. Chain-fruit cholla, jumping cholla; *cholla*; *hanam* (O'odham); *cotéexet* (Seri); *seve'e choa* (Yaqui) (figures 85 and 86)

Shrubby to arborescent chollas, often 1–3 (5) m; trunk well developed and seldom straight, sometimes reaching 20–22 cm in diameter, often with several major branches. (Some are tall enough that you can stand in the shade of a spreading jumping cholla, being careful of spiny joints overhead and fallen joints littering the ground.) Smaller branches, or joints, mostly at the crown or top of the plants; joints green all year, loosely attached. Stem tubercles longer than wide and relatively high. Spines usually obscuring the stem surface, with large, loosely fitting golden-yellow to straw-colored sheaths.

Figure 85. *Opuntia fulgida* var. *fulgida*. Lolita Blanco harvesting fruit from a plant about 2.5 m tall, near El Desemboque San Ignacio, 1980 (Felger and Moser 1985). Photograph by R. S. Felger.

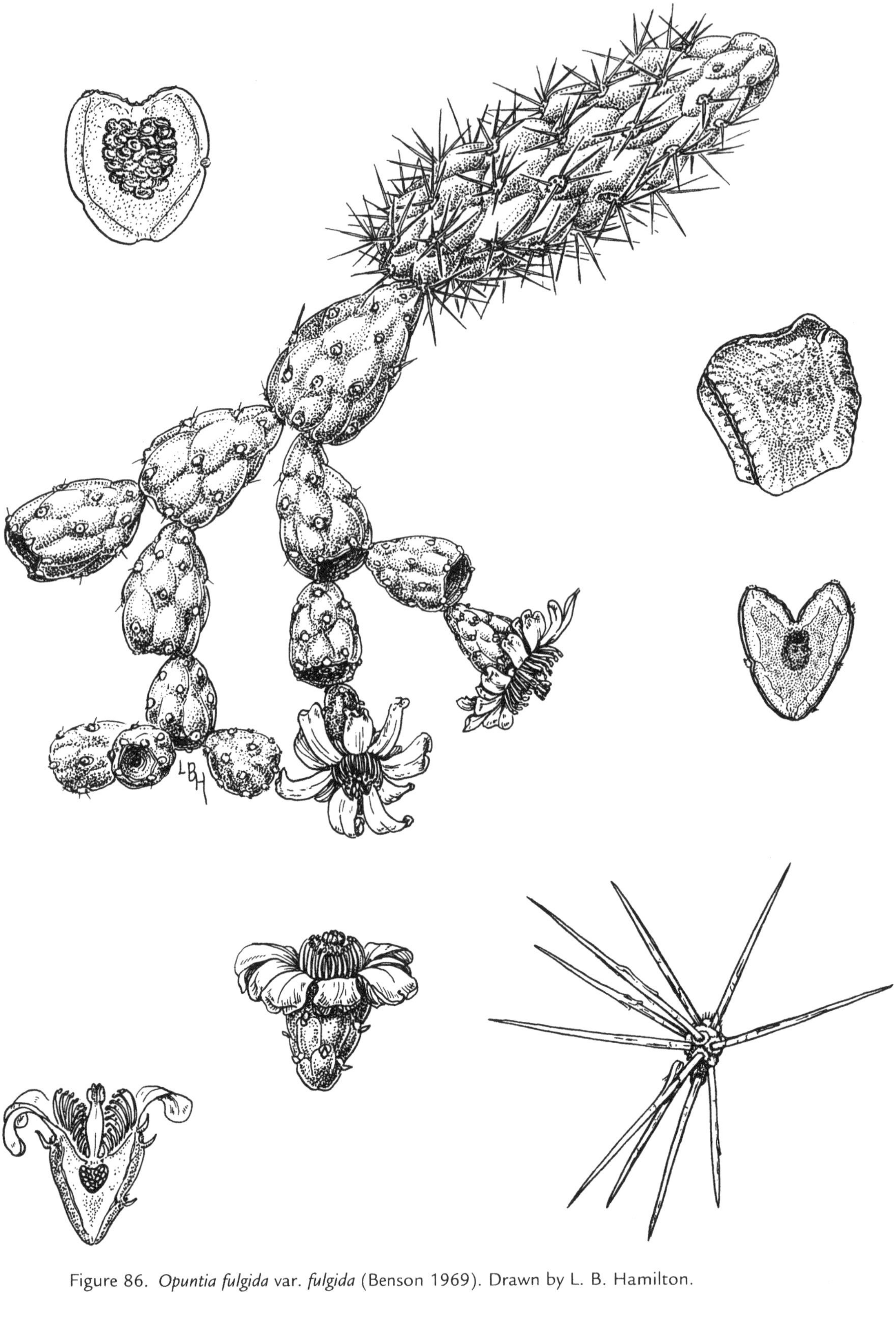

Figure 86. *Opuntia fulgida* var. *fulgida* (Benson 1969). Drawn by L. B. Hamilton.

Flowers often 2.5–5 cm wide, bright purple-pink; larger tepals spreading widely, few and not totally ringing the flower (the flower lacks a full complement of "petals"). Flowering in pulses during warmer times of year, often following summer rains, and also mid-April–early May. In summer the flowers open in the late afternoon, often by 6 PM, and often wither before dawn; in spring they may remain open into the next morning. At night flowers are visited by medium-sized moths and during daylight by medium-sized and large bees. Flower buds form from the previous year's fruit. Fruits often 3–6 cm, mostly obovoid, remaining green and fleshy, proliferating in hanging chains of (3) 6–15 fruits (the longer chains sway with the wind), often persisting for several years. Reproduction by seed is rare; propagation is almost entirely vegetative from fallen joints and fruits, which readily form roots and grow into new plants.

Variety *fulgida*: bajadas, plains, and valleys, and less common on slopes; Sonoran desertscrub; western Sonora, near sea level–975 m (3200 ft), from the Guaymas region northward to southern Arizona. This variety is characterized by large spines with loosely fitting spine sheaths that obscure the stem surface. Large, or arborescent plants, sometimes 3–5 m, occur in western Sonora from the vicinity of El Desemboque San Ignacio northward including the northeastern part of the Pinacate region, to southern Arizona.

The fruits are edible, and the Seri harvested them from specific groves known to have extra large fruits (Felger and Moser 1985). The slightest contact dislodges the spiny joints, with painful results—hence the common name. There is often a litter of fallen, spiny joints beneath the larger plants, which serves as a nursery area for various plants including other cacti such as *Echinocereus*, *Mammillaria*, and sometimes even *Carnegiea*.

This species includes two varieties. Variety *mamillata* ranges from southern Arizona to northwestern Sinaloa but does not extend into the very arid desert of northwestern Sonora. It does not become arborescent and has fewer and smaller spines with tighter-fitting sheaths, making the stem surface readily visible. Both varieties may occur intermixed, but intermediates are surprisingly uncommon. *Opuntia cholla*, endemic to the Baja California peninsula, is closely related and similar in appearance to *O. fulgida*.

***Opuntia thurberi*** [*Cylindropuntia thurberi; Opuntia alamosensis*]. *Cholla, siviri; megueri, seveli* (Guarijío); *sibiri* (Mayo) (figure 87)

Chollas, often 3–5+ m with a rather well-formed trunk. Stems green or gray-green, turning reddish purple during seasons of drought and cooler weather; tubercles 3–4 cm; spines often unequal in length, the longer spines 10–36 mm. Leaves 20–24 mm.

Flowers 6.0–6.5 cm wide; larger tepals bright yellow to yellow-green, approximately 2.5 cm; filaments pale green to yellow-green, the longer ones 12–15 mm; style cream white to pale pink; the stigma lobes 5, pale green. Pericarpel spines slender, readily deciduous. Fruits similar in color to the stems, fleshy, long-persistent, and solitary (sometimes another fruit may grow from a previous fruit), smooth to shallowly tuberculate, often long-pyriform. Flowering March–May.

Slopes, valleys, canyons, and plains; Sonoran desertscrub, thornscrub, and tropical deciduous forest; eastern and southern Sonora, near sea level–1000 m (3280 ft). Common in lowland and foothill thornscrub and tropical deciduous forest, especially in southeastern Sonora, and northward along the eastern margin of the Sonoran Desert to the vicinity of Cucurpe and the type locality at Bacoachi in northern Sonora. Also southwestern Chihuahua and Sinaloa.

The stems are used in making whitewash for walls (see *Heliocarpus attenuatus*, Tiliaceae). The fresh fruits are eaten with salt. In the Río Mayo region the fruits are boiled or prepared as a tea and used as a remedy for diarrhea. The young leaves are eaten fresh from the plant by the Guarijío and noted for their salty flavor.

*Opuntia thurberi* is distinguished from the closely related *O. versicolor* by its usually more arborescent habit, generally longer, thinner, and greener joints, longer and fewer tubercles, and more heterogeneous, fewer, and longer spines.

***Opuntia versicolor*** [*O. thurberi* subsp. *versicolor; Cylindropuntia versicolor*]. Staghorn cholla; *cholla, siviri; hepem ihéem* (Seri); *sevii* (Yaqui)

Shrubs to rarely small trees 4 (5) m; differing from *O. thurberi* as follows: stems generally dark gray-green, often tinged yellow and purplish in dry seasons and winter; tubercles 1.0–2.3 cm; spines often relatively equal in length, the longer spines 9–23 mm. Flowering March–May.

Slopes, valleys, and plains; Sonoran desertscrub and thornscrub; northern and central Sonora, near sea level–915 m (3000 ft): south-central Arizona to the Guaymas region. Tree-sized plants occur toward the northeastern part of the range.

This species replaces *O. thurberi* to the north and west. There is great variation in tepal color (yellow, orange, red, purplish, or greenish) among individual plants in south-central Arizona and perhaps adjacent Sonora. However, these plants may be of hybrid origin. Farther south in Sonora, such as south of Hermosillo, all plants have yellow-green flowers like those of *O. thurberi*.

***Opuntia wilcoxii***. Prickly pear; *nopal, tuna; nopá* (Guarijío); *navo* (Mayo); *naavo* (Yaqui) (figure 88)

Prickly pears, shrub sized and abundant in coastal thornscrub, and commonly arborescent, 2–6 m, with a well-formed trunk and a broad, open crown in tropical deciduous

Figure 87. Young plant of *Opuntia thurberi* (courtesy Raymond M. Turner). Photograph by J. R. Hastings.

Figure 88. *Opuntia wilcoxii*. Near Alamos. Photograph by K. S. Bunnell.

forest of southeastern Sonora. Pads slightly asymmetric, broadly elliptic and narrowed at each end; epidermis finely pubescent or sometimes glabrous. Areoles with tufts of long, tawny yellow glochids and great variation in spines, or many areoles without spines or with 1 stout, spreading yellow-white spine becoming deflexed with age, and sometimes with 1 or 2 additional smaller spines.

Flowers 6–7 cm wide; larger tepals 3.0–3.5 cm, bright yellow, the filaments yellow, the style cream color, the stigma pale yellow. Fruits 4–5 cm, wine purple including the pulp. Flowering (March) April-May, the peak usually late May; fruits ripening in the same season or in summer.

Plains, valleys, and slopes; thornscrub and tropical deciduous forest; southern Sonora, near sea level–455 m (1500 ft).

The nomenclature of this prickly pear is provisional. The description refers to the native, arborescent prickly pear of southeastern Sonora and northern Sinaloa. Some plants have short hairs and glands on the surface of the pads, whereas others are glabrous (perhaps the hairs and glands are sometimes deciduous).

Another population of shrub-sized to small-tree prickly pears occurs near San Javier in east-central Sonora and appear similar to the *O. wilcoxii* of southeastern Sonora. At higher elevations and farther eastward, such as the vicinity of Mulatos in east-central Sonora, there are also similar appearing prickly pears that are occasionally treelike to 5 m. These prickly pears appear to be *O. durangensis* or an allied species. They differ from the tropical deciduous forest tree prickly pear as follows: pads or cladodes with areoles more numerous (about 50% more) and more closely spaced, the spines shorter, more numerous, more uniform, white rather than yellow, and glochids consistently short (not long yellow tufts as in *O. wilcoxii*), and the pubescence much shorter and denser and often restricted to the pericarpel of the flower.

The lowland Guarijío distinguish two kinds of tree prickly pears in the mountains northeast of Mesa Colorado, north of Alamos. One has yellow flowers and reddish fruits consistent with the more widespread and common *O. wilcoxii*, and the other is distinguished by its lighter colored ("white") flowers and fruits.

### *Pachycereus*

Multiple-stem columnar cacti, some very large with a well-developed trunk. Immature (juvenile or sterile) portion of stems with stout spines, the mature (fertile) portions with coalescent areoles and weak, bristly spines or spineless. Flowers large, the ovary and fruits with scales bearing hairs, bristles, and/or spines. Mexico to Guatemala; 5 species.

*Pachycereus pecten-aboriginum* is diploid ($n$ = 11); *P. pringlei* is autotetraploid ($n$ = 22; Pinkava et al. 1973, 1977).

1. Trunk often > 1.5–2 m; fertile (mature) stems with 10–12 ribs; fruits covered with persistent, long, golden spines. ________ ***P. pecten-aboriginum***

1' Trunk usually < 2 m; fertile stems with (12) 13–15 ribs; fruits usually with short, feltlike, tawny, but not golden, spines or sometimes bristly. ________ ***P. pringlei***

***Pachycereus pecten-aboriginum*** [*Cereus pecten-aboriginum*]. *Etcho; chiquí* (Guarijío); *etcho* (Mayo) *utzivo* (Opata); *echo* (Yaqui) (figure 89)

Candelabralike cacti 5–15 m; trunk solitary, straight, and usually well developed, 1.5–2 m or more in height; branches many, erect, relatively slender to moderately thick. Immature stems (30–50 cm below apex), often 18–22 cm in diameter, with 10 or 11 ribs; areoles with few, rigid, sharp, white to gray spines, the central spine single, well developed, spreading to erect, 1.2–9.5 cm × 1.4–3.4 mm near the base, the radials short and spreading. Mature stems 10.5–17.5 cm in diameter, with 10–12 stem ribs; areoles with many bristly, erect, or slightly curved spines in a dense tuft, the spines 0.3–0.5 mm in diameter, the larger ones often 1–3 cm.

Flowers white, 8–10 cm and showy, opening in the evening and closing about midday; tepals white; ovary with dense velvety brown hairs; floral bracts mostly 13–20 mm, linear with long-attenuate tips covering the developing flower buds. Fruits densely covered with long, golden yellow spines; fruits splitting apart at maturity, revealing scant, bright red, firm, and scarcely juicy pulp and shiny black seeds. Seeds 4–5 mm (apparently the largest of all the columnar cactus seeds, at least those in northern Mexico). Flowering January–March; fruiting June–July.

Slopes, valleys, plains, and arroyos; thornscrub, tropical deciduous forest, and the southern edge of Sonoran desertscrub; central and southern Sonora, near sea level–975 m (3200 ft). Pacific slopes of Mexico from Nayarit to Sonora and southwestern Chihuahua, and the Cape Region of Baja California Sur. Populations in southern Mexico may represent a separate species (Bravo-Hollis 1978).

*Etchos* generally overtop thornscrub, but in tropical deciduous forest the canopy generally overtops the *etchos* (Gentry 1942). This is one of the most abundant trees in the subtropical scrub of southern Sonora, especially in the foothills and lower reaches of the Sierra Madre. It is frost sensitive and recommended for cultivation only in nearly frost-free regions. *Etcho* is relatively fast growing (unlike the *cardón*).

The fruit pulp is eaten fresh or cooked and often made into preserves or syrup, and it was formerly used to make wine. There is, however, much less pulp than in *cardón* fruits. The seeds are ground into a grayish, buttery paste. The Mayo made tortillas from ground seeds mixed with some corn flour; *etcho*-seed flour can also be made into pancakes, which have a blueberry-like flavor. The ground seeds were also used as cooking oil (Yetman 1998). To make a hairbrush, the spines are removed from about two-thirds of the fruit, and the remaining spines trimmed to about one centimeter in length. This use suggested the epithet "Indian comb."

***Pachycereus pringlei*** [*Cereus pringlei*]. *Cardón, sahueso; xaasj* (Seri) (figures 90–96)

One of the most massive cacti in the world; often 8–15 m, the trunk short but enormous, often branching within 0.5–3 m of the base with many massive branches. Stems bluish glaucous. Immature stems densely covered by rigid spines like those of young saguaros; mature stems with (12) 13–15 ribs and short, bristly spines or spineless.

Flowers open in the late afternoon or evening and remain open part of the following day, closing earlier on hotter days. Flowers 7–10 cm; tepals white. Outer surfaces of the flowers and fruits with a dense cover of feltlike hairs. Fruits about as large as a medium-sized peach, ripening in early summer; splitting open irregularly at maturity to reveal usually vividly colored crimson-purple to pinkish white pulp filled with small black seeds; fruit pulp sweet and delicious. Spine length and color of the fruit pulp is variable, although an individual plant bears fruit of only one color and general spine length. Seeds 3 mm. Flowering March–May; fruits ripening July.

Arid habitats including rocky slopes, bajadas, and desert plains with sandy soils or partially saline silty soils;

Figure 89. *Pachycereus pecten-aboriginum*. Near Mazatán (Turner et al. 1995). Photograph by J. R. Hastings.

Sonoran desertscrub; western Sonora, mostly near the coast from near Ciudad Obregón northward to Puerto Lobos, and inland to the vicinity of Pitiquito; near sea level–400 m (1310 ft). Also throughout most desert regions of the Baja California peninsula and islands in the Gulf of California.

The Seri classified four types of fruits according to the color of the pulp: red-purple, white, pale orange-yellow, or pinkish white. In addition, they named three kinds of *cardón* according to fruit characters: (1) plants with small fruits, without spines and rather bitter-tasting red-purple pulp, (2) spiny, normal-sized fruits, with sweet pulp of all colors, and (3) fruits not spiny, normal-sized, with sweet pulp of all colors (Felger and Moser 1985).

This cactus is unusual in having trioecious populations with unisexual male plants, unisexual female plants, and bisexual plants. Trioecy occurs near known maternity roosts of the lesser long-nosed bat (e.g., on Isla Tiburón and near Bahía Kino), one of the primary pollinators and seed dispersers for this cactus. Gynodioecy (separate female and

Figure 90. *Pachycereus pringlei*. Near Pozo Coyote, northeast of El Desemboque San Ignacio (Felger and Moser 1985). Photograph by E. T. Nichols.

Figure 91. *Pachycereus pringlei*. Roberto Herrera in 1983 at the base of a cardón well-known to the Seri. The sticks driven into the stem were used to hang meat after men returned from a hunt. About 1 km east of El Desemboque San Ignacio. (Felger and Moser 1985). Photograph by R. S. Felger.

Figure 92. (*top*) *Pachycereus pringlei*. Lesser long-nosed bat feeding from flower. Photograph by M. D. Tuttle (copyright Merlin D. Tuttle, Bat Conservation International).

Figure 93. (*middle*) *Pachycereus pringlei*. Lesser long-nosed bat feeding from fruit. Photograph by M. D. Tuttle (copyright Merlin D. Tuttle, Bat Conservation International).

Figure 94. (*bottom*) *Pachycereus pringlei* with ripe fruit. Photograph by M. D. Tuttle (copyright Merlin D. Tuttle, Bat Conservation International).

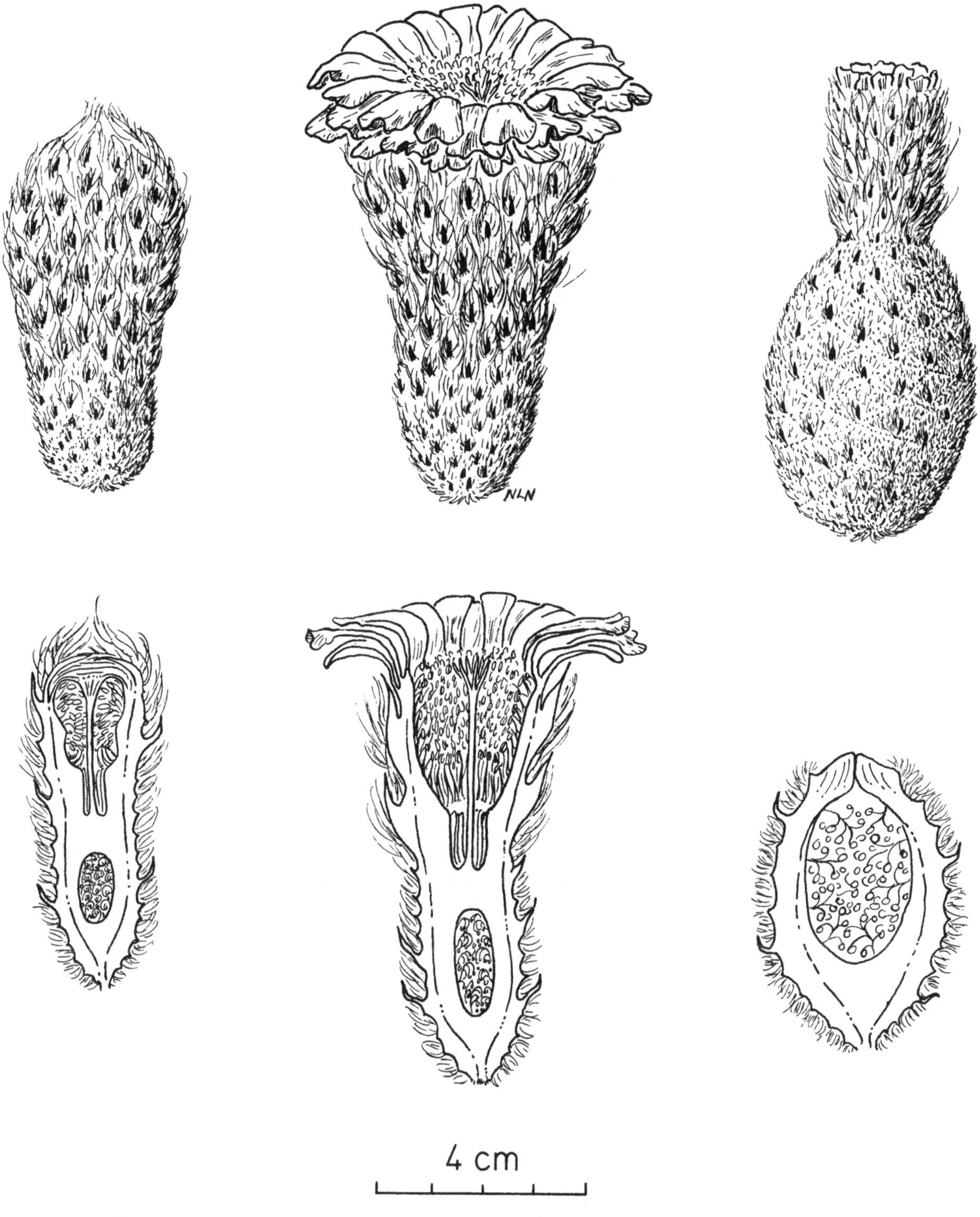


Figure 95. *Pachycereus pringlei*. North of Bahía Kino. Drawn by N. L. Nicholson.

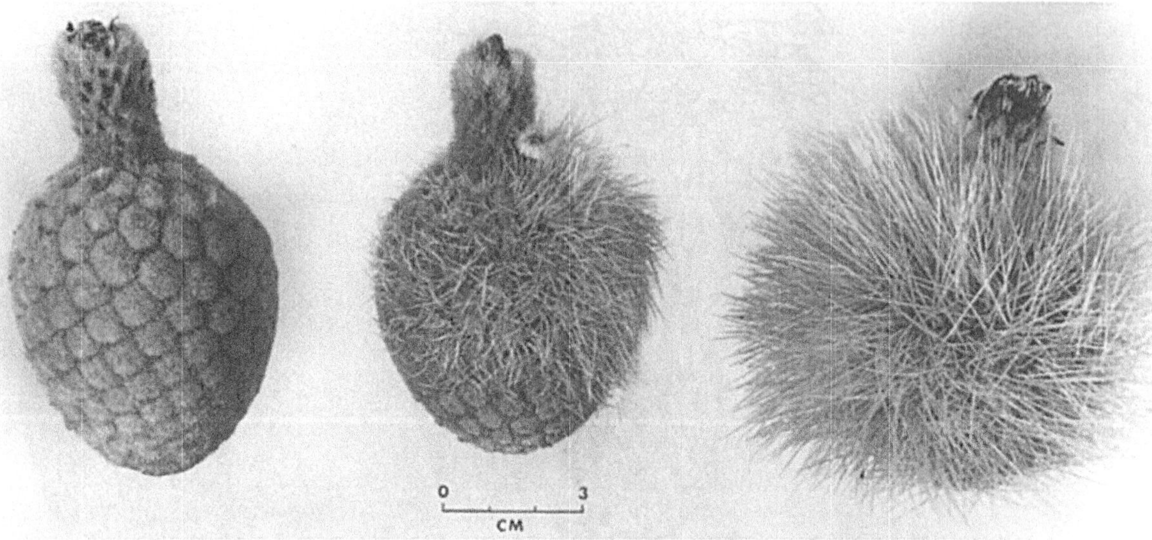


Figure 96. *Pachycereus pringlei*. Variation in spine length and density on fruits. Each fruit is from a different plant in a single population, north of Bahía Kino (Felger and Moser 1985) Photograph by R. S. Felger.

bisexual plants) occurs in regions > 50 km from roosting colonies (e.g., in Baja California Sur) (Fleming et al. 1994; Fleming and Hamrick 1998).

Although generally slow growing, under cultivation cardón can grow considerably faster than the saguaro. It is readily grown from seed, and small plants can be successfully transplanted. This species is frost sensitive. As with the saguaro, stem cuttings rarely form roots. The fruits were harvested by the Seri, the sweet pulp made into wine and the seeds ground into a buttery paste. The seeds were often stored in large ceramic vessels (Felger and Moser 1985).

### *Pilosocereus*

Mexico and the Florida keys to South America; 35 species.

***Pilosocereus alensis*** [*Cephalocereus alensis; Pilocereus alensis*]. Woolly torch; *pitahaya barbona; matacachí* (Guarijío); *aaqui jímsera* "bearded pitahaya" (Mayo) (figures 97 and 98)

Columnar cacti 3–6 (8) m, with numerous, mostly erect to ascending straight branches. Stems often 9–11.5 cm in diameter; ribs 10–13; stem surfaces bluish glaucous. Flowering, upper portions of the stems with a pseudocephalium of dense white hairs forming a "beard" along one side of the stem, the hairs covering flower buds. Flowers pale pink, relatively large, showy, and nocturnal, their bases hidden in the dense hairs of the pseudocephalium. Fruits globose, with a delicate pink or rose-colored spineless rind, and juicy, bright magenta sweet pulp with small black seeds. Flowering June; fruits ripening soon after.

Slopes, canyons, and ledges; tropical deciduous forest; southeastern and east-central Sonora, 230–915 m (750–3000 ft). Sometimes locally common beneath the forest canopy. Northernmost plants grow on south-facing mountain slopes 10–14 km west of Tepoca near Highway 16. Also southwestern Chihuahua to Jalisco.

The fresh fruit is delicious. It is worthy of cultivation in any large cactus garden and readily grows from cuttings or seeds but is highly frost sensitive. In desert regions it thrives best with light shade.

The type collection of *P. alensis* was obtained by Leon Diguet in the Sierra del Alo, Jalisco. It is similar to, but apparently distinct from, *P. leucocephalus*, originally collected near Horcasitas, Tamaulipas (Bravo-Hollis 1978), and not Horcasitas, Sonora, as stated by Britton and Rose (1920).

### *Stenocereus*

Many-branched cacti; stems ribbed and spiny; areoles with red-brown glandular hairs. Flowers large, white inside, pink to red-purple outside, nocturnal and largely bat and moth pollinated. Fruits succulent, spiny, the spine clusters deciduous from ripe fruits; pulp (funiculi) sweet. Arizona to northern South America and the Caribbean; 24 species. Reference: Gibson (1990).

There are two other shrub-sized species in Sonora, *sina* (*S. alamosensis*) and *pitahaya agria* (*S. gummosus*). Both have

Figure 97. *Pilosocereus alensis*. Near Alamos. Photographs by K. S. Bunnell (plant) and M. A. Dimmitt (flowers).

arching, many branched stems; those of *sina* are often rather snakelike, whereas those of *pitahaya agria* tend to be thicker and more rigid. *Sina* has relatively small, pink and red, tubular and slightly asymmetric diurnal flowers probably pollinated by hummingbirds, and smallish, marginally edible red fruits. It is common through much of the lowlands of Sonora except the arid northwest. *Pitahaya agria* has viciously sharp, brittle spines, large, white, and radial nocturnal flowers, and the baseball-sized red fruits are among the most delicious in the world. This species occurs along the coast between Punta Chueca and El Desemboque San Ignacio, on Islas San Esteban and Tiburón, and through most of the Baja California peninsula.

1. Stem ribs 8–10 (12); trunk single, erect and straight, lower branches usually arising above 1.5 m; tropical deciduous forest in foothills and mountains in southeastern Sonora. __________ ***S. montanus***

1' Stems ribs 13–18; trunkless or with few, short, ascending trunks, the lower branches mostly arising below 1 m; widespread. __________ ***S. thurberi***

***Stenocereus montanus*** [*Lemaireocereus montanus*]. *Saguira, pitahaya colorada; sahuí* (Guarijío) (figure 99)

Candelabralike columnar cacti often 6–8 (12) m; well-developed, solitary trunk usually 1.5–2 m in height, with half a dozen or more erect branches. Stem ribs 8–10 (12). Areoles separate, with a nearly circular tuft of red-brown, glandular hairs, the exudate producing dark red to blackish encrustations on the spines. Areoles of trunk and lower limbs with long, stout spines, these areoles often regener-

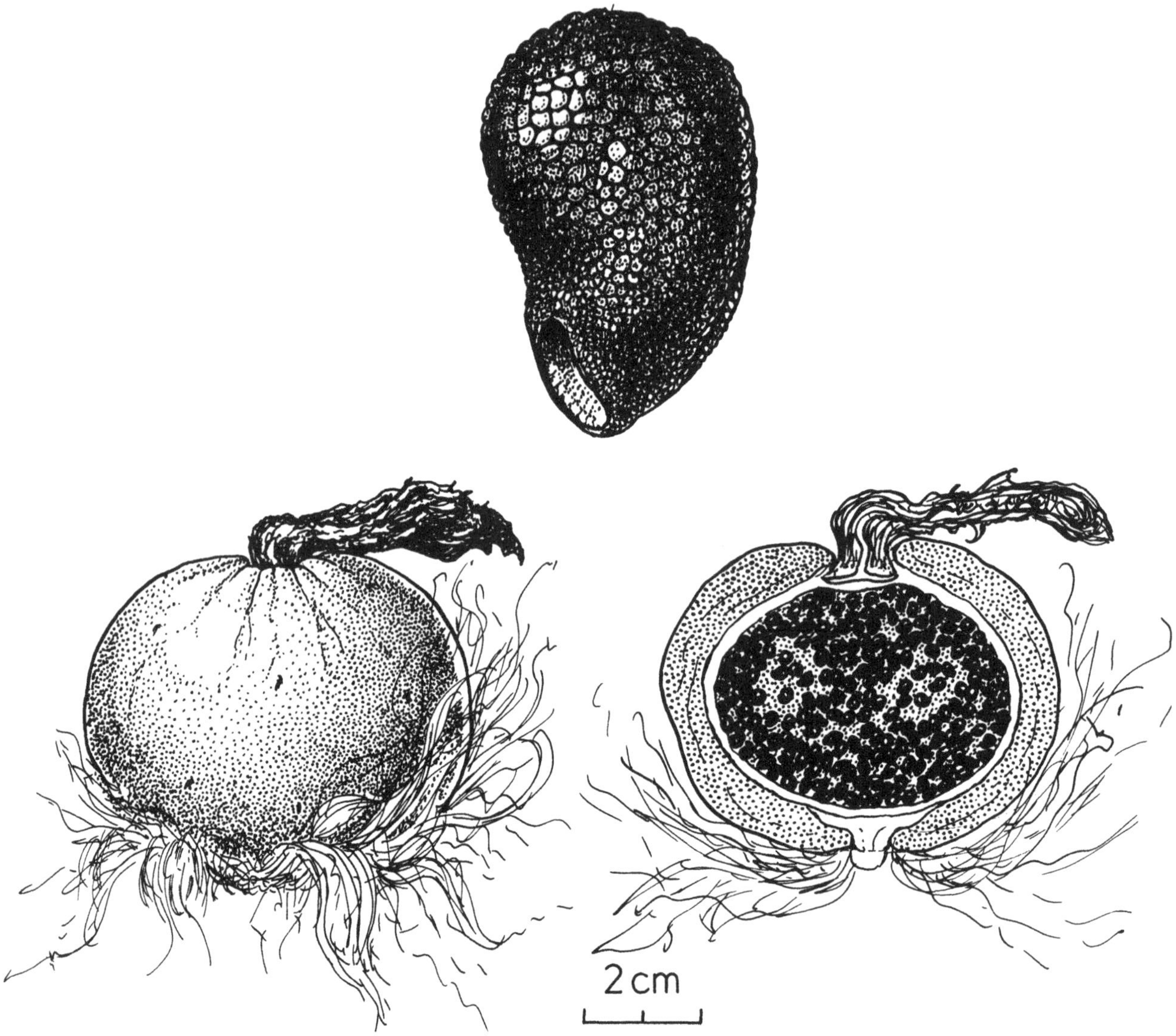


Figure 98. *Pilosocereus alensis*. Fruits and seed. Drawn by L. B. Hamilton.

ating new large spines; upper stems of different plants with either long or shorter spines; major spines (5) 6–12 per areole, sometimes with 1–3 additional small spines; the spines of different lengths, the largest 2–8 cm × 0.8–1.3 mm, needlelike, the dorsal (upper) 1 or 2 larger spine(s) somewhat flattened and twisted.

Flowers nocturnal, white inside, rose colored outside, relatively large, resembling those of *pitahaya dulce* (*S. thurberi*). Fruit skin green to orange; pericarpel scales (bracts) reddish green. Fruits apparently indehiscent, approximately 5 cm in diameter, longer than wide, spiny, the spines falling away at maturity; pulp and seeds edible. Flowering late April–May; fruits ripening June, falling simultaneously and promptly consumed by animals.

Slopes, valleys, and canyons; tropical deciduous forest; southeastern Sonora, 305–1065 m (1000–3500 ft). Also adjacent southwest Chihuahua to Colima.

People in southeastern Sonora recognize and name three kinds of *saguira* according to the fruit color, which is consistent for a given plant: *amarillo* (fruit pulp bright orange), *rojo* (red), and *blanca* (white). They say the *amarillo* is the most common of the three kinds, all three are delicious, the spines of the mature fruits are not troublesome ("se floja la espina cuando madura"), and that *saguira* fruits are more delicious than those of *pitahaya dulce*. As with *tunas* (prickly pear fruits), the fruits are regarded as *potila*. This concept, applied only to cactus fruits, implies they are edible before being fully ripe.

Figure 99. *Stenocereus montanus.* Northeast of Alamos. Photographs by K. S. Bunnell.

Where the forest has been cleared for buffelgrass, solitary saguiras left in the field often have knobby, frost-damaged stems even though frost is infrequent and slight. Saguiras sheltered by a canopy of natural forest, however, show no discernable freeze damage. *Stenocereus montanus* remains poorly studied. This handsome tree cactus is worthy of wider cultivation. As with *Pachycereus pecten-aboriginum*, it is relatively frost sensitive but can be cultivated as far north as Phoenix, Arizona. Stem cuttings readily form roots and are sometimes incorporated into living fences.

The original description of *S. montanus* reports the species as having diurnal flowers, a characteristic contradicted by field observations and ethnobotanical information. *Saguira* can be confused with *etcho* (*Pachycereus pecten-aboriginum*)—they are remarkably similar in general appearance. Features that set *S. montanus* apart are the distinct (noncoalescent) areoles of mature stems with brown, feltlike hairs, slender brown spines, usually fewer stem ribs, nocturnal flowers, shorter, broader floral bracts rounded at the tips (rather than long, slender and pointed), indehiscent, very juicy fruits with bright orange, red, or white pulp, and shorter, deciduous spines. The few-spined areoles give *S. montanus* stems a distinctive appearance, and with a little practice it is easy to recognize this species in the field. It extends into higher elevations than does *etcho* and is restricted in distribution to the upper elevations of the tropical deciduous forest in mountainous regions.

***Stenocereus thurberi*** [*Cereus thurberi*; *Lemaireocereus thurberi*]. Organpipe cactus; *pitahaya dulce*; *mehueri* (Guarijío), *aaqui* (Mayo); *cucuwis* (O'odham); *tutuis* (Pima Bajo); *ool* (Seri); *aaki* (Yaqui) (figures 100–104)

Columnar cacti 3–10+ m, with 13–19 shallow, rounded ribs. Spines slender, at first red-brown, becoming gray to nearly black, spreading in all directions, the longer spines 1–4 (5) cm, straight and needlelike (or occasionally a few spines twisted). Areoles separate, with dark red-brown, glandular hairs, these turning black within one or two years, the exudate producing dark red to blackish encrustations on the spines.

Flowers 7–9 × 6–7 cm, nocturnal, closing shortly after dawn; inner tepals white or pale pinkish lavender with white margins and bases; scales on tube purplish red; nec-

Figure 100. *Stenocereus thurberi*. Southern form with a well-developed trunk. Photograph by K. S. Bunnell.

tar chamber relatively large, its secretion copious. Fruits fleshy, the rind thin, red when ripe; pulp of mature fruit delicious, sweet, juicy, and bright red; spines of ripe fruits readily detaching. Seeds 2.0–2.5 mm, blackish, with a weakly developed ridge along back. Flowering and fruiting mostly in early summer, but flowers recorded March–December.

Slopes, ridges, bajadas, and plains; Sonoran desertscrub, thornscrub, tropical deciduous forest, and occasionally in lower elevations in grassland and oak woodland. Near sea level–1200 m (3935 ft). The most widespread columnar cactus in Sonora, ranging from near Sonoyta and the east side of the Sierra Pinacate in the northwestern part of the state more or less continuously eastward and southward through the lowlands, foothills, and lower mountain elevations. Sinaloa to southwestern Chihuahua and southwestern Arizona, islands in the Gulf of California, and both states of Baja California.

In the thornscrub and tropical deciduous forest of southern Sonora and northern Sinaloa the trunks are well developed, branching occurs well above the ground, and the plants may reach 10 m or more in height. Bats and birds thus have ready access to the flowers and fruits, which are held above or into the upper reaches of the surrounding canopy. Plant size generally decreases, and the trunkless condition is more pronounced northward in open, sparsely vegetated desert areas. Freeze damage is evident on many plants in northern Sonora and southern Arizona. Freezing temperatures often kill the stem tip causing a conspicuous constriction scar, this results in a new near-terminal branch and knobby stems.

*Pitahaya* is one of the most popular desert fruits. The pulp, seeds, and rind are eaten fresh or can be made into wine or preserved by drying. *Pitahaya* is a general term used in Mexico for fruits of various columnar cacti. This word is of Carib or Haitian origin and appeared in print as early as the 1530s (Fernández de Oviedo y Valdés 1959). Organpipe is

Figure 101. *Stenocereus thurberi*. Near Estación Luis, about 70 km southeast of Navojoa, southwestern Sonora. Photograph by D. Yetman.

Figure 102. *Stenocereus thurberi*. Northern form. Photograph by E. T. Nichols.

an attractive landscape subject for warmer desert and semidesert regions. The growth rate is moderate. It can be propagated by seed or stem cuttings and is incorporated into living fences.

## Capparaceae—Caper Family

Herbs, shrubs, or occasionally trees; often strong or foul smelling with watery sap containing mustard-oil glucosides. Leaves alternate. Flowers in racemes with bracts or solitary in axils; bisexual or sometimes unisexual, radial to moderately bilateral. Sepals and petals each ≥ 4, separate or the sepals united, or petals absent. Stamens ≥ 6, as long or longer than the petals. Ovary and fruit usually with a prominent stipe. Fruits fleshy (in Sonora) but sometimes drying capsulelike, 1- or several seeded.

Worldwide, mostly tropical and subtropical, some temperate, most in arid regions; 45 genera, 675 species.

1. Leaves compound. ______________________ ***Crateva***

1' Leaves simple.

  2. Shrubs or small trees with slender, single or multiple trunks; glabrous or with silvery peltate

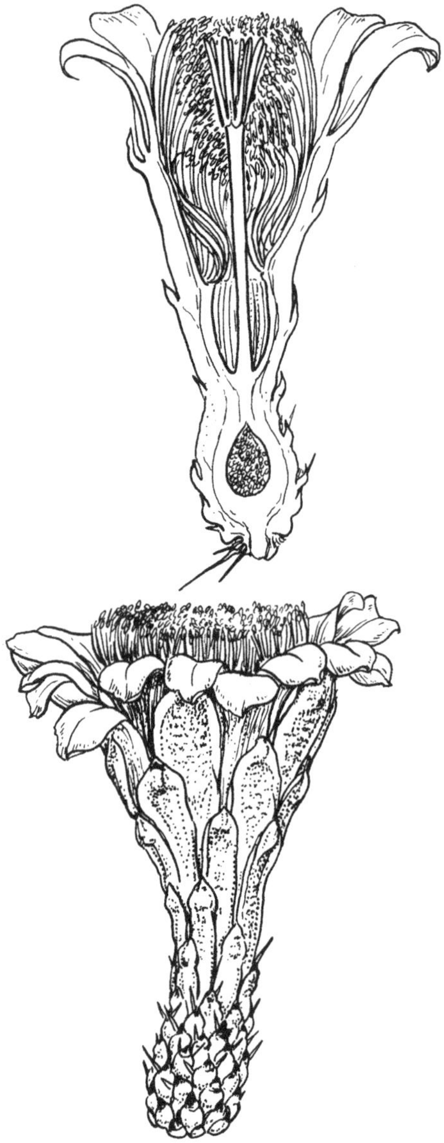

Figure 103. *Stenocereus thurberi*. Drawn by L. B. Hamilton.

scales; leaves 1–3 (5.5) cm; flowers bisexual; petals white. ______ ***Capparis***

2' Small trees with a solitary, thick trunk; pubescence of simple hairs; leaves 4–11 cm; dioecious; petals absent. ______ ***Forchhammeria***

## *Capparis*—Caper

Trees and shrubs. Leaves simple with tough, leathery blades. Flowers bisexual; ovary and fruit on a prominent stipe. Tropical and subtropical regions of both hemispheres; 300 species. References: Cozzo (1946), Hunziker (1984).

Some have edible buds and fruits; capers are flower buds of *C. spinosa*, native to the Mediterranean region.

1. Plants with silvery to brownish scales; leaves linear-oblong; fruiting stipe < 1 cm. ______ ***C. atamisquea***

1' Plants glabrous or glabrate; leaves broadly elliptic to oval; fruiting stipe 6–7 cm. ______ ***C. flexuosa***

***Capparis atamisquea*** [*Atamisquea emarginata*]. *Palo zorillo; juvavena* (Mayo); *cöset* (Seri); *huvak vena* (Yaqui) (figure 105)

Woody shrubs; occasionally trees 5–8 m with several trunks. Twigs straight, stout, rigid, and brittle, often branching at right angles and bluntly thorn tipped. Young stems, lower leaf surfaces, pedicels, sepals, and fruits densely covered with silvery white to somewhat yellowish peltate scales. Leaves 1–2.7 (5.5) cm × 3–6 mm; petioles stout, 1–2 mm, continuous with the prominent midrib; blades linear-oblong, dark green above, dull grayish below, the lateral veins indistinct, the margins entire except the apex sometimes notched; stipules absent.

Flowers solitary or few in leaf axils, on pedicels 4–7 mm, cream white, moderately bilateral; sepals 4, the outer 2

Figure 104. *Stenocereus thurberi*. Photograph by M. D. Tuttle (copyright Merlin D. Tuttle, Bat Conservation International).

much larger; petals 4 (6), approximately 6.5 mm; fertile stamens 6 (7) with filaments approximately 5.5 mm, plus 0–3 staminodes. Fruiting stipe 4–8 mm; body of fruit 8.0–11.3 mm, oval with a small, pointed tip. Ripe fruits splitting to reveal a bright red, fleshy aril, the exocarp and aril with a strong, spicy odor and horseradish taste. Seeds 1 (2 or 3) per fruit, 3.8–5.0 mm, nearly orbicular to broadly kidney shaped. Flowering April–June; fruits ripening in August.

Plains and arroyos; Sonoran desertscrub; western Sonora, near sea level–245 m (800 ft). Northwestern Sinaloa to Organ Pipe Cactus National Monument in southwestern Arizona and Baja California Sur and southern Baja California; also Argentina and Chile.

The flowers attract honeybees, native bees, large tarantula-hawk wasps (*Hemipepsis* sp.), and many other insects. The red arils and seeds are eaten by birds including the house finch and verdin, which are probable seed dispersers.

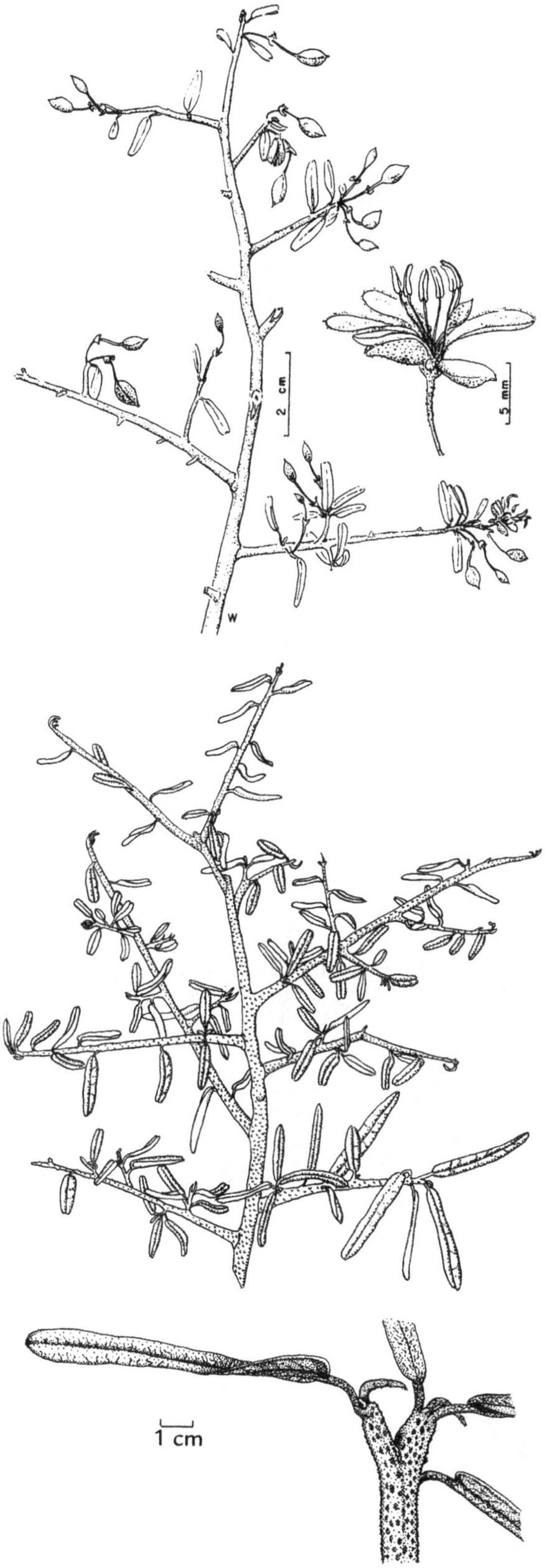

Figure 105. *Capparis atamisquea*. Top panel drawn by I. L. Wiggins (Wiggins 1980); bottom panel drawn by C. M. Marlett.

***Capparis flexuosa.*** *Mimbre del monte, tabelojeca* (Sinaloa); *tabarjeca* (Mayo)

Mostly shrubs; sometimes small trees to approximately 5 m. Glabrous or glabrate, the new herbage and buds minutely puberulent. Apparently evergreen. Leaves 4–7 × 2–4.5 cm; petioles stout, 5–6 mm; blades mostly broadly elliptic to oval, with prominent lateral veins, the tips rounded or blunt to notched; margins essentially entire—very shallowly undulate with closely woven veins forming a margin or border; stipules minute.

Flowers solitary or several at stem tips; pedicels approximately 1 cm. Sepals approximately 8 mm, rounded; petals 2–2.5 cm, obovate, white. Stamens numerous, 5+ cm, cream white, long exserted. Ovary stipe slender, to 5 cm. Fruits indehiscent, the body sausagelike, red-brown, 10.5–16.5 cm on a stipe 6–7 cm. Seeds surrounded by red pulp. Flowering June–July (early August).

Localized and rather rare in washes and low areas in coastal thornscrub at Masiaca and the vicinity of Camahuiroa in southwestern Sonora, near sea level–50 m (165 ft). Widespread in tropical America.

The root has a horseradish-like flavor, and the plant has many medicinal uses in Latin America (Standley 1922).

## *Crateva*—Garlic Pear

Tropics worldwide; six species. This genus includes trees and shrubs with medicinal properties and some with edible leaves and fruits.

***Crateva palmeri.*** *Perihuite* (Sinaloa); *capúsari* (Mayo)

Shrubs to small trees to 8 m, with a spreading crown and light gray bark. Young twigs, petioles, and lower leaf surfaces finely puberulent with white hairs. Leaves apparently drought deciduous; leaflets 3, 5–7 cm, obovate, thin, paler beneath, the upper surfaces green and glabrate with age, the tips acute, the lateral leaflets asymmetric; petioles 3–6 cm and slender; stipules small and deciduous.

Flowers purplish, crowded at the ends of dense, umbel-like racemes; pedicels 2.4–4.5 cm. Perianth yellow-green; calyx 4-lobed, 6.5–7.0 mm; petals 4, 10–17 mm, narrowly oblanceolate with a long, slender claw; stamens often 12, 3.6–4.2 cm, purple, long exserted; ovary on a long pedicel-like stipe. Fruits edible, on a stipe 6.5 cm, the body rounded, fleshy, 3.5–5 cm in diameter. Flowering March–April; fruits ripening June–July.

Arroyo margins in bottomland; tropical deciduous forest and thornscrub; southern Sonora near the Sinaloa border south of Alamos and rare at Masiaca; 50–300 m (165–985 ft). Guatemala and western Mexico.

The fruits are said to be sweet-tasting (David Yetman, personal communication 1998). The closely related, more widespread, and tropical *C. tapia* is well-known for its edible, garlic-scented fruits.

### Forchhammeria

Mexico to Central America and the West Indies; 10 species.

***Forchhammeria watsonii.*** Lollipop tree; *Jito, palo jito; jito* (Mayo), *hi'ito* (Yaqui) (figure 106)

Small lollipop-shaped trees often 5–7 m, unarmed, with a single, thick, smooth, pale gray trunk and a dense crown with remarkably little dead wood. Evergreen or briefly leafless or nearly so at flowering time; new leaves emerge above inflorescences at the time of flowering or fruiting. Leaves sessile or the petioles very short, with small, simple hairs; blades rather thick and blunt-ended, dark green and sparsely hairy above, pale greenish white and densely hairy beneath, 4–11 cm × 8–10 mm, linear to narrowly linear-oblong, the margins entire and distinctly inrolled below; stipules represented by small, deciduous glands.

Dioecious. Racemes 3–7 cm, densely flowered, attracting honey bees and other insects, the pedicels 2–6 mm. Calyx 1.0–1.5 mm, cuplike, shallowly toothed; petals absent. Male flowers yellow; stamens mostly 16–24, 4–5 mm. Female

Figure 106. *Forchhammeria watsonii*. Left panel drawn by I. L. Wiggins; right panel drawn by J. Floyd.

flowers 2–3 mm, maroon, with reduced stamens. Fruits sessile (without a stipe), mostly 10–12 mm, slightly obovoid, purplish to reddish orange, the pulp sweet. Mass flowering March–April; fruiting May–June and sometimes November.

Plains and valleys; best developed on deep, alluvial soils; Sonoran desertscrub and thornscrub; west-central and southwestern Sonora; near sea level–245 m (800 ft). Common from La Palma, south of Hermosillo, to northwestern Sinaloa and in Baja California Sur.

The leaf shape shows adaptation to its arid habitat—seedlings and young plants have narrowly linear leaves, whereas those of the mature tree are much broader. The fruits are eaten boiled with or without sugar. The Yaqui say that only certain, gifted people are able to prepare it so that it will taste good; for others it will be bitter (Felger and Molina unpublished notes). An exceptional tree about 10 m tall near Masiaca in southern Sonora is locally known as *El Jitón* (Yetman 1998). *Jito* occurs in essentially frost-free regions and is highly frost sensitive in cultivation. This unusual tree has ornamental potential for dry, tropical areas. From a distance it might be confused with San Juanico (*Jacquinia macrocarpa*), from which it is distinguished by the thicker trunk and larger, unarmed leaves. Reports of *Forchhammeria* on Isla Tiburón and northeast of Hermosillo (Turner et al. 1995) are undoubtedly based on misidentified sightings of *Jacquinia*.

## Caprifoliaceae—Honeysuckle Family

Herbs to trees. Mostly north temperate and boreal, and montane in the tropics; 15 genera, 400 species.

This family includes honeysuckle (*Lonicera*) with several species in Sonora.

### *Sambucus*—Elderberry

Subshrubby perennial herbs with creeping rhizomes to softwood shrubs and trees. Old and New World, mostly temperate and subtropical or montane in the tropics; 9 species. Reference: Bolli (1994).

The fruits and sometimes the flowers are edible, or the fruits may be toxic. The fruits have variously been used for wines and as a source of blue dye or hair dye, especially by the Romans (Pan was said to dye his hair with elderberries). Some elderberries are cultivated as selected cultivars, and some are used to produce insecticides.

***Sambucus nigra*** subsp. ***cerulea*** [*S. cerulea; S. neomexicana*]. Blue elderberry; *tápiro; sauco* (Yaqui) (figure 107)

Small, relatively short-lived trees; trunks single to mostly multiple, the trunks and limbs thick; wood soft and weak; stems pithy. Bark of young stems silvery gray, becoming brown, scaly and deeply fissured on older branches and trunks. Lenticels longitudinally grouped and conspicuous. Herbage at first pubescent, becoming glabrate or glabrous. Leaves partially to fully deciduous in the summer; new growth and leaves appearing with summer rains and especially in fall and winter. Leaves opposite, odd-pinnate, 8.5–22+ cm; the leaflets 3–7, 4.5–11.5 cm, highly variable, thinner with moist conditions but rather thickish under drier conditions, oval to mostly lanceolate, sometimes narrowly so, the bases unequal, the margins serrate-toothed, the terminal leaflet often largest; stipules absent in this subspecies.

Inflorescences dense, terminal, compound umbel-like cymes 6–17+ cm wide. Flowers cream white, 5.0–6.5 mm wide; sepals very small; corollas 5-lobed; ovary inferior. Fruits rounded, berrylike juicy drupes approximately 5 mm in diameter, dark blue or blackish (when ripe) with a wax-covered, glaucous surface; seeds 3–5. Flowering mostly winter and spring in the desert and into summer at higher elevations. Fruits without a bitter taste, becoming sweet when they turn purple.

Canyon bottoms, riverbanks, washes, streams, and arroyos; grassland, oak woodland, and pine-oak woodland, and occasionally extending into upper elevations of Sonoran desertscrub. North-central and northeastern Sonora, 365–1525 m (1200–5000 ft). Pacific North America: British Columbia, Alberta, and Montana to northern Mexico including Baja California, Chihuahua, Durango, Nuevo León, Sonora, and Sinaloa.

This tree is widely cultivated in northern Sonora, sometimes beyond the distribution of wild populations. It is grown for the berries which are eaten fresh but mostly made into preserves, jelly, and wine, and for its medicinal properties. However, the unripe fruits and herbage can be poisonous (Schmutz and Hamilton 1979).

According to Bolli (1994), *S. nigra* occurs in Europe, the Americas (one subspecies extends to Argentina in the Andes), and Macronesia. Bolli includes six geographically separated subspecies; however, it is sometimes not possible to distinguish them except by provenience. Subspecies *cerulea* is the only member of the genus in southwestern North America including northwestern Mexico. This subspecies shows greater adaptation to a seasonally dry and warm climate than any other member of the genus. The elderberry in Arizona and Sonora has long been known as *S. mexicana*, but according to Bolli (1994) the type of *S. mexicana* is referable to *S. nigra* subsp. *canadensis*.

## Celastraceae—Staff-tree Family

Trees, shrubs, or woody vines; glabrous or with minute hairs. Leaves alternate (in Sonora), simple, deciduous or evergreen, or absent; stipules small and deciduous or absent. Flowers on jointed pedicels, bisexual or unisexual.

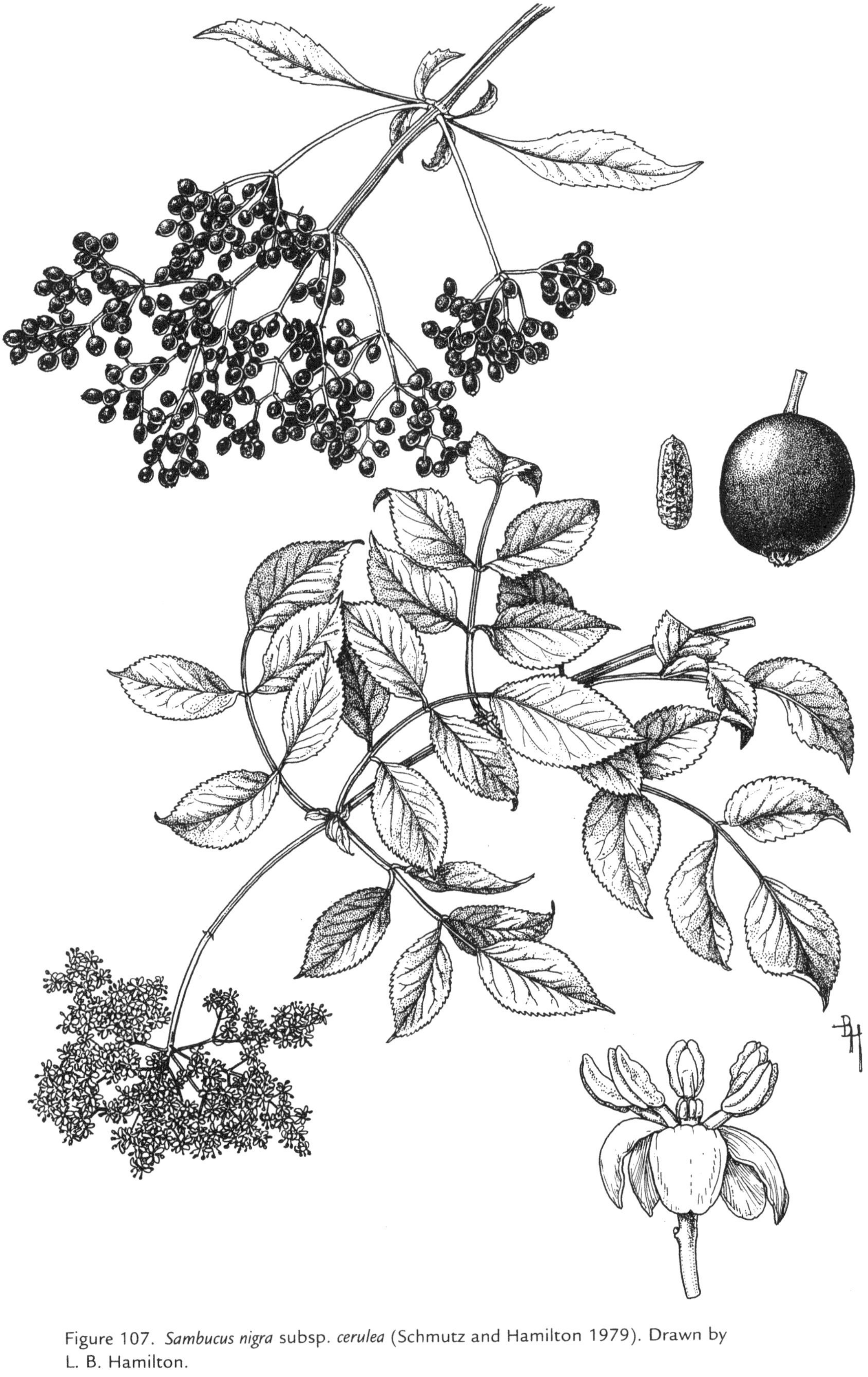

Figure 107. *Sambucus nigra* subsp. *cerulea* (Schmutz and Hamilton 1979). Drawn by L. B. Hamilton.

Calyx lobes 4 or 5, persistent. Petals 4 or 5, the stamens as many as the petals, on a broad, flat disk. Ovary superior. Fruits various, dry, or fleshy, mostly several seeded. Generally in tropical regions; 55 genera, 800 species.

1. Branches leafless, and spinescent. ________ ***Canotia***
1' Branches conspicuously leafy, unarmed.
   2. Leaves broad, thick, and succulent, the margins entire or notched at the tip; fruits leathery, 3-angled, and dehiscent, not winged; coastal. __***Maytenus***
   2' Leaves narrow, thin, and not succulent, the margins toothed; fruits papery, 3-winged, and indehiscent; inland, foothills, and mountains. ____ ***Wimmeria***

*Canotia*

The genus has 1 species.

***Canotia holacantha.*** Canotia, crucifixion thorn (figure 108)

Shrubs or rarely small trees, essentially leafless with well-developed, rigid, hardwood trunks and limbs; trunks and major limbs often twisted, the bark longitudinally shredding. Leaves scalelike, minute, quickly deciduous, present only on fresh, new growth. Smaller branches and twigs pale green, rather rigid, mostly spine tipped; dark areas of

Figure 108. *Canotia holacantha* (Benson and Darrow 1945). Drawn by L. B. Hamilton.

densely crowded, dark red-brown stalked glands conspicuous on stems just above the nodes.

Flowers white, in small clusters, 5-merous, 8–9 mm wide; sepals and petals with slightly ragged margins. Fruits persistent, 1.5–2 cm, at first fleshy, soon becoming woody capsules with 5 carpels splitting apically into awned valves; 5 chambered, each with several seeds. Flowering May–June (August).

Sonoran desertscrub; north-central Sonora, 610–1525 m (2000–5000 ft). In Sonora known only from the foothills southeast of Magdalena de Kino to the nearby Sierra Baviso, where it is locally common on rocky slopes, and another population near Altar. Primarily occurs in central and northern Arizona and extreme southern Utah.

*Canotia* is one of three genera of crucifixion thorns in the Sonoran Desert (see *Castela* and *Koeberlinia*). It somewhat resembles a leafless foothill palo verde (*Parkinsonia microphylla*), from which it is readily distinguished by the persistent capsules, shredding bark, and somewhat different growth form. The report of *Canotia* on Isla Tiburón (Turner et al. 1995) is probably based on a misidentified *Castela polyandra* or *Koeberlinia spinosa*.

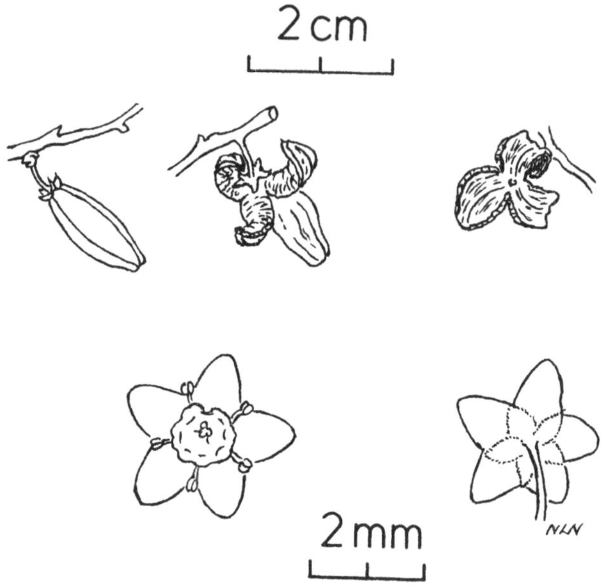


### *Maytenus*

Trees and shrubs. Tropical and warm temperate regions nearly worldwide; 200 species.

***Maytenus phyllanthoides*** [*Tricerma phyllanthoides*]. *Mangle dulce; pasio* (Mayo); *cos* (Seri) (figure 109)

Mound-shaped hardwood shrubs or sometimes small trees 4–6 (8) m, unarmed, densely and sinuously branched, with brittle twigs and smooth, gray bark. Young herbage, petioles, inflorescences, pedicels, and sepals with microscopic, often sparse hairs, otherwise glabrous. Evergreen; leaves 2–6 × 1–3.5 cm, leathery and semisucculent, obovate, the base cuneate (gradually narrowed to the petiole), the tips rounded or sometimes shallowly notched, the margins otherwise entire except the youngest, developing leaves with minute, deciduous marginal teeth, new leaves and shoots often reddish; petioles 2–6 mm; stipules minute, red-brown, and glandular.

Flowers solitary or in small, several-flowered axillary clusters among the leaves. Flowers unisexual or sometimes appearing bisexual, the plants perhaps often dioecious. Flowers pedicelled, 3–4 mm wide, yellow-green, (4-) 5-merous, with the ovary on a fleshy disk and the stamens borne on the disk margin. Fruits 3-angled, 9–12 mm, with 3 chambers before splitting open, the valves reflexing to reveal a sticky, bright red, fleshy, and moderately sweet aril covering 1–3 seeds. Flowering mostly April–November.

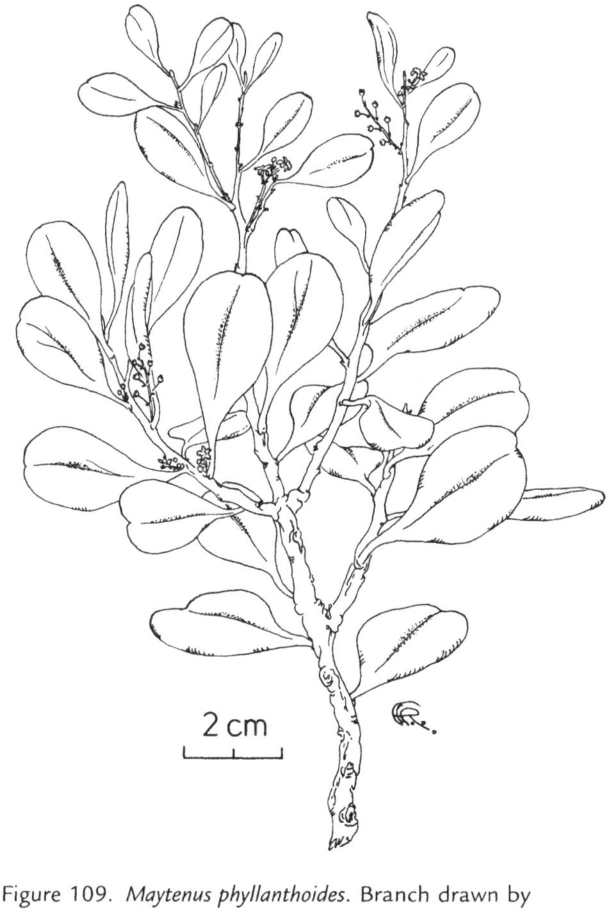


Figure 109. *Maytenus phyllanthoides*. Branch drawn by F. Runyan (Felger and Moser 1985); fruits and flowers drawn by N. L. Nicholson.

Sonoran desertscrub and thornscrub. Coastal Sonora from Estero Sargento and Isla Tiburón southward. Abundant along protected coastal esteros, bays, and inlets; inland margins of mangroves and often flooded by seawater at highest tides and also frequently extending inland from the shore along arroyo beds and low-lying silty flats with salt scrub. Often forming impenetrable thickets. Extensive stands occur on low silty, semisaline flats inland from Bahía Kino and Estero de la Cruz. Also Baja California Sur and Sinaloa.

This species is closely related to *M. texana* of the Gulf of Mexico and Caribbean region and differs by minor characters including pubescence.

In cultivation mangle dulce makes a handsome evergreen screen or hedge. It has survived -6.5°C (20°F) without damage in cultivation in southern Arizona. Because it is tolerant of both saline and dry soils, it should be suitable for coastal dune stabilization and as an ornamental halophyte. This species is sometimes placed in the genus *Tricerma* with about eight species.

### *Wimmeria*

Trees and shrubs. Mexico and Central America; 12 species. Reference: Lundell (1939).

***Wimmeria mexicana*** [*W. confusa*]. *Algodoncillo, papelío; machicari* (Guarijío); *chi'ini* "cotton" (Mayo) (figure 110)

Attractive, unarmed hardwood trees to 8 (11) m, mostly taller than wide; trunk and branches erect to ascending, the crown often spreading and rather sparse, the trunk commonly reaching approximately 30 cm in diameter. Bark smooth, white beneath large, flaking, irregular, gray plates. Young twigs, petioles, flowering branches, and pedicels with minute, short hairs, otherwise glabrous. Leaves tardily drought deciduous, 2–6 cm, petioled; blades lanceolate to elliptic or oblanceolate to obovate, the base tapering and decurrent on the petiole, the margins shallowly toothed; stipules absent.

Inflorescences axillary, 7–11 mm, of small umbel-like cymes on slender peduncles, several flowered or the flowers solitary. Flowers 7.5–8 mm wide, white to cream color, bisexual, on slender pedicels 4–6.8 mm; calyx lobes and petals each 5. Mass flowering (July–October), often in response to heavy autumn rains, attracting hoards of insects, especially bees and flies—the trees can be located by the loud humming of insects. Fruits 1–1.4 cm, about as wide as long, indehiscent, papery, with three broad wings, the stigma and style persistent; one seeded.

Southeastern and east-central Sonora, 305–975 m (1000–3200 ft). Locally common in tropical deciduous forest on rather arid, rocky flats and slopes and sometimes among low-elevation oaks. Oaxaca to Sonora and Chihuahua.

"A small, slender, rather closely branched tree, the bark of which peels off freely in dry seasons, whence the name 'papelio'" (Gentry 1942: 172). When fresh leaves or bark of *Wimmeria* are slowly torn apart, slender fibers are seen. These delicate, cottonlike strands consisting of the pulled-out xylem wall thickenings have earned this tree the name *algodoncillo* (see *Cornus*). The leaves are often chewed by insects, resulting in ragged margins. The wood is used for fence posts, house beams and posts, and fuel. The Guarijío prepare medicinal tea from a mixture of *algodoncillo*, chamomile, and cilantro for "pain in the body" ("malo en el cuerpo, dolor del cuerpo"). Both *Wimmeria* and *Dodonaea* have somewhat similar three-winged papery fruits.

## Clethraceae—Clethra Family

The family has 1 genus. Trees and shrubs. Tropical and subtropical, and some in temperate regions; Americas, Asia, Malaysia, and one species in Madeira; at least 65 species. References: Sleumer (1967), Standley and Williams (1966).

The Clethra family is allied to the Ericaceae.

Figure 110. *Wimmeria mexicana*. Drawn by M. B. Johnson.

***Clethra mexicana*** [*C. lanata*]. *Madroño* (figure 111)

Large shrubs on exposed slopes, or straight-trunked, slender, evergreen trees to 15 m along canyon streams; bark corky, gray, and furrowed. Twigs, petioles, veins of lower leaf-blade surfaces, and inflorescence branches with extremely dense, velvety orange-brown stellate hairs, the leaves glabrate above, the lower leaf surfaces with a dense mat of minute whitish hairs beneath the longer orange-brown hairs. Leaves alternate and spirally arranged, simple, evergreen (at least in riparian canyons), frequently clustered at stem tips, often 5–20 cm, rather thick and broadly obovate, the margins nearly entire to minutely toothed on the distal half; petioles stout, 7–28 mm; stipules absent.

Inflorescences 5.5–14 cm, densely flowered terminal clusters of racemes or few-branched panicles. Flowers white, bisexual, fragrant, on densely hairy pedicels about as long as the flowers, the buds subtended by densely hairy, deciduous bractlets. Calyx 3–4 mm, persistent, densely hairy outside, divided nearly to the base; lobes 5, broadly oval; petals 5, approximately 7 mm, deciduous, separate, white, fringed at the broad tip. Stamens 10; anthers opening by apical pores. Ovary superior, hairy; style 1; stigma 3-lobed; style and stigma rather stout and persistent. Fruit a capsule, nearly round, 3–5 mm in diameter, 3-valved, partly enclosed by the calyx; seeds many, 1.0–1.2 mm, flattened and with a narrow wing. Flowering April–June.

Slopes and canyon streams; oak and pine-oak woodland, often growing with *Alnus, Cupressus, Ilex, Persea*, and *Magnolia*, as well as pines and oaks. The northernmost population was found by Alberto Búrquez in Arroyo El Reparo near Yécora, at about 1100 m, and at El Trigo, 1480 m; east-central and southeastern Sonora, 1065–1675 m (3500–5500 ft). Panama to Sonora and Chihuahua.

Plants from Durango, Jalisco, and southward often have much longer inflorescences.

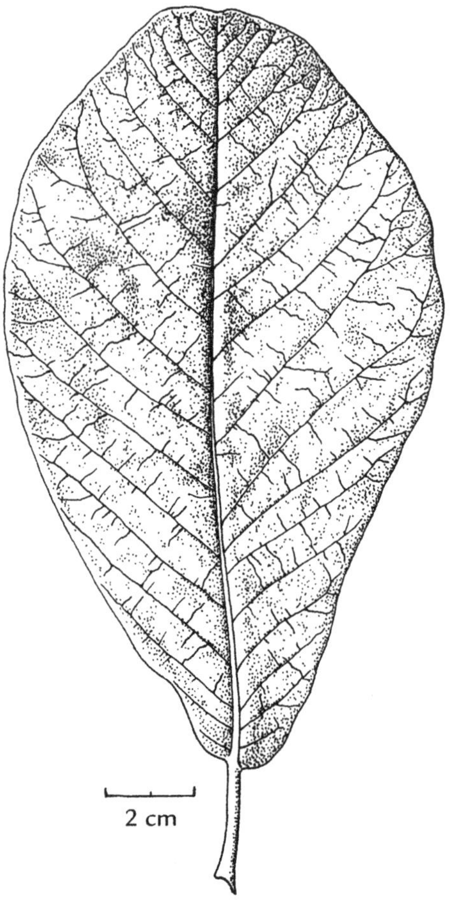


Figure 111. *Clethra mexicana*. Drawn by M. B. Johnson.

## Cochlospermaceae—Cochlospermum Family

Herbs, shrubs, and trees; tropical and subtropical regions of both hemispheres; 2 genera, 15 species. Reference: Poppendieck (1981).

Two species of *Amoreuxia*, herbaceous root perennials locally called *saiya*, occur in Sonora.

### *Cochlospermum*

Trees, shrubs, and subshrubs of dry, tropical regions of both hemispheres; 12 species.

***Cochlospermum vitifolium***. Buttercup tree, rose imperial; *palo barril, rosa amarilla* (Sinaloa); *ciánori* (Mayo) (figure 112)

High-crowned, softwood trees to 16 m; trunk straight and tall, often reaching 45–55+ cm in diameter. Bark smooth; limbs and trunk silvery gray with low, wartlike bumps; twigs reddish. Young growth with yellow sap or juice; leaves, sepals, petals, and filaments with secretory canals containing red liquid. Sparsely pubescent with simple hairs on young growth, major leaf veins and outer surfaces of the sepals. Drought deciduous; leaves present only during the summer rainy season, turning yellow to orange and quickly shed in early fall. Leaves alternate, often 38–42 cm, long petioled; blades often 20–30 cm wide, thin, deeply and digitately 5 (7)-lobed, the lobes with acuminate tips, the margins shallowly serrated; stipules slender and deciduous.

Inflorescences short, the pedicels often 2.5–3.5 cm, stout and red-brown. Flowers 8.5–10 cm wide, produced when the trees are leafless. Outer sepals red-brown, the inner one with a red midstripe; petals, filaments, and style brilliant yellow with small red spots and streaks, and the petals also with red near their bases. Sepals 5, separate, in 2 series, the outer with 2 smaller sepals; margins papery. Petals 5, sepa-

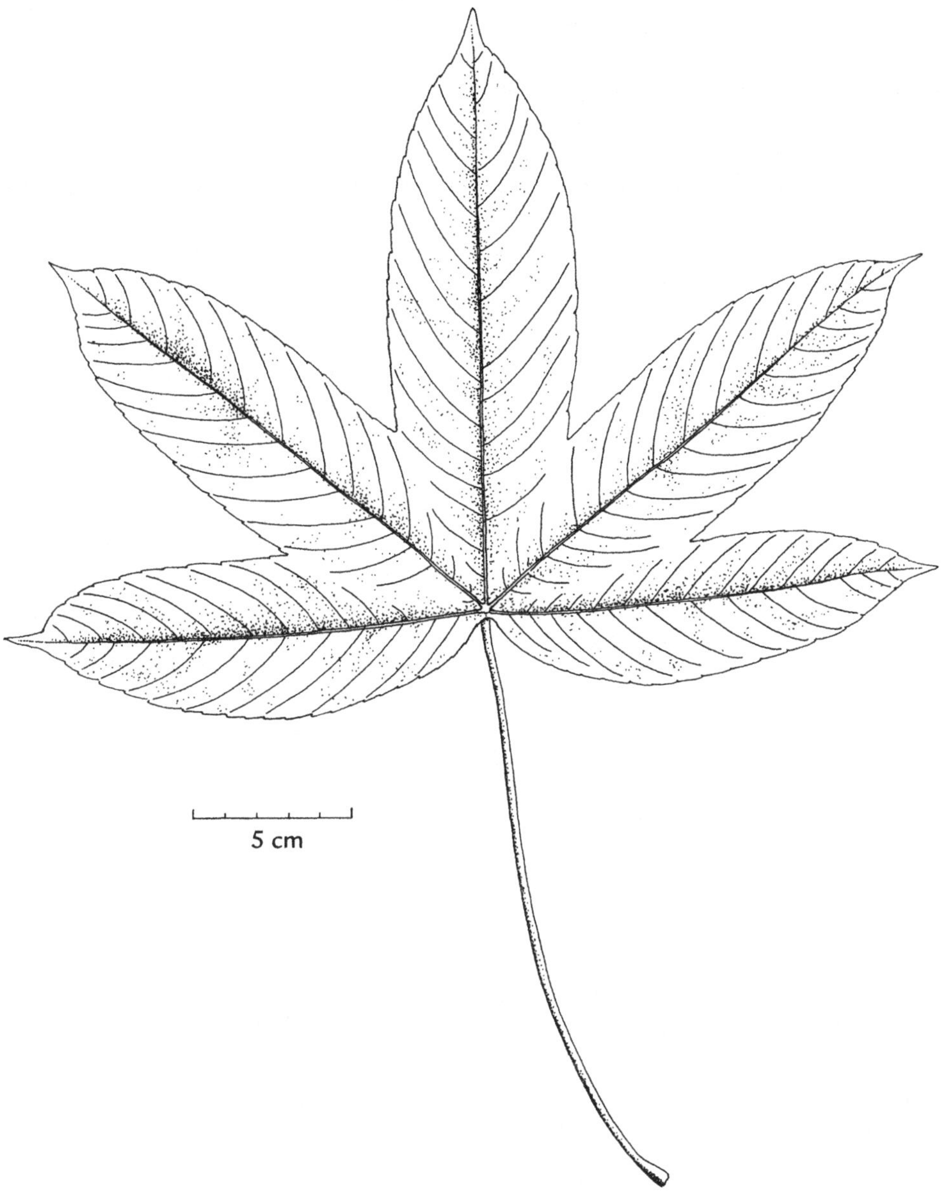


Figure 112. *Cochlospermum vitifolium*. Drawn by M. B. Johnson.

rate, 4–6 × 3.5–5 cm, the tips shallowly notched to cleft or irregularly undulate. Stamens numerous, large; anthers bright yellow, each with a single terminal pore and two minute, triangular basal pores; with age the filaments become red and the anthers orange. Ovary superior, green, and pubescent. Fruit a capsule approximately 10 cm, with woody valves, the seeds cochleate with long, fine, white cottony hairs. Flowering December–March.

Slopes and canyons in tropical deciduous forest; southeastern Sonora, 305–760 m (1000–2500 ft). Bolivia and northern Brazil to Sonora and the West Indies.

The flowers, with their pored anthers, are buzz pollinated. In the morning, fresh flowers are often strewn about on the dry forest floor like cups of gold. Flowering spans about one month or more, often with a peak during the second and third weeks of March. Mature fruits gradually empty as winds liberate the seeds on fluffy parachutes of cottony hairs. The large capsules somewhat resemble those of *Ceiba*, which is not unexpected given that both families are members of the Malvales.

This is a very desirable tree for nearly frost-free regions and is readily propagated by cuttings or seed. In cultivation it can grow as much as 2 m per season. The plants begin flowering at a very early age, when < 2 m. Although generally not cultivated in Sonora, it is grown as an ornamental in tropical regions worldwide, and can be weedy, thriving in disturbed habitats including roadsides. It is used as a living, flowering fence in many tropical areas in Mexico. A form with "double" flowers is common in cultivation. This cultivar has larger flowers than wild trees and does not set seed. It is propagated by cuttings. The sap of this species as well as that of three African *Cochlospermum* species has long been used as a remedy for jaundice.

## Combretaceae—Combretum Family

Tropical trees, shrubs, and lianas; 20 genera, 500 species. Includes important Old World timber trees. The largest genera, *Conocarpus* and *Terminalia*, each have more than 200 species. Reference: Tomlinson (1986).

*Conocarpus erecta* is a large, woody shrub on the shores of the Cape Region of Baja California Sur and in Sonora occurs on the landward side of mangroves at Bahía de Santa Bárbara (west of Navojoa); in Sinaloa it becomes a tree. It is readily distinguished from *Laguncularia* by its alternate leaves and absence of petiole glands. *Terminalia catappa*, sea-almond, is a common cultivated tree in tropical coastal regions globally. Within the family genera that are strictly mangroves are *Laguncularia* and *Lumnitzera*, the latter with two species in Asia.

### *Laguncularia*

The genus has a single species.

***Laguncularia racemosa*.** White mangrove; *mangle blanco*; *posio tosa*, *moyet* (Mayo); *pnaacoj hacáaiz* (Seri); *siari paseo* (Yaqui) (figures 113 and 114)

Shrubs or small trees 3–5+ m, the wood relatively hard. Thick, knobby, branched roots (pneumatophores) often developing from the shallow, horizontal root system. Herbage glabrate; inflorescences and flowers densely pubescent with minute hairs. Twigs terete, with numerous, moderately conspicuous lenticels. Leaves evergreen, opposite and decussate; blades 3.2–8.7 × 1.7–4 cm, mostly broadly elliptic or oval to sometimes obovate, shiny green, brittle, semisucculent, the margins entire; petioles 6–12 mm; stipules none.

Leaves with several kinds of secretory structures: (1) conspicuous pair of circular petiole glands near the leaf blade functioning as extrafloral nectaries, secreting a sweet solution; (2) dotlike or porelike submarginal glands on the lower surface at the junctions of minor veins; in young leaves these seem to function as hydathodes, secreting water or mucilage; (3) microscopic, translucent dotlike glands in small pits excreting salt.

Flowers unisexual or bisexual, the plants dioecious or hermaphroditic. Flowers crowded in terminal and axillary spikelike racemes, or sometimes few-branched panicles, 2.7–9.5 cm. Flowers subsessile, 3–4 mm wide; calyx with 5 small, persistent teeth; petals 5, inconspicuous, greenish white, hairy, and quickly deciduous; stamens 10 in 2 series. Male flowers with a well-developed style but lacking an ovary. Female flowers with a well-developed disk, an inferior ovary, and nonfunctional stamens. Fruits of 1-seeded nutlets 1.4–2.0 cm, ovoid, conspicuously ribbed; the spongy outer portion facilitates dispersal in water. Seedling development begins while the fruit is still on the tree; the seedling emerges rapidly after the fruit falls and becomes stranded in shallow tidal water. Flowering during warmer months; fruits probably ripening 2 or 3 months after flowering.

Saltwater bays, esteros, inlets, and river estuaries in close association with the other two mangroves; coastal Sonora from Estero Sargento and Isla Tiburón southward. Mexico to Panama, Gulf of Mexico and Caribbean to Florida, northern South America, and West Africa.

In the tropics the white mangrove develops into a tree sometimes 20 m in height. The flowers are visited by insects including bees. The northernmost Sonoran white mangroves are sometimes severely frost damaged. White mangrove is most abundant in the tidal zone between the red (*Rhizophora*) and black (*Avicennia*) mangroves.

## Convolvulaceae—Morning Glory Family

Diverse growth forms, largely herbaceous; worldwide, mostly in the tropics; 55 genera, 1600 species.

### *Ipomoea*—Morning Glory

Mostly vines, or sometimes shrubs or trees in seasonally dry tropics. Tropical and warm temperate regions of the world; 600–700 species. References: Austin and Huáman (1996), Austin and Simao Bianchini (1998), McPherson (1981).

The sweet potato is *I. batatas*. The pink-flowered shrub *I. carnea* subsp. *fistulosa* is widely cultivated in Sonora, as are morning glory vines. There are probably more than two dozen native and weedy herbaceous vining species in Sonora. They generally grow and flower with warm weather; the winter-flowering habit of *I. arborescens* is unusual among the Sonoran members of the genus. Morning glory flowers typically open in the morning and fade with midday heat, hence the common name.

The tree morning glories belongs to an unusual assemblage of about 10 species of large shrubs and small trees in the series *Arborescentes* and range from northern Mexico to

Figure 113. *Laguncularia racemosa*. An old, isolated tree, Estero Sargento. (Felger and Moser 1985). Photograph by R. S. Felger.

Peru. The arborescent habit depends on an anomalous type of wood formation. Two shrubby members of this series grow in Sonora: a drought-deciduous shrub 2–3 (4) m in the Sierra el Aguaje north of Bahía San Carlos seems to represent an undescribed species. *Ipomoea chilopsidis* is a trunkless shrub with multiple, stout stems to about 4 m. At its northernmost limits, in southeastern Sonora on Sierra Saguaribo and at Santa Bárbara, it grows on indurated ash soil among lower oaks; it also occurs in nearby southwestern Chihuahua.

***Ipomoea arborescens.*** Tree morning glory; *palo santo, palo blanco*; *tochiyó* (Guarijío); *jútuguo* (Mayo) (figures 115, 116, and 117)

Trees often 6–12 (15) m, with thick trunks (often 50–60 cm in diameter) and limbs; wood fibrous, succulent, very soft, and pithy. Herbage and flowers with milky sap; pubescent, especially the young shoots somewhat woolly with short, seldom straight hairs to 0.7 mm. Drought deciduous; trees usually leafless from fall until the summer rains of the following year. Leaves alternate, simple, 10–27 cm; blades ovate, thin, with short hairs often with a minute swollen base, more sparsely pubescent with age, especially on the upper surface, the base truncate to cordate, the tip pointed, the margins entire; midrib and major veins impressed on the upper leaf surface and raised below, the midrib base (just above the petiole) with a pair of blisterlike glands 1–3 mm, turgid on young, enlarging leaves (these glands are the same color as the midrib and may be difficult to see, especially on older or dried specimens); stipules absent.

Flowers white, showy, and fragrant, on few- to many-flowered clusters or branchlets, usually only one flower in a cluster opening at a time; pedicels much longer than the short peduncle. [These clusters represent a group of reduced 1 (2)-flowered cymes on a reduced shoot that re-

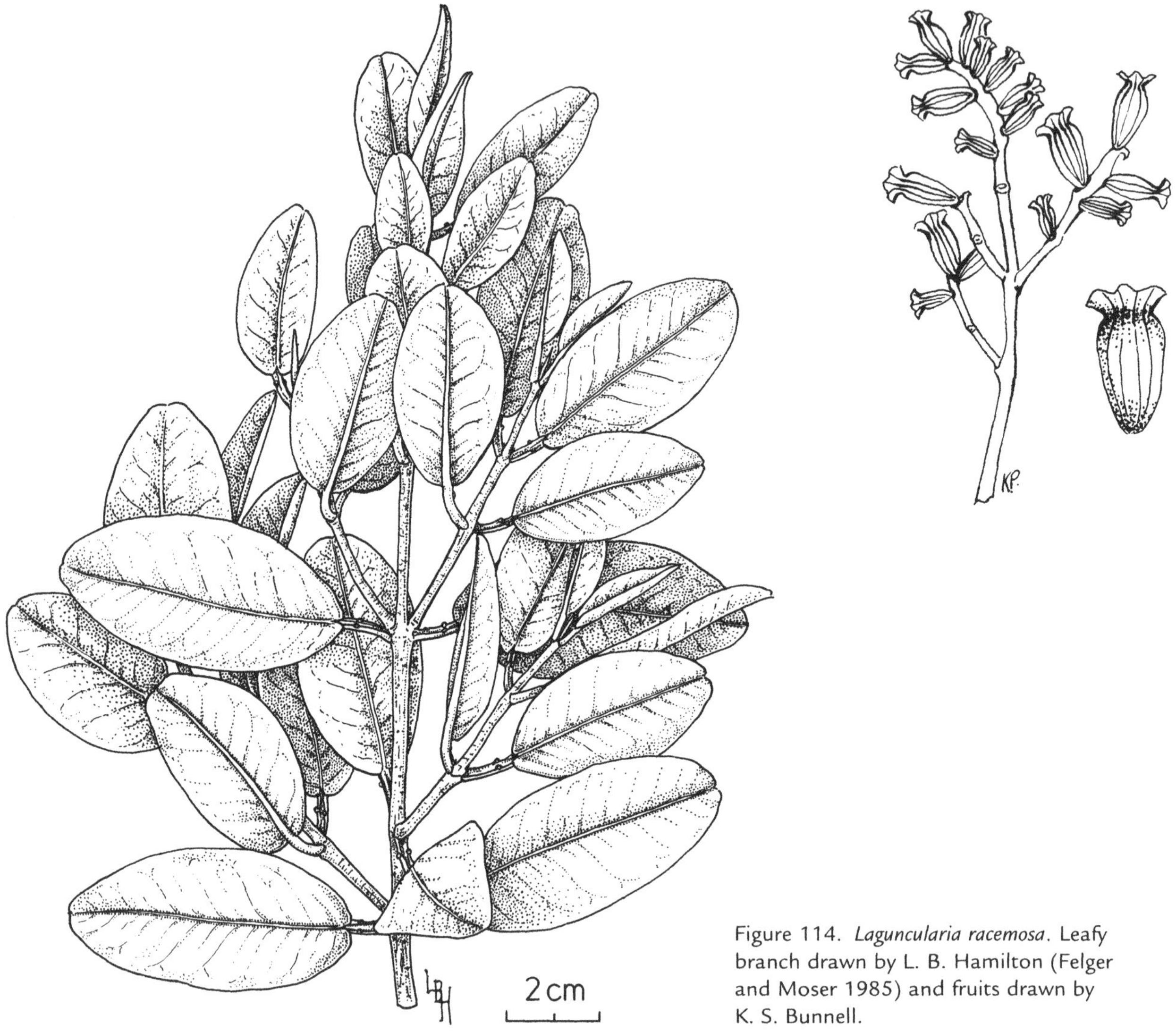


Figure 114. *Laguncularia racemosa*. Leafy branch drawn by L. B. Hamilton (Felger and Moser 1985) and fruits drawn by K. S. Bunnell.

sembles a raceme.] Sepals 5, shorter than the corolla tube, tough, very thick especially toward the base, overlapping, the outer surfaces pubescent to glabrate, the inner surfaces with straight, appressed hairs mostly 0.2–0.4 mm. Corollas slightly twisted in bud, the portion exposed in bud (the interplicae) pale yellow, thick, and forming a star-shaped pattern on the opened corolla, the interplicae glabrate or sparsely pubescent toward their tips, the rest of the limb folded fanlike in the bud and glabrous. Open corollas often 6.5–9.5 cm wide, white, with a thick and semisucculent tube, the throat nearly white to pale yellow-green inside and often with purplish markings; limb white, saucer shaped (rotate) with 5 fused lobes each notched at the apex. Stamens and style about as long as the corolla tube, the filaments, anthers, style, and stigma white. Stamens 5, the filaments attached to the base of the corolla tube, the attachment site of each filament surrounded by a 3–5 mm dense tuft of fleshy, translucent hairs filling the base of the corolla tube and blocking access to the ovary and nectar. Corolla and stamens falling as a unit. Style 1, persistent, the stigma capitate, with 2 rounded lobes (often appearing as 1). Ovary base surrounded by a yellow, ring-shaped nectary. Fruit a capsule with up to 4 seeds. Seeds brown, 9.0–13.5 mm, with dense, silky hairs 12–14 mm on the dorsal margins. Flowering November–March (April) with peak flowering in December and January; fruits ripen May–June.

Slopes, ridges, valleys, and arroyos; Sonoran desertscrub, thornscrub, tropical deciduous forest, and lower elevations in oak woodland (rarely at the lower edge of pine-oak woodland in tropical areas of southeastern Sonora); north-central and northeastern to southern Sonora, near sea level–1220 m (4000 ft). Sonora and southwestern Chihuahua to Oaxaca.

Figure 115. *Ipomoea arborescens*. Near Alamos. Photograph by K. S. Bunnell.

> A spectacular tree 20 to 30 feet high, with smooth white-gray bark like the hide of a hippopotamus. It flowers in winter when leafless, holding a high, thin spread of white corollas like stars against the morning sky. These stars soon fall upon the ground, where the deer eat them. With the summer rains the tree forms a dense foliage, which on the characteristically recurved branches is somewhat plumelike, especially from a distance (Gentry 1942: 213).

The flowers are visited by hummingbirds including the plain-capped starthroat in the vicinity of Alamos and the broad-billed in lowland, desert regions, as well as by hawkmoths (*Hyles lineata* and others). The buds attract both red and black large ants. Farther south this and other species of tree morning glory are commonly bat pollinated (Butanda-Cervera et al. 1978); however, in Sonora hawkmoths are the primary pollinators, and bats have not been seen visiting the flowers (Alberto Búrquez, personal communication 1998). The floral biology is variable and is largely temperature dependent. The flowers commonly open in the late afternoon. During warmer weather the corollas fall the next morning with warming daytime temperatures, but on cooler days they often do not fall until the second night. The trees are self-incompatible (Alberto Búrquez, personal communication 1998).

Especially during drought, burros and other livestock in southeastern Sonora chew on the bark and sometimes proceed to eat the entire trunk. Many trees have trunks scarred from such damage. During drought, the larger branches are chopped up for cattle feed. The young trees have thick roots that were used as food by the Yaqui.

Figure 116. *Ipomoea arborescens*. Drawn by M. B. Johnson.

These trees can grow rapidly (often > 2 m per year) during the hot summer if provided with adequate soil moisture. In cultivation and with ample watering, as well as in certain riparian tropical canyons, leaves may be retained at least through January, when they finally succumb to cold weather. Even under these conditions new leaves do not appear until hot weather in late spring or early summer. In cultivation in Tucson, freeze damage occurs at –4°C (25°F), but recovery can be rapid. The twigs twine if they come in contact with a suitable surface such as a chain-link fence, betraying their relationship to the vining morning glories. Palo santo is readily grown from seed and is a spectacular landscape tree for large areas. The plants do not do well in containers because they soon become pot-bound and then cease active growth. This unusual tree deserves wider cultivation and is already grown as an ornamental around Nairobi and at lower elevations in Kenya (Noad and Birnie 1989).

*Ipomoea arborescens* (Kunth) G. Don is the only truly tree-sized member of the genus in Sonora. There are two taxonomic entities that have been treated as different varieties, or synonyms. However, there is confusion in the

Figure 117. *Ipomoea arborescens*. North of Hermosillo. Photograph by M. A. Dimmitt.

nomenclature stemming from an error in author citation. Gentry (1942: 213) called the lowland Río Mayo foothill populations in southeastern Sonora *I. arborescens* var. *glabrata* (A. Gray) Gentry, based on "*Ipomoea murucoides* var. *glabrata* Gray, Contr. Nat. Herb., vol. 1, p. 107, 1891." Gentry mistakenly cited Gray as the author rather than Rose. The 1891 publication by Rose is based on a collection from Alamos, Sonora. The nomenclature is doubly confused because Gray's var. *glabrata* is a synonym of *I. intrapilosa*, a distinctive species that occurs farther south in Mexico.

Gentry (1942: 214) named the higher elevation or montane populations in southeastern Sonora and adjacent southwestern Chihuahua *I. arborescens* var. *pachylutea* Gentry, and distinguished it from his var. *glabrata* in having "yellowish bark, larger and more pubescent leaves, longer and stouter pedicels, larger and more numerous flowers, larger sepals, and generally heavier inflorescences." Other differences include wood that is apparently not as soft, corollas with a more prominently maroon-purple throat, and pale lavender filaments. Variety *pachylutea* occurs at roughly 500–1220 m (1640–4000 ft) in tropical deciduous forest and the lower oak zones. It is possible that this more "tropical" of Gentry's two varieties is not distinguishable from var. *arborescens* of farther south in Mexico.

Trees of the lowland and northern Sonora populations have conspicuously lighter colored (whiter) bark, and the purple coloration of the corolla tube is much reduced or lacking as compared with var. *pachylutea*. Gentry gave the name var. *glabrata* to the "smaller, less pubescent-leaved form characteristic of the species throughout the foothill regions of southern and central Sonora." The trees that extend into the Sonoran Desert north of Hermosillo are probably the same taxon as Gentry's var. *glabrata*.

Grown side by side in Tucson from seed collected in Sonora, plants of the foothills var. *pachylutea* bear leaves nearly twice as large as those of the lowland, thornscrub-desert var. *glabrata* (*sensu* Gentry 1942). In these plantings var. *pachylutea* tends to branch near the base of the plant with large horizontal branches, whereas plants of var. *glabrata* generally do not share these characters and develop a thick trunk at an earlier age. However, in dense forest near Alamos var. *pachylutea* is an upright tree. The differences between the two Sonoran taxa may warrant recognition at the specific rank. Their relationship with the presumably more widespread and southern var. *arborescens* has not been investigated.

## Cornaceae—Dogwood Family

Mostly trees and shrubs. Mainly in temperate North America and Asia; 14 genera, 100 species.

### *Cornus*—Dogwood

Mostly North America and Eurasia; 50 species. Some species have edible fruits. When fresh *Cornus* leaves are gently torn apart, the primary veins remain connected by delicate fibers. Although apparently diagnostic in north temperate regions, this feature is also present in plants of other families in Sonora (see *Wimmeria*).

***Cornus disciflora.*** (figure 118)

Spreading, symmetrical shrubs or trees to 10 m, oval in outline, with pliant branches and dense foliage. Herbage, outer surfaces of calyx and petals, styles, and fruits with minute, appressed hairs. Leaves opposite, 6–13.5 cm, with fine white hairs, and glabrate above; petioles 0.6–2 cm; blades elliptic with prominent veins, much darker above, the tip acute, the margins entire; stipules absent. Flowers about 2 dozen in dense, capitate clusters on peduncles 3.8–5.0 cm, elongating in fruit. Flowers 6 mm, sessile; calyx 3–4 mm with 4 short lobes, elongating in fruit; petals 4, 3.5–4 mm, thick, and white; stamens 4; ovary inferior, the style exserted approximately 1 mm. Fruits small purple drupes 10–12 mm. Flowering January–March.

Streams in moist canyons; oak woodland and pine-oak woodland; southeastern Sonora, 1350–1675 m (4430–5500 ft). In Sonora known only from Sierra Saguaribo along the Chihuahua border. More common and widespread in nearby southwestern Chihuahua; southward to Central America.

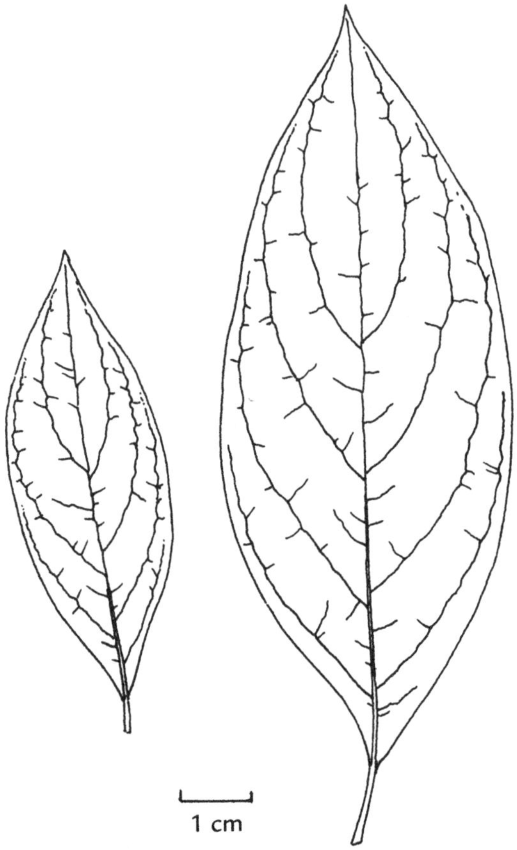


Figure 118. *Cornus disciflora*. Drawn by M. B. Johnson.

## Ebenaceae—Persimmon Family

Trees and shrubs; 2 genera, 485 species.

### *Diospyros*

Evergreen or deciduous trees and shrubs with very hard wood. Tropics and subtropics worldwide, and some in temperate regions; 475 species. Many yield valuable wood and edible fruits including East Indian ebony (*D. ebenum*) and the persimmon of commerce (*D. kaki*) native to Japan and China.

***Diospyros sonorae.*** Sonoran persimmon; *guayparín*; *caguorara* (Mayo) (figure 119)

Trees to 18 m with a trunk to approximately 1 m in diameter and a dense, rounded crown, the bark checkered, dark brown. New growth densely pubescent with small hairs, soon becoming glabrous. Leaves essentially evergreen, alternate, simple, (4.5) 6–14.5 cm, broadly oblong to obovate, firm, glossy green above, dull below, pinnately veined, the midrib and veins raised below, impressed above, the tip rounded to shallowly notched, the margins entire; petioles short and stout, often 1–4 mm; stipules absent.

Dioecious. Calyx and corolla both 5-lobed. Flowers white and green. Male flowers not seen. Female flowers 1–3 in leaf axils, approximately 3 cm wide, the pedicels and outer petal surfaces pubescent with minute hairs, the pedicels 8–12 (20) mm; calyx green, sparsely hairy, the tube very short, the lobes 12–13 × 8–11 mm, oblong to ovate, thick, with rounded tips; corollas urn shaped, approximately 10 mm, white, somewhat fleshy, deciduous, the tube short, the lobes broadly obovate with broadly rounded tips. Ovary superior. Calyx enlarging in fruit, the style persistent; fruits fleshy and indehiscent, globose, 2.5–3.5 cm in diameter, at first yellow and pubescent with short, white hairs, turning blackish, shiny, and glabrous when fully ripe, the pulp soft, blackish brown. Seeds 13.3–15.8 mm, somewhat flattened, several per fruit. Flowering May–June; fruits December–January.

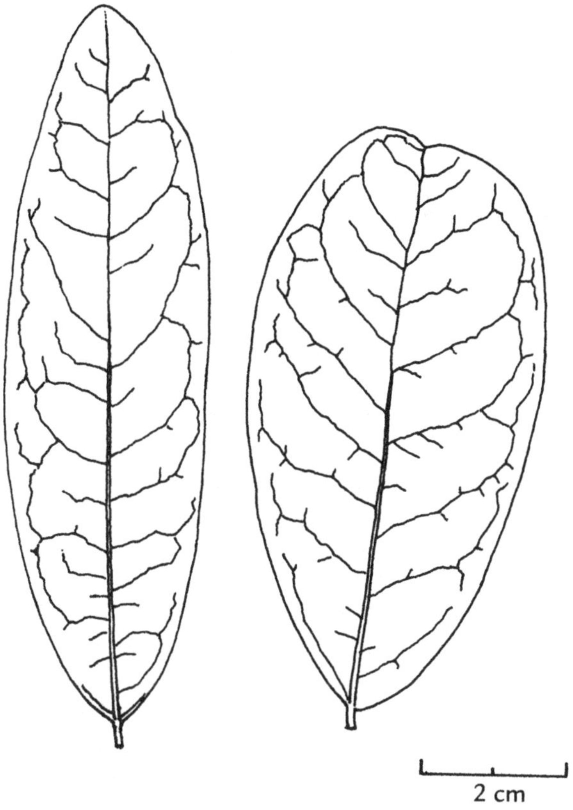


Figure 119. *Diospyros sonorae*. Drawn by M. B. Johnson.

Arroyo and valley bottoms, canyons and coastal plains, mostly localized near habitation, sometimes common in hedgerows; tropical deciduous forest and thornscrub, near sea level–455 m (1500 ft); southern Sonora: Alamos region and southward to Sinaloa, where it is more widespread, farther west at Arroyo Masiaca, and near the coast at Arroyo Camahuiroa and Playa Las Bocas.

The slightly sweet and astringent fruits (the fleshy pericarp) are eaten fresh or cooked. The seeds, parched and ground, are made into atole. The distribution in Sonora suggests naturalization from purposeful introduction. The fruits are relished by frugivorous bats. They carry the fruit to their roosts, and their habit of frequenting porches in Alamos has unpleasant consequences.

## Ericaceae—Heather Family

Mostly shrubs, some perennial herbs and trees, usually in acidic soils, the roots associated with symbiotic mycorrhizal fungi. Worldwide, mainly cool temperate and subtropical regions, and higher elevations in the tropics, absent from deserts; 103 genera, 3350 species.

The family includes manzanita (*Arctostaphylos*), heather (*Calluna*), heath (*Erica*), azalea and rhododendron (*Rhododendron*), and blueberry and cranberry (*Vaccinium*). Several genera of subshrubs and shrubs including manzanita occur in the mountains of north-central and eastern Sonora.

### *Arbutus*—Madrone, *Madroño*

Handsome trees, often with thick bases, the bark peeling or flaking in copper-colored thin sheets or plates. Pubescent and often with some glandular hairs or glabrous, especially with age. Leaves alternate, simple, and evergreen; blades somewhat tough, often with a slightly asymmetric base; petioles and midribs usually reddish; stipules absent.

Inflorescences terminal, usually shorter than the leaves, of one or more racemes or few-branched panicles. Flowers fragrant (at least in *A. arizonica*, often strong with an odor somewhat like chocolate), bisexual, on rather stout pedicels subtended by a bract enclosing 2 bractlets. Calyx small with 5 broad lobes, almost hidden by the ballooning of the corolla base. Corollas creamy white, with a broad bell- or urn-shaped tube, inflated basally and surmounted by 5 small, partially overlapping reflexed lobes. Stamens 10, emerging from beneath a green, 5-lobed nectary disk and enclosed by the baglike corolla tube, the filaments slender above and expanded below into a swollen base covered with slender, white hairs covering the ovary and nectary disk. Anthers bulbous and opening by pores (the pores appear subterminal but like many members of the heath family, the anthers invert during development so that they mature in an upside-down position), each anther with two fleshy, wormlike recurved awns or spurs. Ovary superior, of 5 carpels, glandular-roughened (papillose), the style columnar with a capitate, green stigma slightly protruding from the corolla tube. Fruit a semifleshy, red-orange, rounded, edible berry, the surface granular-warty (papillate); seeds mostly 15–25 (3–5 per chamber or carpel).

Western United States to Central America and the Mediterranean region; 14 species. Reference: Sorenson (1995).

The urn-shaped corollas and fleshy-awned anthers are unique among Sonoran trees. Silk pouches of the communal larvae of *Eucheira socialis*, a pierid butterfly, are prominent in madrones of the Mexican highlands at least as far north as the vicinity of Yécora. The larvae venture out at night from their often large pouches to feed on the leaves, sometimes defoliating the trees. In many areas of Mexico people have made use of the larvae for food and the silk for paper and fabric (Kevan and Bye 1991).

1. Bark of trunks and major limbs rough textured (checkered), brown to pale gray. ____________ ***A. arizonica***

1' Bark of trunks and major limbs smooth, cream colored to bright coppery or reddish. _______ ***A. xalapensis***

***Arbutus arizonica.*** Arizona madrone; *madroño*

Small to medium-sized trees or occasionally 15–20 m. Smaller branches and twigs usually with checkered, flaking, reddish bark; bark of larger limbs and trunks conspicuously checkered; bark on trunks of larger trees becoming like that of oaks. Young herbage minutely and often sparsely pubescent, the older leaves glabrate. Leaves 7.5–12 × 2–3.5 cm; petioles prominent and reddish; blades lanceolate to narrowly elliptic, glossy, with tapered bases, the margins entire to sometimes serrated even on the same tree or branch. Inflorescences 3–7 cm. Calyx lobes 2–2.5 mm with somewhat membranous margins; corollas 5.5–7 mm. Fruits approximately 1 cm in diameter. Flowering May–June.

Slopes, ridges, cliffs, and canyons; oak and pine-oak woodland; north-central and eastern Sonora; 1220–2440 m (4000–8000 ft). Jalisco to southeastern Arizona and southwestern New Mexico.

Mearns (1907) reported madrone trees 20 m tall with enormous trunks on the west side of Sierra San Luis. Similar large trees occur on moist canyon slopes east of Yécora, in nearby Chihuahua and on the Sierra de Alamos. At least in our region, *A. arizonica* differs from *A. xalapensis* by its distinctive growth habit with straighter branches, different bark characteristics, and usually smaller leaves, flowers, and fruits. *Arbutus arizonica* tends to extend into drier, more temperate habitats than does *A. xalapensis*, although they often grow intermixed.

***Arbutus xalapensis*.** Smooth madrone; *madroño blanco* (figure 120)

Small to medium-sized trees to approximately 15 m; new bark smooth and mottled cream white, turning copper colored or reddish with age. Twigs, leaves, and inflorescences with white hairs; or with age the leaf blades glabrate. Leaves 8–12 (16) × 2.3–4.5 (8) cm; petioles 1.5–4 cm; blades elliptic to oval or ovate, dull to glossy above, the base tapered to mostly broad and truncate, the margins entire or occasionally serrulate. Inflorescences 4–7 cm. Calyx lobes 2.5–3.5 mm, with membranous, ciliate margins; corollas 6–8 mm. Fruits 1–1.5 cm in diameter. Flowering March–April.

Slopes, ridges, and canyons; pine-oak woodland. Mountains in eastern Sonora, from the Sierra el Tigre and Sierra Huachinera southward, 1370–2285 m (4500–7500 ft). Also southern New Mexico and western Texas to Guatemala; not in Arizona.

The leaves are generally larger and trunks and limbs often more contorted than those of *A. arizonica*, giving the appearance of an overgrown manzanita.

## Erythroxylaceae—Coca Family

Trees and shrubs with hard wood. Tropics mostly in the New World; 4 genera, 260 species.

### *Erythroxylum*

Pantropical, mostly in the New World; 250 species.

Cocaine is extracted from the leaves of several South American species, especially *E. coca*.

***Erythroxylum mexicanum*.** *Mamoa; momo ogua* (Mayo) (figure 121)

Locally common, densely branched hardwood shrubs or trees to 5+ m, with spreading, stout, straight, stiff branches; glabrous. Leaves ultimately drought deciduous, alternate, simple, often 3–8 cm; blades thin, broadly obovate, entire, uniformly green on both surfaces, rounded to shallowly notched at tip; petioles approximately 2 mm; stipules about as long to shorter than the petioles, scalelike, with fringed margins.

Flowers in small sessile clusters on nodes below the leaves on short twigs; pedicels slender. Flowers white, 5–8 mm, subtended by minute bracts; calyx lobes 5; petals 5 with basal appendages and soon falling; stamens 10, exserted, the filaments united below in single series, the ovary superior. Fruit a 1-seeded drupe approximately 1 cm, ellipsoid, with bright red fleshy pulp; sepals and dried stamens persistent at the base. Flowering May–June; fruits ripen June–September.

Slopes, arroyos, and valleys; tropical deciduous forest; southeastern Sonora, 180–610 m (600–2000 ft). Also Oaxaca to southwestern Chihuahua and northward on the Atlantic drainage to Tamaulipas.

The leaves of this species apparently lack significant narcotic alkaloids. The plants are usually leafless during dry seasons and become leafy with summer rains. In cultivation, however, new leaves may appear as early as March. As with other species in the genus, the seeds retain viability for only a short time. It is an attractive plant worthy of cultivation in hot, near frost-free climates.

## Euphorbiaceae—Spurge Family

Plants of extremely diverse growth habits, from minute herbs to large trees, many with diverse succulent growth forms, especially in dry regions of Africa. Many with milky sap. Leaves of Sonoran tree genera simple, alternate, often with small or quickly deciduous stipules. Flowers unisexual; plants monoecious or dioecious. Various flower parts often reduced or absent; perianth often inconspicuous or absent, the sepals and petals separate or united below, or petals absent. Ovary superior, often 3-chambered; seeds 1 or 2 per chamber. Styles usually 3, simple or branched; stigmas often 3. Fruit a capsule with the segments separating elastically from a persistent column (in Sonora except *Drypetes*). Seeds often with a basal appendage (caruncle).

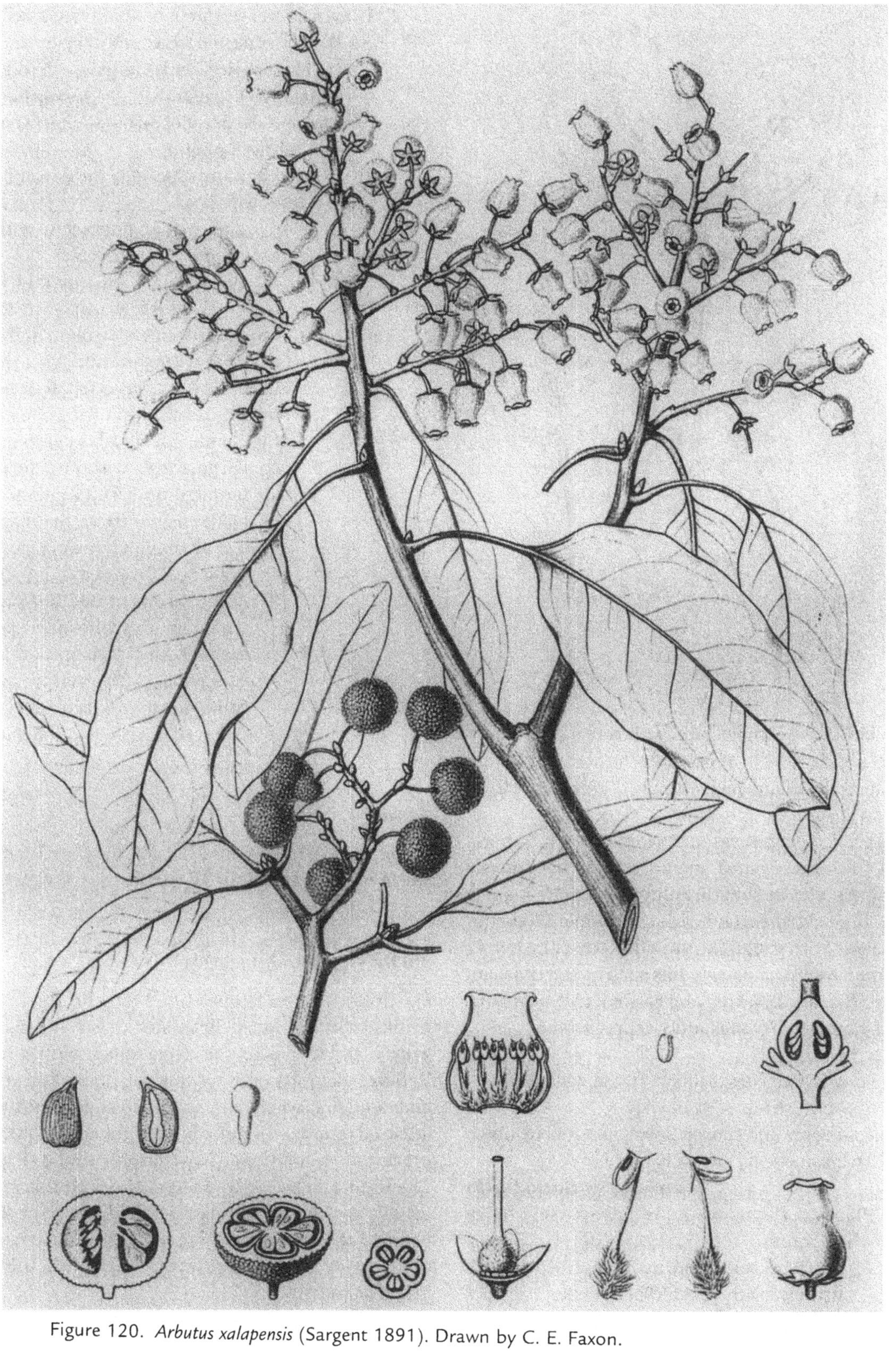

Figure 120. *Arbutus xalapensis* (Sargent 1891). Drawn by C. E. Faxon.

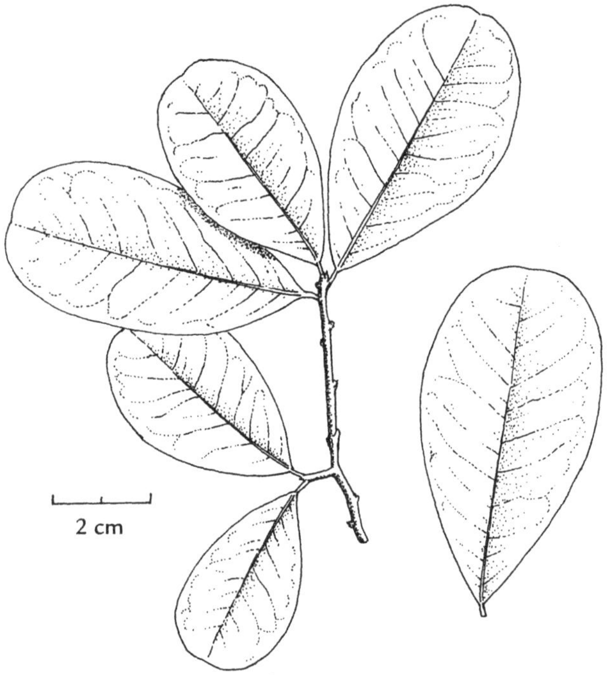


Figure 121. *Erythroxylum mexicanum*. Drawn by M. B. Johnson.

Worldwide, mostly tropical and subtropical; 300 genera, 7500 species, 144 in Sonora. References: McVaugh (1961), Steinmann and Felger (1997), Webster (1975, 1994).

This is the fourth largest vascular plant family in Sonora. *Euphorbia*, the largest genus, includes many herbaceous and shrubby species in Sonora; cultivated species include poinsettia (*E. pulcherrima*) and various ornamental shrubs and succulents with cactuslike growth forms. The family includes numerous economic plants such as the Brazilian rubber tree (*Hevea brasiliensis*) and cassava (*Manihot esculenta*). Many members of the family are poisonous.

1. Trunk beset with spines; fruits > 10 cm wide. *Hura*
1' Trunk not spiny; fruits < 3 cm wide.
  2. Leaves deeply and conspicuously palmately lobed.
    3. Twigs and leaves densely hairy. ***Jatropha malacophylla***
    3' Plants glabrous or the leaves sparsely hairy when young.
      4. Leaf lobes with entire margins, except the tips usually with a hairlike bristle; capsules smooth or at least not prickly. ___ ***Manihot***
      4' Leaf lobes with serrated margins, not bristle tipped; fruits burlike with herbaceous prickles. ___ ***Ricinus***
  2' Leaves entire, serrated, or shallowly lobed.
    5. Bark of trunk and larger limbs papery and exfoliating in sheets; leaf blade margins with conspicuous, usually stalked glands. ___ ***Jatropha cordata***
    5' Bark not papery, not exfoliating; leaf blade margins without glands.
      6. Herbage and especially the flower buds with lepidote scales. ___ ***Croton***
      6' Plants glabrous or pubescent with simple hairs, without lepidote scales.
        7. Leaves 1.5–3.5 cm, mostly broadest above the middle (obovate to oblanceolate), the tips often blunt and rounded; desert and thornscrub. ___ ***Adelia***
        7' Leaves ≥ 5 cm, broadest at or below the middle (lanceolate, ovate, or elliptic), the tips pointed; tropical deciduous forest.
          8. Large trees, without latex; young herbage and fruits pubescent; leaf blades asymmetric, the margins entire or shallowly wavy; fruits ovoid, not lobed, 1-seeded. ___ ***Drypetes***
          8' Large shrubs to medium-sized trees, with copious white latex sap; plants including fruits glabrous; leaf blades symmetrical, the margins serrated; fruits mostly 3-lobed, (2) 3-seeded. ***Sebastiania***

### *Adelia*

Tropical and subtropical regions of the New World with a few in warm deserts; 10–12 species, 5 or 6 in Mexico.

***Adelia virgata.*** *Ona jújugo* (Mayo) (figure 122)

Mostly shrubs, sometimes to 5+ m with thick, multiple trunks; without latex. Branches mostly upright, rigid, woody, and wandlike, the twigs sometimes thorn tipped. Herbage glabrate, or especially the nodes and vegetative buds and fruits with short, simple hairs. Short-shoot leaves fascicled (clustered), quickly drought deciduous, and appearing after rains much like those of ocotillo (*Fouquieria*), 15–38 mm, obovate to oblanceolate, pinnately veined, usually dark green and gland-dotted, entire or shallowly notched at tip; long-shoot leaves larger, to 6 cm, entire or occasionally 3-lobed; stipules absent but with tufts of hairs in leaf fascicles.

Dioecious. Flowers green, small, without petals. Capsules 2- or 3-lobed, 11–12.5 mm wide, two- or three-seeded; seeds rounded, 5–5.5 mm in diameter. Flowering mostly during warmer months.

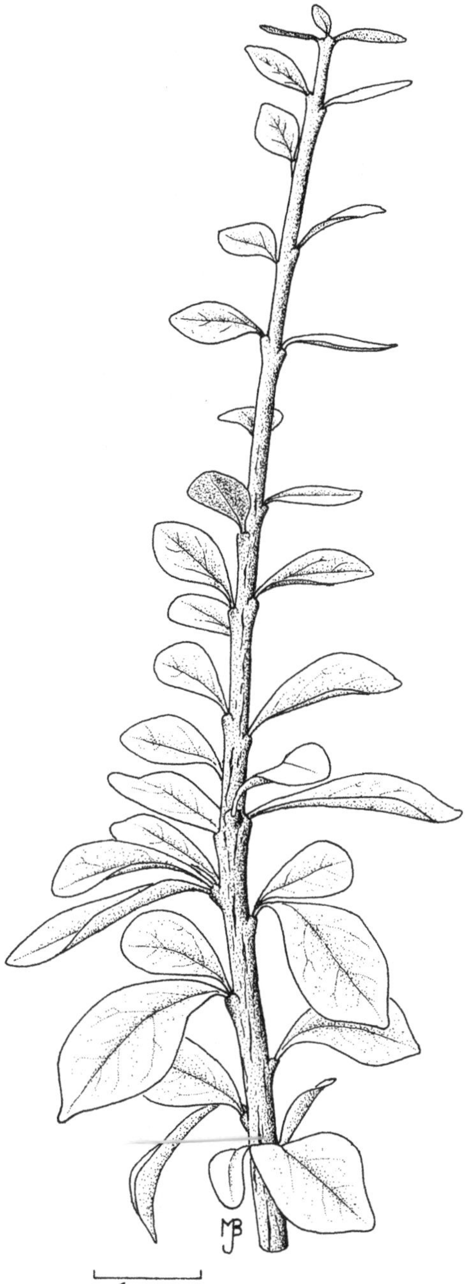


Figure 122. *Adelia virgata*. Drawn by M. B. Johnson.

Valleys, plains, canyons, and arroyos; Sonoran desertscrub at the southern margin of the desert and thornscrub; 10–350 m (30–1150 ft). Vicinity of Bahía San Pedro and San Carlos southward to western Sinaloa and Baja California Sur. Tree-size plants occur in the interior of Sierra el Aguaje north of Guaymas.

*Adelia obovata*, a shrub of northwestern and central Sonora, differs from the closely related and perhaps conspecific *A. virgata* mainly by its more pubescent leaves, and it does not reach tree size. Their taxonomic status remains unresolved (Steinmann and Felger 1997).

### *Croton*

Herbs, shrubs and trees. Warm regions of the world, especially the tropics with centers of diversity in Mexico and Brazil; 800 species, 14 in Sonora.

The herbage is generally toxic, and the plants are not eaten by livestock. Croton oil is obtained from *C. tiglium*. The tropical "croton" grown in gardens in southern Sonora and elsewhere is *Codiaeum variegatum*.

***Croton fantzianus.*** *Vara blanca; sejcó* (Guarijío) *cuta tósari* (Mayo) (figure 123)

Small trees 5–7 m, with slender, straight trunks often 6–10 cm in diameter (on Sierra de Alamos occasionally to about 10 m with the trunk about 20 cm in diameter), the wood hard and durable; sap not milky. Bark slightly rough, mottled white and gray, exfoliating in small, thin flakes, the twigs with prominent lenticels. Sepals (buds) and young herbage densely clothed with silvery, lepidote scales partially falling away from the upper surfaces of older leaves, the scales with short to long rays (hairs), especially the central ray. Stipules 3–9 mm, subulate to linear. Leaves, especially the larger ones drought deciduous, 3.5–20 × 2–10.5 cm; petioles 2–5 cm; blades broadly ovate to oval, often cordate at the base, darker green above, conspicuously palmately veined with 5–9 veins, rounded to obtuse or acute at tips, the margins entire.

Monoecious. Inflorescences of axillary spikelike racemes 8–37 mm with 18–30 crowded flowers, the buds rounded, the lower 1–3 flowers female, the others male; flowers cream colored, subsessile, the pedicels mostly ≤ 0.5 mm in length. Male and female flowers with 5 calyx lobes and 5 petals, the calyx lobes overlapping, shorter than the petals; petals 1.3–2.5 mm, oblong-elliptic with ciliate margins and often a few lepidote scales on the outer surface. Male flowers approximately 5 mm including the 15–16 exserted stamens. Female flowers larger, the ovary 3-chambered, clothed with velvety, yellow-white, long-rayed stellate hairs; styles 3, united at the base, dark purple-brown, spreading, ultimate style branches irregularly divided, thick and with 24 or more slender ultimate divisions. Capsules 6–8 mm, 3-lobed and 3-seeded, rounded, relatively smooth with long-rayed stellate hairs. Seeds 6.5–8.5 mm, shiny brown to gray, with a white caruncle. Flowering and fruiting November–March.

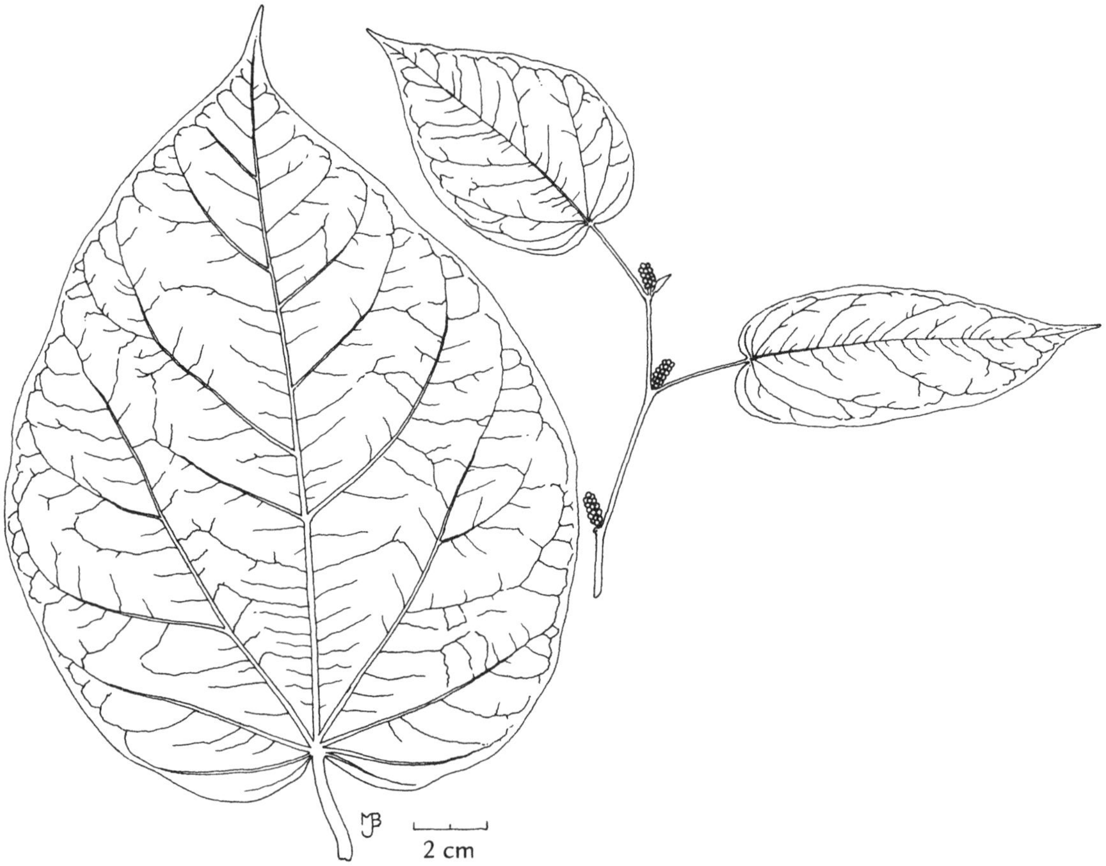


Figure 123. *Croton fantzianus*. Drawn by M. B. Johnson.

Rocky dry hills, valley plains, and canyon slopes in tropical deciduous forest; southeastern Sonora, 300–700 m (985–2300 ft), in the vicinity of Alamos and nearby areas.

During the long dry seasons, variously from early November through May, the plants generally possess well-formed, conspicuously rounded flower buds that remain dormant until favorable moisture conditions occur. This species belongs to section *Eluteria*, which comprises about one dozen New World species, distinctive from other New World crotons by a combination of lepidote scales and well-developed petals in the female flowers (McVaugh 1961, Webster 1993). *C. fantzianus* is the only member of this section in Sonora and is the only truly arborescent *Croton* in the state. This species is described from Nicaragua and occurs sporadically northward along the Pacific lowlands to Guerrero. There is a gap in the known distribution between Guerrero and Sonora and northern Sinaloa.

Large quantities of the slender trunks are cut mostly for tomato stakes and sometimes also for fences. About every week two large trucks leave Alamos fully loaded with thousands of *vara blanca* poles destined for use in agricultural fields in southern Sonora, Sinaloa, and Baja California Sur. When cut back the plants usually recover and develop multiple stems, attaining a shrubby aspect, but the regrown stems are not considered commercially desirable. The trees, once common in the Alamos region, are now scarce. The slender poles are also used for *portal* roofing and local wattle and daub house walls.

### *Drypetes*

Tropical regions worldwide; 200 species, 20 in the New World tropics and 4 in Mexico.

*Drypetes* is unique among the Euphorbiaceae in having mustard oils.

***Drypetes gentryi.*** *Cortopico, palo verde; joyarí* (Guarijío) (figure 124)

Large, hardwood trees attaining heights of 20–30+ m; without latex. Bark smooth and mottled gray, with age

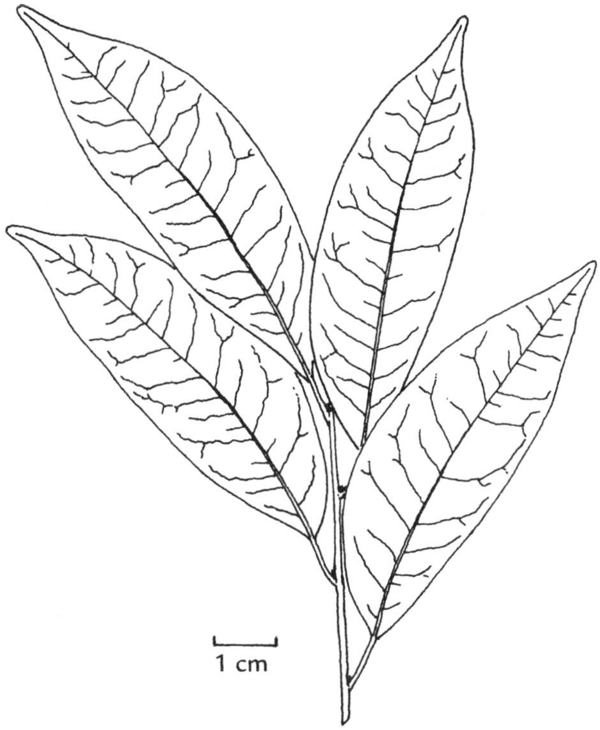


Figure 124. *Drypetes gentryi.* Drawn by M. B. Johnson.

checkered in patches on older branches and fissured on the trunk; larger trunks irregularly fluted. Leaf buds and twigs with short hairs, becoming glabrate. Evergreen with a dense crown; leaves 5–11 cm, somewhat leathery, petioled; blades slightly glaucous, lanceolate to ovate, elliptic or oval, pinnately veined, often asymmetric especially toward the base, the apex acute, the margins entire to shallowly wavy; stipules reddish brown, minute, and scalelike.

Dioecious. Inflorescences of small, axillary clusters; flowers and pedicels densely pubescent; flowers inconspicuous: male flowers white; female flowers 2.5–3.0 mm, sepals green to yellow-green, pubescent, shorter than the ovary, the ovary tawny brown and densely pubescent, the stigmas olive green, fleshy, prominent and caplike, the pedicels 2–4 mm. Fruits of 1-seeded drupes, 11–12 mm, somewhat ovoid and not lobed, gray green, scruffy-velvety with short hairs; seeds without a caruncle. Flowering October–March; fruits March–May.

Riparian tropical deciduous forest; southeastern Sonora, 310–1065 m (1015–3500 ft). In deep, moist canyons such as Arroyo Gochico, Arroyo Verde, at Choquincahui, on Sierra de Alamos, and narrow side canyons of the Río Guajaráy. Often grows with other large tropical, mostly evergreen riparian trees such as *Aphananthe, Ficus* spp., *Sideroxylon tepicense*, and *Trophis*. Also Colima and Jalisco to Sinaloa and tropical canyons in southwestern Chihuahua.

Chachalacas feed on the fruits. This is one of the taller trees in Sonora, and the only large euphorbiaceous tree in the state. These trees somewhat resemble some of the large wild figs; the leaves and fruits are rather similar in size and color to those of *Ficus pertusa. Drypetes* is readily distinguished by its lack of milky latex, often asymmetric leaves, and one-seeded, pubescent fruits.

### *Hura*

Tropical regions of the Americas; two closely related species.

***Hura polyandra*.** Cannon-ball tree; *haba, haba del indio.*

Glabrous trees to 15 m with a spreading canopy, heavy foliage, and milky sap; trunk thickly studded with stout spines. Stipules covering the leaf bud, soon deciduous. Leaves drought deciduous; petioles about as long as the blades; blades often 10–20 cm, relatively thick and somewhat leathery, broadly heart shaped, the margins with gland-tipped teeth, the leaf tip short-acuminate.

Monoecious. Flowers without petals; male flowers sessile and crowded in long-peduncled, thick spikes; female flowers solitary, axillary. Capsules > 10 cm in diameter, shaped like a flattened pumpkin with about 20 carpels arranged like the segments of an orange, and explosively dehiscent (violently flinging the seeds as far as 14 m). Seeds 2.5–3 cm, somewhat disk shaped and resembling a fava bean (*haba*).

Tropical deciduous forest in central Sinaloa to Nicaragua. Occasionally cultivated in southwestern Sonora.

A grove of these trees reported to occur near Navojoa is probably derived from planted trees (Alberto Búrquez, personal communication 1995). The caustic and poisonous sap has been used as a fish poison, and the seeds have been used to poison coyotes and other animals. The wood is soft and easy to carve. It can be propagated by cuttings or by seed.

### *Jatropha*

Shrubs, trees, or sometimes herbaceous perennials or annuals, with clear or colored (but not milky) latex. Stipules often modified into glands. Male and female flowers each with 5 sepals and 5 petals. Male flowers with somewhat bell-shaped corollas with a broad tube and reflexed (curled) lobes; stamens in two whorls, the filaments united below in a column with 5 basal glands. Female flowers larger and on thicker peduncles and pedicels than the male flowers, 1 or 2 at a node, the sepals enlarging in fruit. Mostly tropical and subtropical, many in semiarid to arid regions, largely in the Americas, also Africa to India; 150 species.

Probably most species are toxic, containing toxalbumins, alkaloids and other substances. Several shrubby species are widespread and common in Sonora. *Jatropha cinerea* is a common shrub in western Sonora and occasionally becomes nearly arborescent in southern Sonora. Another shrub species in western Sonora, *J. cuneata*, is used by the Seri to make baskets. Three other species occur in Sonora, two shrubs and one root perennial.

1. Bark peeling in papery sheets during dry season; glabrous. ______ ***J. cordata***

1' Bark not peeling; herbage velvety-pubescent. ***J. malacophylla***

***Jatropha cordata***. Sonoran tree-limberbush; *copalío, torote papelío, miguelito; guajpé* (Guarijío); *to'oro, mo'mo* (Mayo); *vaas* (Pima Bajo); *jay sapo* (Yaqui) (figure 125)

Glabrous trees to approximately 10 (13) m, usually taller than wide with a narrow, upright habit, the crown often relatively small; wood soft and semisucculent, the smaller branches flexible. Trunk well developed, the bark exfoliating in thin, papery sheets, especially during dry seasons; twigs of current season reddish. Sap watery, exuding clear and amber colored from wounds of larger limbs and trunk and drying sulfur yellow. Leaves drought deciduous, present only during the summer rainy season, yellowing and quickly shed in September or early October; leaf blades 3.2–8.5 × 2.4–7.0 cm, heart shaped to ovate, the tips obtuse to moderately acuminate, the margins with conspicuous tack-shaped stalked glands, otherwise nearly entire to more often with small, shallow gland-tipped teeth; petioles 1.4–7.2 cm; stipules represented by stalked glands.

Flowers white to pink. Male flowers in few- to many-flowered cymose inflorescences 3–10 cm; peduncles and pedicels slender; fresh sepals 1.5–2.5 (4) mm, united at base, the lobes overlapping, ciliate and occasionally some with glandular margins; corollas 6–8 mm, the tube rose-pink or sometimes white, the lobes 1 mm, white; stamens 8, basal glands larger than the anthers. Female flowers 1 (2) at a node; pedicels 1.5–3 mm and relatively thick, the flowering sepals 2.5–7.0 mm, the margins with tack-shaped glands like those of the leaves. Fruiting pedicels enlarging to 7–18 mm, and the sepals enlarging to 2 cm. Seeds (1) 2 or 3, round-ovoid, 10–12 × 7–9 mm, mottled or uniformly light to dark brown. Flowering (June) July–August.

Slopes, plains, and valleys; Sonoran desertscrub, thornscrub, tropical deciduous forest, and occasionally lower elevations in oak woodland. Across southern Sonora northward to the Guaymas-San Carlos region, and farther north inland to the vicinity of Cucurpe and Río Bavispe in north-central and northeastern Sonora; near sea level–1000 m (3280 ft). Also southwestern Chihuahua and Sinaloa to the state of México.

*Jatropha cordata* is frost sensitive, and its northern distribution is limited by freezing temperatures. The northernmost populations are multiple-stemmed shrubs showing evidence of repeated freeze damage.

The papery-barked trunks of this tree resemble those of some *Bursera* species, and they can be difficult to distinguish when leafless. The paper-bark *Bursera* species have pungently aromatic sap, but although *J. cordata* has a distinctive odor, it is neither strong nor pungent. The Mayo

Figure 125. *Jatropha cordata*. Leaf of young shoot on right. Drawn by M. B. Johnson.

and Yaqui have used the wood to carve Pascola masks. As with *Bursera*, the soft wood is easy to carve and does not split on drying. It can be propagated from cuttings or grown from seeds. The trunks and larger branches are sometimes used for posts to make living fences. This plant is an important host for the larvae of *Rothschildia cincta*, a large saturniid moth. Yaqui and Mayo men use the cocoons of this moth to make the *teneboim* leg-rattles used in ceremonial dances.

***Jatropha malacophylla*** [*J. platanifolia*]. Sycamore-leaf jatropha; *sangrengado*; *jeyó* (Guarijío); *sa'apo* (Mayo) (figure 126)

Large multiple-stem shrubs or rarely small trees to 5 m with 2 to several thick trunks to 14 cm in diameter and a spreading crown; bark smooth and gray. Cuts to the main stems produce copious, blood-colored sap, the younger stems yield clear sap that in a day or so turns red-brown. Stems and twigs rather thick, semisucculent, not very flexible. Herbage densely to sparsely velvety-pubescent. Leaves emerging with first summer rains and falling quickly at the end of the rainy season. Leaves 12–20 × 10–15 cm; petioles 4–9 cm; blades pale green and sycamorelike, palmately lobed with 3 larger and 2 smaller lobes and few smaller, coarse teeth; minute, deciduous, red-brown stalked glands in the position of, and apparently representing, the stipules.

Inflorescences of short, thick-stalked fascicles of cymes 1–2 cm including the flowers, each cyme with 1 or 2 female flowers and several male flowers. Male flowers 8.5–9.5 × 7–8 mm, greenish white to white; sepals separate, mi-

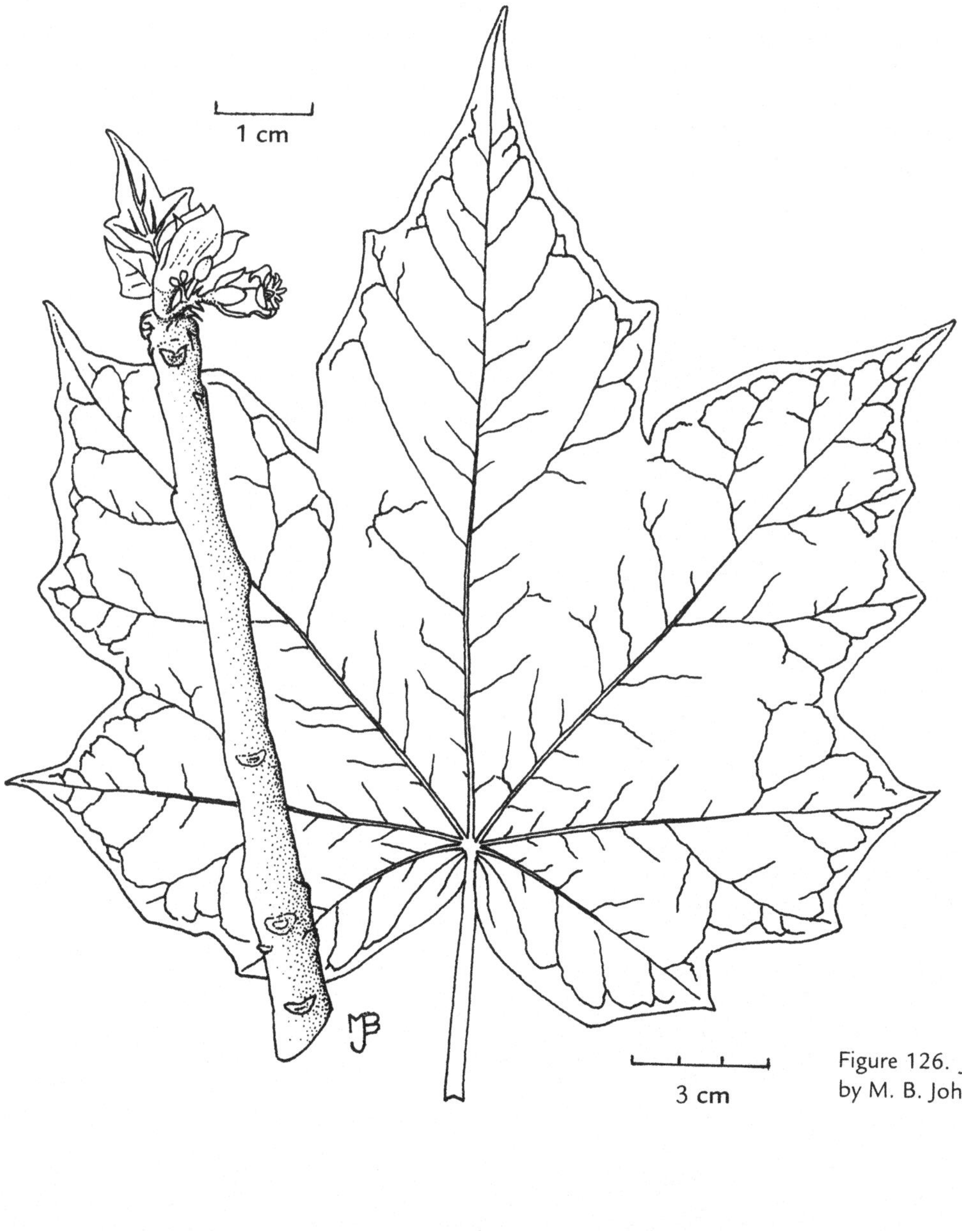


Figure 126. *Jatropha malacophylla*. Drawn by M. B. Johnson.

nutely ciliate, otherwise glabrous or sparsely pubescent, shorter than the petals, subequal, not overlapping, entire or with few shallow teeth (or occasionally longer than the petals and unequal like the female sepals); corolla lobes 4.5 mm, reflexing to expose the stamens; stamens 10, with pale yellow conical glands at the base of the staminal column; inner surface of petals and the "top" (inner side) of anthers with a tuft of white, hispid hairs overtopping the anthers and guarding the entrance to the flower. Female flowers approximately 20 mm across; sepals green, unequal, broadly ovate, the margins with few shallow teeth to nearly entire, enlarging to 2–2.5 cm and persistent on fruits; petals white, with spreading white hairs inside; style and stigmas green; styles 3, each with a 2-branched stigma. Pedicels of male flowers slender, 3–6 mm; pedicels of female flowers 1–3 mm, thick, elongating to 3.5 cm on fruits. Capsules 2 cm, 3-seeded. Seeds 10.5–13.5 mm, ovoid, blackish with a thin, whitish caruncle. Flowering May–June with emerging leaves and also later in summer; fruits July–September.

Widespread and common, slopes and canyons, often in dry, rocky habitats; tropical deciduous forest; east-central and southeastern Sonora, 260–1000 m (855–3280 ft). Also Oaxaca to Sinaloa and southwestern Chihuahua. In Sonora mostly in the Municipio de Alamos; the northern record near Santa Ana in the Municipio de Yécora. In cultivation it is highly frost sensitive.

### *Manihot*

Herbaceous perennials to small trees with soft wood, usually (in Sonora) with milky sap. Leaves large, with long petioles, the blades palmately deeply parted; stipules linear-threadlike, quickly deciduous. Monoecious. Inflorescences of terminal racemes or panicles; female flowers below the male flowers. Sepals petal-like, 5, united below in male flowers and nearly separate in female flowers; petals absent. Capsules 3-seeded. Seeds mottled gray and brown, with a caruncle.

Tropical and subtropical regions of the New World; 100 species, 20 in Mexico and 5 in Sonora. Reference: Rogers and Appan (1973).

This genus includes *M. esculenta*, cassava or *yuca*, one of the most important tropical root crops. The fresh roots of some, or perhaps all, species contain cyanogenic poisons.

1. Bark gray, not papery; primary leaf lobes often with secondary lobes, the apex acute. ______ ***M. aesculifolia***

1' Bark bronze colored, peeling in thin, papery sheets; primary leaf lobes without secondary lobes, the apex broadly obtuse to truncate or emarginate. _____ ***M.*** sp.

### ***Manihot aesculifolia.***

Glabrous shrubs and rarely small trees 5+ m with a well-developed trunk. Bark gray-brown, not exfoliating. Petioles 8–17 cm; leaf blades about as broad as long, 15–40 cm wide, bicolored, moderately glaucous below, lobes (5) 7 or 9, obovate, often with secondary lobes, the tips acute and sometimes with a short bristle. Leafy and flowering only during summer rainy season.

Inflorescences of diffuse panicles or racemes 12–25 cm. Calyx of male flowers approximately 1 cm, whitish and glaucous outside, moderately yellow inside. Capsules approximately 15 mm, rounded, the surfaces rough and glaucous. Seeds 8.5–12 mm, the margins moderately angled.

Mostly canyons and rocky slopes, tropical deciduous forest; southeastern Sonora, 200–450 m (655–1475 ft). Southward to Panama.

### ***Manihot*** sp. (figure 127)

Shrubs or small trees to 7 m, with thick, burseralike multiple trunks and limbs, or sometimes single trunked. Bark exfoliating in large, thin, coppery red, papery sheets. Stems relatively thick; the first-season twigs often to 1.8 cm in diameter. Glabrous or sparsely pubescent with short hairs. Petioles 7.5–37.5 cm; leaf blades bicolored, moderately glaucous below, 10.5–25 × 14.5–30 cm; lobes (7) 9, obovate, the larger (middle) lobe 10–20 × 4–8 cm, the tips obtuse to sometimes rounded or notched with a slender, hairlike bristle 5–18 mm. Leafy only during summer rainy season.

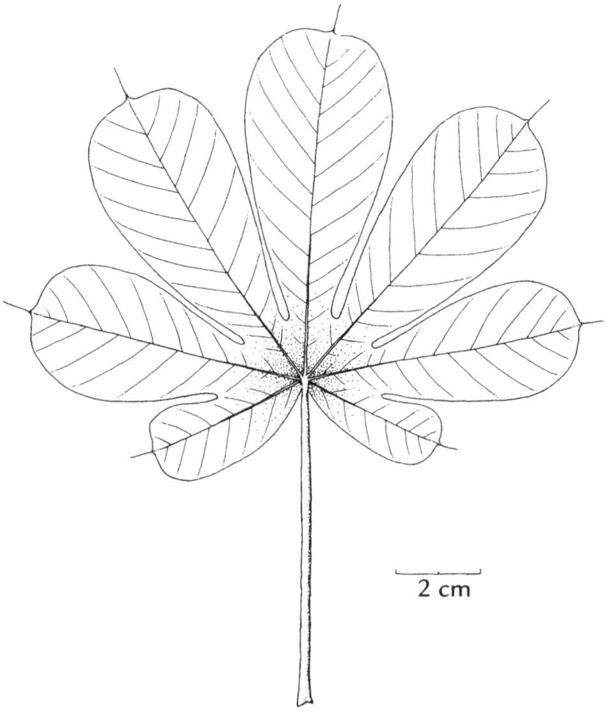


Figure 127. *Manihot* sp. Drawn by M. B. Johnson.

Flowers yellow-green, in panicles. Capsules mostly 18–20 mm, rounded, the surfaces rough and glaucous. Seeds 12–13.5 mm, the margins moderately angled. Flowering July–September.

Canyons and slopes; tropical deciduous forest; east-central and southeastern Sonora, 200–760 m (655–2500 ft). Especially common on some mountainsides and canyon slopes, such as steep slopes and riverbanks along or near the Río Mayo between Chorijoa and Mesa Colorado and northward to the vicinity of El Novillo. Also in Sinaloa.

The identity of this unusual manihot is not resolved. It does not seem to fit into any of the six known species of the truly arborescent manihots in the section *Foetidae*, all from Mexico.

### *Ricinus*

The genus has 1 species.

***Ricinus communis.*** Castorbean; *higuerilla*; *yorimaqui* (Guarijío); *kevenia* (Yaqui) (figure 128)

Shrubs or sometimes arborescent to 5–6 m, annuals or living several years; glabrous; extrafloral nectaries often present at the nodes. Stems hollow, becoming solid and woody below; bark smooth; sap clear and watery. Stipules large, united into a sheath and covering the buds. Petioles 12–30 cm; leaf blades often 20–35+ cm and about as wide, peltate, palmately lobed, the lobes large, pointed, and with gland-tipped, toothed margins.

Inflorescence a dense terminal panicle; male flowers below. Petals absent; female flowers small, with red stigmas; each male flower with repeatedly branched filaments bearing as many as 1000 anthers. Capsules 2–3 cm, burlike with firm but herbaceous prickles, explosively dehiscent. Seeds 3 per capsule, 1.5–2.5 cm, mottled, oblong, with a conspicuous caruncle. Flowering in warmer months.

Widely planted or semicultivated and sometimes weedy or naturalized along streams, canyons, and arroyos or washes and disturbed habitats including field margins, canal banks, and roadsides; Sonoran desertscrub, thornscrub, tropical deciduous forest, and oak woodland. Nearly statewide, near sea level–1065 m (3500 ft). Native to the Old World and naturalized in warm regions worldwide.

The entire plant is poisonous; 1–3 seeds can be fatal to a child, and 2–8 to an adult; death from ingestion may occur in 1–12 days. The seeds are the source of castor oil. Many cultivars have been developed, including dwarfed varieties selected for indehiscent capsules that facilitate commercial production. A red-leaved cultivar grown as an ornamental is occasionally found along roadsides and in washes.

### *Sebastiania*

Glabrous woody shrubs and trees with milky latex sap. Monoecious. Inflorescences terminal, slender, and spike-like. Petals absent; male flowers small, many, the stamens yellow; female flowers green, one to several, larger than and below the male flowers.

Mostly in tropical regions of the Americas, the majority in Brazil; 100 species.

*Sebastiania bilocularis* (*Sapium biloculare*), *hierba de la flecha*, is a common, many-stemmed Sonoran Desert shrub. It is widespread in the lowlands of Sonora. In the more tropical regions bordering the southeastern margin of the desert it occasionally becomes 5 m. Secretions of the inflorescence of *S. bilocularis*, at least, fluoresce strongly under black light.

1. Petioles without glands. ____________ ***S. pavoniana***

1' Petioles bearing conspicuous glands near the blade.

  2. Leaves < 1.5 (2.5) cm wide; petioles < 1 cm; leaf margins often with glands; ovary mostly 2-chambered with 2 styles. ____________ ***S. bilocularis*** (not treated in text)

  2' Leaves usually > 3 cm wide; petioles ≥ 2 cm; leaf margins without glands; ovary 3-chambered with 3 styles. ____________ ***S. cornuta***

***Sebastiania cornuta.*** *Hierba de la flecha*; *jagüéy* (Guarijío) (figure 129)

Large shrubs or sometimes small trees to 8 m, the branches erect-ascending; bark smooth, gray. Leaves evergreen to tardily drought deciduous; stipules moderately persistent, 1.5–3.5 mm, membranous, broadly triangular to ovate, with ragged margins; petioles (1) 2–3.5 cm, with 1 or 2 conspicuous pairs of glands just below the blade; blades (5) 9–17 × (1.5) 3.5–5.5 cm, lanceolate to ovate or nearly elliptic, shiny green, thickish, the margins minutely serrated, the leaf tips mostly acuminate.

Inflorescences (4.5) 6–9 cm. Styles 3; ovaries and capsules 3-lobed and presumably 3-seeded. Flowering March and July–September.

Scattered populations along riverbanks, canyon bottoms, and slopes; tropical deciduous forest; southeastern Sonora, 225–600 m (740–1970 ft). Mostly in the Río Mayo region from near Guajaray and Mesa Colorado eastward. Also southwestern Chihuahua and southward to Nayarit.

The bark has been used to poison fish, and the white sap was used to poison arrows. The Sonoran plants have been called *Sapium appendiculatum* (now treated as

Figure 128. *Ricinus communis* (Schmutz et al. 1968). Drawn by L. B. Hamilton.

*Sebastiania appendiculata*), a species from farther south in Mexico.

**Sebastiania pavoniana** [*S. palmeri; S. pringlei*].
Mexican jumping bean; *brincador, hierba de la flecha; jogüé* (Guarijío); *túbucti* (Mayo) (figure 130)

Large shrubs to slender, erect trees to 12 m; trunks to 35 cm in diameter; bark smooth, grayish; branches and twigs slender. Leaves tardily drought deciduous, (6) 8–17 × 1.5–5.5 cm; petioles 0.7–1.8 cm; blades elliptic to ovate-elliptic, shiny green, rather thin, the margins minutely serrated; stipules 0.5–0.8 mm, brown, membranous, soon deciduous. Inflorescences 4–6 cm. Styles 3; capsules 3-lobed, 3-seeded. Flowering June–August.

Slopes, arroyos, and canyon bottoms; tropical deciduous forest; southeastern and east-central Sonora, 275–925 m (900–925 ft). The northernmost populations are in the

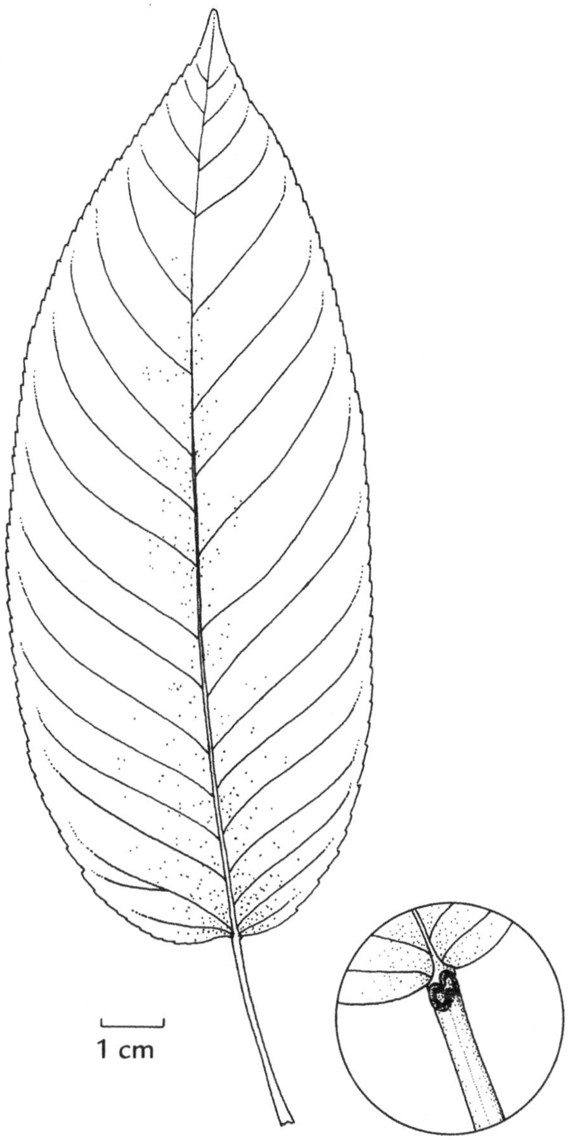


Figure 129. *Sebastiania cornuta*. Drawn by M. B. Johnson.

vicinity of San Javier and Tepoca. San Luis Potosí and Jalisco to Sonora and Baja California Sur.

The pulverized bark was used for poisoning fish. Infestations of the larvae of a tortricid moth, *Cydia deshaisiana*, cause the seeds to "jump." Alamos has been a center of the jumping bean business. The short-lived jumping beans, which are still sold, were widely marketed as novelties that fascinated children in pre-electronic times. The seeds of related trees in other parts of the world are host to moth larvae and become jumping beans. *Sebastiania pringlei* and *S. palmeri* are host to jumping bean moths elsewhere in the American tropics (Webster 1967). *Spirostachys africana* and *Sapium ellipticum* [*Shirakia elliptica*] provide the African versions of these motile seeds (Palgrave 1977).

## Fabaceae (Leguminosae)—Legume Family

Plants of highly varied growth habits: annual or perennial herbs, shrubs, trees, or vines, many with nitrogen-fixing bacteria in root nodules. Unarmed or armed with thorns, prickles, or stipular spines. (Species characterized as being armed may also bear some unarmed stems, or some individual plants in a population may be unarmed.) Leaves and branches usually alternate. Stipules usually present, often well developed and persistent, sometimes spinescent. Leaves mostly once or twice pinnately compound, often with pulvini at the base of petioles, pinnae, and individual leaflets. (The pulvini are swollen areas, often dark or differently colored, that swell or contract to open or close the leaf parts, often in response to water stress and diurnal cycles. In some cases pulvini are present but not functional). Extra-floral nectary glands present on leafstalks (petiole and rachis) of most mimosoids but absent from the other two subfamilies among Sonoran tree species. Leaflets with pinnate venation, or palmate in *Bauhinia*, the margins entire or the apex sometimes notched or with a projection.

Flowers usually bisexual. Calyx usually 5-lobed. Petals mostly 5, often separate. Stamens 5 or 10 to numerous. Pistil single, superior, simple (one carpel), the ovules attached along one side. Fruit usually a pod, very diverse, including the following characteristics or combinations in our region:

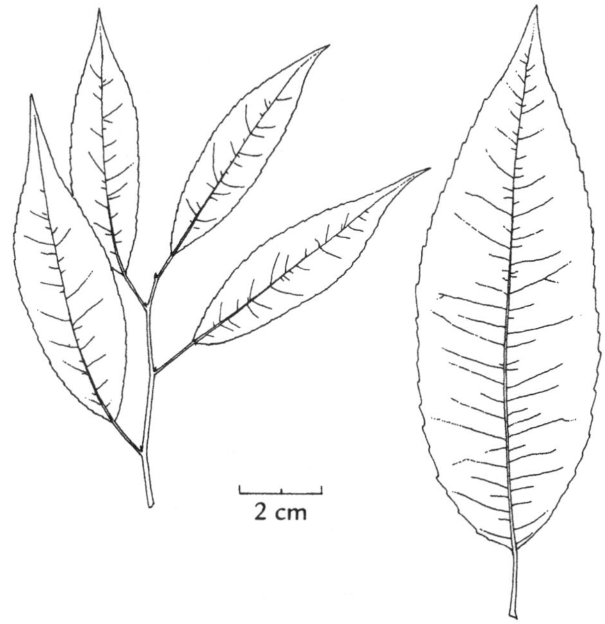


Figure 130. *Sebastiania pavoniana*. Drawn by M. B. Johnson.

Indehiscent:

- Mesocarp or pulp mealy and carbohydrate-rich, attracting vertebrate herbivores that act as dispersal agents (e.g., *Prosopis*).
- Mesocarp none, the pods dry (e.g., *Eysenhardtia*).
- Pods breaking into thin, one-seeded, wind-dispersed segments (e.g., *Mimosa palmeri*).

Dehiscent:

- Pods with fleshy arils, the valves partially separating but not elastically (e.g., *Pithecellobium*).
- Pods dry, the valves separating partially to fully along both sutures but not elastically and not coiling, often gradually or tardily dehiscent (e.g., *Leucaena, Lysiloma*).
- Pods dry, explosively dehiscent upon drying, flinging the seeds away from the parent plant, the valves suddenly and elastically coiling (e.g., *Caesalpinia palmeri*), or pods sometimes gradually and loosely coiling and seeds gradually falling (e.g., *Coursetia*).

Seeds 1 to many, often very long lived; seed coat often hard and impervious to water and air (scarification or nicking the seed coat thus enhancing germination); often preyed upon by diverse bruchid beetles.

Worldwide: 650 genera, 16,500 species, at least 350 species in Sonora. The largest and most widespread family of trees in Sonora, with 51 tree species recorded in the state. Three distinctive subfamilies sometimes treated as separate families:

**Caesalpinioideae**—Senna or caesalpinioid subfamily. Mostly trees and shrubs, sometimes vines or herbs; seldom nitrogen fixing. Leaves once or twice pinnate, rarely reduced to one leaflet. Flowers often large and showy, bilaterally symmetrical but often not conspicuously so, one petal unlike the others. Stamens usually ≤ 10, mostly separate. Generally tropical and subtropical and insect pollinated. *Bauhinia, Caesalpinia, Conzattia, Haematoxylum, Parkinsonia, Senna.*

**Mimosoideae**—Mimosoid subfamily. Largely trees and shrubs, mostly nitrogen fixing. Leaves twice pinnate in Sonora; leafstalks mostly with conspicuous, usually crateriform or cup-shaped nectary glands. These extrafloral nectaries variable and often diagnostic; usually raised and green when fresh, the center often convex and viscid with exudate and later becoming sunken. The nectar is presumed to attract ants that attack herbivores, but the relationship is not well established and perhaps other functions are involved (see *Leucaena*).

Flowers radial except the pistil, mostly small, sessile or nearly so, and close together in spikes, heads, or racemes. Stamens 10 to numerous, often much longer than the petals and forming the showy part of the flower, the anthers small and often with a deciduous gland at the tip. Seeds often with a horseshoe- or U-shaped groove on each side or face. Mostly tropical and subtropical. The flowers, with their exserted stamens and style, are often pollinated by unspecialized pollen and nectar feeders that crawl over the surface of the inflorescence, or by fluttering or hovering insects or birds. *Acacia, Albizia, Chloroleucon, Havardia, Leucaena, Lysiloma, Mimosa, Pithecellobium, Prosopis.*

**Papilionoideae**—Pea or papilionoid subfamily. Annual to perennial herbs, vines, shrubs, or trees, mostly nitrogen fixing. Leaves pinnately or digitately compound or rarely reduced to one leaflet. Flowers bilaterally symmetrical (pea-like). Upper petal (banner or standard) flanked by two lateral petals (wings), the two lower petals coalesced into a boat-shaped structure (keel) enclosing the stamens and style. Stamens mostly 10, with 9 filaments united into an open sheath surrounding the ovary and the 10th filament partially or wholly separate from the others. Tropical to temperate. Often pollinated by bees. *Brongniartia, Coursetia, Diphysa, Erythrina, Eysenhardtia, Lonchocarpus, Olneya, Piscidia, Platymiscium, Psorothamnus, Robinia.*

Tree legumes form the major perennial biomass of the desert and subtropical regions of Sonora and many other dry regions of Mexico. Many are nitrogen fixing, probably providing the major source of nitrogen for desert-subtropical ecosystems. The family includes numerous economic plants including beans, soybeans, peas, peanuts, alfalfa, and clover. Legume trees have featured prominently in the economies of the people of Sonora, from ancient to present times [e.g., ironwood (*Olneya*), mesquite (*Prosopis*), and *guamúchil* (*Pithecellobium*)].

References: Barneby (1989), Isely (1998), Johnson (1988), McVaugh (1987), Polhill and Raven (1981), Turner and Busman (1984).

1. Palo verde trees, the bark of branches and most limbs conspicuously green for several years; flowers caesalpinioid, yellow. ______ ***Parkinsonia***

1' Not palo verde trees, the bark not green or usually green only for 1 or 2 years; flowers various.

  2. Trees leafless or the leaves simple or appearing simple.

    3. Leaves deeply bilobed, > 4 or 5 cm wide. ______ ***Bauhinia***

    3' Leaves not bilobed, < 1.5 cm wide.

      4. Trees wispy, unarmed, with peeling, papery white bark; western and central Sonora south of Caborca. ***Acacia willardiana***

4' Trees densely branched with spinescent twigs, the bark not peeling, not papery white; in Sonora northwest of Puerto Peñasco. ________ ***Psorothamnus***

2' Leaves compound (once or twice pinnate).

5. Leaves with 3 leaflets.

6. Twigs and petioles with short, thick spines; leaves alternate. ________ ***Erythrina***

6' Unarmed; leaves opposite or 3 per node. ________ ***Platymiscium***

5' Leaves with more than 3 leaflets.

7. Leaves once pinnate.

8. Plants entirely unarmed.

9. Leaflets 3–10 cm.

10. Leaflets apiculate (with a short projection at the tip); pods dehiscent and the valves coiling.

11. Stipules 1–3 cm, green and leafy, crescent-shaped, soon deciduous; pods not glandular, not constricted between the seeds. ________ ***Brongniartia***

11' Stipules to 0.5 cm, slender and straight (not leafy), brown, persistent; pods glandular, constricted between the seeds. ________ ***Coursetia***

10' Leaflets not apiculate, the tips rounded, blunt, or notched; pods indehiscent or tardily dehiscent, the valves not coiling.

12. Leaflets appearing glabrous or glabrate (hairs minute and inconspicuous); leaflets 2–4 cm, the lateral veins inconspicuous; flowers lavender or purple; pods tardily dehiscent, flat, and thin. ________ ***Lonchocarpus***

12' Leaves obviously pubescent; leaflets 3–12 cm, the lateral veins conspicuous; flowers white and pink or yellow; pods indehiscent, winged or moderately compressed, and thick.

13. Leaves odd-pinnate, silvery; flowers white, pink, and reddish; pods 3.5–9 cm, 4-winged. ***Piscidia***

13' Leaves even pinnate, green; flowers yellow; pods > 20 cm, not winged. ________ ***Senna***

9' Leaflets < 3 cm.

14. Pods papery and inflated (bladderlike), or the trunk and stems corky winged. ________ ***Diphysa***

14' Pods not papery or inflated, or the stems not corky winged.

15. Pods 9–11 mm, single seeded; leaves aromatic and glandular. ________ ***Eysenhardtia***

15' Pods > 15 mm, usually multiple seeded; leaves not aromatic, not glandular.

16. Leaflets apiculate (with a short projection at the tip); flowers white with yellow and pink; pods to 6 cm, glandular, the valves coiling. ________ ***Coursetia***

16' Leaflets not apiculate; flowers lavender to purple; pods (6) 8–18 cm, not glandular, the valves not coiling. ________ ***Lonchocarpus***

8' At least some branches armed with spines, thorns, or spinescent twigs.

17. Trunk and stems uniquely corky winged. ________ ***Diphysa suberosa***

17' Trunk and stems not corky winged.

18. Trunk and limbs fluted–sculptured; bark smooth; leaves even-pinnate; leaflets 4–8, broadest above middle, notched at tip, with conspicuous lateral veins; flowers caesalpinioid, yellow. ***Haematoxylum***

18' Trunk and limbs not fluted, rounded in cross section, the bark rough on the trunk and older limbs; leaves odd-pinnate (or sometimes even-pinnate in *Olneya*); leaflets 6–19, mostly broadest at or below the middle, usually not notched at tip, the lateral veins inconspicuous; flowers papilionoid, pink or lavender.

19. Leaflets < 2 cm; desert and thornscrub. ________ ***Olneya***

19' Leaflets 2.5–6 cm; montane, oak and conifer zones. ________ ***Robinia***

7' Leaves twice pinnate (the pinnae even pinnate).

20. Leaves with 1 or 2 pairs of pinnae.

21. Leafstalks without nectary glands; flowers caesalpinioid. ________ ***Caesalpinia***

21' Leafstalks bearing one or more nectary glands (*A. farnesiana* often without glands); flowers mimosoid.

22. Each pinna with 2–6 leaflets.

23. Each pinna with 4 or 6 leaflets. ________ ***Acacia pringlei***

23' Each pinna with 2 leaflets. ________ ***Pithecellobium***

22' Each pinna with more than 6 leaflets.
24. Inflorescences spicate (more than twice as long as wide).
25. Stamens numerous; pods conspicuously flattened, dehiscent (sometimes tardily so), without mesocarp (not pulpy). __________ ***Acacia* (*A. greggii, A. willardiana*)**
25' Stamens 10; pods terete to moderately flattened or strongly coiled, indehiscent, with at least some mesocarp (pulp). __________ ***Prosopis***
24' Inflorescences capitate (about as wide as long).
26. Trees unarmed; bark on trunks and limbs mostly whitish to light gray and smooth; rim or margin of pods separating from valves; in Sonora only at Bahía San Pedro and Cañón las Barajitas. __________ ***Lysiloma candidum***
26' Trees armed with spines; bark on trunks and limbs not whitish and entirely smooth; rim or margin of pods not separating from valves; widespread.
27. Spines straight; pods indehiscent, with abundant mesocarp. __________ ***Acacia farnesiana***
27' Spines curved; pods tardily dehiscent, without mesocarp.
28. Bark rough and shredding with age; spines (prickles) internodal; pods often curled or curved, usually moderately constricted between some or all seeds. __________ ***Acacia occidentalis***
28' Bark smooth with transverse ridges or spines; spines at nodes; pods straight, mostly not constricted between seeds. __________ ***Havardia sonorae***
20' Leaves usually with 3 or more pairs of pinnae.
29. Leafstalks with one or more conspicuous nectary glands.
30. Stamens 10, separate. __________ ***Leucaena***
30' Stamens numerous, > 10, separate or united.
31. Stamens separate. __________ ***Acacia***
31' Stamens united.
32. Stipules leafy (often soon deciduous); valves of pods separating from a conspicuous rim. ***Lysiloma***
32' Stipules not leafy; valves of pods not separating from the rim.
33. Unarmed; leaves 10–27 cm, the bark smooth. __________ ***Albizia***
33' Mostly armed with persistent spines or thorns on at least some branches; leaves 1.5–14.5 cm (if > 10 cm, then the bark mottled or peeling in irregular plates).
34. Bark peeling or flaking in plates or strips, conspicuously mottled; armed with straight thorns to 25 mm; leaflets 5–50 mm; pods curved, curled, or twisted. ***Chloroleucon***
34' Bark uniformly gray, not peeling or flaking; armed with catclaw-like curved spines to 3 mm; leaflets 2–7 mm; pods straight. __________ ***Havardia***
29' Leafstalks without nectary glands.
35. Flowers caesalpinioid.
36. Leaves with 2–4 pairs of pinnae. __________ ***Caesalpinia***
36' Leaves with 10–14 pairs of pinnae. __________ ***Conzattia***
35' Flowers mimosoid.
37. Spines at nodes (stipular), terete or dorso-ventrally flattened; flowers yellow, in globose heads. ***Acacia* (*A. farnesiana, A. pennatula*)**
37' Spines (prickles) internodal, laterally compressed; flowers pink-purple, in cylindrical spikes. ***Mimosa***

### *Acacia*

Mostly trees and shrubs, often armed with spines; paired stipular spines, or prickles scattered on the stem between nodes or unarmed. Leaves twice pinnate; leaflets often small and numerous; leafstalks often with nectary glands. Flowers mimosoid, small, usually yellow or white, crowded in spikes, spikelike racemes, or heads; stamens numerous and separate; anthers small, often with a terminal gland. Pods highly variable.

Largely subtropical to tropical, many in arid and semiarid regions; 1200 species, largely in warmer parts of Australia, Africa, and the Americas. References: Clarke and Seigler (1989), Isely (1969), Lee et al. (1989), Seigler and Ebinger (1988).

After *Astragalus*, *Acacia* is the world's largest legume genus. The Sonoran tree acacias have strong, hard wood. There are four shrub species in Sonora: *A. angustissima* with several varieties, *A. crinita*, *A. millefolia*, and *A. neovernicosa* (see *A. constricta*).

1. Glabrous or glabrate; unarmed.
  2. Bark not papery and peeling; leaves with 4–8 pairs of pinnae; petiole less than half as long as the leaf. _ ***A. coulteri***
  2' Bark papery and peeling; leaves with 1 (2) pair(s) of deciduous pinnae; petiole making up all or most of the leaf stalk. ________ ***A. willardiana***
1' Often pubescent; at least some branches armed with spines or prickles.
  3. Armature of unpaired, internodal recurved and laterally compressed prickles ("spines") like small rose thorns.
    4. Large shrubs to small trees or sometimes medium-sized trees; leaves often 1.5–3.5 cm, with 2 or 3 pairs of pinnae; flowers in cylindrical, spikelike inflorescences. ________ ***A. greggii***
    4' Medium-sized to large trees; leaves often 2.5–6.5 cm with 2–4 (5) pairs of pinnae; flowers in rounded, headlike inflorescences. ________ ***A. occidentalis***
  3' Armature of usually paired (stipular) spines at nodes, the spines straight, terete, or channeled and boat shaped.
    5. Leaves with 7 to many pairs of pinnae.
      6. Spines flattened or the larger ones boat shaped; leaves with 7–25 pairs of pinnae. ______ ***A. cochliacantha***
      6' Spines oval in cross section; leaves with 20–50+ pairs of pinnae. ________ ***A. pennatula***
    5. Leaves with 1–6 pairs of pinnae.
      7. Leaves with 1 pair of pinnae, the larger leaflets 7–45 mm. ________ ***A. pringlei***
      7' Leaves with 2–6 pairs of pinnae, the leaflets 1–7 mm.
        8. Pods indehiscent, dark-brown to blackish, thick, and not constricted between seeds. ___ ***A. farnesiana***
        8' Pods dehiscent or not, red-brown or pale colored, slender and relatively thin or flattened, and constricted between the seeds.
          9. Pods dehiscent, viscid with red-brown glands; mostly north of the Guaymas region. ____ ***A. constricta***
          9' Pods indehiscent, finely pubescent with sub-appressed hairs at least when young; uncommon, southwestern Sonora. ________ ***A.*** aff. ***pacensis***

*Acacia cochliacantha* [*A. cymbispina*]. Boat-spine acacia; *guinolo* (vicinity of Alamos), *chírahui*; *sinalá* (Guarijío); *chírahui* (Mayo, vicinity of Alamos), *chírajo* (Mayo); *koowi tami* 'pig teeth' (Yaqui) (figure 131)

Vase-shaped shrubs or small trees, 5–6 (8) m, with many straight branches; twigs with numerous, small lenticels. Bark dark brown, relatively smooth. Herbage glabrate; new herbage with sparse, short, white hairs. Spines stipular, often 1–5 cm, paired, at first red-brown and flexible, becoming whitish and firm, dorso-ventrally flattened, the larger ones boat shaped (unique among the flora of northwestern Mexico). Foliage gradually drought deciduous, leafing out with summer rains; leaves 5–15 (21) cm; petioles with a raised nectary gland; pinnae 7–25 pairs; leaflets 1.5–3.5 mm and numerous.

Flowers yellow-orange, in rounded heads. Pods 7.5–17 cm × 8–11 mm, dark red-brown, thick and moderately compressed, straight to slightly curved, indehiscent with pulpy mesocarp. Flowering June–September (November).

Slopes, valleys, plains, arroyos, and disturbed habitats; thornscrub, tropical deciduous forest, margins of Sonoran desertscrub, and occasionally in grassland. Widespread in Sonora except the eastern edge and northwestern part of the state, near sea level–975 (1700) m (3200 [5575] ft). The highest elevation records are along highways; probably also spread by cattle. Baja California Sur, and Sonora and Chihuahua to Veracruz and Chiapas.

This is one of the most abundant woody plants of central and southern lowland Sonora and Sinaloa. Larger, tree-sized chirahuis with well-developed trunks occur in tropical canyons, mostly in southeastern Sonora. This species quickly colonizes abandoned milpas and other disturbed areas including buffelgrass pastures in tropical deciduous forest, often forming impenetrable spiny thickets. In northern Sonora it is often severely damaged by winter freezing; the northernmost population occurs between Cucurpe and Sinoquipe. Along the coast it extends no farther north than the Guaymas region.

It is an excellent source of firewood, and the wood has been used for construction. The pods are eaten by burros and mules, and formerly by the Guarijío people (Gentry 1942). Burros often pull off long strips of bark to eat. In fall the dry hillsides in southeastern Sonora are often red-brown due to masses of ripening pods.

***Acacia constricta*****.** Whitethorn acacia; *mezquitillo*; *gidag* (O'odham) (figures 132 and 133A)

Usually shrubs, sometimes small trees to 5 m. Bark smooth, red-brown, becoming gray with age and flaking. Spines stipular, reaching 1–3 cm, paired, straight, and terete, usually white, or sometimes absent. Herbage glandular, usually viscid and sprinkled with mostly golden-colored glands, sparsely to moderately pubescent with small, white hairs, becoming glabrate. Leaves winter deciduous and tardily drought deciduous; new leaves appearing in spring; leaves 2–7 cm; leafstalks channeled; the petioles with a prominent nectary gland; pinnae 3–6 pairs; leaflets many, 1.5–3 mm.

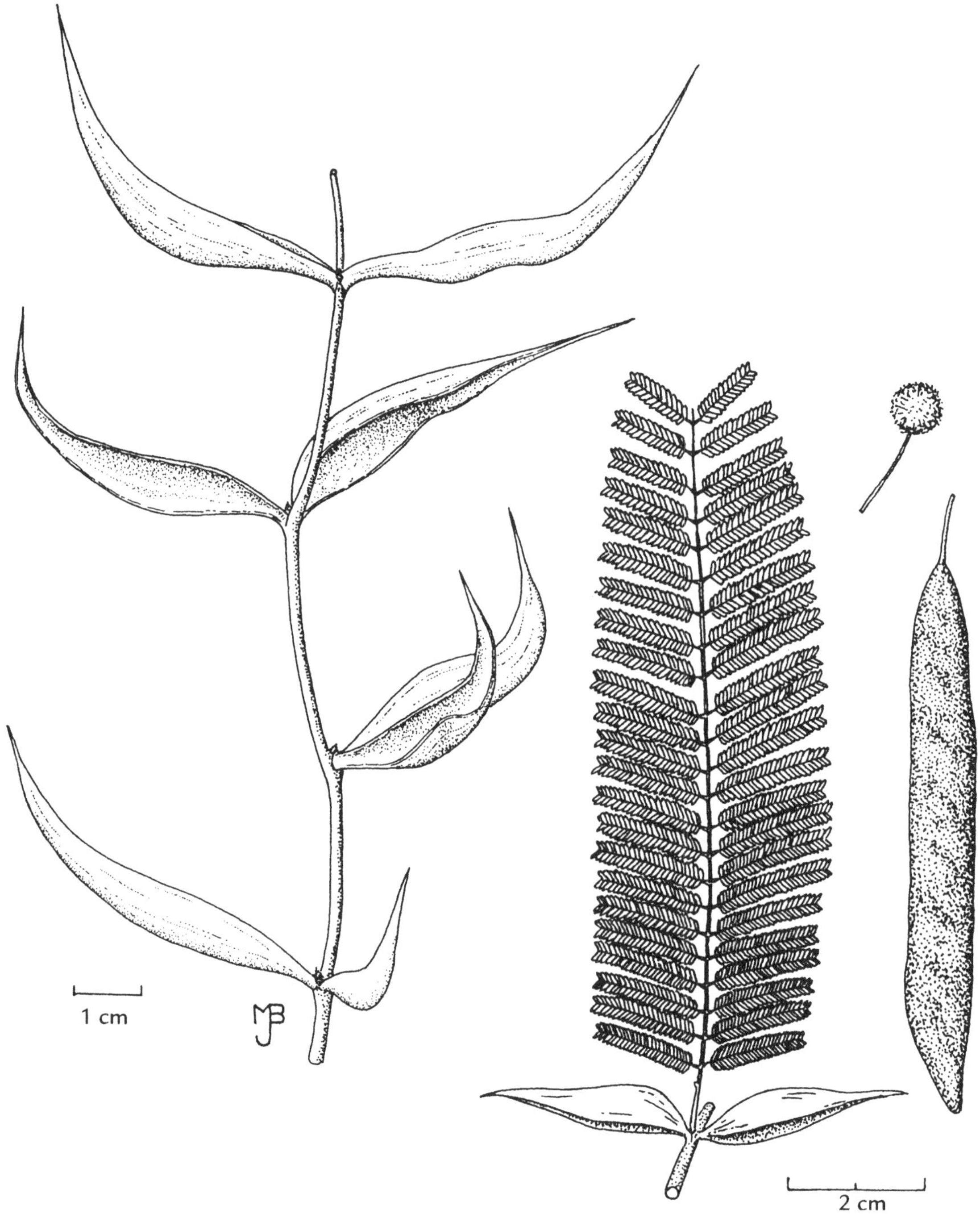


Figure 131. *Acacia cochliacantha*. Drawn by M. B. Johnson.

Flowers bright yellow, in rounded heads. Pods 4.5–13.5 cm, 4–6 mm maximum width, reddish with glistening viscid glands, moderately flattened, constricted between seeds, dry, gradually dehiscent, the valves separating partially to fully. Flowering in warmer months, especially with summer rains.

Slopes, valleys, bajadas, plains, and arroyos; Chihuahuan and Sonoran desertscrub, thornscrub, and grassland. Widespread in Sonora but not in the southeast nor extreme northwest, and generally not south of the Guaymas region; near sea level–1220 m (4000 ft). Baja California and Arizona to Texas, through much of arid and semiarid Mexico to Puebla but apparently not south of Sonora on the west coast.

To avoid the danger of a late spring frost, Sonora and Arizona farmers often waited until the whitethorn leafed

Figure 132. *Acacia constricta* (Benson and Darrow 1945). Drawn by L. B. Hamilton.

out before planting corn. Mesquite occasionally makes a mistake and leafs out before the last frost, the whitethorn is more reliable.

The closely related Chihuahuan whitethorn (*A. neovernicosa*, = *A. constricta* var. *vernicosa*) replaces *A. constricta* in Chihuahuan desertscrub in northeastern Sonora. *Acacia neovernicosa* is a spindly, sparingly branched shrub, 3–4 m, with generally smaller, more viscid leaves, usually fewer pinnae, and more densely fascicled leaves on short spur branches. *Acacia constricta* is tetraploid, while *A. neovernicosa* is diploid (Isely 1969).

The southern record in Sonora is from Navojoa (Gentry 1942), and Turner et al. (1995) include a record for northwestern Sinaloa. Larger trees in southern Sonora attributed to this species are likely to be *A. farnesiana*, or perhaps *A.* aff. *pacensis*. Records from northwestern Sinaloa, southern Sonora, and Baja California Sur (Turner et al. 1995) should be examined for their relationship with *A. pacensis*.

***Acacia coulteri***. *Guayabillo*; *maguó* (Guarijío); baihuío (Mayo) (figure 133B)

Small trees 5–8 (10) m, the bark dark gray, generally smooth but occasionally with small papery, peeling patches; unarmed and glabrous or very sparsely pubescent. Canopy feathery and bright green during summer rains; leaves drought deciduous; leaflets often falling before the rachillas. Stipules bristlelike, 1–3.5 mm. Leaves 8.5–21 cm; petioles often bearing an elevated nectary gland; pinnae 4–8 pairs; leaflets many, 3.5–10 mm, oblong and acute at the tip.

Flowers yellow-white, in dense cylindrical spikes 2.5–6.5 cm. Pods 6–12.5 × 1.8–2.6 cm, flattened and with a conspicuous rim, gradually dehiscent. Flowering (April) June–October.

Slopes and canyons; thornscrub and tropical deciduous forest; central and southern Sonora, 30–610 m (100–2000 ft). Often forms extensive stands on rocky slopes at the desert-thornscrub ecotone in east-central Sonora. Sonora to Oaxaca and northward in eastern Mexico to Nuevo León and Tamaulipas.

This attractive tree is recommended for landscaping and tolerates winter freezing as far north as Tucson. *Guayabillo* produces excellent wood for house construction, fence posts, and firewood and has been heavily exploited in southeastern Sonora. It might be confused with *Lysiloma divaricatum*, especially during dry seasons when both are leafless. *Acacia coulteri* is distinguished by slightly larger leaflets, the body of the pod does not separate from the rim, and the stipules are slender and bristlelike rather than leafy. Farther south, such as in Oaxaca, it is a large tree with papery, peeling bark.

***Acacia farnesiana*** [*A. minuta* subsp. *densiflora*; *A. smallii*]. Sweet acacia; *huizache*, *vinorama*; *cu'cá* (Guarijío); *cu'uca* (Mayo); *ston kudur(i)* (Pima Bajo); *poháas camoz* (Seri); *kuka* (Yaqui) (figure 133C)

Shrubs to small trees, generally < 10 m, or rarely 15–18 m especially in settlements in the lower Yaqui and Mayo river valleys; pubescent, the leaves becoming sparsely hairy with age. Spines reaching 2–4.5 cm, stipular, paired, white, straight, and terete. Foliage evergreen to tardily drought deciduous and winter deciduous toward the northern part of the state. Leafstalks channeled; leaves 2.5–5.3 cm, sometimes with a dark nectary gland on the petiole or rachis; pinnae 2–4 pairs; leaflets many, 3–7 mm.

Flowers bright yellow-orange, in globose heads, with a strong and far-reaching sweet fragrance. Pods 4–7.5 cm, indehiscent with pulpy mesocarp, straight or slightly curved, approximately 1 cm thick and terete; seeds crowded in two rows. Flowering November–April.

Slopes, canyons, valleys, plains, and arroyos; Sonoran desertscrub, thornscrub, tropical deciduous forest, and grassland. Sometimes spreading as a shrub along roadsides into oak and pine-oak habitats in east-central Sonora to 1900 m. Statewide except the northwest and northeast corners; 30–1065 (1850) m [100–3500 (5000) ft]. Especially common in disturbed habitats such as roadsides and agricultural and heavily grazed areas. Perhaps not native to North America.

This species is extensively cultivated as an ornamental tree in southern Arizona and to a lesser extent in Sonora. A rather variable species, originally restricted to the warmer regions of the American tropics, now naturalized and weedy in some warm regions around the world. An essential oil from the flowers was an important ingredient in perfumes.

In northern Sinaloa and Sonora this acacia is generally called *vinorama*, whereas *huizache* is the name usually applied to it in southern Sinaloa and elsewhere in Mexico. *Acacia farnesiana* may be difficult to distinguish from *A. constricta* if pods are not available: herbage of *A. farnesiana* lacks the golden glands usually present on *A. constricta*, and in cultivation *A. farnesiana* is more robust with larger leaves than *A. constricta*.

***Acacia greggii*** [*A. greggii* var. *arizonica*] Catclaw acacia; *uña de gato*; *'u:paḍ* (O'odham); *tis* (Seri); *hu'upa keka'ala* (Yaqui) (figures 133D and 134)

Irregularly branched shrubs or trees often 2–8 m or more; trunk usually crooked; heartwood reddish; bark rough and dark gray. At least some branches armed with small, sharp, recurved, and usually laterally compressed prickles scattered along the stems between the nodes. Young twigs,

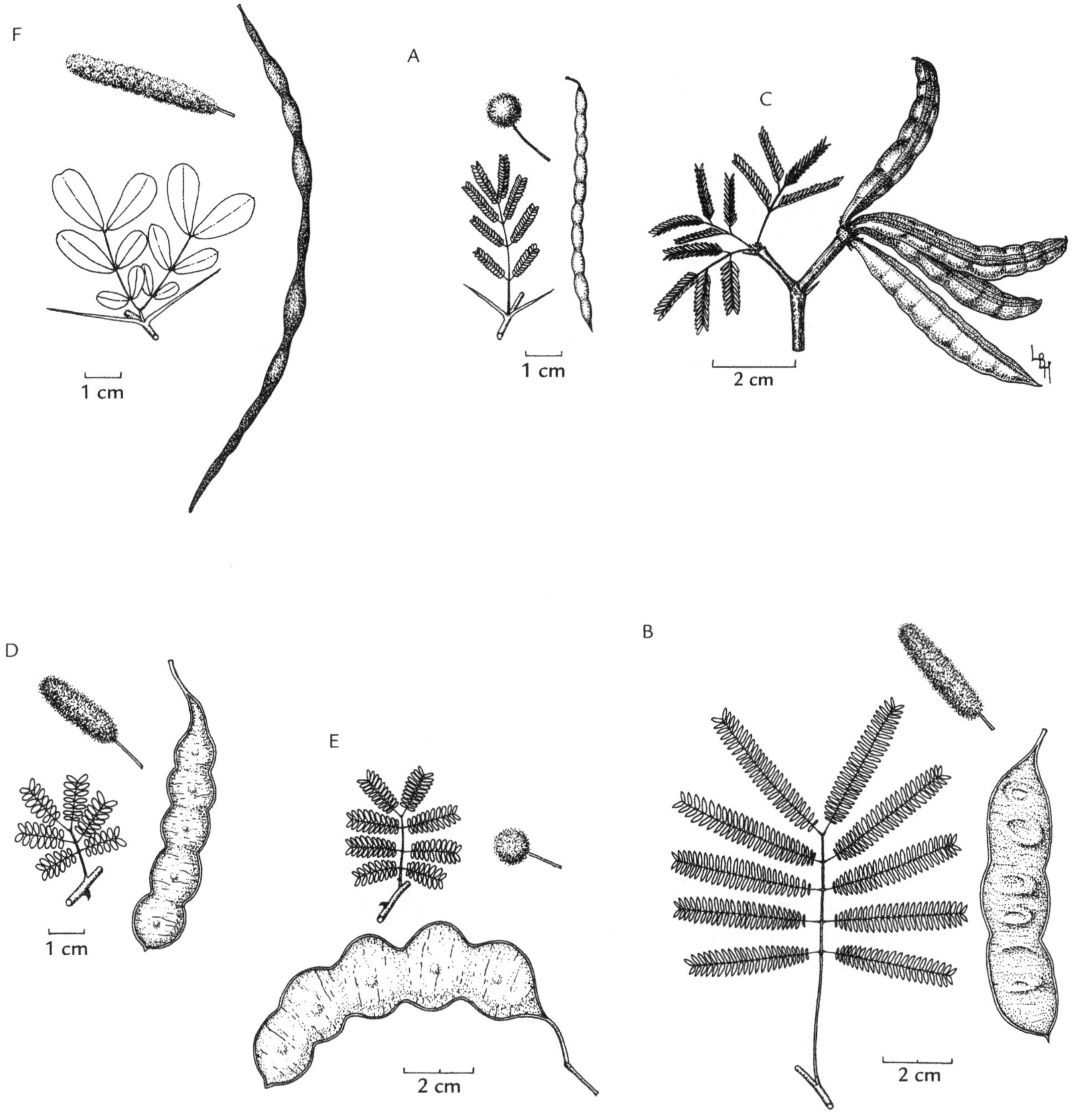


Figure 133. *Acacia*. (A) *A. constricta*, (B) *A. coulteri,* (C) *A. farnesiana* (drawn by L. B. Hamilton), (D) *A. greggii*, (E) *A. occidentalis*, (F) *A. pringlei* subsp. *californica*. A, B, D, E and F drawn by M. B. Johnson.

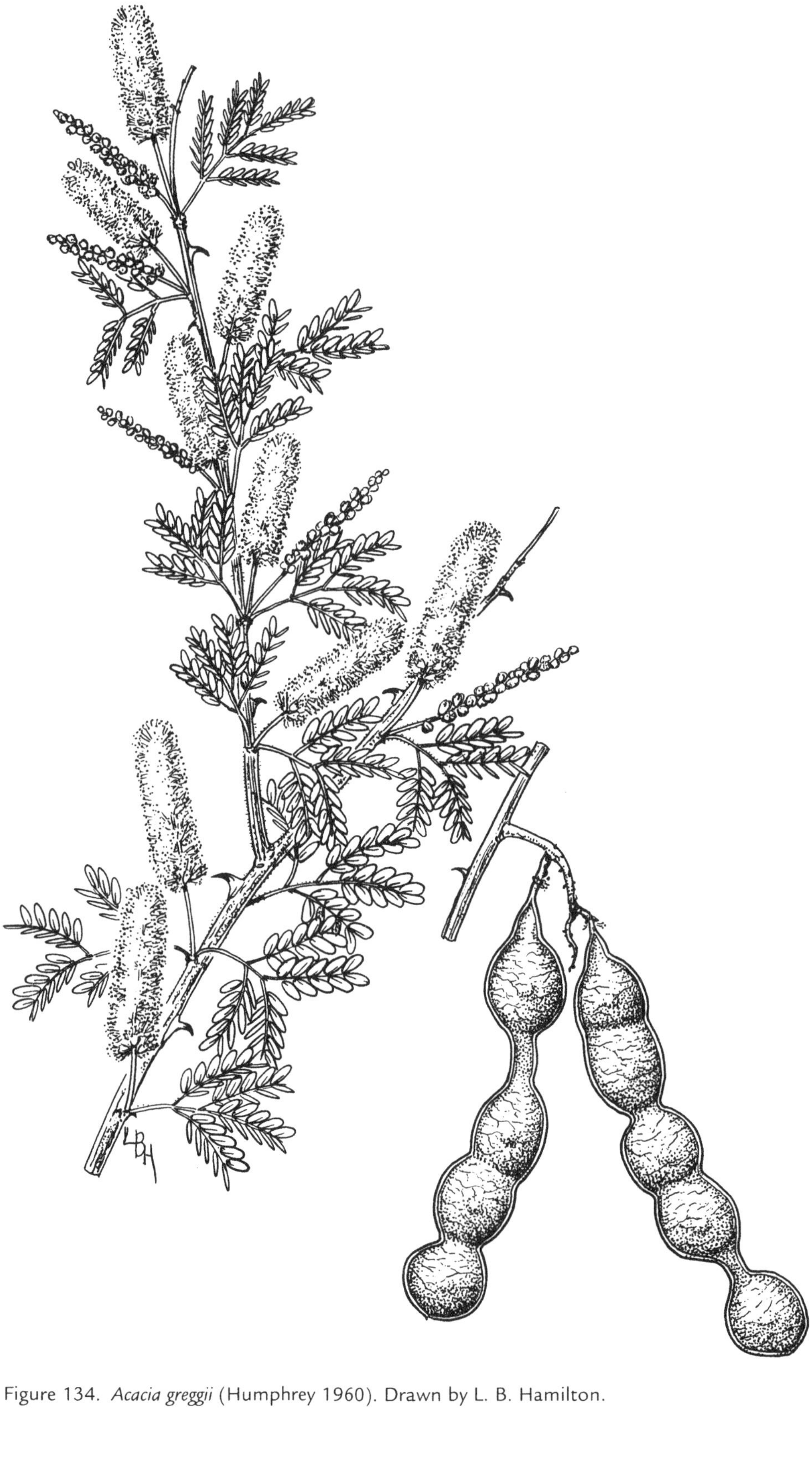

Figure 134. *Acacia greggii* (Humphrey 1960). Drawn by L. B. Hamilton.

leaves, and inflorescences moderately to densely pubescent with short hairs. Leaves tardily winter deciduous, often (1.5) 2.5–3.5 cm; petioles with a prominent nectary gland; pinnae 2 or 3 pairs; leaflets 3.5–6 (8) mm, 4–6 (7) pairs per pinnae.

Flowers cream colored, fragrant, in dense, cylindrical spikelike racemes, the flowering portion often 1.8–3 cm. Anthers minute (ca. 0.1 mm), without a glandular appendage. Pods (3.5) 6–15 × 1–2+ cm, flattened, ribbonlike, usually curved or curled, constricted between some seeds but usually not all seeds, reddish to tan, dry and indehiscent or tardily semidehiscent (the whole pod usually falls and is chewed by rodents). Flowering April–June; pods in June.

Slopes, canyons, flats, and arroyos; Sonoran and Chihuahuan desertscrub, and grassland; northern and central Sonora, near sea level–1525 m (5000 ft). Common along desert washes; widespread in western Sonora, apparently not ranging south of the Guaymas region. Across the northern tier of Mexican states, Durango, northern Baja California Sur, and Texas to southwestern Utah and southeastern California.

Isely's (1969) delimitation of *A. greggii* into three varieties is not entirely satisfactory, and the keys and distributional patterns are confusing. The Sonoran plants are var. *arizonica* Isely, characterized by densely pubescent leaves, but because "the leaflets of part of Gregg's original gathering are puberulent, a taxonomic distinction is difficult to sustain" (Barneby 1989: 7).

***Acacia occidentalis.*** Tree catclaw; *tésota* (northern Sonora), *teso* (Alamos region and Mayo); *te'sá* (Guarijío) (figure 133E)

Trees to 15 m with spreading canopies to at least 15–18 m across; well-developed and often straight trunk occasionally to nearly 1 m in diameter; heartwood reddish; bark rough and dark gray. At least some branches armed with small, sharp, recurved, usually laterally compressed prickles scattered along stem internodes. Herbage glabrate. Leaves tardily drought and winter deciduous, but often with at least some foliage remaining, 2.5–7 cm; petioles with a prominent nectary gland; pinnae 2–4 (5) pairs; leaflets 2.5–8.5 mm, (4) 6–14 pairs per pinna.

Flowers cream colored, in rounded heads, highly fragrant, the scent like that of orange blossoms. Pods 7.5–15.5 × 2–2.5 cm, flattened, ribbonlike, usually curved or curled, constricted between most to all seeds, reddish to tan, dry, indehiscent or tardily semidehiscent. Flowering February–April.

Mostly valleys and arroyos; from the southern and eastern margins of the Sonoran Desert to grassland, thornscrub, and tropical deciduous forest; north-central and northeastern to southern Sonora; near sea level–1200 m (3935 ft). Best developed in deep sandy soil along larger arroyos and river valleys; on hillsides seldom taller than 6 m but with a distinct trunk and tree form. Otherwise known only from northern Sinaloa.

This handsome tree is recommended for cultivation and is occasionally grown in southern Arizona. The wide-crowned canopy produces pleasant, filtered shade. The hard, dark wood is used for many purposes. Young nursery stock has survived at least −7°C (18°F) without damage. It is similar and undoubtedly closely related to *A. greggii* but differs in having a single, well-developed trunk forming a symmetrical tree, leaves often with an additional pair of pinnae, and a rounded inflorescence.

***Acacia*** aff. ***pacensis.*** *cochirepa* (Mayo)

Large shrub or small tree to approximately 5 m with large, straight, terete spines, bright yellow flowers in globose heads, and slender, indehiscent pods. Resembling *A. constricta* but often appearing more like *A. schaffneri* of the Chihuahuan Desert in north-central Mexico with its relatively thick and seemingly fewer, somewhat spreading branches and denser and darker green foliage.

This rather uncommon acacia has recently been found in the lowland thornscrub of southwestern Sonora. It may be *A. pacensis*, known for certain only from Baja California Sur. *A. pacensis* differs from *A. constricta* in having small bracts near the apex of the peduncle and indehiscent, pubescent pods (Rudd and Carter 1983).

***Acacia pennatula*** subsp. ***pennatula.*** Feather acacia; *algarrobo, chírahui, palo garrobo, algarrobo espino* (Sinaloa); *yepohuecha* (Guarijío) (figure 135)

Mostly large shrubs, sometimes small trees to approximately 7 m. Bark light to dark brown or red-brown, finely to coarsely granulated. Young twigs, leafstalks, inflorescences, and peduncles with dense, short, spreading, white to yellow hairs; leaflets sparsely hairy. Twigs mostly with painfully sharp, stout, paired stipular spines reaching 1 (1.8) cm and oval in cross-section. Leaves nearly evergreen or winter deciduous, 8.5–25 cm; petioles often with a nectary gland; pinnae often 20–62 pairs; leaflets 1.4–2.3 mm, numerous.

Flowers bright yellow, fragrant, in rounded heads often 1.5 cm wide. Pods (5) 7.5–13.5 × 1–2 cm, indehiscent with mealy mesocarp, thick and moderately flattened, dark red-brown. Flowering May–October.

Slopes and valleys; upper thornscrub, upper tropical deciduous forest, oak woodland, and lower pine-oak woodland; eastern Sonora; 305–1525 m (1000–5000 ft). Widespread, especially in the lower oak zone and its ecotone with thornscrub or tropical deciduous forest, and often abundant in disturbed habitats; the northernmost record is from Cumpas. Subspecies *pennatula* ranges from north-

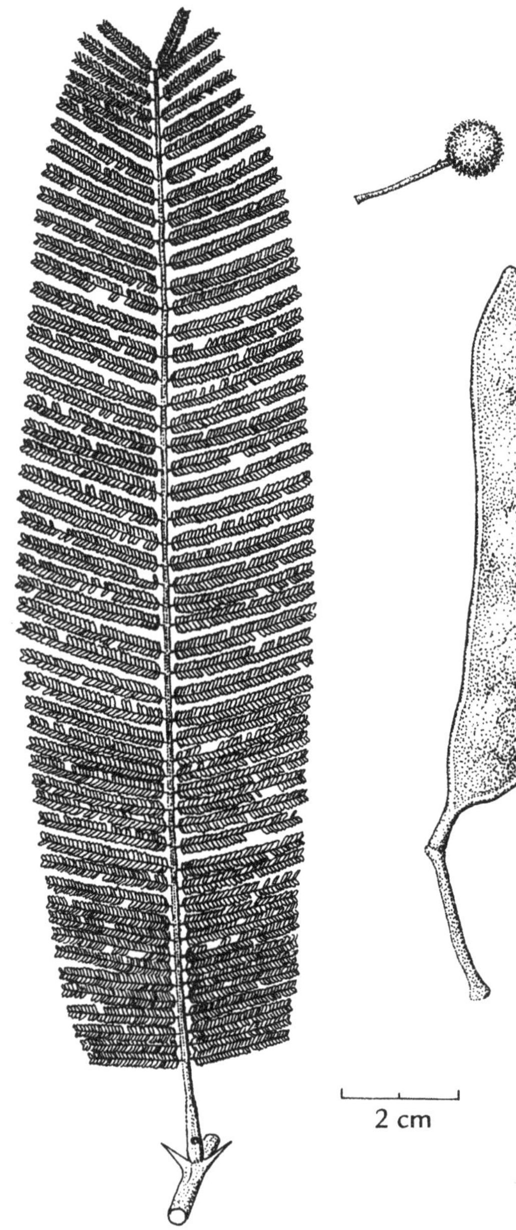


Figure 135. *Acacia pennatula* subsp. *pennatula*. Drawn by M. B. Johnson.

ern Mexico to northern South America. Subspecies *parvicephala* occurs from southern Mexico to northern South America (Seigler and Ebinger 1988).

This acacia is sometimes cultivated in southern Arizona. The attractive, feathery foliage and dark bark are distinctive. It is damaged at temperatures of −4°C (25°F). The Guarijío ground the pods including the seeds for flour when they did not have maize, and the bark was used to dye leather (Gentry 1942). In Sinaloa the plant is used as a "tónico sexual para burros."

***Acacia pringlei*** subsp. ***californica*** [*A. californica*]. *Chicorai* (Yaqui) (figure 133F)

Large shrubs or small trees sometimes 6–8 (12) m, often with multiple trunks. Spines stipular, the larger ones 3–5 cm, paired, white, terete, and straight. Herbage with small, short hairs. Foliage mostly evergreen or partially and tardily drought deciduous. Leaves with one pair of pinnae, each with 2 or 3 pairs of leaflets, the terminal pair usually largest; leaflets often obovate to elliptic, sometimes ovate, the larger leaflets 7–30 (45) mm; petioles with a prominent nectary gland.

Flowers cream white, in cylindrical spikes 2–6.5 cm. Pods 9–20 cm, red-brown, slender, and constricted between the seeds (resembling a slender pod of *Parkinsonia aculeata* or a very long pod of *A. constricta*), dry and tardily dehiscent. Flowering (March) May–June.

Slopes, valleys, canyons, arroyos, and coastal plains; Sonoran desertscrub and thornscrub; west-central, east-central, southeastern, and probably southwestern Sonora, near sea level–670 m (2200 ft). Also extreme northwestern Sinaloa and the Cape Region of Baja California Sur. Subspecies *pringlei* is widely disjunct in eastern and southern Mexico from Tamaulipas to Yucatán and Oaxaca.

The dark-green foliage contrasts strikingly with the white spines. The leaflets are the largest of the Sonoran acacias. The distribution in Sonora is highly fragmented, with many isolated, often small, local populations. Perhaps it was once more widespread, as much of the probable former range has been converted to agriculture, and the wood is hard and suitable for fuelwood. The single record from Sinaloa is a specimen collected by Gentry (#7327: University of Arizona Herbarium) from "the Sonora-Sinaloa boundary below Agiabampo. April 17, 1945. Coastal desert scrub [=thornscrub]; open plain with calcareous clay soil." Most of this coastal plain has been converted to agriculture, and presumably this acacia also occurred (or still occurs) in southwestern Sonora.

***Acacia willardiana.*** *Palo blanco*; *cap* (Seri); *nawi'o* (Yaqui) (figures 136 and 137)

Thin, wispy trees to 7 m, the trunk and limbs slender, the wood hard and flexible, the twigs and smaller branches slender and often drooping; bark whitish, papery and peeling away in sheets during dry seasons; unarmed and glabrous except sometimes with few, scattered, minute, white hairs. Leaves 10–27 cm, bearing a dome-shaped nectary gland between the pinnae; the pinnae 1 (2) pair(s); leaflets 4–15 pairs per pinna; the pinnae soon deciduous leaving the flattened leafstalk to function as a phyllode.

Flowers pale yellow, in dense cylindrical spikes 2–7 cm. Pods 8–18.5 × 1.2–1.5 cm, flat, straw colored, dry, and tardily dehiscent; seeds light brown, disk-shaped, germi-

Figure 136. *Acacia willardiana*; Pedro Comito on the east side of Isla Tiburón (Felger and Moser 1985). Photograph by R. S. Felger.

nating readily without scarification. Flowering February–June and October.

Rocky habitats on bajadas, arid slopes, cliffs, ridges, and arroyos; Sonoran desertscrub and thornscrub. Topolobampo in northwestern Sinaloa through western Sonora to the Cerros Anacoretas (south of Caborca), and Islas San Pedro Nolasco and Tiburón; near sea level–915 m (3000 ft).

The phyllodes, which resemble those of many Australian acacias, make this species unique in the New World. In addition, the peeling white bark and absence of spines are notable among Sonoran acacias. It has no obvious close relatives, but it may be more closely related to *A. coulteri* than to any other acacia. Palo blanco is readily grown from seed and thrives in cultivation in southern Arizona, although it is somewhat frost sensitive. It should be brought into cultivation in Sonora and other, hot, dry regions.

### *Albizia*

Mostly trees, the flowers mimosoid. Warm, mostly tropical regions, largely in the Old World; 150 species, 19 native to the Americas. Reference: Barneby and Grimes (1996).

The Old World *A. julibrissin* is widely cultivated in temperate regions including northern Sonora. Its cold-tolerance is unusual for the genus. Another Old World tree, *A. lebbeck*, is grown in relatively frost-free regions of the state.

***Albizia sinaloënsis.*** *Joso, palo joso; hooso* (Yaqui) (figure 138)

Handsome, unarmed trees to 20+ m with a high bole and smooth yellow-white bark, the trees generally taller than broad but sometimes developing a spreading canopy. Sometimes spreads by root sprouts. Twigs and leaves densely hairy, or the leaflets glabrate above. Foliage deciduous in late spring; new leaves appearing in late spring and early summer. Stipules 0.7–1.7 mm, subulate, and deciduous. Leaves 10–27 cm; pinnae (3) 4–8 pairs; leaflets (5) 7–16 mm, 10–21 pairs per pinna; petioles usually with a nectary gland near the base and sometimes additional glands along the rachis.

Flowers sessile in globose heads; heads solitary on 1–5 fascicled axillary peduncles 2–3.5 cm. Calyx 2.5–3 mm, campanulate, green, the outer surfaces densely pubescent with short, white hairs, the inner surface glabrous, the

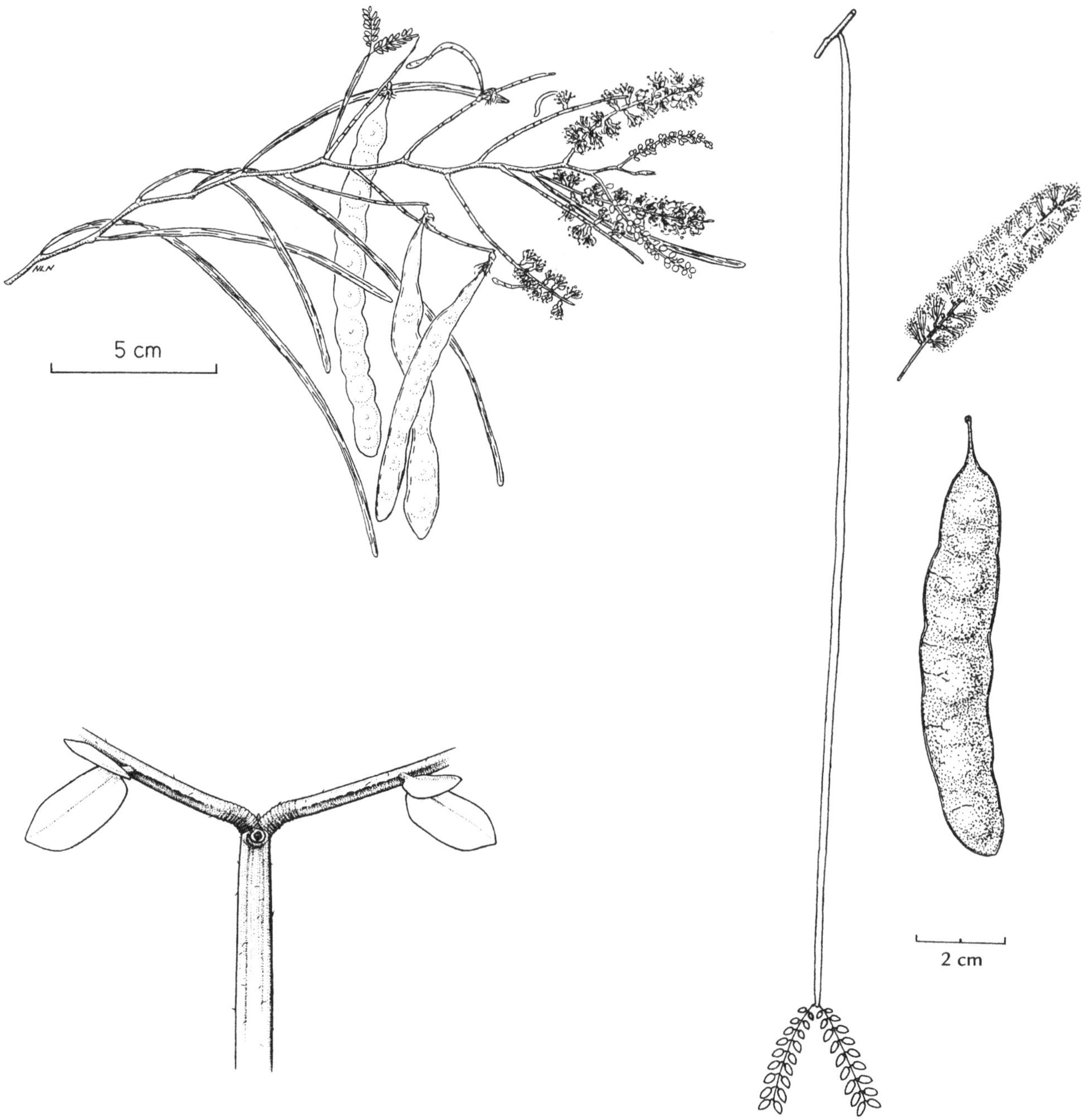


Figure 137. *Acacia willardiana*: flowering and fruiting branchlet (drawn by N. L. Nicholson), leaf, pod, and single inflorescence (drawn by M. B. Johnson), and petiole gland (drawn by A. K. Jakob).

lobes 0.8–1 mm, obtuse. Corollas 4.5–5 mm, cream colored, drying yellow-green, glabrous or sparsely hairy inside with scattered minute glands on the tube and white hairs densest toward apex, the lobes 1.1–1.5 mm. Stamens numerous, the filaments united below into a tube about as long as the corolla tube, the anthers 0.2 mm, without glands. Pods 7.5–16 cm, broadly linear, flattened, acute at both ends, velvety with minute hairs, straw colored, dehiscent, the margin ribbed, the apex with a slender apiculate tip; seeds 7–9 mm, 4–6 mm wide, flattened, smooth, cream colored to light brown. Flowering May; pods in winter.

Streams and banks of larger arroyos and river valleys and river bottom settlements in coastal thornscrub and tropical deciduous forest; east-central and southern Sonora; 30–305 m (100–1000 ft); the northernmost grove

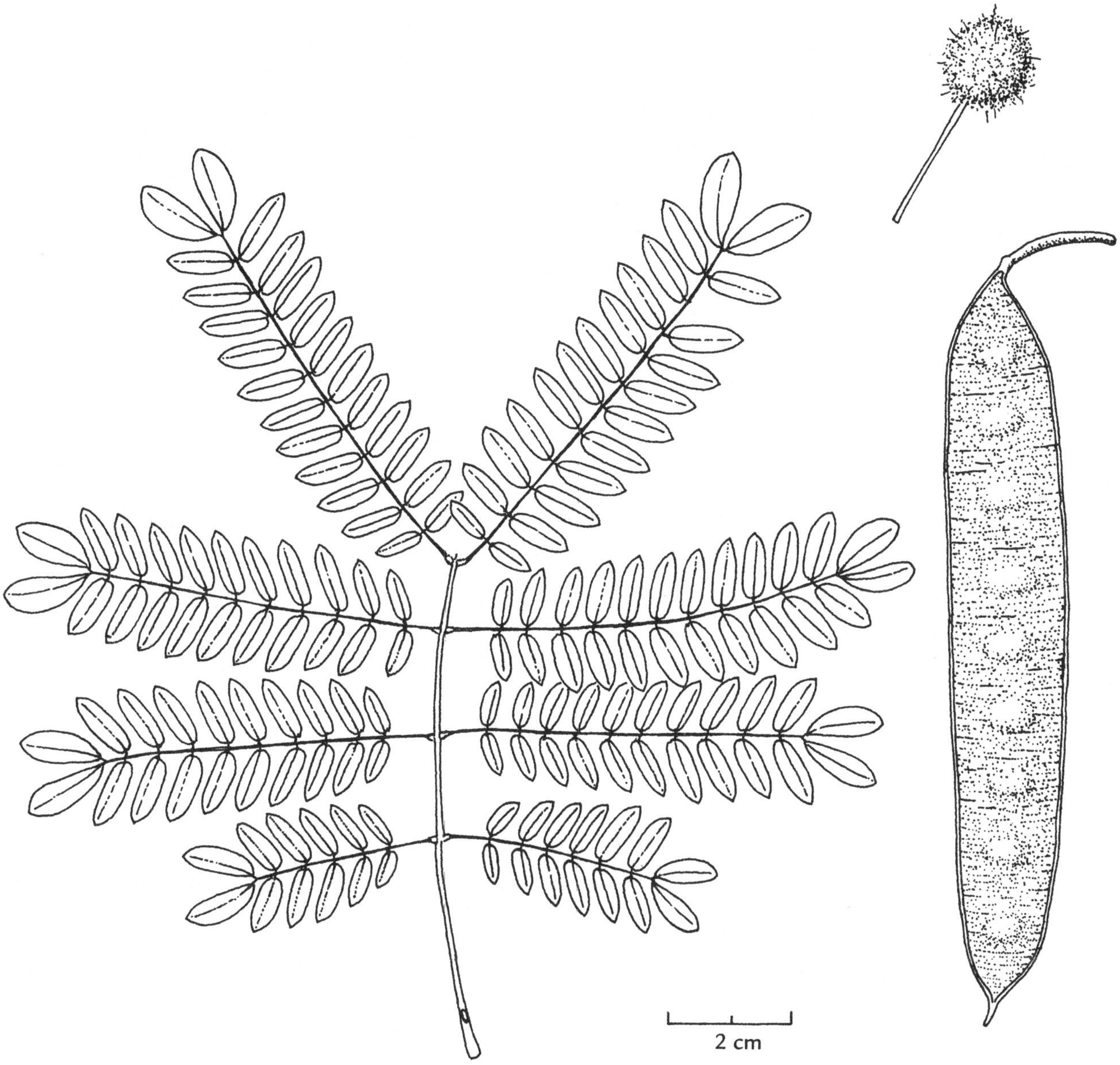


Figure 138. *Albizia sinaloënsis*. Drawn by M. B. Johnson.

at Arroyo Suaqui Grande (Tecoripa) in the Río Yaqui drainage near Highway 16 in east-central Sonora. Southward to Esquinapa in southern Sinaloa.

In Sonora these trees often grow in large groves near settlements and have long been cultivated or protected in river valley villages such as at Potam. The trees occasionally show freeze damage. The bark is sometimes used for tanning but gives a lighter color to leather and is less desired than bark of mauto (*Lysiloma divaricatum*). This fast-growing tree is worthy of wide cultivation in tropical regions with ample soil moisture.

With the exception of *A. sinaloënsis*, the New World *Albizia* species have dimorphic flowers. Barneby and Grimes (1996: 216) indicate that *A. sinaloënsis* "seems in most respects a form of *A. tomentosa*" adapted to drier conditions. *Albizia tomentosa* ranges from Central America northward along the Pacific slope of Mexico to Jalisco.

### *Bauhinia*

Mostly trees and shrubs. Flowers caesalpinioid, often showy, sometimes with unisexual flowers. New and Old

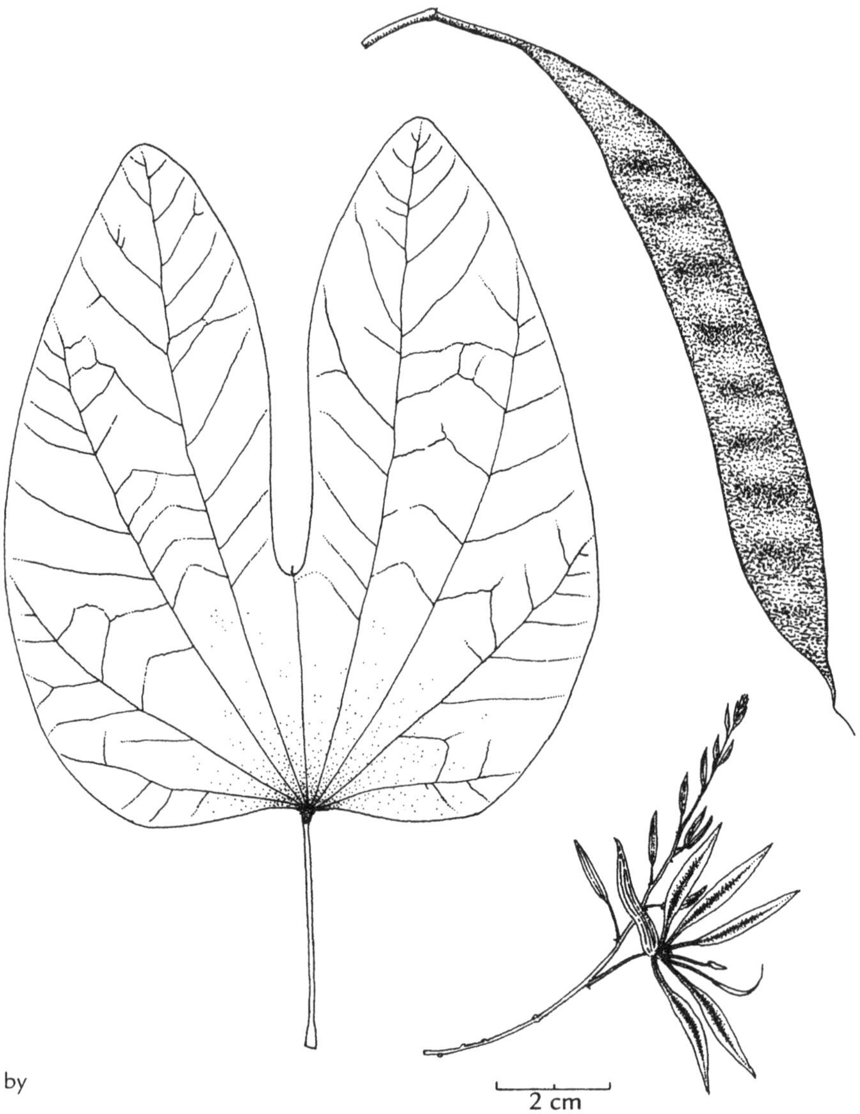


Figure 139. *Bauhinia pringlei*. Drawn by M. B. Johnson.

World tropics and subtropics with a few in deserts; 300 species. Reference: Wunderlin (1983).

The Asian *B. variegata*, called orchid tree because of its large, attractive flowers, is widely cultivated in the lowlands of Sonora and rarely occurs as a volunteer along roadsides.

***Bauhinia pringlei***. *Palo piojo blanco* (figure 139)

Unarmed, slender trees to 8 m with smooth, white bark. (When rubbed or scuffed the bark turns brilliant yellow-green; other local trees also show the same phenomenon.) Herbage with short hairs when young, soon glabrate. Foliage luxurious during summer rainy season and apparently facultatively drought deciduous. Leaves simple; petioles 2.5–6.5 cm; blades often 5.5–23.5 × 5.5–21 cm, deeply bilobed, shaped like a bovine hoof print, palmately veined and glabrate except along veins; stipules 1.2–2.0 mm, green with reddish tinges, narrowly triangular, soon withering, each stipule base associated with a turgid, bright lime-green nectary 0.9–1.7 mm. [The stipules are paired, one on each side of the pulvinus at the petiole base. The associated extrafloral nectary attaches to the scalelike base of each stipule. These nectaries exude viscous, clear liquid and attract small ants. Although he did not identify these structures as nectaries, McVaugh (1987) pointed out that they are commonplace in the genus and homologous with prickles in other *Bauhinia*.]

Flowering when leafless, or from leafless branches; inflorescences of usually several loosely flowered racemes 7.5–19 cm at ends of twigs, each raceme with several or more flowers. Flowers bisexual, or sometimes unisexual, male flowers without a pistil, or the female flowers with small staminodes only. (This species is apparently the only legume in Sonora with unisexual flowers.) Pedicels, outer calyx surface, filaments of sterile stamens, and claws of petals with short to loosely woolly white hairs. Pedicels slender, 9–15 mm. Calyx 2 cm, the sepals separate below, fused above, spathelike and reflexed, white with broad, maroon stripes. Petals 3.0–3.7 cm × 3–5.5 mm, white with a pink-purple midstripe, narrowly lanceolate with a slender claw. Filaments white below, maroon above; functional stamen 1 (2), 17–20 mm; sterile stamens (8) 9, 5–7 mm. Pods 13–15 × 1.3–1.6 cm, flattened, narrowed below to a relatively long stipe, explosively dehiscent, flinging the seeds with considerable force. Seeds 6.5–8 × 6.4–8.3 mm, 3.0–3.7 mm thick, irregularly shaped, mostly quadrangular or polygonal to almost orbicular in outline and compressed, orange-brown, smooth and glossy. Flowering December–March.

Extreme southeastern Sonora, 380–390 m (1245–1280 ft), the northernmost trees about 30 km southeast of Alamos near the road to El Zapote. Locally common on well-drained, fairly level ground and hills. (This habitat supports a rich assemblage of tropical deciduous forest trees and shrubs. The clearing of forest for buffelgrass is rapidly diminishing the Sonoran population.) Much more common and widespread in nearby north-central Sinaloa and southward in western Mexico to Guerrero, Zacatecas, and the state of México.

This magnificent tree deserves wide cultivation. It is expected to be frost tender but should thrive at least as far north as Hermosillo and perhaps in the warmer areas of southern Arizona.

### *Brongniartia*

Trees and shrubs. Flowers papilionoid. Western Texas and Sonora to the Andes; 50 species. Several shrub species in Sonora. Reference: Dorado (1992).

**Brongniartia alamosana.** *Palo piojo, vara prieta; sejcó* (Guarijío); *cuta nahuila* (Mayo) (figure 140)

Unarmed shrubs or small, low-branched, spreading trees to 6 m. Bark smooth, pale gray-green with conspicuous pale-colored lenticels; bark at base of young plants with deeply furrowed corky ridges. Herbage at first densely silky-haired; older leaflets sparsely pubescent; twigs and leafstalks remaining hairy. Stipules 1–3 cm, leafy, crescent shaped, and deciduous. (The stipules are the largest of the Sonoran tree legumes.) Leaves drought deciduous, 14.5–26 cm, odd-pinnate; leaflets 9–15, 3–6.5 cm, oval, often with an apiculate tip.

Inflorescences racemelike (see below). Flowers attractive, dark maroon-purple, becoming blackish inward surrounding a bright yellow-green center resembling an eye spot, the banner 2 × 2 cm. Pods 4.5–6.0 cm, flat, with a sharp terminal point, explosively dehiscent, the valves coiling loosely; seeds approximately 1 cm. Flowering May–August.

Slopes and arroyos; thornscrub and mostly tropical deciduous forest; southeastern and east-central Sonora, 60–915 m (200–3000 ft). Also northern Sinaloa to Guerrero.

"In the warm, dry days of late fall, the tree stands naked in a leafless forest, when the pods burst open through the warm hours of the day and there is heard a sound as of the popping of guns, as it shoots out its seeds" (Gentry 1942: 141). This plant flowers in profusion just before leafout and sparingly through the summer rainy season. Plants in cultivation are self-fertile; freeze damage occurs below −4°C (25°F). The long, slender branches are cut for poles used in construction. However, the Mayo common name is a derogatory term meaning "good for nothing" because the wood is weak and not regarded as good for fuel or other uses.

Most members of this genus including the Sonoran species have a specialized inflorescence: "The flowers are produced in what appear to be sessile fascicles (usually of two or three flowers, or the flowers solitary) at potentially leafy nodes, usually of newly developing branchlets" (McVaugh 1987: 298).

### *Caesalpinia*

Mostly trees and shrubs, unarmed (the four native Sonoran tree species) or with prickles or "thorns." Leaves twice pinnate; leafstalks often with small, stalked glands. Flowers caesalpinioid, in terminal or axillary racemes. Pedicels jointed (the portion below the joint is the actual pedicel). Filaments hairy toward their bases. Pods highly variable. New and Old World tropics and subtropics with a few in deserts; 125 species. Reference: Lewis (1998).

Two shrub species are widely cultivated in Sonora. Red bird-of-paradise (*C. pulcherrima*), native to the Caribbean and widely naturalized in eastern Sonora, has large, brilliant red-orange flowers. Yellow bird-of-paradise (*C. gilliesii*), native to South America, is one of the most common cultivated shrubs in the region and often becomes weedy. *Caesalpinia cacalaco*, a large shrub or small tree in lowland Sinaloa, is cultivated in southern Sonora, where it occasionally becomes feral in roadside ditches. It has attractive yellow flowers and numerous large conical "thorns" along the stems and trunks.

Various members of this genus are used for lumber, dyes, medicines and ornamentals. During the Middle Ages, wood of the Old World *C. sappan* was an important article of commerce in Europe. The wood yields a red dye and was long known as brazil, brasil, or brésil. Some South American caesalpinias, primarily *C. echinata*, became a valuable

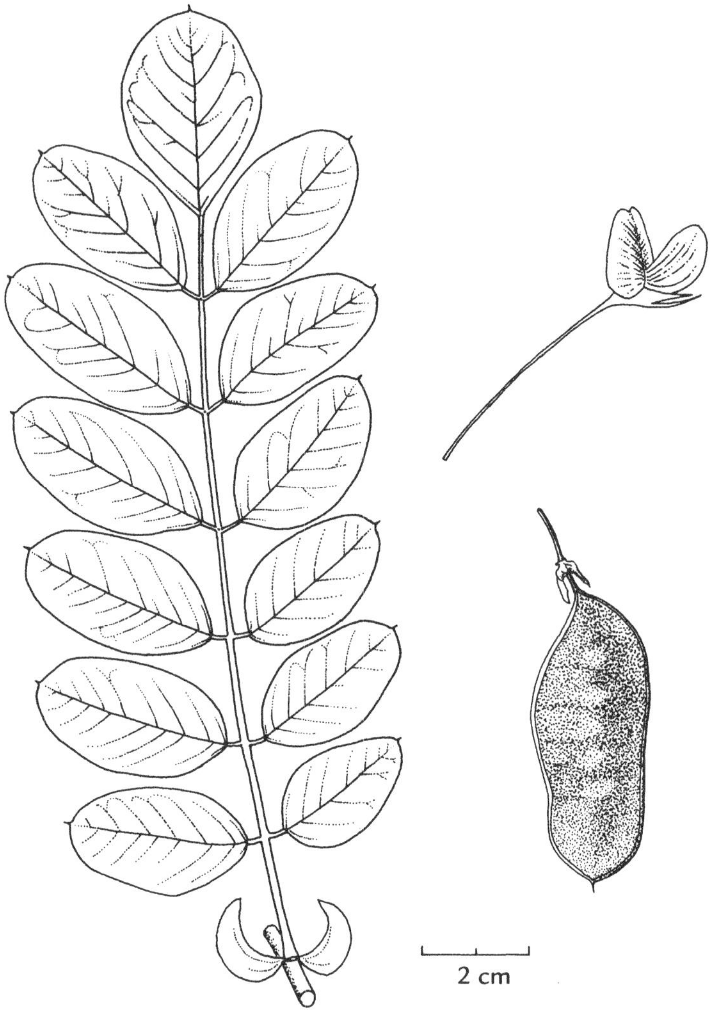


Figure 140. *Brongniartia alamosana*. Drawn by M. B. Johnson.

dye source during colonial times and the Republic of Brazil received its name from its forests of these dyewood trees (see Record and Hess 1943 and *Haematoxylum*).

1. Leaflets 3.2–10 cm; lower calyx lobe (sepal) about three times longer than the others. ________ ***C. platyloba***

1' Leaflets to 3 cm; calyx lobes all similar, or the lower one less than twice as long as the others.

  2. Pods indehiscent, thick, woody, and semipersistent. ***C. sclerocarpa***

  2' Pods explosively dehiscent, strongly compressed, the valves firm but not woody, quickly falling.

    3. Leaflets > 1.5 cm; pods 6–8 cm. ***C. caladenia***

    3' Leaflets < 1.5 cm; pods < 5 cm. __ ***C. palmeri***

***Caesalpinia caladenia.*** *Palo piojo blanco; guetapochi* (Guarijío), *jícamuchi* (Mayo) (figure 141)

Shrubs or slender, single-trunk trees to 6 (13) m. Bark whitish and smooth on trunks, reddish on twigs becoming gray, with conspicuous whitish lenticels. Racemes, pedicels, and calyces densely pubescent with glandular and nonglandular hairs, otherwise glabrous or glabrate. Stipules absent. Leaves drought deciduous, 15–35 cm; leafstalks often with stalked glands between the pinnae; pinnae 2–4 pairs and often also with a terminal pinna; leaflets 1.5–3.0 cm, ovate to obovate, 3–5 pairs per pinna. Flowers when leafless or nearly so; racemes 8–28 cm. Flowers bright yellow, 2.5–3 cm wide; calyx lobes all similar. Pods 6–8 cm, flat, hairy, conspicuously dotted with red-brown, stalked glands, explosively dehiscent. Flowering December–April.

Slopes and valleys; thornscrub and tropical deciduous forest; southeastern and east-central Sonora: the Alamos region westward toward Navojoa and northward to south of Moctezuma; 245–525 m (800–1715 ft). Lewis (1998) reports that this species is apparently endemic to Sonora, although populations at least in northern Sinaloa are probably conspecific.

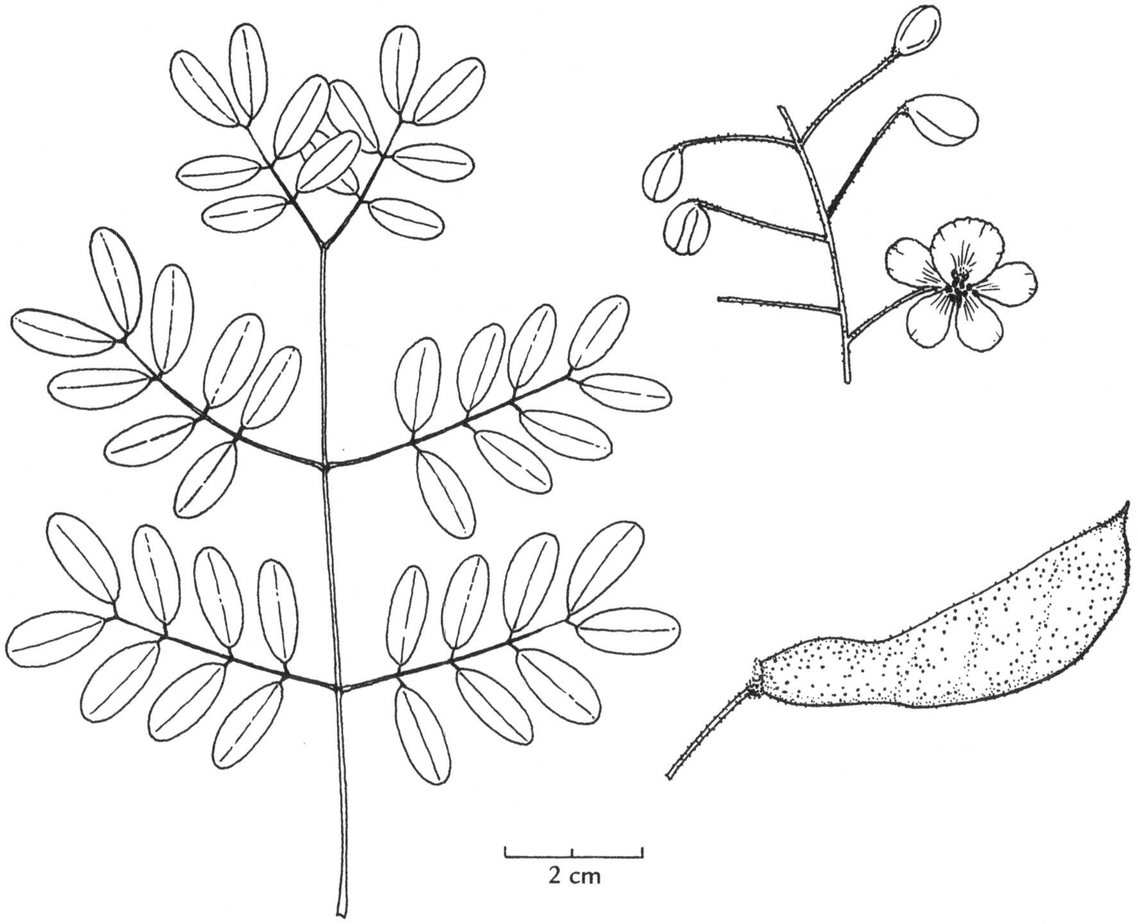


Figure 141. *Caesalpinia caladenia*. Drawn by M. B. Johnson.

*Caesalpinia caladenia* is closely related to *C. acapulcensis*, known from Jalisco to Michoacán, which some authors have treated as a synonym of *C. caladenia*. "However, the situation is further complicated by the intergradation of *C. caladenia* with *C. standleyi* and the intergradation of *C. caladenia* and *C. palmeri*" (Lewis 1998: 62). *C. standleyi* occurs in Sinaloa, Nayarit, and Jalisco.

**Caesalpinia palmeri.** *Palo piojo; jícamuchi* (Mayo); *hap oácajam* (Seri); *kume'a ouwo* (Yaqui) (figure 142)

Usually shrubs, occasionally small, slender trees to 8.5 m with multiple or single trunks to 14 cm in diameter. Bark gray with conspicuous raised lenticels (hence the Mexican vernacular name), the twigs red-brown. Herbage glabrous or the young leaves often with white hairs. Stipules absent. Leafstalks with small, stalked glands between the pinnae. Leaves gradually drought deciduous, 3.5–14 cm; pinnae 3–5 pairs plus a terminal pinna; leaflets (2.5) 6–10 (14) mm, elliptic to oval, oblong, oblanceolate, or obovate, (1) 2–4 pairs per pinna.

Flowering axis, pedicels, calyces, ovary, and style densely pubescent with short, white hairs. Racemes densely flowered and short, or few-flowered clusters on short shoots, the pedicels jointed just below the flower. Flowers bright yellow, portions fading red-orange, often 18–20 mm wide. Calyx 8–10 mm, maroon, the lower lobe irregularly 3-parted, slightly larger to about the same size as the other lobes. Pods 2.8–4.8 cm, flat, hairy, dotted with stalked red-brown glands, explosively dehiscent. Flowering various seasons, often with a peak March–April.

Hills, plains, and valleys; western and central Sonora from 31°N and southward; near sea level–625 m (2050 ft). Widespread as a shrub in Sonoran desertscrub and thornscrub, becoming trees only in dry regions of tropical deciduous forest, especially between Alamos and the Sinaloa border. Also in Sinaloa. In cultivation in southern Arizona, this plant is hardy to at least −6.5°C.

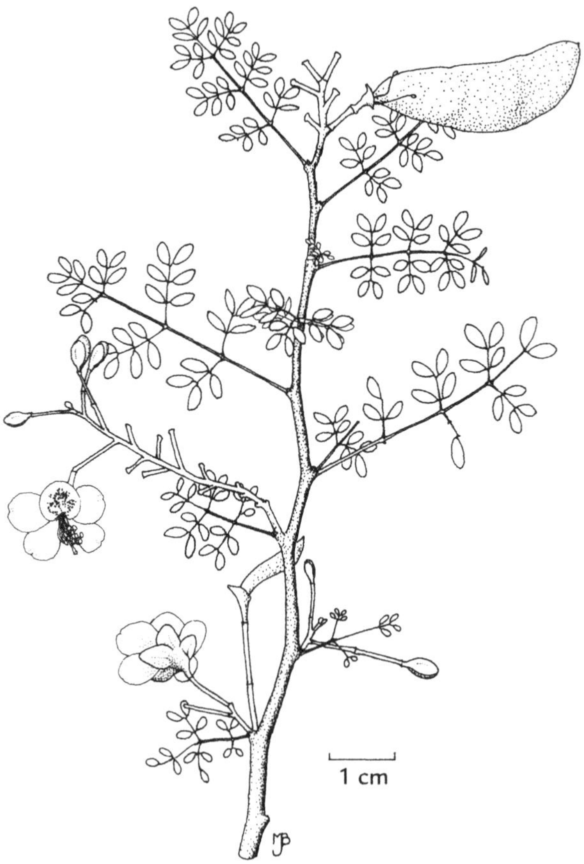


Figure 142. *Caesalpinia palmeri.* Drawn by M. B. Johnson.

***Caesalpinia platyloba.*** *Palo colorado, arellano* (Sinaloa); *hueraquí* (Guarijío); *cuta síquiri* (Mayo) (figure 143)

Trees to 7 (12+) m, with very hard wood. Bark smooth, dark lead-gray with conspicuous small, white lenticels, with age flaking away in irregular pieces. Young herbage and inflorescences, especially the buds, densely pubescent with short, brown, velvety hairs; mature herbage glabrate. Stipules linear, deciduous. Leaves drought deciduous, 27–36 cm; the pinnae 2, 3, or sometimes 4 pairs; leaflets mostly 3–10 cm, broadly elliptic to ovate, 5–7 pairs per pinna.

Flowers bright yellow; calyx hairy; lobes yellow with glandular-fringed margins, the lower lobe 9–10 mm, about 3 times longer than the others, boat shaped and cupping the stamens; 4 petals 8–10 mm, the banner larger than the others. Pods 6–10.5 × 2.7–3.6 cm, more or less oval, flattened, red-brown to tan, with a conspicuous rim and short stipe, indehiscent and dry, developing in summer and persisting at least until March. Flowering June–July.

Slopes and canyons; thornscrub and tropical deciduous forest; southern and east-central Sonora, 10–760 m (30–2500 ft). Also southwestern Chihuahua and Sinaloa to Oaxaca and Tamaulipas, and disjunct in Yucatán. The strange lower calyx lobe is similar to that of *C. pumila*, a Sonoran Desert shrub.

In both Sonora and Sinaloa the wood is prized for construction, and fence posts are said to last 100 years. The wood is used as a substitute for that of amapa (*Tabebuia*). The posts often take root and grow into trees, forming living fences. Larger palo colorado trees have become scarce in many areas, especially the Alamos region, due to serious overharvesting, although technically the trees are protected by law. In Tucson this tree has shown promise as a landscape subject and tolerates at least several degrees of frost.

***Caesalpinia sclerocarpa.*** *Ébano, palo freno; tubchi* (Mayo) (figure 144)

Trees 7–12 (15) m. Trunk thick and well developed, to 3 m and 50–75 cm in diameter. Bark at first smooth, mottled gray, white, and red-brown, with age curling and peeling away in large flakes; lenticels light brown. Glabrous except minute yellow-brown hairs on inflorescences, pedicels, and sepals. Stipules deciduous, often 2.5–3 mm, thin, with fringed margins. Leaves drought deciduous, often 8–12 cm, with 2 or 3 pairs of pinnae plus a terminal pinna; leaflets 12–20 mm, mostly broadly elliptic to oval or obovate, 3 or 4 (5) pairs per pinna.

Racemes often 5–10 cm on peduncles 1–2 cm. Flowers yellow with red spots, approximately 1.5 cm wide. Pods 4–7 × 1.5–2.5 cm, more or less oval, woody, glabrous, approximately 1 cm thick, moderately flattened, convex, red-brown to tan, with a conspicuous rim and a short stipe, indehiscent, developing in summer and persisting at least until the following March. Seeds approximately 7 mm, brown, several per pod. Flowering at the end of the summer rains.

Arroyos in plains of coastal thornscrub. Extreme southwestern Sonora; 15–30 m (50–100 ft) elevation in the vicinity of Agiabampo and Camahuiroa, growing with *Capparis flexuosa, Diospyros sonorae, Guazuma ulmifolia*, and *Lysiloma watsonii*. Southward in the lowlands of western Mexico to Oaxaca, mostly in tropical deciduous forest.

The wood is highly prized and excellent for roof beams.

### *Chloroleucon*

Trees and shrubs. Flowers mimosoid. Dry tropical America; 10 species. Reference: Barneby and Grimes (1996).

***Chloroleucon mangense*** var. ***leucospermum*** [*Pithecellobium leucospermum; P. undulatum; Chloroleucon undulatum*]. *Palo fierro, palo pinto; huorijata* (Guarijío); *ooseo sutu* "lion's claws" (Mayo); *pinto uus* (Pima Bajo) (figure 145)

Trees 5–8 m, with very hard wood and often spreading branches; some branches with paired, straight, stout, sharp

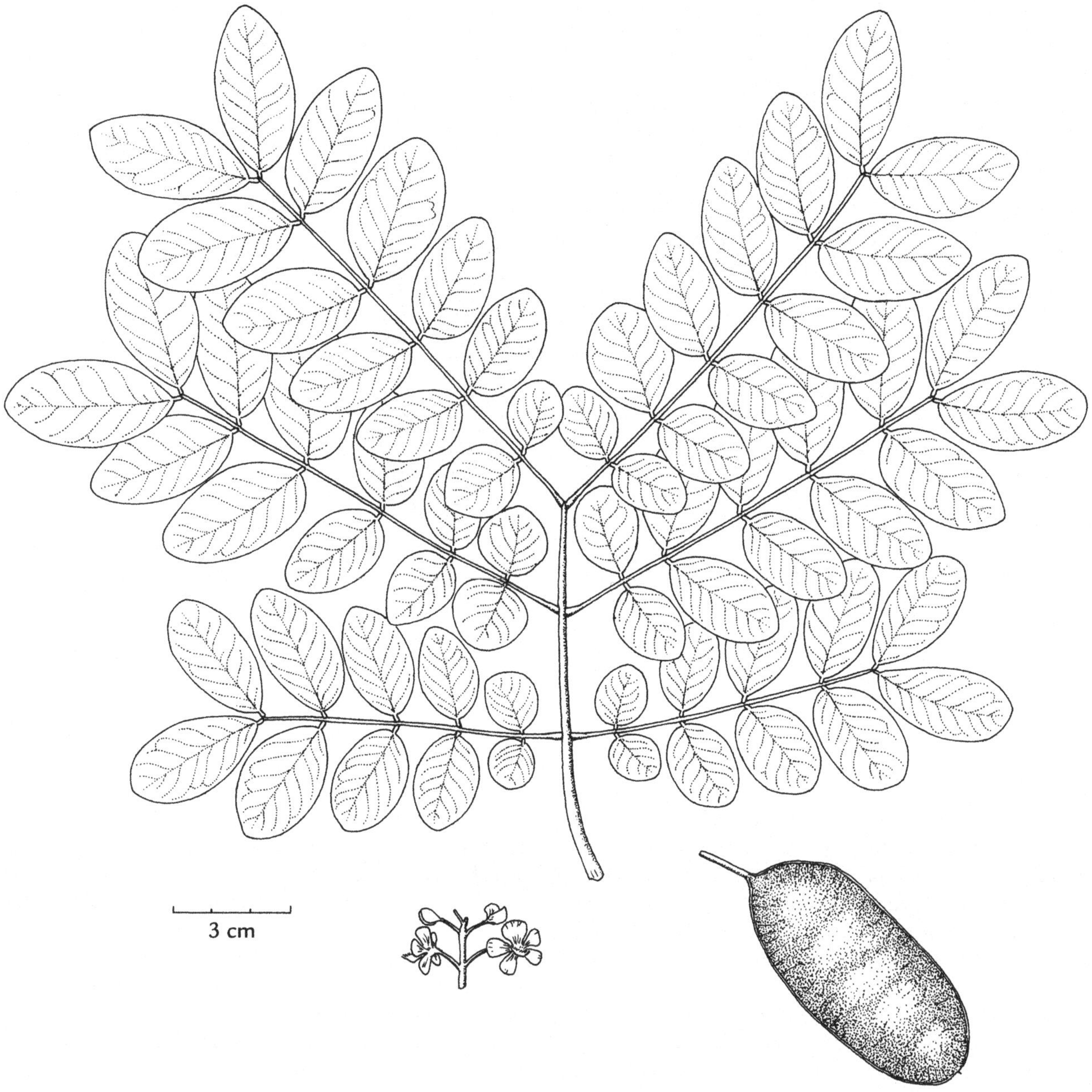


Figure 143. *Caesalpinia platyloba*. Drawn by M. B. Johnson.

thorns to 2.5 cm and often bearing lenticels. (The thorns appear stipular but actually are modified sterile peduncles). Bark smooth, mottled gray, white, and brown during dry seasons and greenish during the summer rainy season, peeling in large flakes and curled, longitudinal strips (one common name reflects that the trunk is "pinto"). Leaflets glabrate, the herbage otherwise mostly with short, whitish hairs. Dry-season buds covered by broad, thin, and striated scales. Stipules mostly lacking, when present to approximately 10 mm, slender and thin, striated, and quickly deciduous (falling long before the leaf is fully enlarged). Leaves drought deciduous, 6.5–14.5 cm, with 3–6 pairs of pinnae; leaflets 5–13 mm, 5–12 pairs per pinna, often with a prominent nectary gland on the petiole and smaller ones on the rachis and rachillas.

Flowers in axillary heads on slender peduncles 1.0–2.5 cm, the fruiting peduncles thick and to 3.5 cm. Flowering with the newest, emerging leaves or just before leafout; flowers cream white. Pods 13.5–27.5 × approximately 1 cm, linear, moderately flattened, curved, sessile, moderately constricted between the seeds, dry, tardily dehiscent, the valves partially separating. Flowering June–July (January).

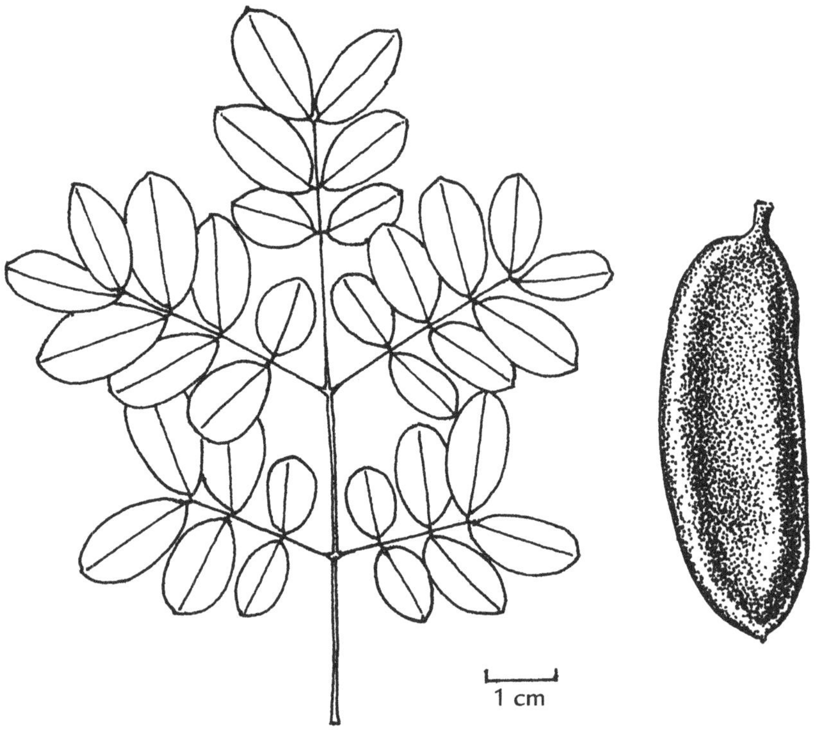

Figure 144. *Caesalpinia sclerocarpa*. Drawn by M. B. Johnson.

Slopes, canyons, valleys, plains, and arroyos; tropical deciduous forest and occasional in thornscrub; southeastern and east-central Sonora; occasionally in extreme southwestern Sonora; 45–745 m (150–2445 ft). The northernmost populations occur between San Javier and Tepoca. Cape Region of Baja California Sur, Sonora, and southwestern Chihuahua to Central America. This species ranges from northern Mexico to South America and the Caribbean and includes six varieties, but only one occurs in Mexico.

The forked branches are made into the three-pronged *horqueta* to support the large clay *ollas* (water vessels) seen on Sonoran patios. The trunks are often hollow, holding water or providing habitats for lizards and other animals.

### *Conzattia*

Large trees. Flowers caesalpinioid. Mexico; probably 2 species.

**Conzattia multiflora** [*C. arborea*]. *Joso de la Sierra, palo joso; sejhuahuí* (Guarijío); *joso* (Mayo); *navío* (Sinaloa) (figure 146)

Unarmed, straight-boled overstory trees 12–25 m; trunks tall and as slender as 50–60 cm in diameter, with a high, open canopy, the wood pungent-smelling, weak, and brittle (*vidriosa y quebradiza*). Bark whitish and smooth except for rough, orange lenticels. Stipules absent but with unusual, pubescent glandular areas in the position of stipules. Emerging leaves densely silky-haired, becoming glabrate. Leaves present only during the summer rainy season, 32–70+ cm, even-bipinnate; 10–14 pairs of pinnae; leaflets 15–25 mm, oblong with a small mucronate tip, 12–21 pairs per pinna; without nectary glands. Inflorescences of many crowded racemes 9–22 cm, emerging at stem tips with new leaves. Flowers bright yellow; petals approximately 1 cm. Pods 4.5–10 cm, flattened, red-brown, persistent for several months and ultimately dehiscent, the valves not coiling. (From a distance the large clusters of reddish pods have the appearance of flowers.) Flowering June–July.

Steep slopes and ridges; southeastern Sonora; 200–1200 m (655–3935 ft). Tropical deciduous forest in western Mexico; Sonora, and southwestern Chihuahua to Oaxaca and Puebla.

Individuals or small stands are widely scattered in mountainous areas. Apart from the palms, this species has the largest leaves of any Sonoran tree. It towers over the low cover of scrubby foothill trees as the tallest member of the hillside forest. At the height of the dry season the brown of the sere hillsides is punctuated by yellow patches of flowering conzattias.

The Sonoran plants have often been identified, apparently incorrectly, as *C. sericea*. The latter species is reported to have fewer pinnae (three or four pairs) and fewer leaf-

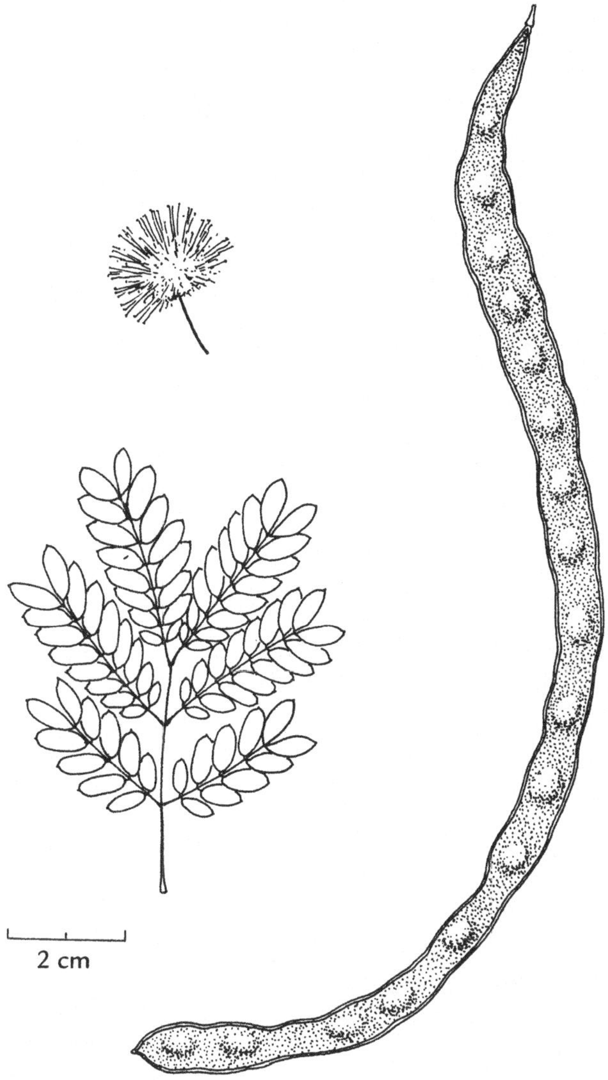


Figure 145. *Chloroleucon mangense* var. *leucospermum*. Drawn by M. B. Johnson.

lets (ca. nine pairs and sericeous on both surfaces) than *C. multiflora*.

### *Coursetia*

Trees, shrubs, and perennials or rarely annual herbs. Flowers papilionoid. Southwestern United States to Argentina; 38 species. Reference: Lavin (1988).

**Coursetia glandulosa.** [*C. microphylla*] *Sámota, samo prieta; samó* (Guarijío); *causamo* 'mountain samo' (Mayo); *hehe ctoozi* (Seri); *cau samo* (Yaqui) (figure 147)

Many-stemmed large shrubs or somtimes treelike, 2–7 m, in southeastern Sonora, with several slender trunks, the wood flexible and hard. Bark smooth and gray. Young herbage densely pubescent with short, appressed hairs, and becoming sparsely pubescent with age. Stipules subulate, to approximately 5 mm. Leaves gradually drought deciduous (winter deciduous in northern Sonora), even-pinnate; leaflets 10–22, 0.5–4 cm, thin, elliptic to oval, the midrib extending into an apiculate point. Summer rainy-season leaves luxuriant, sparsely pubescent, reaching 13.5 cm with larger leaflets; spring dry-season leaves fewer, densely pubescent, often 1–2.5 cm with smaller leaflets.

Flowering branches, pedicels, calyx, ovary, and pods densely pubescent with stalked glands, the glands glistening yellow-translucent when fresh. Flowers single or 2–10 in axillary racemes to 3 cm, subtended by small, deciduous bracts, pealike, 10–16 mm; calyx red-brown, densely glandular; banner petal white, often with a faint red tinge, partially reflexed, the wing petals pale yellow, enclosing the keel petals. Pods 3–7.5 cm, tan, glandular-sticky, constricted between the seeds, dry, dehiscing gradually or popping open, the valves loosely coiling; seeds 1–6. Often flowering January–February in southern Sonora and (February) March–April (May–November) in northern Sonora; often with mass flowering after rains.

Slopes, ridges, valleys, canyons, and arroyos; Sonoran desertscrub, thornscrub, tropical deciduous forest, and lower elevations in grassland. Through much of Sonora, except extreme northwestern part of the state; near sea level–1150 m (3770 ft). Southern Arizona and Chihuahua to Oaxaca, and Baja California Sur. In desert regions densest on north and east-facing slopes and at higher elevations.

In cultivation it is hardy to about −5°C (23°F). The stems are sometimes encrusted with orange-colored lac produced by the ant-tended scale insect *Tachardiella*. The lac, known as *goma* or *gomilla de Sonora*, was used by native Americans as an all-purpose adhesive (Felger and Moser 1985). The plants have some horticultural potential, primarily for the attractive foliage and flowers.

### *Diphysa*

Unarmed woody shrubs or small trees, or the twigs sometimes spinescent. Stipules deciduous. Leaves glabrate, odd-pinnate; leaflets mostly subopposite. Flowers papilionoid, bright yellow, in short axillary inflorescences, with a conspicuous, promptly deciduous to persistent small bract at the apex of the pedicel. Pods indehiscent, dry, often inflated (the thin exocarp separates from the hard endocarp to form lateral bladders), the base narrowed to a stipe. Mexico to northern South America; 15 species.

There are four species in Sonora. *D. thurberi*, a small shrub in mountains in the eastern part of the state, and *D. racemosa*, a shrub with gummy-viscid pubescence in

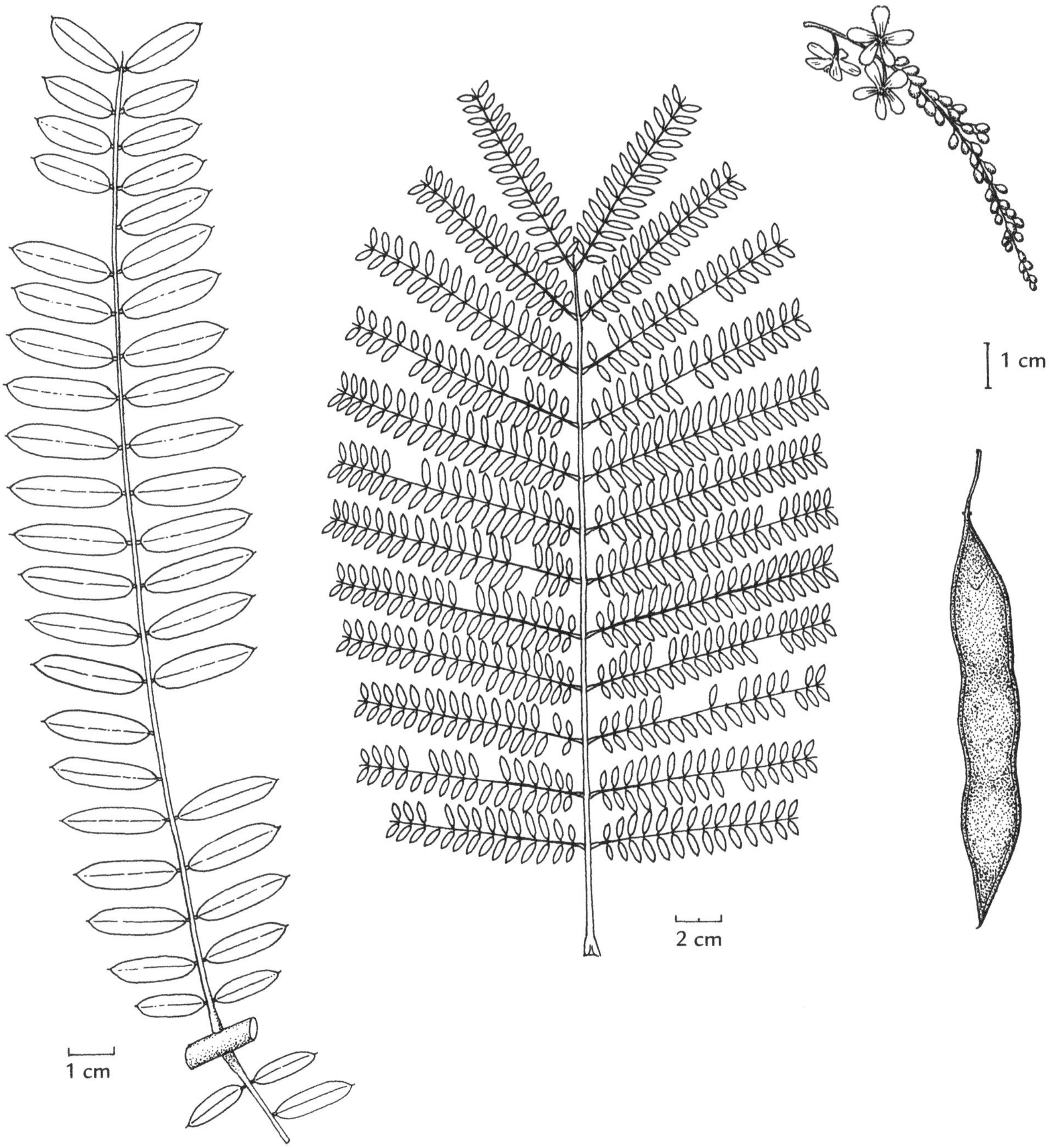


Figure 146. *Conzattia multiflora*. Drawn by M. B. Johnson.

thornscrub and tropical deciduous forest in southeastern Sonora.

1. Trunk and branches not corky; leaflets glandular punctate on both surfaces. ____________ ***D. occidentalis***
1' Twigs and branch corky winged, the trunk also corky; leaflets glandular punctate only on upper surfaces. ***D. suberosa***

***Diphysa occidentalis.*** *Güiloche, palo nahuila; jusiro* (Guarijío); *guicobo* (Mayo) (figure 148A)

Shrubs in desert regions, and small trees to at least 6 m in thornscrub and tropical deciduous forest; very hard wood; usually with several trunks. Bark smooth, gray. Leaves drought deciduous, 4.5–15.5 cm; leaflets often 10–15 per leaf, 9–16 mm (longer during the summer rainy season

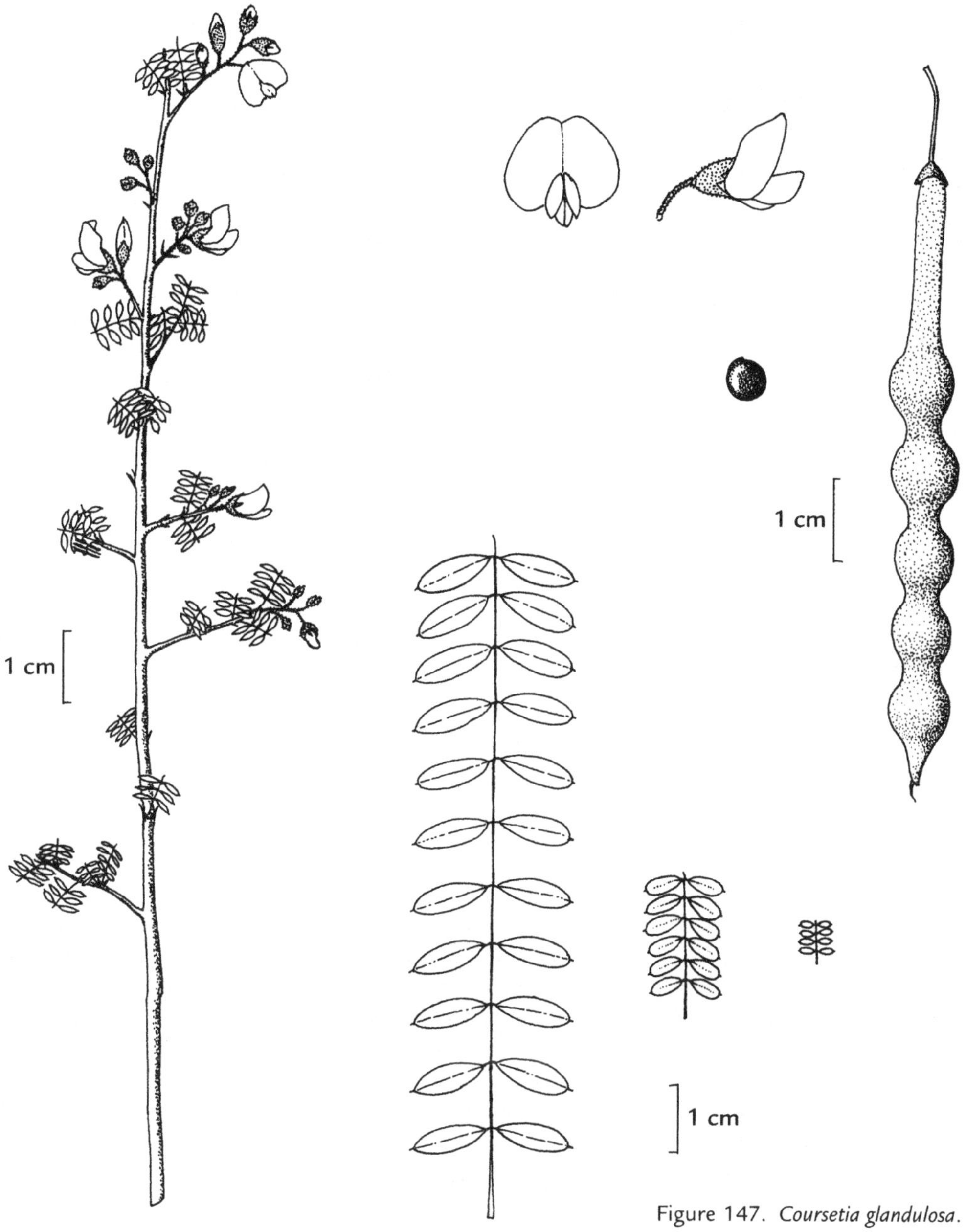


Figure 147. *Coursetia glandulosa*. Drawn by M. B. Johnson.

and much shorter during dry seasons), oblanceolate to lanceolate or oblong, both surfaces minutely glandular punctate, the tip often with a minute projection (mucro).

Flowers approximately 2 cm wide, bright, clear yellow and showy, the ground sometimes littered with fallen flowers. Pods 4.5–7.5 (8.5) × (1.0) 1.5–2.0 cm, inflated, puffy-papery (membranous), mostly ripe in late summer. Peak flowering with summer rains and sometimes mid-winter to April.

Slopes, valleys, canyons, and arroyos; Sonoran desertscrub, thornscrub, and tropical deciduous forest; north-central (east of Cucurpe) to southern Sonora; near sea level–760 m (2500 ft). Southward in western Mexico to Guerrero.

The conspicuously inflated, thin walled, straw-colored pods are unique among Sonoran trees.

***Diphysa suberosa***. *Corcho; juyapori* (Guarijío) (figure 148B)

Slender, erect shrubs or trees, 2–8 m, with a thick trunk and small crown. Bark dark gray, with conspicuous, irregular, interrupted thick corky ridges or wings often 5–7 cm deep. (The four wings of each growth pulse are staggered at a slightly different angle from those of the previous one. Even the small branches have conspicuous corky-winged bark.) Twigs often branching at right angles, the tips often spinescent. Leaves 4.5–8 cm; leaflets 7–15,

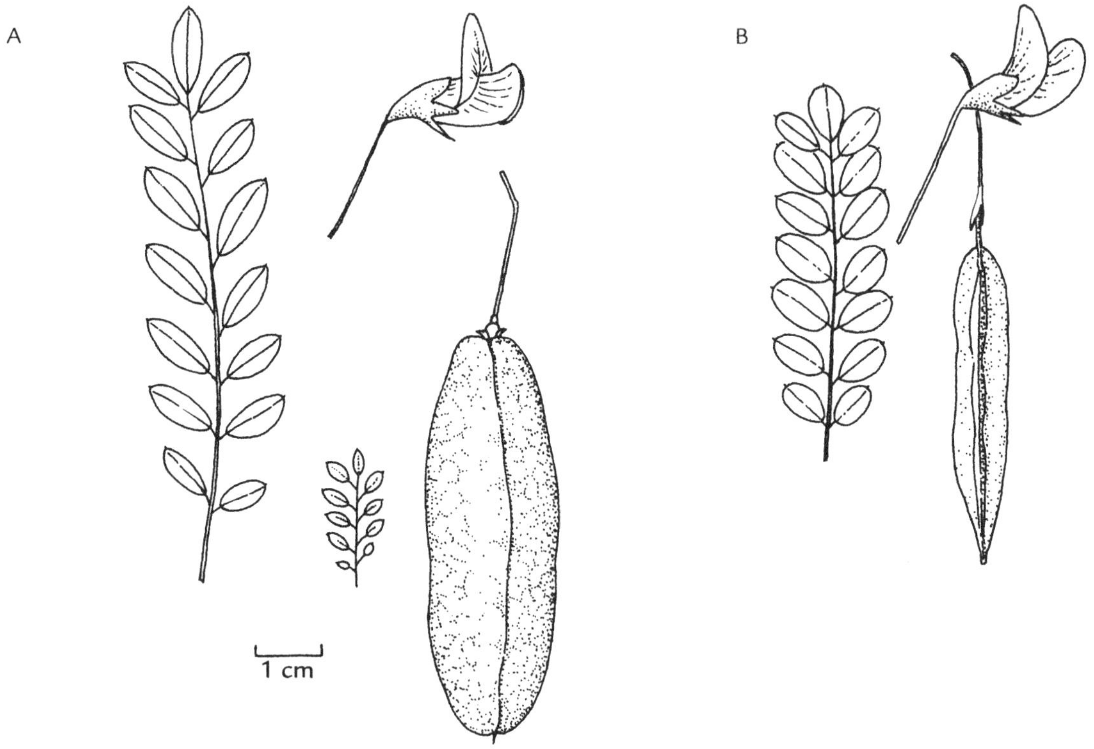


Figure 148. *Diphysa*. (A) *D. occidentalis*, (B) *D. suberosa*. Drawn by M. B. Johnson.

7–17 mm, obovate, with minute glands on the upper surfaces only.

Flowers yellow, approximately 2 cm wide. Pods 3.2–6.7 × 0.8–1.2 cm, with a short stipe, dark brown, persistent, pendent on slender pedicels, moderately papery-puffy and inflated. Flowering March–July.

Slopes, valleys, and arroyos, mostly on indurated ash; thornscrub, tropical deciduous forest, lower elevations in oak woodland, and the margin of Sonoran desertscrub; central and southeastern Sonora; 60–1065 m (195–3500 ft). Also southwestern Chihuahua and southward in western Mexico to Oaxaca.

The larger, tree-sized plants occur in limited areas in the upper reaches of tropical deciduous forest and adjacent oak zones; this species is fairly common in these habitats in the Guarijío country of southeastern Sonora. It is apparently localized and uncommon over much of the rest of its range in Sonora.

### *Erythrina*

Trees and shrubs with soft wood. Flowers papilionoid. Tropical and subtropical regions; 110 species. References: Bruneau (1996), Conn and Snyder-Conn (1981), Sherbrooke and Scheerens (1979), Toledo (1974).

Most New World species have tubular flowers adapted to hummingbird pollination; the banner petal usually long and narrow, and the wing and keel petals much shorter. Most of the Old World and a few New World species have more open flowers adapted to pollination primarily by passerine birds. The flowers of various species of *Erythrina* in Mexico are cooked and eaten.

***Erythrina flabelliformis*****.** Southwest coral bean; *chilicote, pionilla; guaposi* (Guarijío); *jévero* (Mayo); *xloolcö* (Seri) (figure 149)

Shrubs in northern Sonora, and trees in central and southern Sonora to 10 m, with a small, high crown, and a well-formed, erect trunk often 60–100 cm in diameter, the major branches erect to ascending. Bark in mature forest trees rather smooth, often coppery orange; lenticels conspicuous, the lenticular area often enlarging to become blunt protuberances on the trunk and larger branches, these sometimes developing into stout, conical, blunt spines to 2.5 cm. Wood very soft and light colored. Seedlings and young plants with a swollen caudex. Bark of younger branches smooth, light brown to yellow-gray, becoming dark gray to brown with irregular vertical white stripes ("stretch marks") marking the expansion zones.

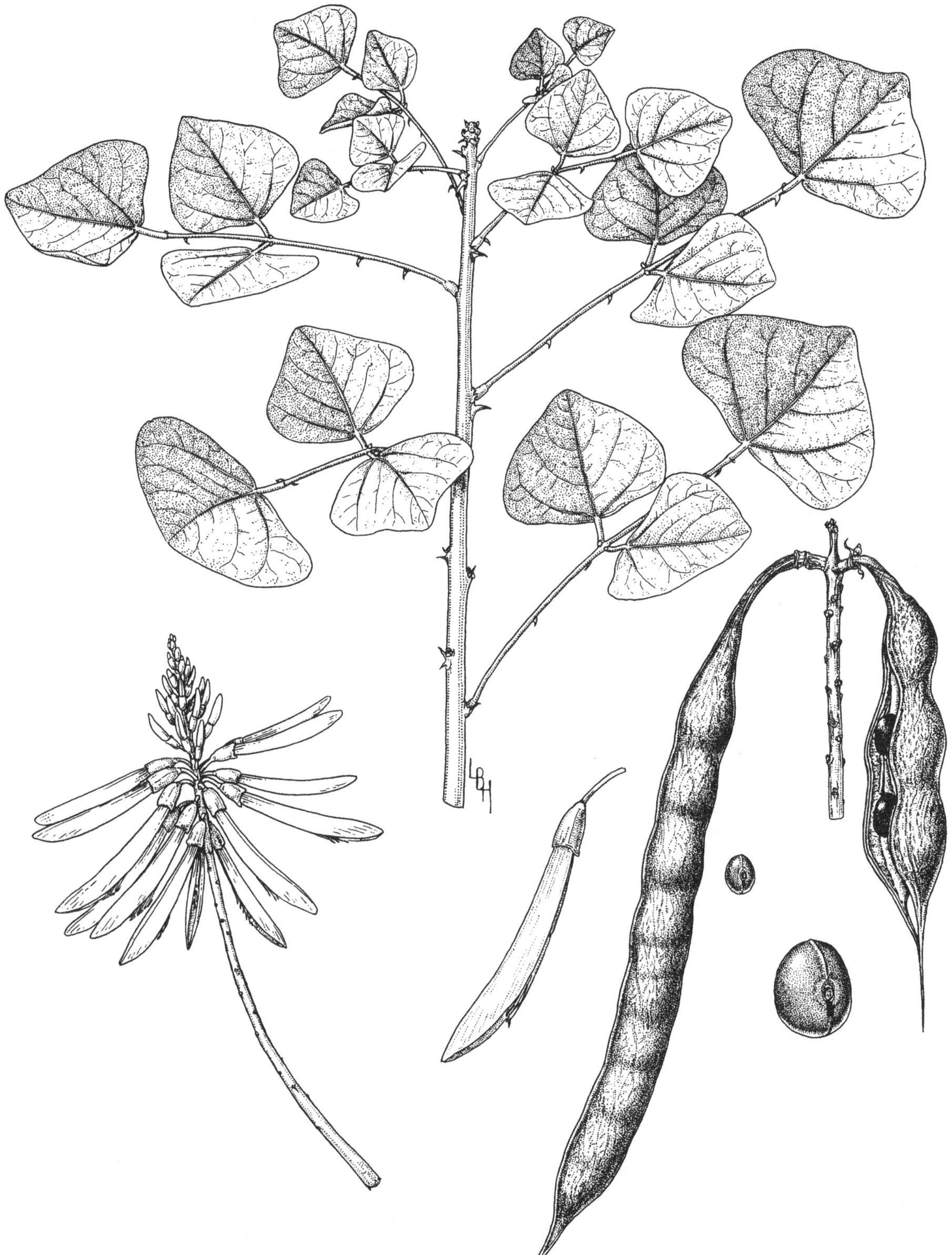


Figure 149. *Erythrina flabelliformis* (Schmutz and Hamilton 1979). Drawn by L. B. Hamilton.

Stems relatively thick, at least some branches with solitary, short, stout, sharp, and recurved spines below the leaf nodes; these spines often persisting on the trunk and becoming broad, transverse, and decurved. Young herbage, calyx, ovary, and young pods densely woolly-pubescent, becoming glabrate. Leaves appear in early summer, often with the first rains, light green, and quickly yellowing and falling with drought in September. Leaves 15–25 cm; leaflets 3, stalked, 3.5–7.5 × (2.8) 5.8–11 cm, broadly ovate to deltoid; leafstalks sometimes with small prickles and with nectary glands between the leaflets.

Inflorescences of terminal and axillary racemes 7.5–21.5 cm. Calyx short and spathelike, splitting on one side; with a conspicuous green nectary gland exuding nectar from the tip of the bud, turning dark and becoming less functional after the flower opens. Banner petal scarlet-red, narrow, to 5.5 cm, the keel and wing petals reduced. Pods 13–26.5 cm, somewhat woody, terete, constricted between seeds, commonly persisting for at least several months, eventually splitting to reveal the seeds, which gradually fall away. Seeds 1–12, approximately 15 mm and about two-thirds as wide, bright red, or occasionally yellow, orange, or light brown; immature seeds porcelain white. Flowering May–July, when leafless or with the first new leaves; pods ripen June–November; seeds germinate with summer rains of the next year.

Gentle slopes to ridges, cliffs, canyons, and arroyos; tropical deciduous forest, thornscrub, Sonoran desertscrub, grassland, and oak woodland. Widespread in Sonora except the extreme northwest, 10–1830 m (30–6000 ft). Western Mexico from Michoacán to southeastern Arizona and extreme southwestern New Mexico, and Baja California Sur.

Coral bean is frost sensitive, and in cultivation the stems are killed at –4°C (25°F). Plants in the northern populations, or at high elevations in the south, grow among large rocks that serve as heat reservoirs. Stems projecting above the rocks are often freeze-killed, resulting in multiple-stem shrubs with massive underground caudices. Farther south, such as southeastern Sonora, where frost is light or absent, *chilicote* develops into a distinctive tree and is not restricted to rock habitats. The extrafloral (calyx and leaf) nectaries produce fructose- and glucose-dominant exudate. Ants attracted to these sweets patrol the plants, protecting them from herbivores. In contrast, the floral nectaries produce sucrose-dominant nectars attractive to hummingbirds (Sherbrooke and Scheerens 1979).

Erythrinas are recommended as landscape trees at lower elevations in Sonora. They are easily grown from seeds and stem cuttings made during the active growing season. The soft, easy-to-carve wood does not split on drying and has been used as a substitute for cork (stoppers for bottles and gourds), to make stools and toys, and by the Mayo, Opata, and Yaqui for dance masks. Young green pods are cooked and eaten by the Gaurijío and Mayo, and the Guarijío have used the logs for rafting across rivers.

### *Eysenhardtia*

Flowers papilionoid. Arizona to Texas southward to Central America; 11 species. Reference: Lang and Isely (1982).

***Eysenhardtia orthocarpa.*** Kidneywood; *palo dulce; pahuió* (Guarijío); *baiguo* (Mayo) (figure 150)

Unarmed slender-stem shrubs, sometimes small trees to 6 m; trunk to 15 cm in diameter; bark gray to light tan, becoming moderately fissured and separating into strips or irregular plates, giving a shaggy appearance to older limbs and trunks. Leaves tardily winter and/or drought deciduous (the leaflets often falling first, leaving the semipersistent leafstalks), odd-pinnate, 5–15.5 cm; leaflets 11–55, 6–25 mm, oblong, the lower surfaces dotted with conspicuous brown glands, giving young foliage a pungent tangerine-like fragrance when crushed, later like green mango. (This species and *Psorothamnus spinosus* are the only aromatic tree legumes in Sonora. Also see *Senna atomaria*.)

Inflorescences of densely flowered terminal and axillary racemes 3.5–17.5 cm. Flowers white, 10–12 mm, sweet scented. Pods 9–11 mm, 1–seeded, crowded, reflexed, flattened, brown, dry, indehiscent, and persistent for several months. Flowering April–September.

Slopes and canyons; Sonoran desertscrub, thornscrub, tropical deciduous forest, grassland, and oak woodland; Sonora except northwest one-third of state, 10–1645 m (35–5400 ft), and to 2000 m in southwestern Chihuahua. Sinaloa and Chihuahua to Arizona and southwestern New Mexico.

Two varieties are sometimes recognized. Variety *tenuifolia*, distinguished by glabrous rather than minutely hairy leaflets, smaller leaves, leaflets, and corollas, and longer stipules, replaces var. *orthocarpa* at lower elevations in desertscrub and thornscrub of western Sonora and Sinaloa. *Eysenhardtia orthocarpa* is replaced southward by the closely related *E. polystachya*.

The plants are browsed by the white-tail and mule deer. In cultivation in southern Arizona it can grow 1–1.5 m per year with adequate irrigation, and is hardy to –8° C (17°F). It is well suited as a patio tree.

### *Haematoxylum*

Flowers caesalpinioid. Three species in the American tropics and subtropics and one in Africa.

The New World trees (*H. brasiletto* and *H. campechianum*), known as logwood or *brasil*, were a major source of red and black dye and were one of the major exports during colonial times, and continued to be important well into the twentieth century. Commercially there was no distinction

Figure 150. *Eysenhardtia orthocarpa* (Benson and Darrow 1945). Drawn by L. B. Hamilton.

between these two species. The wood was also known as campeachy, or peachwood (from Campeche, Mexico, one of the principal ports of shipment) (see Record and Hess 1943 and *Caesalpinia*).

*Brasil chino* is a spiny, undescribed *Haematoxylum* shrub species endemic to southeastern Sonora and nearby northern Sinaloa.

***Haematoxylum brasiletto.*** Brazilwood; *brasil; huichilaco, huichachago* (Guarijío); *júchajco* (Mayo); *huchahko* (Yaqui) (figures 151–153)

Shrubs or trees to 10 (12) m, with deeply sculptured, anastomosing fluted ridges on the trunks and limbs; wood very hard with reddish heartwood; smaller branches often zigzag. Bark gray, becoming moderately rough textured, often mottled with whitish to grayish banding of lichens. Twigs often thorn tipped or a single, straight, rigid thorn above the leaf node, or some branches thornless, the leaves often fascicled at the nodes. Glabrous except newest growth with sparse, forward-pointing, wormlike, yellowish appressed hairs. Stipules small and bristlelike. Leaves 2.5–8.3 cm, drought deciduous, sometimes partially twice pinnate but mostly pinnate with 2–4 pairs of leaflets, the leaflets 7–28 mm, broadest above the middle (cuneate to obcordate), notched at apex, with many well-marked parallel veins.

Inflorescences of short axillary racemes. Flowers showy, bright yellow and red-lined at base of the banner, 8–20 mm wide. Pods 1.8–5 cm, oval to elliptic, flattened, tough and somewhat papery, tardily dehiscent by a longitudinal suture along the flat surface, probably wind disseminated, the young pods reddish or green and straw colored at maturity. Seeds 1–4, very flat, 2-lobed. Flowering at various seasons in response to rain.

Slopes, ridges, plains, canyons, and arroyos; thornscrub, tropical deciduous forest, and margins of Sonoran desertscrub; central and southern Sonora; near sea level–885 (1065) m [2900 (3500) ft]. Mexico including Baja California Sur, Sonora, and Chihuahua to Colombia and Venezuela.

The stems begin to develop the unique fluted ridges at about 2.5 cm in diameter. The enlarging stems produce additional ridges and form living sculptures, as if they had melted and then solidified. Brasil is recommended for landscape use in hot, semiarid climates. It is distinctive as a specimen plant or small patio tree or as a spinescent security hedge. Twig die-back occurs at –4°C (25°F), and the plants are severely damaged at –6.5°C (20°F).

Figure 151. *Haematoxylum brasiletto*; near Alamos. Photograph by K. S. Bunnell.

"The wood is excellent fuel and burns with a greenish flame. It is also used in the construction of houses, corrals, etc. A light-red dye or ink and also rosary crosses are made from the deep-red heartwood" (Gentry 1942: 131). The wood burns even when freshly cut and is locally prized as firewood. It has a long and important history for medicinal purposes (Standley 1922).

### *Havardia*

Trees with strong, hard wood; branches mostly armed with paired, rigid, and persistent stipular spines. Leaves twice pinnate, the petioles (in Sonora) bearing a nectary gland. Flowers mimosoid, with strong, sweet fragrance. Pods straight, broadly linear, and flattened, gradually dehiscent. Mostly dry tropical and subtropical North America, Texas to Central America; 5 species. Reference: Barneby and Grimes (1996).

1. Trees with a well-formed trunk and high crown; leaflets 5–7 mm; flower heads solitary or in small racemes with slender peduncles; flowers short-pedicelled. ***H. mexicana***

1' Multiple-stem large shrubs or small trees; leaflets 2–3.5 mm; flower heads solitary on stout peduncles; flowers sessile. ______ ***H. sonorae***

**Havardia mexicana** [*Pithecellobium mexicanum*]. Mexican ebony; *palo chino; huahuichó* (Guarijío); *chino* (Mayo and Spanish) (figure 154A)

Upright trees to 12 (15) m, forming a high, spreading canopy. Bark smooth and gray, with age becoming thick, fissured, and dark colored. Twigs slender and wiry, the spines paired, moderately curved, painfully sharp, often 1–3 mm. Herbage with short, white hairs; nectary gland just below lowest pair of pinnae. Leaves 2.5–6 (8.5) cm, dull gray-green; pinnae 2–6 pairs, leaflets 5–7 mm, 5–10 pairs per pinna.

Flower heads approximately 1.5 cm in diameter, axillary and solitary, or in small terminal racemes; flowers greenish white. Pods 5.4–14 cm, dull brown. Seeds 6–10 mm. Flowering March–May; pods ripen May–June.

Deep soils of river floodplains and large arroyos; Sonoran desertscrub, thornscrub, and tropical deciduous forest; north-central and northeastern Sonora southward to Sinaloa (Comanito, 30 km north of Pericos); 60–1000 m (195–3280 ft). Northern limits within 40 km of the Arizona border in the floodplain of the Río Magdalena and the Altar Valley. Also in the Cape Region of Baja California Sur. In northern Sonora, often forms sparse gallery forests along drainageways with *Acacia occidentalis* and *Prosopis velutina*.

The bark is used for tanning leather and the wood for furniture and stills for making *bacanora* (a mescal- or tequilalike agave liquor). The wood is strong and used for house construction, especially beams, and in southern Sonora the beautiful red heartwood is used for crafts. It should be suitable as a street and yard tree. The upright habit and high canopy produce light, filtered shade. The trees do poorly in rocky soils. Cultivated plants have survived –10°C (14°F) without damage in southern Arizona.

**Havardia sonorae** [*Pithecellobium sonorae*]. Sonoran ebony; *jócono; palo gato* (Sinaloa); *jócona* (Mayo); *wo' kohna* (Yaqui) (figure 154B)

Multiple-trunk or multiple-stem, thicket-forming shrubs or trees mostly 3–7 m (exceptional trees to 12 m occur near San Javier). Pubescent throughout. Painfully sharp, paired stipular spines to 3 mm, broad-based, terete, down-curved, and persisting to form transverse rings or bands on the gray bark. Leaves 1.5–4.5 cm; pinnae 1–4 pairs; leaflets 2–4 mm, 8–24 pairs per pinna.

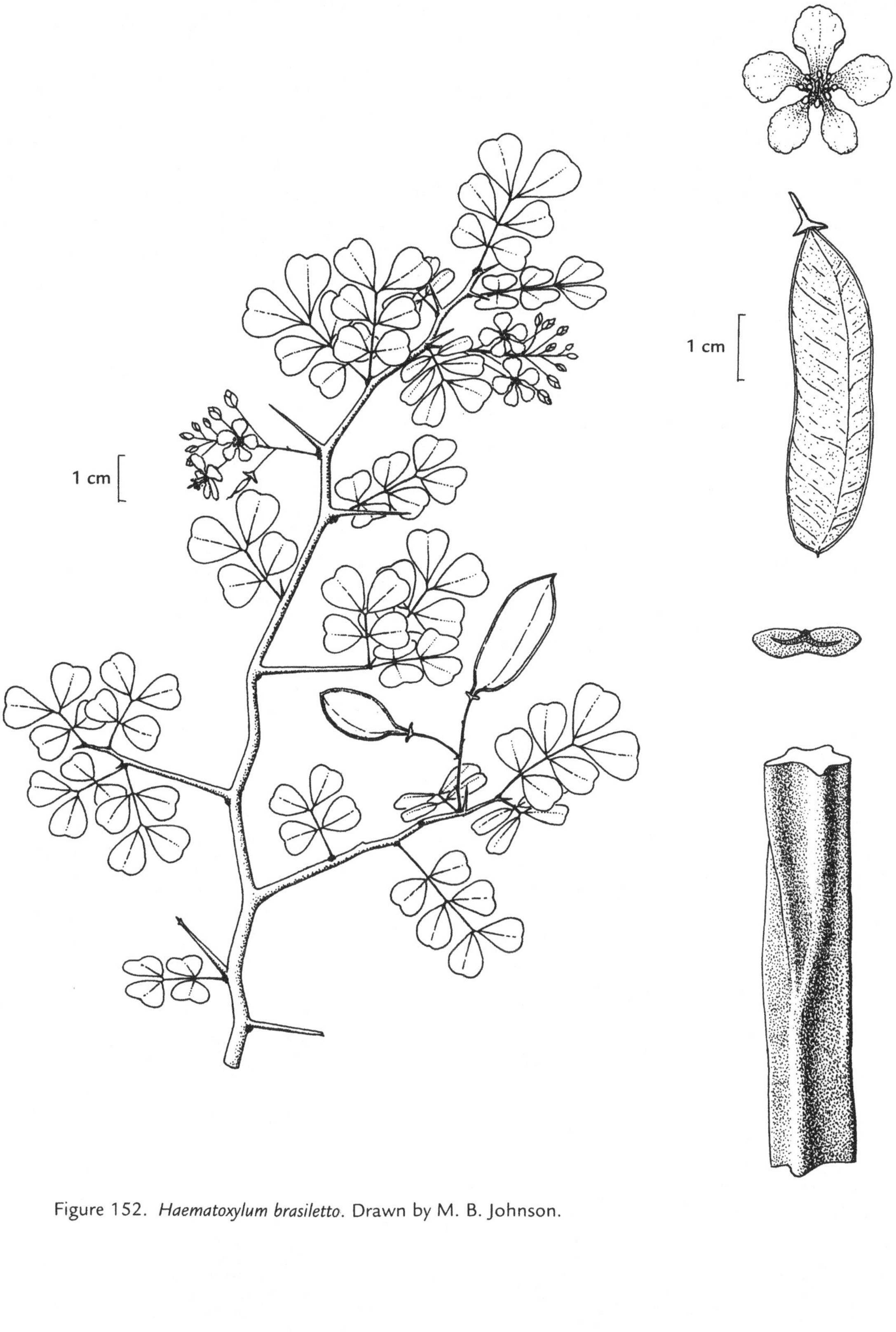


Figure 152. *Haematoxylum brasiletto*. Drawn by M. B. Johnson.

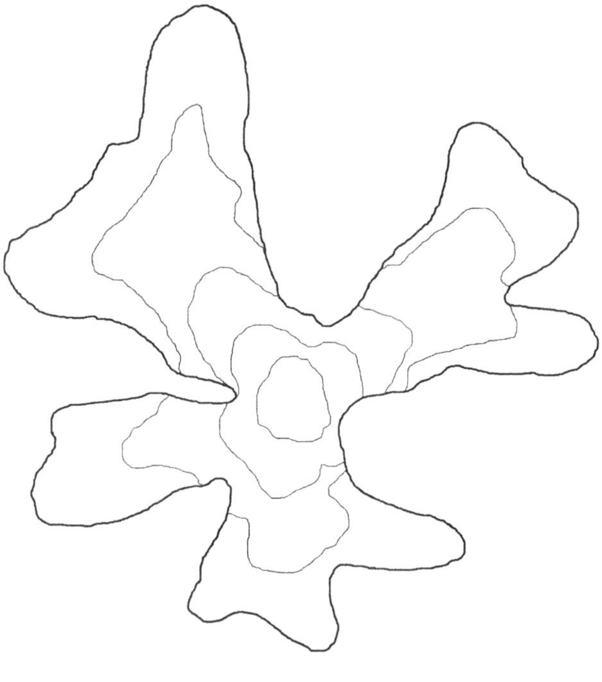

Figure 153. *Haematoxylum brasiletto*. Cross-section of trunk. Drawn by A. Búrquez.

Flowers white, the heads solitary on stout, axillary peduncles. Pods 5–10.5 cm, light colored. Seeds 9–13 mm. Flowering (April) June–August; pods ripening in fall.

Plains, bajadas, arroyos, and sometimes rocky slopes; Sonoran desertscrub, thornscrub, and drier margins of tropical deciduous forest; south-central and southern Sonora, near sea level–455 m (1500 ft). Essentially a thornscrub and desert-edge species; west-central Sinaloa northward in Sonora to the Guaymas region and inland to the vicinity of Tesopaco.

### *Leucaena*

Unarmed shrubs and trees with twice-pinnate leaves; leafstalks channeled, the petiole bearing a prominent nectary gland. [The extrafloral leafstalk glands produce copious nectar exudate, but no relationship between ants and these nectaries has been demonstrated (Hughes 1998)]. Flowers mimosoid, sessile, small, white to cream-white (in Sonora), densely crowded on globose heads; stamens 10, anthers glandless. Pods pendulous, clustered, broadly linear, flat, with a stipe and a thickened rim, dehiscent while still on

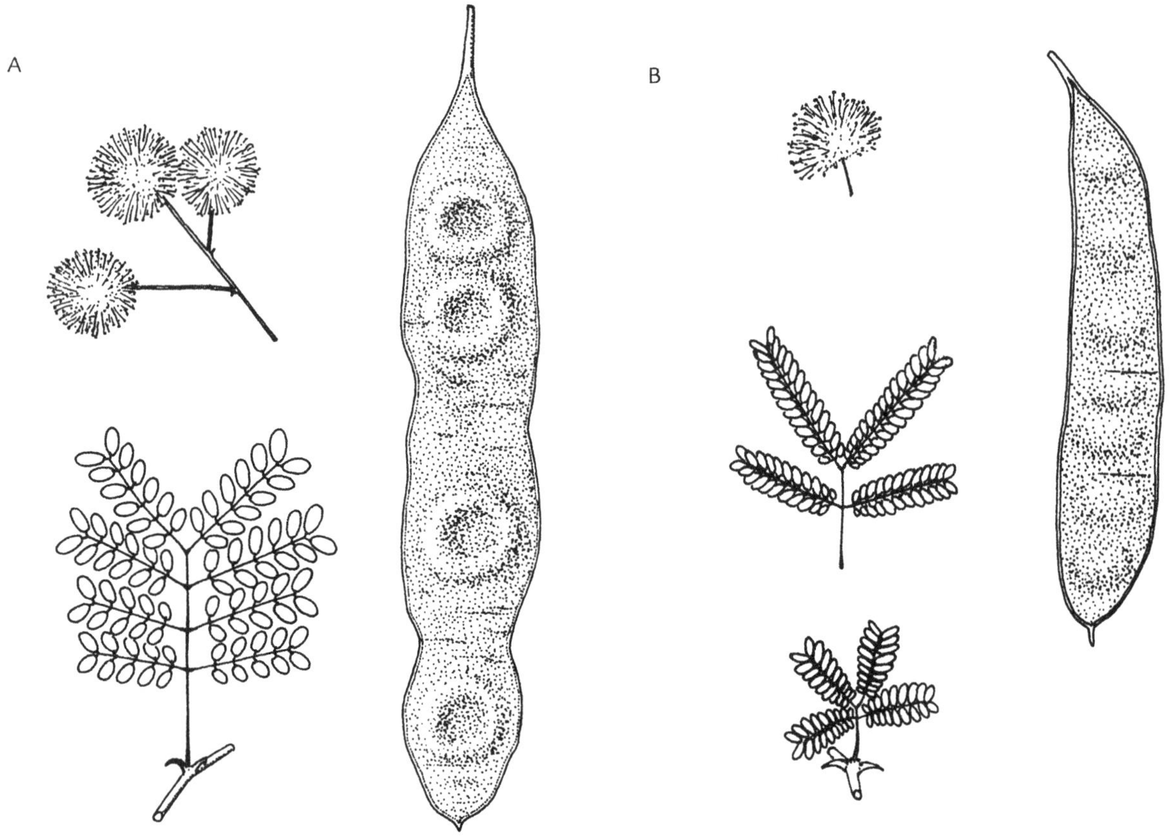


Figure 154. *Havardia*. (A) *H. mexicana*, (B) *H. sonorae*. Drawn by M. B. Johnson.

the tree, the valves separating along both sutures (in Sonora).

Tropical and subtropical regions from southern United States to northern South America, mostly in Mexico and seasonally dry regions; 22 species. References: Hughes (1998), Zárate (1994).

*Leucaena* includes various hybrids and species with edible green pods and seeds; but others have seeds toxic to non-ruminant livestock. Several cultivated and domesticated species provide major economic resources (e.g., *L. esculenta* and *L. leucocephala*).

1. Larger leaflets > 3 cm, 3–7 pairs per pinna. ***L. lanceolata***

1' Leaflets < 2 cm, 12 or more pairs per pinna.
   2. Petiole nectary gland 1.0–1.5 mm in diameter, peg-shaped and cylindrical; pinnae 13–22 pairs per leaf, the leaflets > 30 pairs per pinna; pods 9–13 mm wide. ***L. involucrata***
   2' Petiole nectary 2–3 mm long, crateriform and elliptic; pinnae 5–8 pairs per leaf, the leaflets 12–19 pairs per pinna; pods 18–26 mm wide. ***L. leucocephala***

***Leucaena involucrata.*** *Vara blanca* (figure 155)

Small, multistem trees 2–5 (8) m, with an open, irregular spreading crown; bark smooth and gray. Herbage moderately glabrous or sparsely hairy. Stipules 4–5.9 mm, persistent. Leaves drought deciduous, 15–20+ cm; pinnae (13) 16–22 pairs; leaflets (32) 40–51 pairs per pinna, 4–6 mm, linear, asymmetric at base; nectary on the petiole near the first pair of pinnae, 1.0–1.5 mm in diameter, cylindrical, raised, and peg shaped. Flowers pale cream-white. Pods 11–18 cm × 9–13 mm. Apparently flowering with summer rains; pods ripen September–December.

This enigmatic species was described by Zárate (1994) from specimens collected in the vicinity of El Novillo, along the Río Yaqui drainage in the mountains southeast of Hermosillo. "In the area of El Novillo, it occurs in dry thorn scrub forest where it appears to be restricted to rocky calcareous outcrops. Associated trees include species of *Acacia, Lysiloma, Mimosa, Desmanthus, Piscidia, Tecoma, Rhus, Alvaradoa, Dodonaea,* and *Yucca*" (Hughes 1998: 161). It is otherwise known only from a 1941 collection by Gentry on Sierra Surotato in northern Sinaloa. Hughes (1998: 162) attributed the Sinaloa collection to *L. involucrata* "solely on leaf characters, given the lack of flowers and immaturity of the fruits on that collection. This material differs from the Sonoran collections in being more pubescent." The absence of any other records seems strange. In its leafless condition, however, it might easily be passed by, and from a distance plants on a mountain or hillside might be mistaken for *Lysiloma*.

***Leucaena lanceolata*** var. ***lanceolata*** [*L. microcarpa; L. palmeri; L. sonorensis*]. *Batachín, batillo* (Hughes 1998), *guaje; guasigua* (Guarijío), *hüique* (Mayo) (figure 156)

Slender, single-trunk trees often 6–13 m (or regrowth from cut stumps resulting in multiple-stem shrubs), the major branches erect-ascending; bark gray; or uncommon shrubs 3–4 m at the northernmost localities. Wood soft, the trunks and limbs readily breaking in strong winds. Young twigs, leafstalks, and leaflet bases moderately hairy, becoming glabrous or nearly so. Stipules 1.7–2.3 mm, triangular, broad to narrow, the margins ciliate, soon deciduous. Leaves 23–33 cm; pinnae 3–7 pairs; leaflets 3.3–5.0 cm, lanceolate to ovate or elliptic, 4–7 pairs per pinna. Leafstalks with a prominent raised nectary gland just below the first pair of pinnae, this gland functional on young leaves and soon becoming hard, laterally compressed, and somewhat volcano shaped; also smaller nectary glands below the terminal leaflets and sometimes below other pinnae and leaflets. Flowers white or cream-white. Pods (10) 12–23 × 1.7–2.2 cm, red-brown, glabrous or velvety with minute white hairs and glands. Flowering August–September.

Slopes, canyons, and arroyos; thornscrub and tropical deciduous forest; southeastern Sonora and locally in east-central Sonora (to ca. 8 km south of Tepoca as a shrub); 240–720 m (790–2360 ft); mostly in widely scattered, often small populations and not common. Pacific drainages of Mexico including southwestern Chihuahua to Chiapas and disjunct in central Veracruz and the Cape Region of Baja California Sur. Also var. *sousae* on the Pacific coast of Michoacán, Guerrero, and Oaxaca.

***Leucaena leucocephala** subsp. **glabrata** [*L. glabrata*]. White lead tree, lead tree; *guaje, guacho* (figure 157)

Small trees, occasionally to 15+ m in cultivation. Glabrate or sometimes with sparse, short, white hairs. Stipules 1.5–2.5 mm, triangular with a subulate body and thin, closely appressed, ciliate winged margins. Leaves bright green, 20–31 cm; pinnae 5–8 pairs; leaflets 1.3–1.8 cm, linear-oblong to lanceolate, often moderately asymmetric and curved, 12–19 pairs per pinna; nectary glands on leafstalks between the first pair of pinnae and variously elsewhere on leafstalks and pinnae. Flowers white to yellowish; self-fertile. Pods 12–23 × 1.8–2.6 cm, green to reddish, drying brown. Flowering nonseasonally, especially April–June.

Roadsides, ditches, fields, arroyos, and settlements; Sonoran desertscrub, thornscrub, and tropical deciduous forest; near sea level–720 m (2360 ft).

Since the latter part of the twentieth century, this tree has been widely grown in much of lowland Sonora, and to a lesser extent in the warmer parts of southern Arizona.

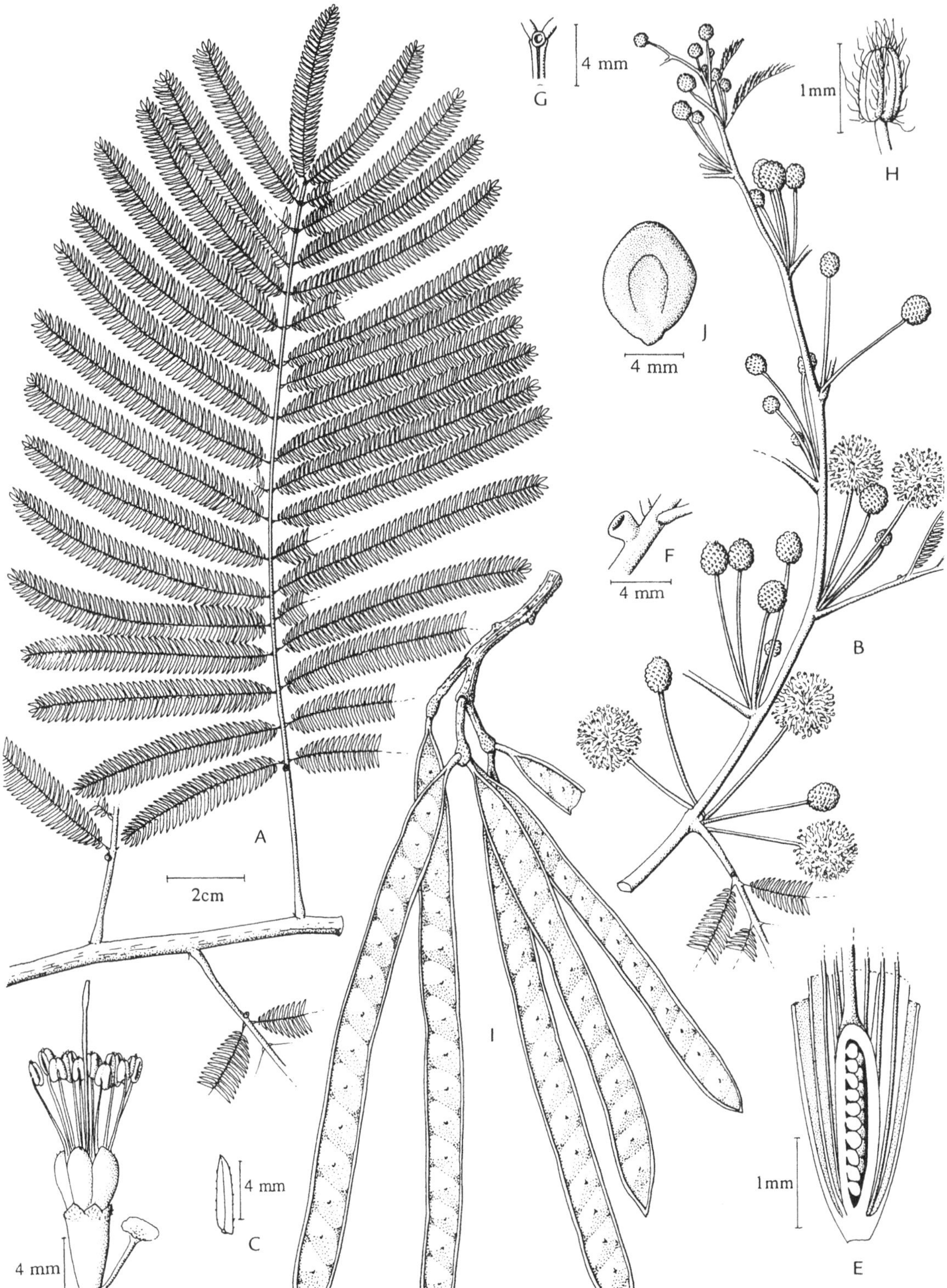


Figure 155. *Leucaena involucrata.* (A) Leaf, (B) flowering shoot, (C) leaflet, (D) flower, (E) longitudinal section of flower, (F) petiolar nectary, (G) rachis tip nectary, (H) anther, (I) pods, (J) seed. (Hughes 1998). Drawn by R. Wise.

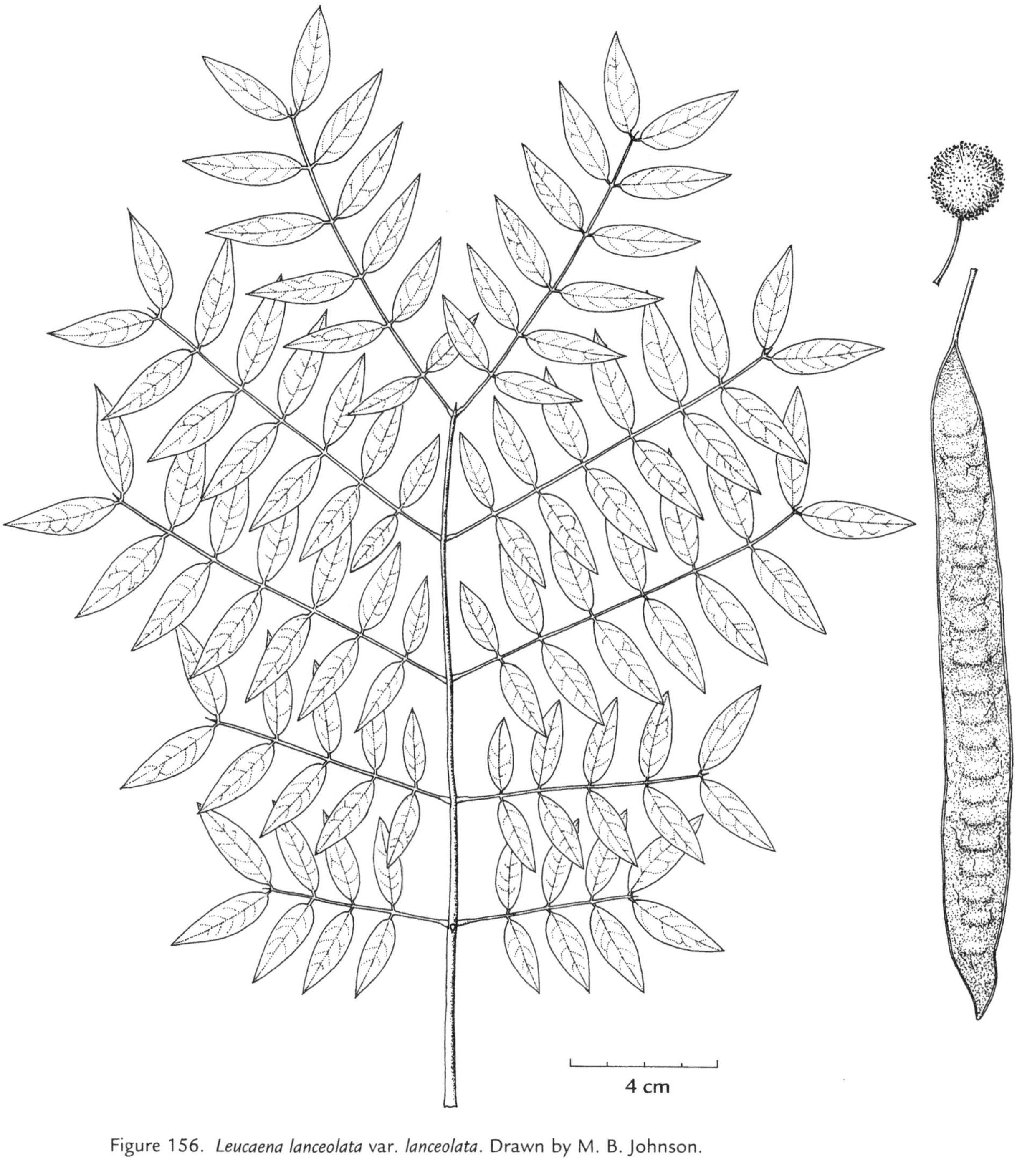


Figure 156. *Leucaena lanceolata* var. *lanceolata*. Drawn by M. B. Johnson.

During this time it has become weedy in urban areas, along roadsides, and other disturbed habitats and is apparently spreading rapidly. It is highly frost sensitive, and its northern limit is probably determined by freezing weather as well as drought.

This tree has long been grown through much of Mexico and may have been cultivated and selected in Mexico in pre-Columbian times. The natural distribution remains unknown. The tender green seeds and pods are important in traditional southern and central Mexican cuisine but are little used in Sonora.

It "has been widely introduced outside Mexico and Central America only in the last few decades, but its active promotion in tropical reforestation means that it is now

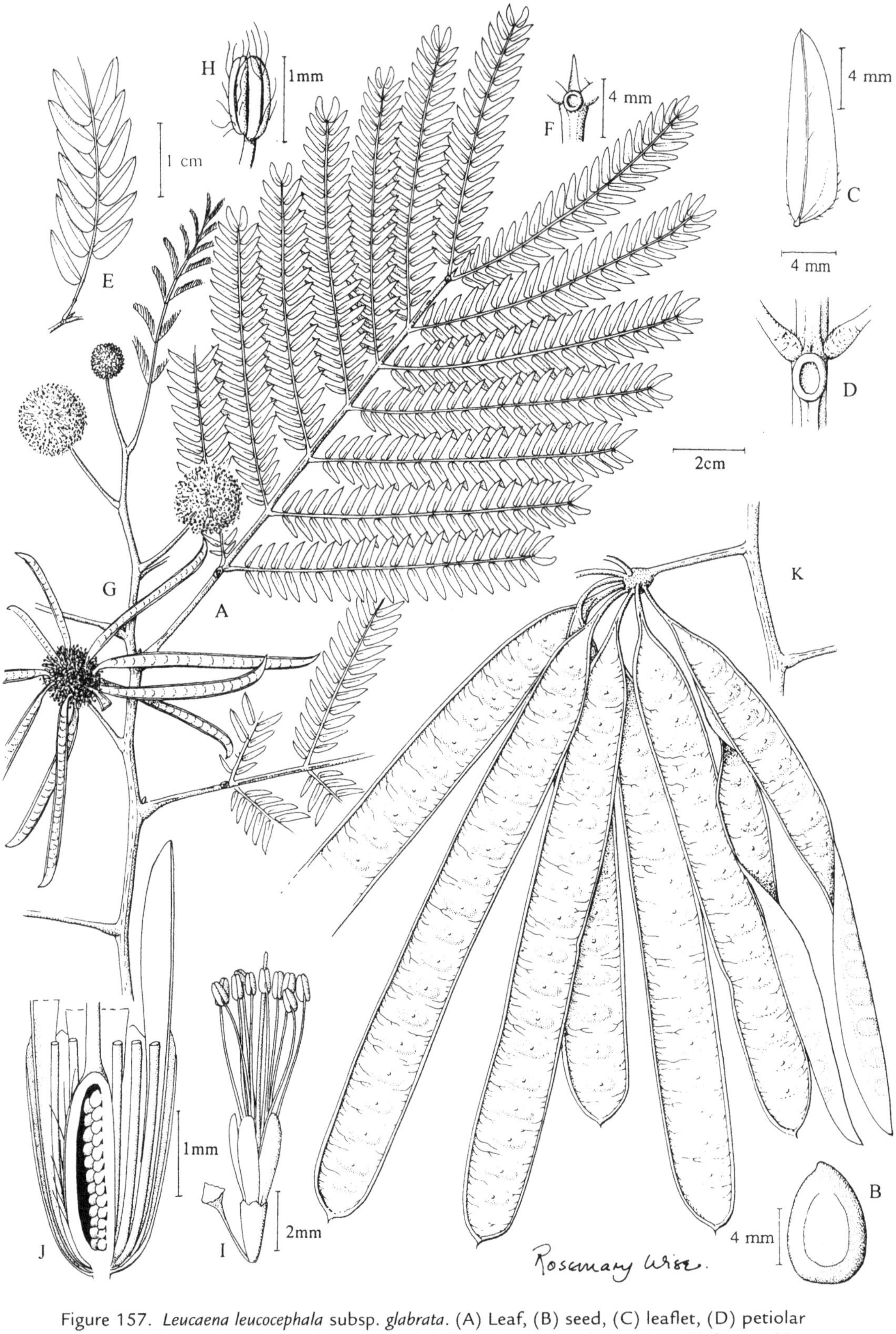


Figure 157. *Leucaena leucocephala* subsp. *glabrata*. (A) Leaf, (B) seed, (C) leaflet, (D) petiolar nectary, (E) inset pinna, (F) rachis nectary, (G) flowering shoot, (H) anther, (I) flower, (J) longitudinal section of flower, (K) pods. (Hughes 1998). Drawn by R. Wise.

in cultivation pantropically" (Hughes 1998: 114, 117). This fast-growing, nitrogen-fixing tree has become a major multipurpose agroforestry crop in tropical regions worldwide. It is used for fuel, food for livestock and people, green manure, medicine, pulp, shading for crops and people, erosion control, and soil improvement. Many horticultural selections or cultivars are known. In many areas it has become a serious invasive weed.

Hughes (1998) recognized two other subspecies, both naturally occurring, from southern and southeastern Mexico and northern Central America.

### *Lonchocarpus*

Flowers papilionoid. Trees, shrubs, and woody vines. American and west African tropics; 150 species.

Various species of *Lonchocarpus* are used as fish poisons, sources of the insecticide rotenone, lumber, medicine, dyes, and beverages.

***Lonchocarpus hermannii*** [*Willardia mexicana*]. Venus tree; *nesco, palo piojo; japiró* (Guarijío) (figure 158)

Slender trees 5–15 (17) m, usually with a high, sparse crown, the trunk to 25+ cm in diameter, with vertical invaginations, the bark gray and finely granular with low, horizontal ridges. Twigs and leaves with small, inconspicuous, white hairs. Stipules absent. Leaves drought deciduous, odd-pinnate, 7–17 cm: leaflets 9–17 per leaf, 2–4 cm, elliptic, the margins moderately cupped downward (but not revolute) during drought.

Flowers in racemes 4–14 cm; lavender or pale pinkish purple, approximately 2 cm, pealike, producing a strong, sweet fragrance. Calyx cup shaped with 5 short teeth. Pods 5–18 cm, flattened, dry, tardily dehiscent; seeds 10–11.5 mm, flat, kidney shaped, uniformly tan to red-brown. Mass flowering but of short duration at the height of the dry season, May–June.

Slopes and canyons; primarily tropical deciduous forest, and uncommon in thornscrub of the coastal plain in the vicinity of Navojoa, and northward in foothills to Sierra San Javier. East-central and southern Sonora east of 109°45'; 50–800 m (165–2625 ft). Also southwestern Chihuahua and Sinaloa to Jalisco.

The flowers attract both native bees and the honeybee. Honey made from the nectar is reported to be poisonous. The wood is used for house beams. The roots served as fish poison, and in Sinaloa the bark has been used to kill parasites of cattle and horses (Standley 1922). The leafless, flowering trees stand out among the gray hillsides during the arid foresummer. The attractive foliage and spectacular purple flowers make it worthy of cultivation in near frost-free regions.

### *Lysiloma*

Unarmed shrubs or trees. Stipules leafy and conspicuous, generally quickly deciduous. Leaves twice pinnate, the leafstalks usually with a conspicuous nectary gland. Flowers mimosoid, white or cream white, crowded in peduncled globose heads or subglobose racemes; stamens numerous. Pods flat, linear-oblong, stalked (with a stipe) gradually dehiscent, the margins forming a thickened, often semipersistent rim separating from the rest of the pod. Tropical and subtropical; Caribbean and Florida, and southern Arizona to Central America. 8 species. Reference: Barneby and Grimes (1996).

The three Sonoran species are distinctive and easily identified by their different branching architecture, appearance of the foliage, and especially by differences in the bark and trunk. They are, however, somewhat difficult to distinguish in a simple, dichotomous key.

1. Bark on trunk and major limbs whitish or light-gray, smooth; leaves usually glabrous or glabrate; stipules mostly smaller than the leaflets; leaflets 12–20 mm; Sonora only in a few coastal canyon-valleys north of San Carlos. ________ ***L. candidum***

1' Bark on trunk and major limbs gray to brown, often dark, rough at least at base; leaves often sparsely pubescent; stipules mostly larger than the leaflets; leaflets 2–9 mm; widespread.
   2. Trunks or major limbs numerous, usually straight-ascending; bark usually gray, smooth except on older wood; flowers sessile. ____ ***L. divaricatum***
   
   2' Major limbs and trunks usually few, often crooked; bark usually dark brown to blackish, rough, and soon becoming checkered; flowers short-pedicelled. ________ ***L. watsonii***

***Lysiloma candidum***. *Palo blanco* (figure 159A)

Slender, graceful trees 3–10+ m, with one to several tall trunks; bark smooth and whitish gray with a chalky bloom (fluorescing under black light), and often with a high, thin crown. Glabrous or the young herbage and especially the inflorescences and buds with small, white hairs. Leaves evergreen or nearly so, 5–14 cm; pinnae (1) 2 or 3 pairs; leaflets 12–20 mm, linear-oblong to oblong or obovate, mostly 5–17 pairs per pinna; leafstalks with a nectary gland between the lowermost pair of pinnae and sometimes between the others.

Flowers sessile, in globose heads. Valves of pods 9–13 × 2.0–2.5 cm, red-brown especially when young, on stipes 1.0–1.7 cm. Seeds 8–9 mm. Flowering mostly March–May.

Sonoran desertscrub in west-central Sonora; near sea level–100 m (330 ft). Otherwise in Baja California Sur and

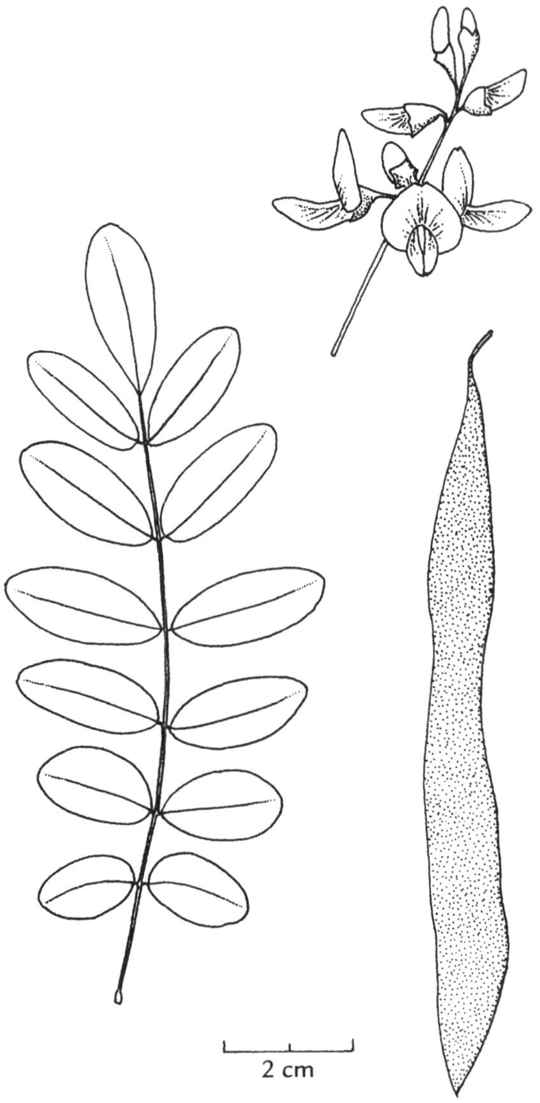


Figure 158. *Lonchocarpus hermannii*. Drawn by M. B. Johnson.

the southern margin of Baja California. In Sonora known only from Bahía San Pedro and nearby Cañón las Barajitas north of San Carlos; locally common on gravelly-sandy soil of broad, braided washes and floodplains of major drainages coursing into the sea.

This handsome tree is easily grown from seed. It is generally too frost sensitive to be grown in Tucson. The white bark and high crown of lacy blue-green foliage make it a desirable street and garden tree for hot, dry climates.

***Lysiloma divaricatum*** [*L. microphyllum*]. *Mauto; ma'a* (Guarijío); *mayo, mayo cuta, jeseri* (Mayo); *cmapöjquij* (Seri); *vamyo* (Yaqui) (figures 159B and 160)

Multiple-stem shrubs in desert regions to graceful trees mostly 4–10 m in thornscrub and tropical deciduous forest. (In some canyons, especially in the Guarijío region, reaching 15–20 m, the trunk diameters 35–43 cm with a high solitary bole or a few erect-ascending trunks or limbs.) Bark brown to gray, becoming rough with age, and on trunks and major limbs of larger trees furrowed and checkered, and flaking away in thick, woody curls or plates. Wood hard and flexible, red-brown beneath the bark. Leaflets glabrate (often with a few hairs near the base), the young herbage, leafstalks and rachillas, inflorescences, and calyces with short hairs; leafstalks with raised nectary glands between the first and the uppermost pairs of pinnae. Leaves drought deciduous, most or nearly all of the foliage commonly shed in early fall, (4) 8–12 cm; pinnae (4) 6–9 (10) pairs; leaflets 2.0–6.5 mm, numerous.

Flowers sessile, in globose heads. Pods 8–13 cm, dark brown, ripening in late summer and fall, and semipersistent. Seeds 7–10 mm, brown to blackish, thinner than those of *L. watsonii*. Flowering June–November.

Slopes, ridges, valley, arroyos, and canyons; Sonoran desertscrub, thornscrub, and tropical deciduous forest; eastern and southern half of Sonora and Isla Tiburón; near sea level–915 m (3000 ft). Northwestern Costa Rica to northern Mexico in Sonora, Baja California Sur, Tamaulipas, and San Luis Potosí.

This is the dominant tree throughout the tropical deciduous forest of Sonora. The semipersistent pods form a significant part of the *mojino* (dry season) color of the tropical deciduous forest. As firewood it is considered to be on a par with mesquite and is used in bakeries and for home heating and cooking. The bark is the most commonly used and preferred tanning agent in southern Sonora. Mauto is one of the favored woods for house construction in southeastern Sonora, including posts, main beams, and roofing cross-supports. Large quantities of poles, or *varas*, are cut and sold as grape stakes when forests are cleared for buffelgrass pasture, and the larger limbs are used as fence posts that are said to last 15 years in the ground. The trees are being overharvested. This fast-growing tree should be investigated for agroforestry.

***Lysiloma watsonii*** [*L. thornberi; L. watsonii* subsp. *thornberi*]. Feather tree; *tepeguaje; machaquí* (Guarijío); *machahuí* (Mayo) (figures 159C and 161)

Many-stemmed shrubs 1–3 m near the northern and upper elevational limits where freeze damaged, and trees 5–13+ m farther south, forming a well-developed but irregular and crooked trunk and spreading crown. Bark on larger limbs and trunk dark-colored, rough, cracking off in large pieces. Young herbage, leafstalks, rachillas, inflorescences, and calyces pubescent with short hairs; leaflets usually glabrous or glabrate. Evergreen or semideciduous unless the leaves are freeze killed, otherwise the old leaves usually remain until leaf emergence in spring, April in southern Sonora and May near the Arizona border; new foliage bright light green, later turning dark green like mesquite foliage. Leaves 10–15 cm; pinnae 5–10 pairs; leaflets 3–9 × 1.2–2.7 mm, numerous, slightly glaucous; petioles with relatively large, raised nectary glands.

Flowers fragrant, cream-white, short pedicelled, in rounded, subcapitate racemes. Pods 12–24.5 × 1.5–2.7 cm, dark brown, often blackish. Seeds 8–10 mm, somewhat disk shaped, rather thick, and brown. Flowering April–May.

Slopes, canyons, and arroyos; thornscrub, tropical deciduous forest, oak woodland, and occasionally in grassland and margins of Sonoran desertscrub; eastern two-thirds of Sonora and less frequent in west-central region (e.g., mountains north of San Carlos); near sea level–1435 m (4700 ft). Sinaloa and Durango northward to western Chihuahua and the Rincon Mountains in southern Arizona.

The bright green foliage makes the trees conspicuous among the otherwise dry, brown landscape of the late spring dry season. This tree is extensively cultivated in southern Arizona. The dense, fernlike foliage provides shade and screening and lends a tropical aspect to the landscape. It is relatively frost sensitive, sustaining twig damage at –4.5°C (24°F). The strong, durable wood has been used for construction and wooden articles. It is a favored wood, in parts of southeastern Sonora second only to that of *Platymiscium*, because it is strong but not too hard to work and durable, and is said to "outlast all." Forked posts for sliding-beam gates are often made of tepeguaje. The seeds, ground as flour, were an important food in the mountains of southeastern Sonora (Gentry 1963). The bark is chewed or prepared as a mouth rinse for diseases of the gums and teeth. The Guarijío use the bark in a mixture to clean wounds (see *Bursera lancifolia*).

Exceptional *tepeguaje* trees, sometimes to 20+ m, occur in tropical riparian canyons along the Río Guajaráy. One tree in a montane canyon above Bavícora in the upper Río Mayo drainage has an estimated height and spread of 30 m. One of its limbs supports a large epiphytic *Ficus cotinifolia*.

### *Mimosa*

Trees and shrubs; flowers mimosoid. Mostly in tropical and subtropical regions, 480 species, mostly in the New World. Thirteen other species in Sonora, spiny shrubs and one perennial herb. Reference: Barneby (1991).

***Mimosa palmeri***. *Chopo*; *cho'opó* (Guarijío); *cho'opo* (Mayo) (figure 159D)

Slender or dense briarlike shrubs or trees to 7 m. Trunk to 20+ cm in diameter; bark fissured with longitudinal, interconnected blackish, rough, peeling strips surrounding smoother, lightly colored interstices. Pubescent throughout including the spines (prickles), with glandular and nonglandular hairs, the glandular hairs often red, tack-shaped, and mostly on the herbage. Stems armed with internodal, single to subopposite prickles to 8 mm, laterally compressed, sharp, recurved, and broad based. Foliage drought deciduous, often leafless or nearly so by December. Stipules 3–4.5 mm, slender, hornlike, and herbaceous. Leaves 6–15 cm; pinnae the 8–13 pairs and opposite to sometimes subopposite, the pulvinus of each pinna surmounted by a pair of hornlike glands 0.3–0.5 mm; leaflets numerous, 3–4 mm, ovate to oblong; leafstalks without nectary glands.

Flowering profusely on many slender spikes 5–12 cm in large, terminal, paniclelike branches. Flowers pink-purple; stamens 10, united at base; anthers glandless. Pods in dense clusters, 3–10.5 cm × 3.0–3.5 mm, beaked, densely pubescent with glandular and nonglandular hairs, constricted between the seeds, separating into one-seeded indehiscent segments, with a cordlike, semipersistent rim. Seeds 4.0–5.0 mm, obovate, compressed, smooth, and light brown. Flowering (April) July–October.

Slopes and valleys; thornscrub, tropical deciduous forest, and sometimes among lowermost oaks; southeastern and east-central Sonora; 150–1000 m (500–3000 ft). Sonora to Mazatlán, Sinaloa.

*Chopo* yields the best wood for making charcoal, which has led to serious overharvesting, and it is also prized for tool handles and fence posts. It is accorded legal but ineffective governmental protection. The bark is chewed to harden the gums and for ailing teeth, and the Guarijío drink tea made from the bark as a remedy for cough. The most extensive populations of larger plants are in remote areas at upper elevations of tropical deciduous forest in the Guarijío lands in the Río Mayo drainage. Elsewhere it tends to be localized. The northernmost population known to us is on Sierra San Javier, at the uppermost margin of thornscrub close to the lowermost oaks. These trees sometimes form heavy trunks. Cultivated trees near Yuma, Arizona, may grow more than 1.2 m per year and form trunks more than 12 cm in diameter within five years. It is not a very attractive tree, but may have value for agroforestry.

### *Olneya*

Flowers papilionoid. The genus is monotypic; its closest relatives include *Coursetia*, *Genistidium*, *Peteria*, and *Robinia*. Reference: Lavin and Sousa (1995).

***Olneya tesota***. Desert ironwood; *palo fierro*; *ejéa* (Mayo); *ho'idkam* (O'odham); *ho'idkam* (Pima Bajo); *comítin* (Seri); *ehea* (Yaqui) (figures 162 and 163)

Trees often 7–8.5 m, long lived and relatively slow growing, with one to several massive trunks and extremely

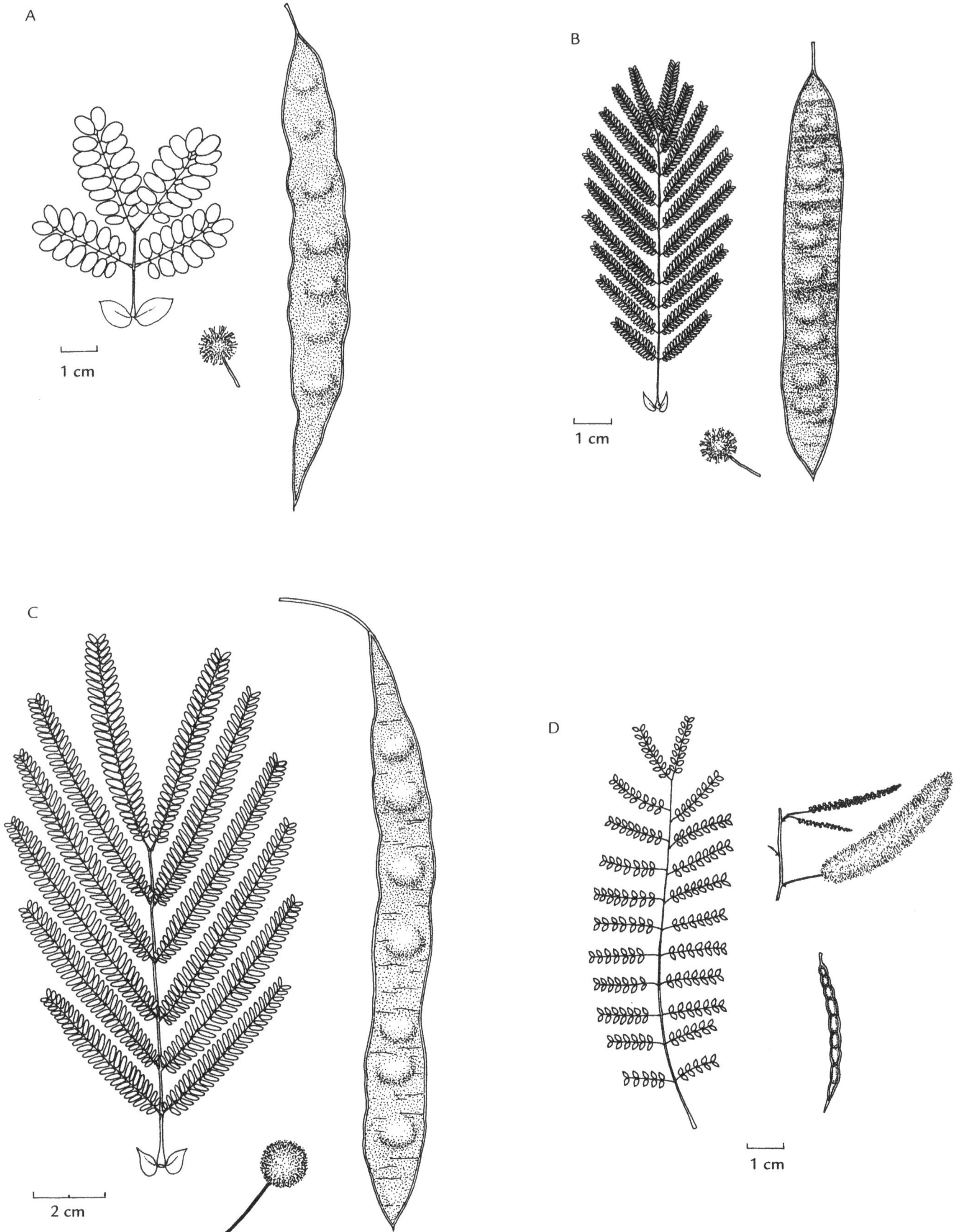


Figure 159. *Lysiloma* and *Mimosa*. (A) *Lysiloma candidum* (unusually small leaf), (B) *L. divaricatum*, (C) *L. watsonii*, (D) *Mimosa palmeri*. Drawn by M. B. Johnson.

Figure 160. *Lysiloma divaricatum*. Canyon near Rancho San Pedro, Sierra Sutucame, Río Mayo region. Juvenile *Opuntia wilcoxii* and *Pachycereus pecten-aboriginum* grow as epiphytes. Photograph by R. S. Felger.

hard, fine-grained wood (it does not float in water and burns with a hot flame); bark gray and shredding. Herbage, inflorescences, and flowers densely pubescent with short, relatively coarse whitish hairs; leaflets sometimes sparsely and minutely pubescent (appearing glabrous except under magnification). Spines (3) 4–11 mm, seemingly stipular, single or paired, sharp, straight or slightly curved, or absent from some twigs. Foliage dense, gray-green; leaves 2.8–6+ cm, tardily drought deciduous, odd- or even-pinnate; leaflets 6–16 per leaf, (7) 10–20 mm, often slightly asymmetric, the tip entire or notched, and sometimes with a minute apiculate projection.

Inflorescences of short, dense, racemes or panicles. Flowers pealike, approximately 1.5 cm; calyx purple-brown, the lobes spreading, the petals mostly pink to lavender or purple, varying from tree to tree. Banner petal with a white, succulent callus at the base of the blade, a chartreuse nectar guide above the callus, a whitish band highlighting the nectar guide, and the flat blade of the banner folding back from the nectar guide later in the day. Pods generally 2–4 cm, thick and nearly terete, dark red-brown, with glandular and nonglandular hairs, tardily dehiscent (often after falling). Seeds 7–8 mm, 1 or 2 (3–5) per pod. Mass flowering, attracting myriads of bees, usually late April–May in northern Sonora, the first flowers often in late March in southern Sonora; pods ripening June–July.

Slopes, valleys, plains, and arroyos; Sonoran desertscrub and thornscrub; western and central Sonora; near sea level–885 m (2900 ft). Common across most of the Sonoran Desert. Southeastern California to the Cape Region of Baja California Sur and southwestern Arizona to southwestern Sonora. Essentially endemic to the Sonoran Desert, extending southward beyond the desert in the lower Río Mayo and Arroyo Masiaca drainages of southwestern Sonora and lowlands of the Cape Region of Baja California Sur.

Ironwoods are largest and most numerous along arroyos and large washes in the drier regions of the desert and less common on rocky slopes. Great stands occur on Isla Tiburón. It is one of the few desert trees that offers worthwhile shade in the desert, and many animals and smaller plants find shelter beneath the limbs. Ironwood often grows with mesquite and blue palo verde to form spinescent xeroriparian gallery forests.

Abundant rainfall and mild winters result in massive flowering, but flowering does not occur every year, and during any given mast year not all of the trees produce flowers. The trees defoliate just before flowering, while adjacent nonflowering trees tend to retain their foliage. New leaves are produced with the summer rains, after the pods fall. Many of the great ironwood stands in Sonora have been plundered for charcoal production and replaced with buffelgrass pasture (Nabhan and Carr 1994). The wood takes a beautiful polish and is in great demand by the Seri Indians and other artisans for sculpture.

### *Parkinsonia*

Large shrubs to medium-sized trees with green bark and relatively soft wood. Twigs armed with paired stipular spines or with a spinescent tip. Short shoots very reduced. Leaves twice pinnate with one or two pairs of pinnae, petioled or the pinnae sessile; leaflets small; leaflets and leaves drought deciduous, the leaflets often falling independently.

Flowers caesalpinioid, in reduced to open axillary racemes; banner petal yellow or whitish and often with orange spots or flecks, often fading to orange or orange-flecked, the other petals yellow. Stamens 10, separate, the filaments yellow and hairy at the base; anthers orange. Pods dry, tardily and partially dehiscent or indehiscent; seeds 1 to several.

Warm to tropical desert and semiarid regions; 10 species. The greatest number of species occur in the Sonoran Desert. References: Carter (1974), Dimmitt (1987), Hawkins (1996), Hawkins et al. (1999).

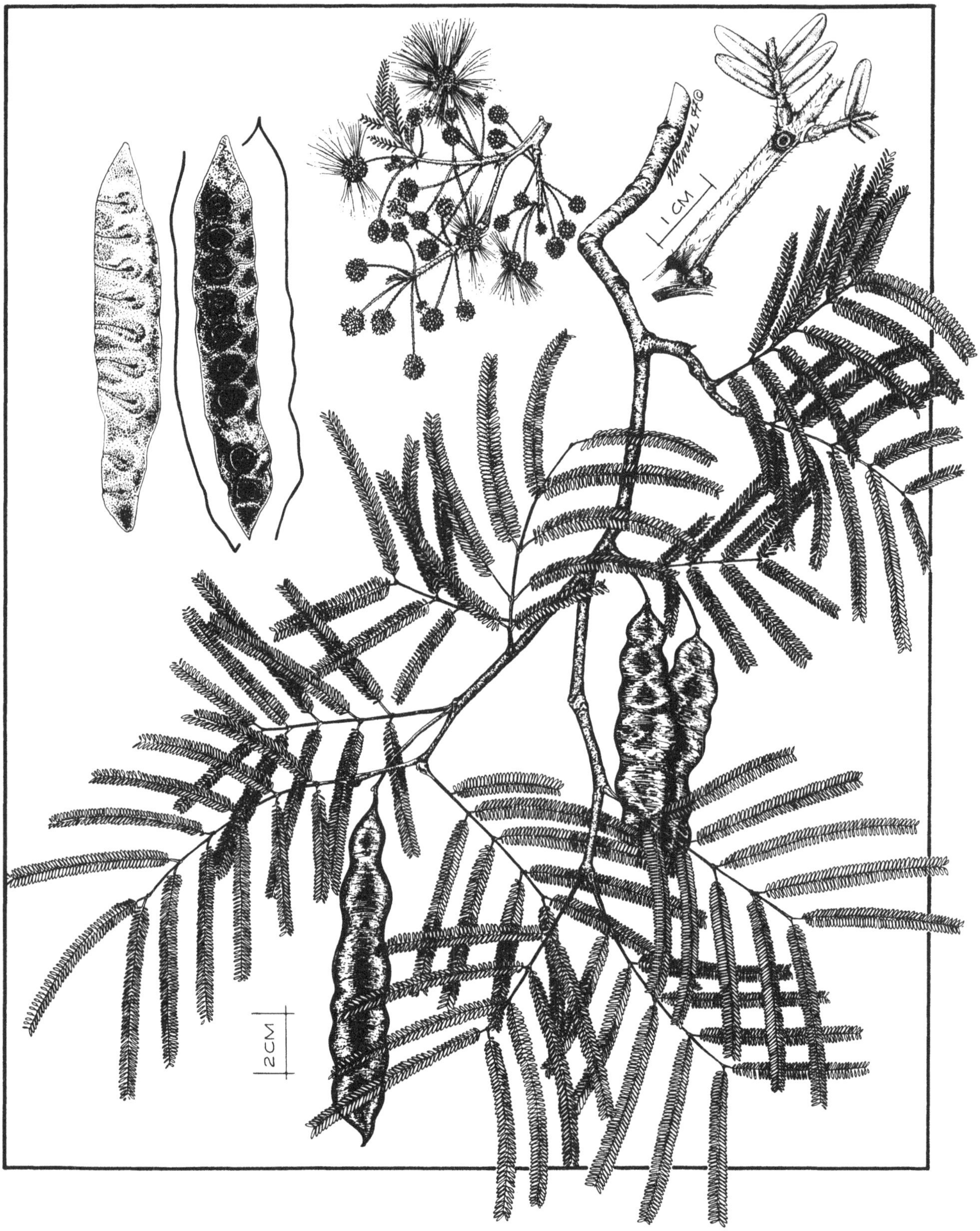


Figure 161. *Lysiloma watsonii*. Note leafstalk gland, upper right. Drawn by N. E. Weaver.

Figure 162. *Olneya tesota*. Vicinity of El Desemboque San Ignacio. (Felger and Moser 1985). Photographs by R. S. Felger.

For more than a century, *Cercidium* and *Parkinsonia* have variously been treated as one or two genera. The work of Hawkins (1996, Hawkins et al. 1999) supports the case for a single, monophyletic genus. *Parkinsonia sensu stricto* includes three species in southern and eastern Africa and the pantropical *P. aculeata*. *Cercidium* ranges from southwestern United States to Argentina.

Various hybrids are encountered in Sonora and elsewhere. A magnificent horticultural selection of a hybrid found in Tucson, called *Cercidium* cv. "Desert Museum," involves hybridization of *P. aculeata*, *P. florida*, and *P. microphylla* (Dimmitt 1987).

1. Leaves > (10) 20 cm. ______________ ***P. aculeata***
1' Leaves < 7 cm.
   2. Twigs spinescent at tip; axillary spines absent; petiole absent; leaflets mostly 1–3.3 mm. _ ***P. microphylla***
   2' Twigs not spinescent at tip; solitary or paired axillary spines often present; leaves petioled; leaflets mostly 2–10 mm.
      3. Leaves with 1 (rarely 2) pair of pinnae, each pinna with 2–4 pairs of leaflets. ___ ***P. florida***
      3' Leaves with one or two pairs of pinnae, each pinna with 4–8 pairs of leaflets.
         4. Spines 1–2 at each node, stout; petioles 4–21 mm; leaflets 4–13 mm; petals including banner bright yellow, or the banner often with orange spots. _____ ***P. praecox***
         4' Spines none or 1 and slender; petioles to 3 mm or none; leaflets 2–6 mm; petals pale yellow, except the banner often white. __________________ ***P. ×sonorae***

***Parkinsonia aculeata*.** Mexican palo verde; *bagote, retama, guacaporo, huacapori, guachaporo, junco* (Sinaloa); *guacapora* (Guarijío); *bacaporo* (Mayo); *oobjam* (Tohono O'odham); *ooba'igam* (Pima Bajo); *vaka'apo, vaka'apoa, vaka'aporo* (Yaqui); *snapxöl* (Seri) (figure 164)

Green-barked trees with a well-developed trunk. Youngest growth with small, white hairs, soon glabrate. Long-shoot leaves often bearing a pair of sharp, stiff stipular spines, which often break off or fall away; leafstalk modified into a stout, stiff, sharp spine often 1–3 cm, bearing near its base 1 or 2 (3) pairs of long, slender, pinnae (each pinna resembles an individual leaf) with conspicuously swollen and flexible pulvini at their bases. Pinnae with flattened, stringy, strap-like rachises (actually rachillas), (10) 20–45 cm; leaflets small and numerous, quickly drought deciduous. (After the leaflets have fallen, the rachillas, which resemble phyllodes, are tardily drought or winter deciduous, but the leafstalk-spine is persistent.) Short-shoot leaves mostly spineless.

Inflorescence a raceme, often (5) 10–16 cm, few flowered. Flowers yellow and showy, 27–34 mm wide; sepals and

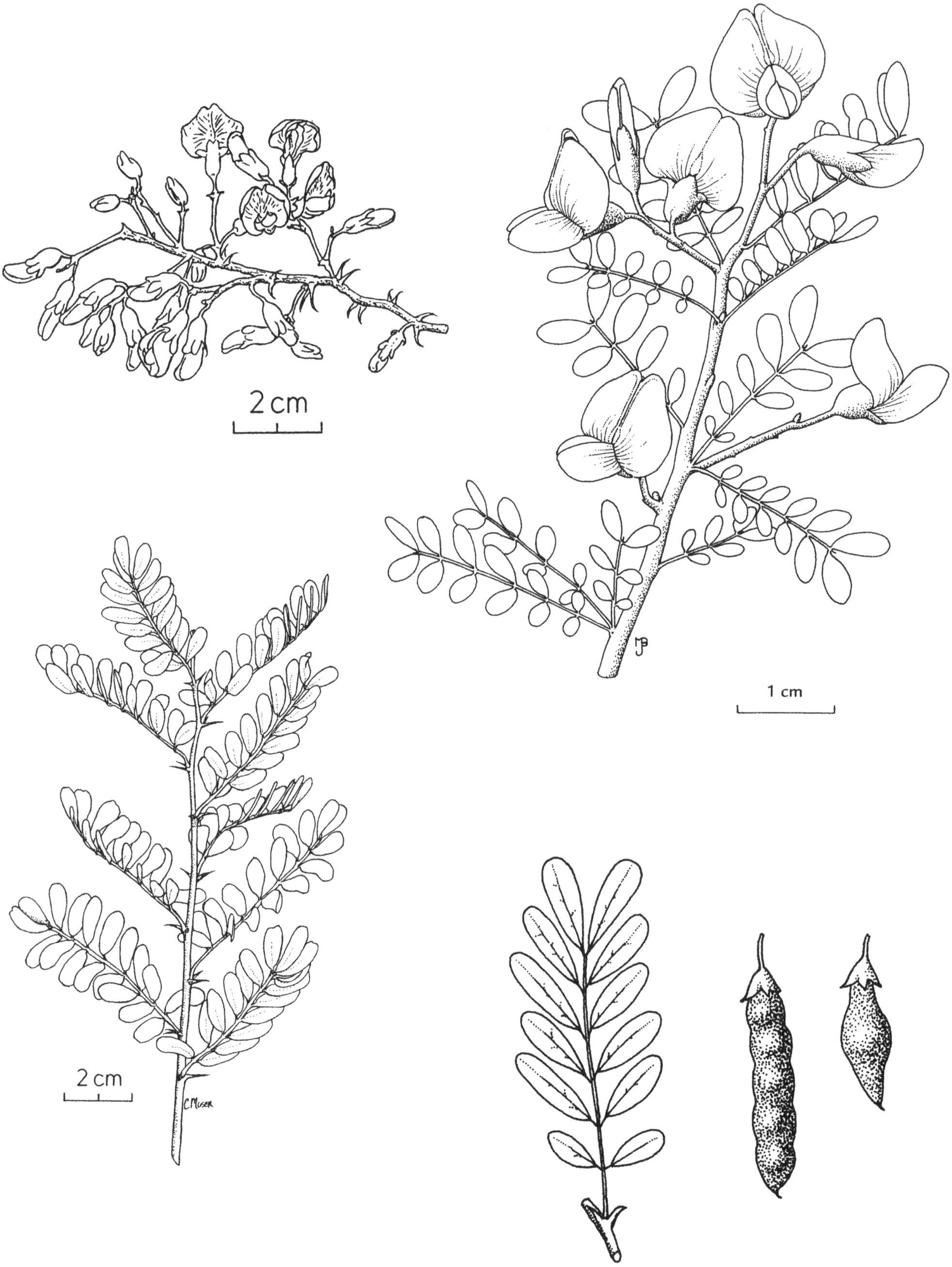


Figure 163. *Olneya tesota*. Drawn by C. M. Marlett and M. B. Johnson.

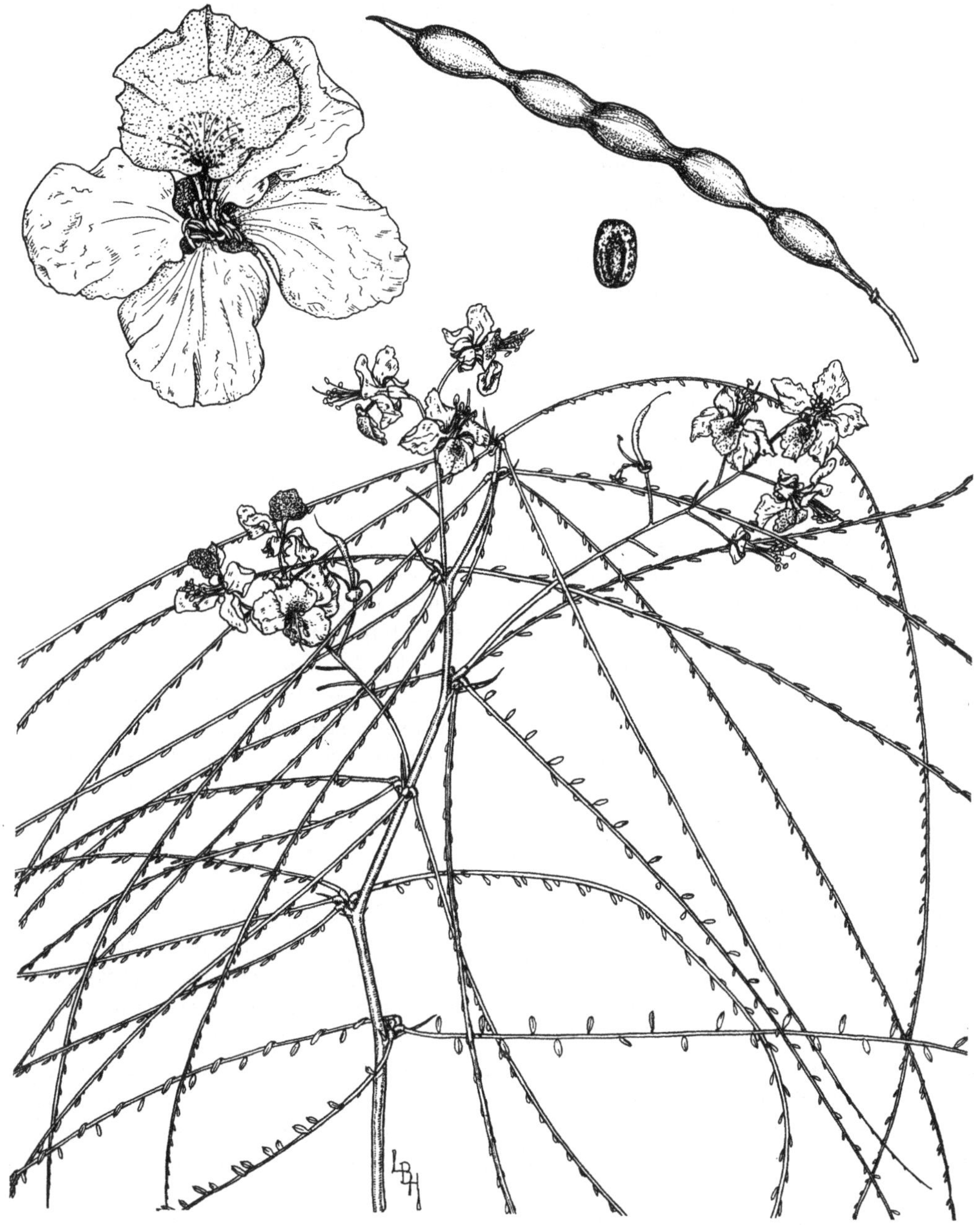

Figure 164. *Parkinsonia aculeata* (Benson and Darrow 1945). Drawn by L. B. Hamilton.

petals bright yellow, the sepals reflexed, the banner at first with red-orange spots or flecks, sepals and banner turning red-orange with age. Pods indehiscent or tardily semidehiscent; seeds few. Mass flowering April and sporadically summer–fall.

Plains, arroyos, and especially in disturbed places; most habitats to oak woodland, and occasionally roadside among pines (e.g., near Maycoba). Apparently spreading in Sonora as natural vegetation gives way to disturbed habitats. One of the most common cultivated trees in the region, often weedy and sometimes naturalized through much of Sonora, near sea level–1370 (1600) m (4500 ft [5250] ft). Especially common along irrigation canals and arroyo or river banks in southern Sonora.

This species is now almost pantropical in arid subtropical and tropical regions in many parts of the world following introduction as an ornamental, hedging, and shade tree, and subsequent invasion and naturalization. The original distribution is difficult to determine because of the extensive, naturalized weedy populations, but it is probably native to Central America and southern Mexico.

The pods can float for up to two weeks in fresh water, which seems to account for its weediness along canals and riverbanks (Hawkins et al. 1999). The Guarijío drink *guacaporo* tea as a remedy for cough and flu.

Occasional hybrids with *P. praecox*, known as ***P. ×carterae***, have been found in the Guaymas-Empalme region and northeastern Sonora, as well as scattered localities elsewhere in Mexico and in Ecuador (Hawkins et al. 1999).

***Parkinsonia florida*** [*Cercidium floridum* subsp. *floridum*. *Parkinsonia torreyana*]. Blue palo verde; *palo verde*; *caaro* (Mayo); *ko'okomaḍk*, *kalisp* (O'odham); *ziij*, *iiz* (Seri); *te'owe* (Yaqui) (figure 165A)

Large shrubs to trees often 7–10 m in height, with a well-developed trunk. Herbage, young limbs and twigs with a bluish cast, especially in the northern part of the geographic range, becoming less pronounced southward. Leaves short-petioled, pinnae 1 (rarely 2) pair; leaflets 5–9 mm, usually 3 or 4 pairs per pinna.

Flowers 18–25 mm wide. Calyx green to yellow-green, the lobes reflexed. Petals and filaments bright yellow, the banner base often with few, small red-orange spots or flecks. Pods 3–7 (10) cm, flattened, not constricted between seeds. Blazing yellow masses of flowers in spring, peak flowering usually in late March and early April (about two weeks in advance of flowering of *P. microphylla*). Flowering February–May (sparse flowering sometimes in November); pods usually ripen May–June.

Mostly bajadas, plains, and arroyos; Sonoran desertscrub, thornscrub, and lower elevations in grassland. Widespread across the western two-thirds of Sonora; near sea level–1220 m (4000 ft). In the desert generally best developed along major washes and riverine floodplains; northwestern Sinaloa to Arizona and southeastern California.

This is one of the few trees in the Sonoran Desert that offers substantial shade from the summer sun. Characteristically a xeroriparian tree, it often grows along desert washes with ironwood (*Olneya*) and mesquite (*Prosopis* spp.). Relatively fast growing, it is grown from seed as a landscape tree in southern Arizona. Cultivated trees are sometimes severely damaged by infestations of a mite that causes abnormal "broom" growth. The young, green seeds are edible and taste much like green peas.

*Cercidium floridum* subsp. *peninsulare* occurs in Baja California Sur and on the south end of Isla Tiburón. It is distinguished by its generally larger leaflets and denser and more persistant pubescence of larger hairs, and there are reported differences in size of the inflorescences and stoutness of pedicels. The flowers of *P. florida* are much brighter yellow than those of *P. microphylla*. Putative hybrids with *P. microphylla* have been found in the bottom of some of the craters in the Pinacate region and elsewhere in the Sonoran Desert (Jones et al. 1998).

***Parkinsonia microphylla*** [*Cercidium microphyllum*]. Foothill palo verde; *palo verde*; *ke:k cehedagǐ* (O'odham); *ziipxöol* (Seri); *wo'i voa'am* (Yaqui) (figure 165B)

Common small trees or large shrubs, usually branching near base from a short, thick trunk. Bark smooth and green with a waxy appearance. Twigs without spines, but ending in a sharp point. Leaves lacking a petiole; pinnae 2, 1–5 cm; leaflets 1–3.5 (6) mm, nearly orbicular, 4–8 (10) pairs per pinna, the individual leaflets and ultimately the leafstalk drought deciduous. (Because the leaves lack petioles, each sessile pinna might be confused for a single, once-pinnate leaf.)

Flowers pale yellow and white, (10) 12–18 mm wide. Calyx tube green to greenish yellow, the lobes pale yellow, spreading to moderately reflexed (but not as strongly as in *P. florida*). Banner petal white, fading pale yellow and orange flecked or not, the other petals pale yellow. Filaments pale yellow. Pods 3–9.5 × approximately 1 cm, constricted between the seeds, the apex often extending into a slender seedless snout. Mass flowering March–May; fruits ripening May–June.

Slopes, ridges, canyons, bajadas, and arroyos; Sonoran desertscrub; northwestern and central Sonora; near sea level–1220 m (4000 ft). Endemic to the Sonoran Desert: widespread and common in the western two-thirds of Sonora south to the Guaymas region, southwestern Arizona, both Baja California states, and southeastern California along the Colorado River.

In cultivation this tree is very slow growing, making it commercially undesirable. Unlike the blue palo verde (*P. florida*), the foothill palo verde usually grows away from washes, and the smaller, paler flowers have a white banner.

***Parkinsonia praecox*** subsp. ***praecox*** [*Cercidium praecox* subsp. *praecox*]. *Brea*, *palo brea*; *hupués* (Guarijío); *choy* (Mayo); *mas* (Seri); *choiy* (Yaqui) (figure 165C)

Spreading trees, 4–8 m, the major branches commonly nearly horizontal; bark pale pea-green. Herbage sparsely to densely pubescent with small, white hairs. Stipular spines absent, solitary, or sometimes paired, reaching 10–18 mm. Leaves sessile or with petioles to 25 mm; the pinnae 1 or 2 pairs, 1.5–4.1 cm; leaflets 4–13 mm, 4–9 pairs per pinna.

Flowers yellow with red spots, 16–20 mm wide. Calyx tube green, the lobes yellow and reflexed. Petals bright yellow, the banner with basal red-orange spots or flecks.

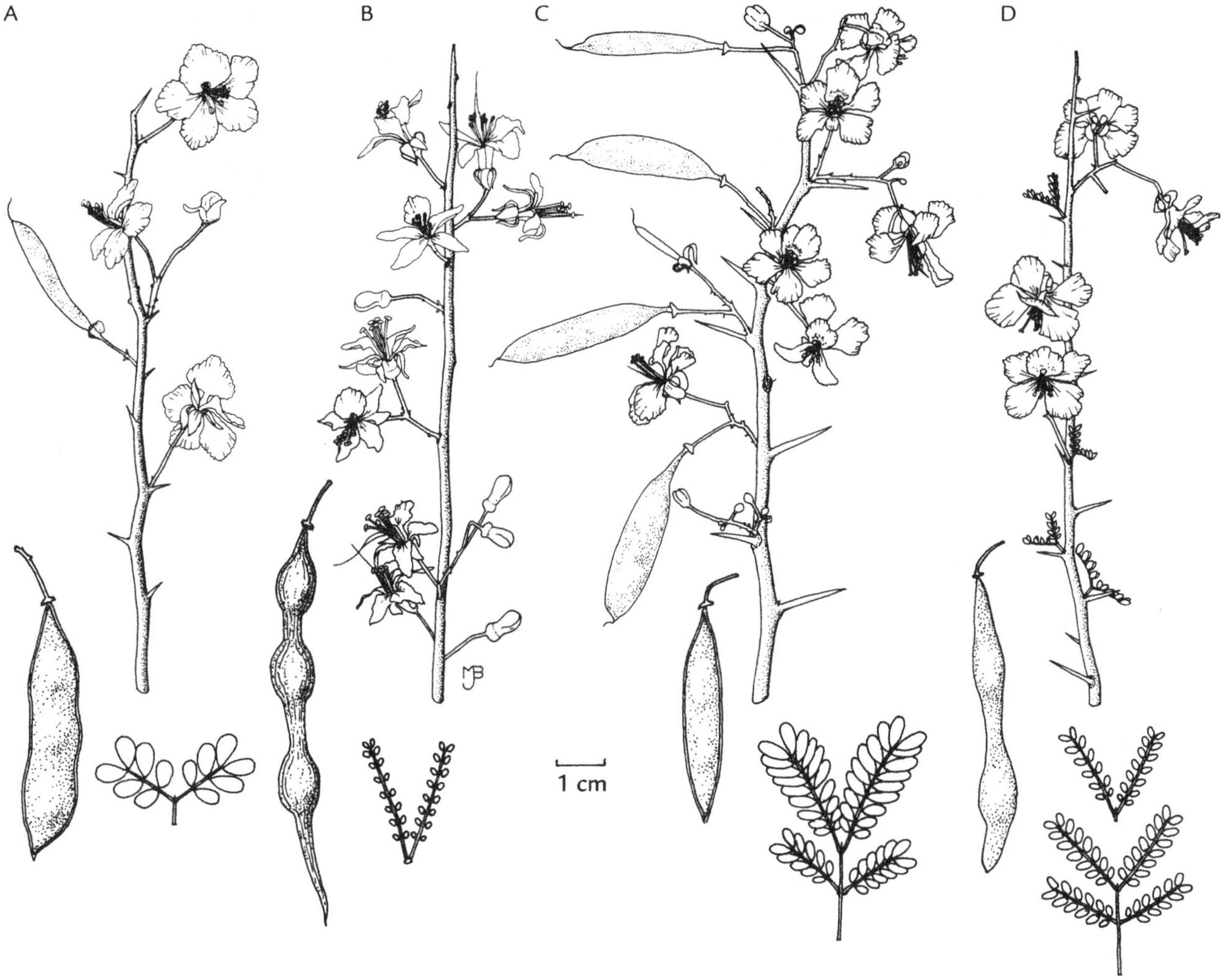


Figure 165. *Parkinsonia*. (A) *P. florida*, (B) *P. microphylla*, (C) *P. praecox*, (D) *P.* × *sonorae*. Drawn by M. B. Johnson.

Pods 3–6 cm, linear-oblong, flattened, not constricted between the seeds. Mass flowering February–June.

Slopes, valleys, plains, and arroyos; Sonoran desertscrub, thornscrub, tropical deciduous forest, and lower elevations in grassland. Widespread, north-central and northeastern Sonora southward; near sea level–1065 m (2495 ft). Also Baja California Sur, and southwestern Chihuahua and Sinaloa to Oaxaca and South America. At least one other subspecies worthy of recognition occurs in South America.

The pea-green bark, open, spreading growth habit, and blue-green foliage are distinctive. The flowers are about the same size and color as those of *P. florida*, but the markings on the banner are more conspicuous. Brea thrives on overgrazed lands and seems to be extending its range in northern Sonora. "The common name, *palo brea*, is derived from the fact that the waxy substance coating the bark, after being scraped from the branches and melted by heat, is used as a 'gum' for gluing together leather objects and furniture; thus it is used just like *la verdadera brea*" (Carter 1974: 30). It is used as a landscape tree in southern Arizona and is generally hardy to about –6.5°C (20°F).

***Parkinsonia* ×*sonorae*** [*Cercidium* ×*sonorae. C. molle*]. Sonoran palo verde; *palo verde* (figure 165D)

Large shrubs or trees to 6 m. Leaves sessile or with petioles to 10 mm; pinnae 1 or 2 pairs, 1.5–3.2 cm; leaflets 2.2–6 mm, 4–7 pairs per pinna. Young herbage moderately pubescent with short, white hairs.

Flowers 15–20 mm wide. Calyx tube green, the lobes yellow and reflexed. Petals yellow (slightly lighter than those of *P. praecox*, but brighter than those of *P. microphylla*), the banner paler toward the center with red-orange spots or flecks. Pods 3.5–8 cm, flattened, slightly constricted between the seeds, the apex often extended into a slender snout. Flowering March–May.

Slopes, valleys, plains, and arroyos; Sonoran desertscrub; central Sonora; 30–490 m (100–1600 ft).

This natural hybrid between *P. microphylla* and *P. praecox* occurs in Sonora and Baja California, where the ranges of the parent species overlap. It is occasionally cultivated in southern Arizona. The plants are neither as slow growing nor as cold hardy as *P. microphylla*.

### *Piscidia*

Trees and shrubs; flowers papilionoid. Florida and Mexico to northern South America; seven species. The indehiscent, corky, four-winged pods are unique among Mexican legumes. Reference: Rudd (1969).

***Piscidia mollis.*** Fish-poison tree; *palo blanco; jopoé, cu'utu sánami* (Guarijío); *cuta tosari, jo opo* (Mayo); *ho'opo, hopo* (Yaqui) (figure 166)

Unarmed trees to 12 m with dense, rounded crowns; bark pale gray to whitish, smooth on young branches, with age becoming rough and deeply fissured. Herbage densely pubescent with silky, white hairs. Stipules approximately 1.5 mm, drying as the leaf emerges and quickly deciduous. Leaves evergreen, 12–21 cm, odd-pinnate; leaflets (7) 9–13, 3.0–8.0 (10) cm, ovate to broadly elliptic.

Inflorescences 4–15 cm, of densely flowered racemes or mostly panicles. Flowers showy, the corollas 15–17 mm, white with mostly pink markings, the banner with an indistinct green eye spot, the keel pink-magenta. Pods 3–10.5 cm including a slender stipe and apiculate tip, indehiscent and distinctly corky with four wide, longitudinal wings. Flowering (May) June–July.

Gentle slopes, valleys, and arroyos; thornscrub and tropical deciduous forest; central and southern Sonora; 15–610 m (50–2000 ft). Also in northern Sinaloa.

Mass flowering occurs just before and with the first summer rains. Flowering trees are leafless, and the silvery, new leaves emerge with the flowers. Individual trees may not flower every year. Nonflowering trees are generally evergreen and can be found among the leafless, flowering trees (see *Olneya*). The leaves contain toxic compounds which have been used as fish poison (Gentry 1942). Fish intoxi-

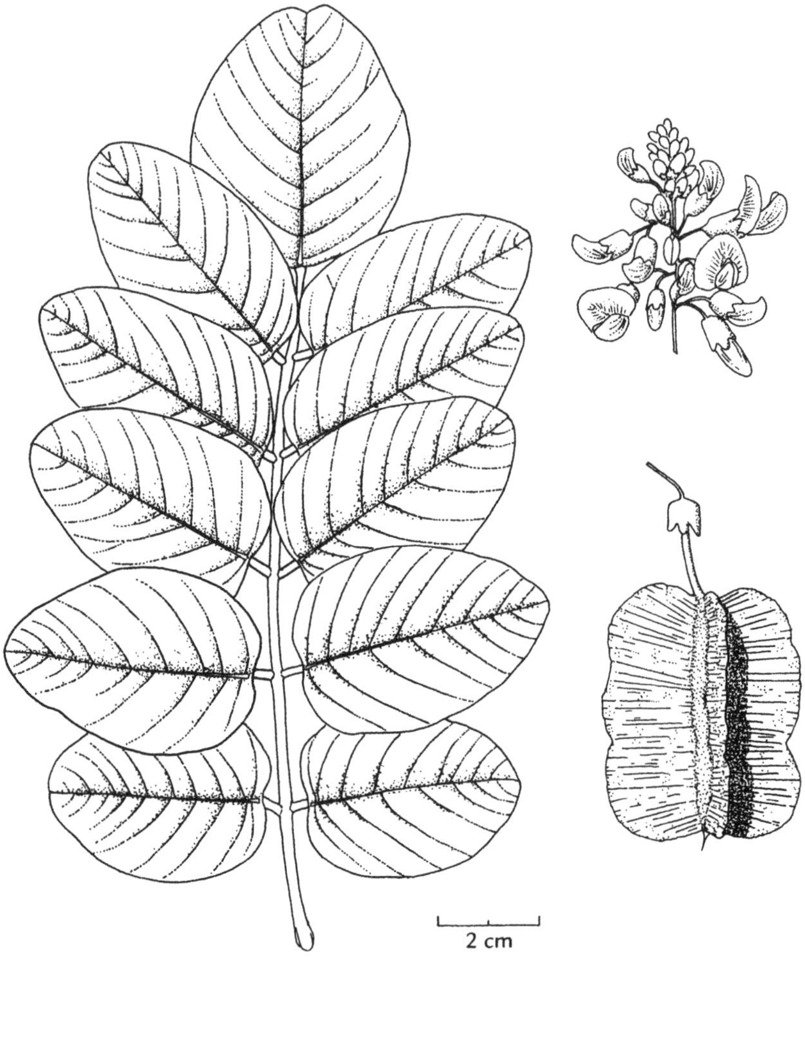


Figure 166. *Piscidia mollis*. Drawn by M. B. Johnson.

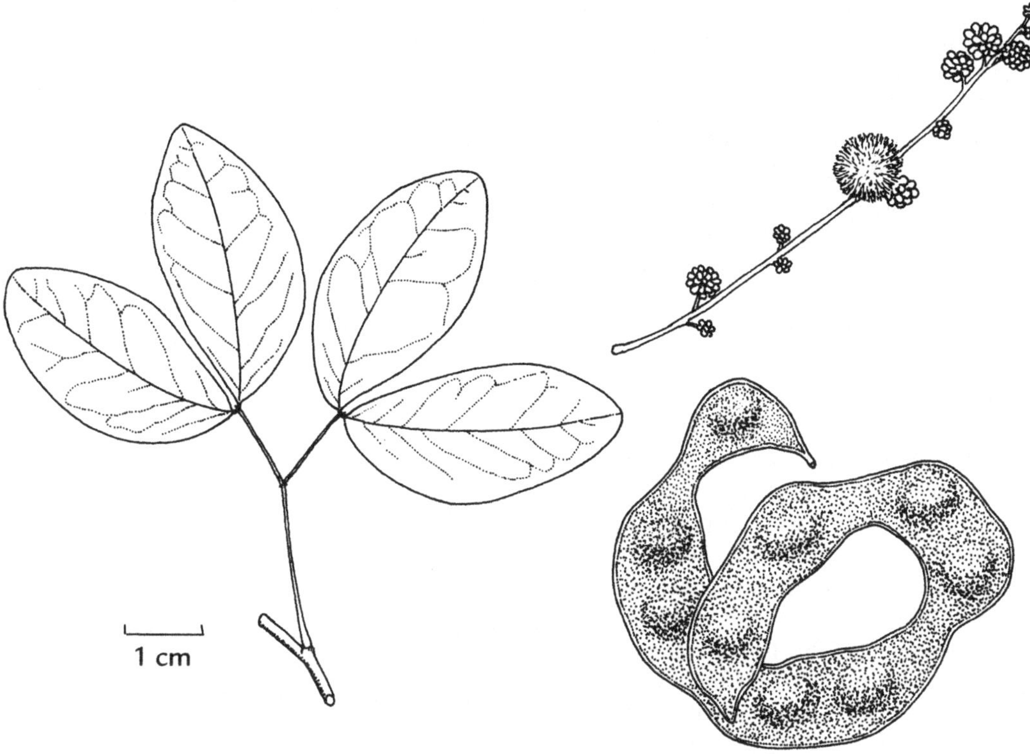


Figure 167. *Pithecellobium dulce.* Drawn by M. B. Johnson.

cated by this plant will apparently recover and are safe to eat. From a distance, the dusty blue foliage and form give the tree an oaklike appearance. It is recommended as an ornamental tree for subtropical regions. In cultivation freeze damage occurs at –4° to –6.5°C (20°–25°F).

### *Pithecellobium*

Trees and shrubs; flowers mimosoid. Pods coiled—a diagnostic feature, called "monkey's earring" in some parts of the tropics and reflected in the scientific name. Tropical and subtropical America; 18 species. Reference: Barneby and Grimes (1997).

One other species in Sonora, *P. unguis-cati*, on the coastal plain in the southwestern corner of the state (Friedman 1996), resembles *P. dulce* but is a shrub (farther south it often becomes a small tree) and differs in part by having subracemose rather than capitate flower heads, and flower heads solitary rather than paired or several at a node. Several other trees species in Sonora, often included in *Pithecellobium*, are treated as *Chloroleucon* and *Havardia*. *Ebenopsis confinis* (*Pithecellobium confine*), on Isla Tiburón and the adjacent Sonora coast, is a shrub to 3+ m, with stout, rigid, hardwood branches and unique, very thick, woody, blackish, and dehiscent pods.

***Pithecellobium dulce.*** Manila tamarind; *guamúchil; maicochini, maicochún* (Guarijío); *maco'otchini* (Mayo); *camótzila* (Seri); *mako'ochiini* (Yaqui) (figure 167)

Broad-crowned trees 10–30 m with relatively soft wood; bark smooth and gray, with age becoming fissured, checkered, and exfoliating in thick pieces on the trunk and lower limbs. Pubescence of slender, spreading, white hairs, especially on young growth, leafstalks, inflorescences, and flower buds; leaf blades glabrous or essentially so, or sparsely pubescent, especially on major veins and margins. New growth often reddish to yellow-green. Stipules developing into paired straight ascending spines, persisting to form horizontal, ringlike ridges on the bark. Leaves with one pair of pinnae; leaflets 2 per pinna, 2–7 cm, broadly oval to ovate; leafstalks channelled (grooved), bearing a raised, cylindrical to obconic nectary at summit of the petiole and smaller ones below the leaflets.

Flowers crowded in heads, these on racemose flowering branches often 10–40 cm. Perianth yellow-green to green; calyx 1.5–2.0 mm with short lobes; petals 3.5–4 mm; stamens cream white, to about 1 cm. Pods loosely coiled, 13–22 cm (if straightened), 12–15 mm maximum width, moderately flattened, often constricted between seeds, partially to tardily dehiscent, the valves twisting to reveal the seeds embedded in an edible, sweet, and spongy aril, "the seeds remaining attached and pendulous by their long stalks, the distal part of the stalk developed as a fleshy white or colored [pink] aril, which renders the black seed more conspicuous. Birds are the most likely agents of dispersal" (Tomlinson 1980: 205). Flowering December–March (June); fruiting May–June.

Arroyos, streams, and riverbanks; thornscrub and tropical deciduous forest; southern and east-central Sonora; sea level–1000 m (3280 ft). Often naturalized in riparian habitats, and probably not native to Sonora. Widespread in Mexico, Central America, and northern South America. Widely cultivated and naturalized in many arid to semiarid tropical and subtropical regions around the world.

*Guamúchil* is one of the most commonly planted trees throughout the lowland regions of Sonora northward to San Luis at the Río Colorado. It was "carried by the Spaniards in colonial times to the Philippines, whence the tree was carried to India, where it was first described and named botanically" (McVaugh 1987:234). It is esteemed as a shade tree and for its edible fruit. The sweet, spongy pulp (aril) of the pods is made into a refreshing drink or eaten fresh. A long bamboo fruit-gathering pole is often seen leaning against a tree at harvest time. The pole has a transverse stick tied to it and resembles a columnar cactus fruit-picking pole. The fruits are sold in local markets. In southeastern Sonora the bark is made into tea taken as a remedy for stomach pain. Cattle and burros relish the leaves, fresh or fallen and dried, and there is invariably a trim line to the reach of livestock. The bark is used in the tanning of hides, especially in Sinaloa, but in Sonora is considered less desirable than other tanning barks because it stinks. The trees are frost sensitive and therefore generally are not grown in Arizona.

Extensive gallery forests of these trees grow in the Guarijío country north of Alamos along the Río Mayo and its tributary, the Río Guajaráy. These groves occur on sandy riverbanks, the trees often of a single age class, possibly having germinated during a flood. In addition, they are planted. Manuel Rodriguez Tejada, a Guarijío man, carries a little gourd containing seeds of this tree and plants them where he thinks there should be a guamúchil tree, such as along the banks of the Río Mayo above Mesa Colorado. The trees generally reach full size in about 20 years or less and seem to live about half a century.

### *Platymiscium*

Trees or shrubs; flowers papilionoid. Tropical America; 20 species, 3 in Mexico.

The genus is unusual among the legumes in having opposite and whorled leaves. Some are important timber trees.

***Platymiscium trifoliolatum.*** *Palo santo; tampicerán* (Mayo, from *ta'ampisa*, "numbs your teeth") (figure 168)

Trees 10–15+ m, glabrous, with strong, hard wood; bark dark gray-brown, rough, and deeply furrowed or checkered. Stipules mostly broadly oblong, often 9–10 mm, green and thickish, soon deciduous, with an abaxial median area of short, thick, glandular hairs. Leaves deciduous in spring before flowering, turning bright yellow before falling; opposite on twigs and mostly whorled (3 per node) on vigorous new growth, 36–42 cm; leaflets 3, shiny green, 12–24 × 5–15 cm, mostly broadly ovate, the apex long acuminate, the terminal leaflet largest.

Inflorescences of slender racemes 7–22 cm, clustered at stem tips with the tender, emerging new leaves. Flowers 10–12 mm, on slender pedicels, and subtended by 2 thick, maroon bracts. Calyx dark maroon, 4.0–4.5 mm, with broad, blunt teeth; banner petal maroon outside, the inner surface of the banner, wing, and keel petals bright yellow-orange. Pods indehiscent, probably 6–7 cm, elliptic, flattened; seeds 1, compressed, probably 2 cm. Flowering March–April.

Localized in moist, tropical deciduous forest of slopes, canyons, and arroyos; southeastern Sonora, 305–940 m (1000–3085 ft). Also Sinaloa, Nayarit, Jalisco, and probably southwestern Chihuahua.

The leaflets are the largest of any legume in Sonora, and the opposite or whorled leaves are unique among legumes of western Mexico. It is the most prized of all lumber trees, although it is rather scarce in Sonora, and is used for house building, heavy gate posts, and molds for making *peloncillos* (brown sugar cones). The tree is called *palo santo* because it flowers during *Semana Santa* (Easter week); the fallen yellow flowers often carpet the ground.

### *Prosopis*

Trees or shrubs (in Sonora); most stems armed with spines. Leaves winter deciduous, sometimes tardily so; brilliant green new foliage appearing with warm weather in March and April; the new shoots often fast growing. Leaves (in Sonora) twice pinnate with 1 pair of opposite pinnae (jugate) or 2 pairs (bijugate); leaflets many; petioles with a nectary gland between the lower pair of pinnae; pulvini well developed.

Flowers mimosoid, small, whitish to yellow or yellow-orange, numerous in densely flowered cylindrical racemes or spikes. Calyx shallowly toothed. Stamens 10 in 2 series and separate; anthers with a small terminal gland. Pods indehiscent, straight or coiled, usually with carbohydrate-rich mesocarp.

Hot, arid, and semiarid regions; 41 species in the Americas, plus 4 from Africa to India. References: Bell and Castetter (1937), Benson (1941), Burkart (1976), Felger

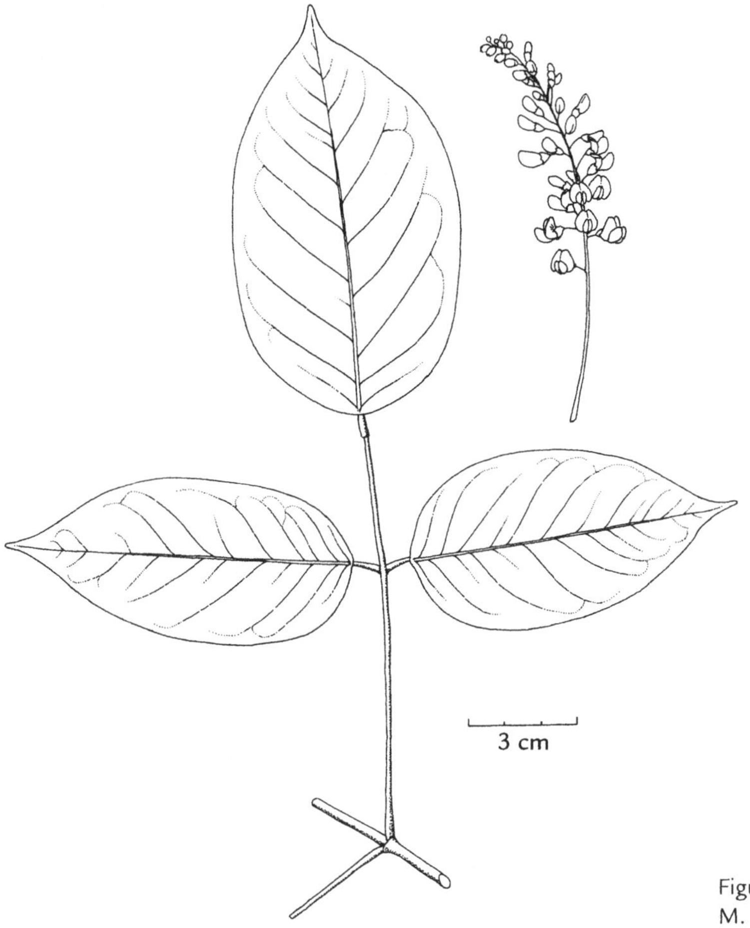


Figure 168. *Platymiscium trifoliolatum*. Drawn by M. B. Johnson.

(1977), Johnston (1962a), Rzedowski and Calderón de Rzedowski (1988), Simpson (1977).

The distinctive screwbean is in section *Strombocarpa* and the mesquites are in section *Algarobia*. They are discussed separately below.

1. Pods tightly coiled; flowers sessile, bright yellow, the petals united below: section Strombocarpa.
   ***P. pubescens***

1' Pods straight or sometimes somewhat curved but not tightly coiled; flowers on pedicels approximately 0.5 mm, cream colored to dull yellow or yellow-orange, the petals separate: section Algarobia.

   2. Leaves mostly bijugate; herbage hairy.
      ***P. velutina***

   2' Leaves mostly jugate (at least on short shoots); herbage glabrous or hairy.

      3. Leaves and twigs usually finely pubescent, sometimes becoming glabrous; fresh flowers orange-yellow; pods with little and bitter mesocarp or without mesocarp; lowland and coastal areas from Tastiota to Navojoa.
         ***P. articulata***

      3' Leaves and twigs usually glabrous (sometimes pubescent in northwestern Sonora); fresh flowers whitish or cream colored; pods usually with ample, sweet mesocarp; widespread.
         ***P. glandulosa***

## Section **Algarobia—Mesquite**

Trunks and larger limbs often crooked or irregular; the wood hard, heavy, and strong although somewhat brittle, and resistant to decay. Bark dark brown to blackish, irregularly checkered to shredding. Amber-colored sap, oxidizing black, oozes from wounds beneath the bark, and white to pale yellowish gum oozes from deeper wounds. Seedlings and young plants rapidly developing deep roots [the trees may put down roots as deep as 20 m or more, with the record being 80 m—the deepest known roots in the world (Solbrig and Cantino 1975)]. Paired or sometimes single thorns (apparently modified branches) developing from short shoots in leaf axils, the thorns straight, terete, rigid, and sharp, often especially large on juvenile and salt-stressed plants.

Racemes several on short shoots or axillary on long shoots. Profuse flowering April–May and irregularly to continuously through early fall when there may be a secondary flowering peak. Pods mostly ripe in early summer and sporadically through the fall, or a second crop in fall. Pods straight to curved (rarely loosely coiled), linear, thick, moderately compressed, often slightly to conspicuously constricted between the seeds, and usually hanging in clusters of several or more; often 5000 or more pods per tree. Seeds encased in a tough, leathery endocarp drying hard and bony; cotyledons bright green even while the seed still inside the pod.

The pods, with their usually sweet mesocarp, are eaten by a wide variety of animals, and germination is usually enhanced after seeds pass through an animal. Although large Pleistocene herbivores with which the mesquites probably co-evolved are gone, cattle have filled the niche as major agents of seed dispersal. Coyotes and other animals also eat the fruit and aid in seed dispersal. The seeds can remain viable for four decades if dry and enclosed in their endocarps (Martin 1948).

Along with the pines, the mesquites are the most economically important trees in Sonora. The populations and sheer biomass of the mesquites are (or were) probably greater than any other trees in northwestern Mexico. Great mesquite forests grew along all major rivers, deltas, and valley bottoms. The wood has been the preferred cooking fuel in desert regions of southwestern North America since ancient times. Most of the great mesquite bosques (forests), or *mezquitales*, have been seriously overharvested or destroyed for fuelwood and charcoal. Stumps left by woodcutters usually regenerate, but today's charcoal-makers destroy the entire plant. Much, or perhaps most, of the finest farmland in Sonora occupies former mesquite bosques.

The pods provided a major food resource for native Americans. There is great potential for the domestication and farming of mesquite for flour-production, and mesquite will probably become an important global agronomic crop. There is much variation in sweetness and flavor of the pod mesocarp. Traditional uses for mesquite included shelter, weapons, tools, cordage, fiber, dye, cosmetics, and medicines, both in North and South America (D'Antoni and Solbrig 1977, Felger 1977, Felger and Moser 1985).

About 24 species of *Algarobia* occur in South America and 6 occur in North America. One species, *P. juliflora sensu stricto*, bridges the gap between the Mexico/North American and South American species. It is a tropical species, extending northward on the Pacific Coast of Mexico to coastal mid-Sinaloa. This species has relatively thin (herbaceous) leaflets, inedible (bitter) fruit pulp, and very thick peduncles and pedicels. The *Algarobia* seem to be in an active state of evolution, and species boundaries are often poorly defined. *Prosopis articulata* and *P. glandulosa* are sympatric in the Guaymas region, and often grew intermixed along the shores of the harbor. *Prosopis glandulosa* and *P. velutina* broadly intergrade in areas of contact.

*Mezquite chileno*, "Chilean" or South American mesquites, are among the most widely planted ornamental trees in desert regions of southern Arizona and are likewise widely cultivated in lowland regions of Sonora. These trees appear to be hybrids mostly involving *P. alba* and *P. chilensis*, and perhaps *P. flexuosa* and *P. nigra*.

***Prosopis articulata*.** Bitter mesquite; *mezquite amargo; sanéa, juupa* (Mayo); *hu'upa* (Yaqui) (figure 169)

Large shrubs to small trees resembling *P. velutina*. Young stems and leaves finely pubescent, often glabrous with age. Leaves of short shoots jugate and often relatively short; those of long shoots jugate and sometimes also bijugate, and often larger. Leaflets 6.8–10 × 1.8–3.0 mm.

Flowers yellow-orange to yellow; anthers yellow to reddish. Pods with little and bitter mesocarp, the pods otherwise resembling those of the other Sonoran mesquites. Flowering primarily April–May (August).

Mostly valleys, plains, and arroyos; Sonoran desertscrub and coastal thornscrub. Near sea level–60 m (195 ft). Widespread through much of the Baja California peninsula, especially the southern half, and otherwise known from western Sonora from near Tastiota to the vicinity of Navojoa.

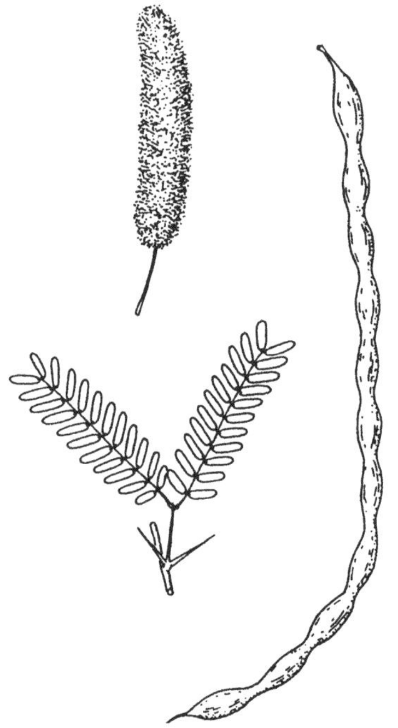

Figure 169. *Prosopis articulata*. Drawn by M. B. Johnson.

*Prosopis articulata* seems most closely related to *P. velutina*, with which it apparently intergrades; perhaps they are not distinct species.

***Prosopis glandulosa*** var. ***torreyana*** [*P. juliflora* var. *torreyana*]. Western honey mesquite; *mezquite; jupa'ará* (Guarijío); *juupa* (Mayo); *kui* (O'odham); *kúi* (Pima Bajo); *haas* (Seri); *hu'upa* (Yaqui) (figure 170A)

Woody shrubs to trees 12+ m. Twigs and leaves generally glabrous. Leaves with 1 pair of pinnae, each with (6) 11–19 pairs of leaflets; individual leaflets about 5 times longer than wide, 4.5–31 × 1.3–6 (8) mm (smaller sizes from drought stressed plants). Thorns often 1–4 cm, at first dark, fading whitish.

Racemes often (4) 6–9 cm. Flowers pale yellow. Flowering primarily April–June (September). Pods often 10–20 cm × 7.5–10 mm maximum width; mesocarp usually sweet and well developed.

Slopes, valleys, plains, arroyos; Sonoran desertscrub, Chihuahuan desertscrub, thornscrub, tropical deciduous forest, and grassland. Near sea level–1370 m (4500 ft). Abundant in the lowlands nearly statewide, and generally replaced inland in the Sonoran Desert by *P. velutina*. Variety *torreyana* ranges from Sinaloa, Chihuahua, and Baja California Sur to western Texas, southwestern Utah and southern California.

Western honey mesquite occurs in many habitats in Sonora but is best developed along drainageways, riverine floodplains, and low-lying, coastal plains with fine-textured soils. Vast bosques of *mezquital* once covered the coastal plains from near Bahía Kino to Tastiota and south of Guaymas. In the Chihuahuan Desert of northeastern Sonora and adjacent regions, this mesquite develops as a shrub with a large, subterranean root-caudex.

Two geographic races of *P. glandulosa* are generally recognized. Variety *glandulosa* occurs east of the continental divide from Texas to Kansas and northeastern Mexico to the Chihuahuan Desert region. The area of overlap and intergradation between the two varieties is extensive. Variety *glandulosa*, at least from Texas, is much more cold hardy than var. *torreyana* from lowland Sonora.

Western honey mesquite is distinguished from *P. velutina* by its jugate leaves and relatively narrow, often large, and rather widely spaced leaflets. *Prosopis glandulosa*, especially the western var. *torreyana*, and *P. velutina* are closely related. Populations from the vicinity of the Río Colorado and the western part of the Gran Desierto dunes in northwestern Sonora are glabrous or glabrate and readily fall into the usually accepted concept of *P. glandulosa* var. *torreyana*. Farther eastward and at slightly higher elevations, approaching the domain of *P. velutina* and encompassing the Pinacate region, the mesquites are intermediate, having the jugate leaves of *P. glandulosa* but the pubescent leaves and twigs characteristic of *P. velutina*. There seems to be a tendency for these intermediate populations to have proportionally broader (and shorter) leaflets to the east, where they approach *P. velutina*—so that except for the jugate leaves one could place them with *P. velutina*. Perhaps these two mesquites should be treated as subspecies.

***Prosopis velutina*** [*P. juliflora* of western American authors, not *P. juliflora* (Swartz) de Candolle; *P. juliflora* var. *velutina*]. Velvet mesquite; *mezquite*; *kui* (O'odham); *hu'upa* (Yaqui) (figure 170B)

Woody shrubs or trees to 15+ m. Similar to *P. glandulosa* var. *torreyana* but the herbage and inflorescences always hairy, most or all leaves on a plant bijugate, and the leaflets proportionally broader, not attaining as large a size, usually more closely spaced, and tending to average more leaflets per pinna. (Many leaves on an individual velvet mesquite may be jugate, especially new leaves in spring and during drought, but at least some bijugate leaves are nearly always present.) Leaflets roughly 3 times longer than wide, often 6–13 × 2–3.5 mm, 14–24 pairs per pinna.

Flowers yellow or cream colored. Peak flowering late March–May and sporadically summer–early fall. Pods mostly in early summer and variously through the fall.

Slopes, canyons, valleys, and arroyos; Sonoran desertscrub, thornscrub, grassland, and lower elevations in oak woodland; northern and central Sonora; near sea level–1370 m (4500 ft). Arizona, northern Sonora southward to the vicinity of Guaymas, and extreme southwestern New Mexico and northwestern Chihuahua.

This is one of the most common trees and shrubs in Arizona upland of the Sonoran Desert and desert-grassland regions of Sonora and adjacent southern Arizona. It is largest and best developed along drainageways, although it also occurs on bajadas, rocky slopes, and other habitats. Large mesquite trees were common along the floodplains of the major drainageways in the region. The combination of bijugate leaves, pubescent herbage and inflorescences, and shorter and proportionally broader leaflets distinguish this mesquite.

## Section **Strombocarpa**

***Prosopis pubescens.*** Screwbean, screwbean mesquite; *tornillo; kujel, kuwidculis* (O'odham) (figure 171)

Large, woody shrubs or trees to 6 m; bark smooth and gray, becoming shaggy on larger trunks. Winter deciduous, leafing out in spring. Spines 2–9 mm, stipular, paired, stout, straight, terete, and white. Young herbage including spines with short, sparse, white hairs, becoming sparser with age. Leaves 3–6 cm, jugate or occasionally some leaves bijugate; leaflets 4.5–13 mm, 4–9 per pinna.

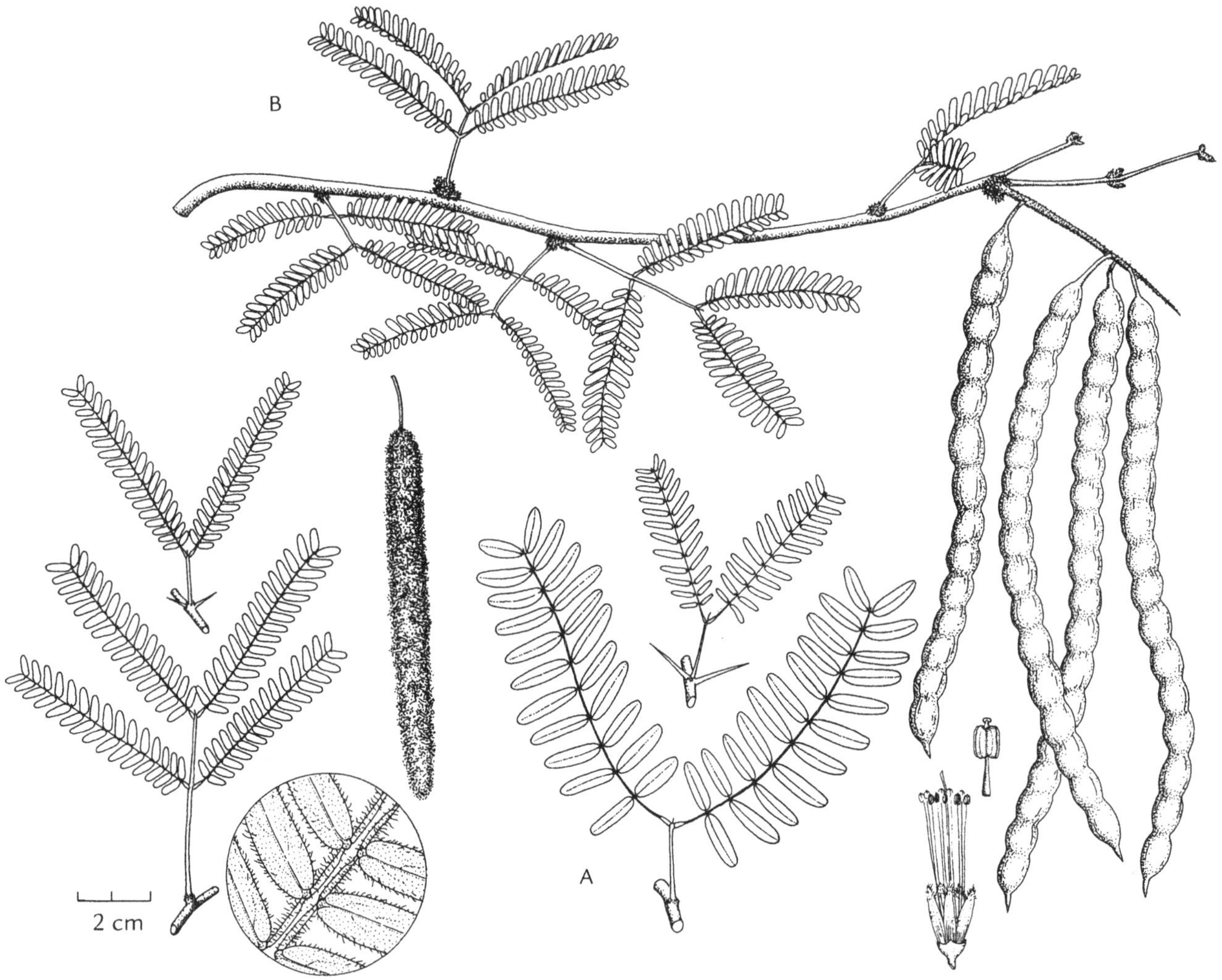


Figure 170. *Prosopis*, section *Algarobia*. (A) *P. glandulosa* var. *torreyana*, (B) *P. velutina*. Drawn by M. B. Johnson.

Flowers sessile, bright yellow, in spikes 3–4.5 cm. Ovary densely pubescent and coiled. Pods likewise tightly coiled, in clusters of (3) 6–20+ pods; individual pods cylindrical, usually 3–4 cm. Flowering May and sporadically summer–fall; pods ripen mostly during summer months (some pods usually persistent).

Sandy, often alkaline or partially saline soils along streams and washes, waterholes, and oases; Sonoran desertscrub; northwestern Sonora; near sea level–245 m (800 ft); mostly at pozos near Bahía Adair, Laguna Prieta, Ciénega de Santa Clara, and the margins of the Río Colorado where it was once abundant, along the Río Sonoyta, Quitobaquito, and Quitovac. Also northeastern Baja California and southern California to southwestern Utah and Texas and northern Chihuahua. The range of this species has shrunk as rivers have become dry or altered due to human activity (Turner et al. 1995).

The sweet mesocarp of the pods were used as food, the wood for fuel and house construction, and the roots and bark for medicine (Bell and Castetter 1937).

### *Psorothamnus*

Flowers papilionoid. North American deserts; 9 species. Reference: Barneby (1977).

The smoke tree is the only arborescent member; the others are shrubs with pinnately compound leaves.

**Psorothamnus spinosus** [*Dalea spinosa*]. Smoke tree (figure 172)

Erect shrubs or trees to 6+ m, with hard wood and well-developed trunks. Bark smooth and gray, with age often somewhat fissured and rough on the trunk. Nearly every

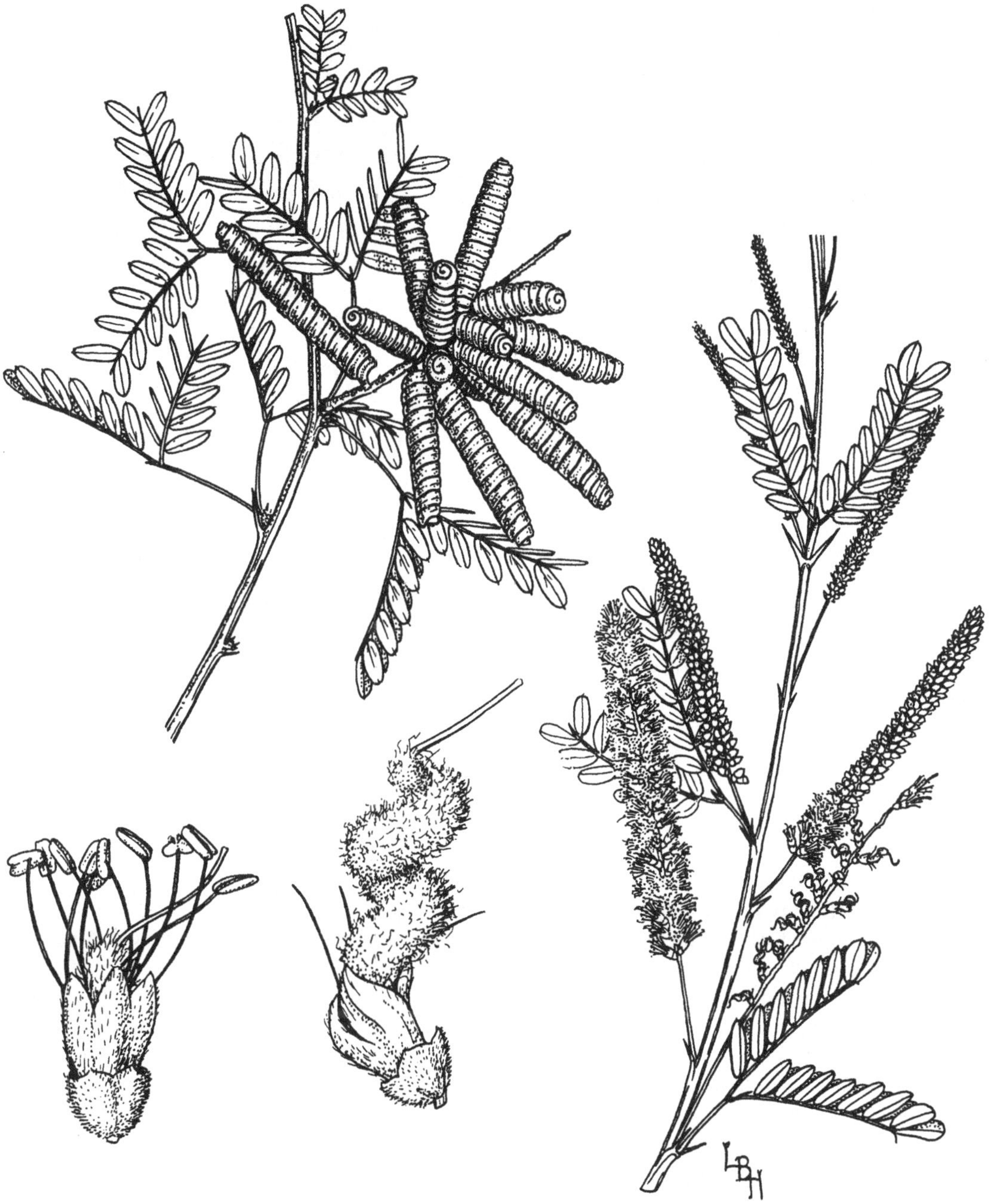


Figure 171. *Prosopis pubescens* (Benson and Darrow 1945). Drawn by L. B. Hamilton.

part of the plant dotted with orange glands containing aromatic resins. (Crushed, fresh twigs have a fruity aroma. This species and *Eysenhardtia* are the only aromatic tree legumes in Sonora.) New growth appears during hot weather. Branches repeatedly, stiffly, and closely forked; twigs silvery to bluish gray, ending in a sharp point, and with dense, short, appressed hairs. Stipules ≤ 1 mm, deciduous. Seedlings and young plants usually leafier (but also drought deciduous) with leaves reaching 20–65 mm. Trees mostly leafless, except the young growth often with few, very sparse, and quickly deciduous leaves reduced to a single leaflet often 5–25 mm, oblanceolate, bluish gray, thickish, and almost succulent. (This unifoliolate leaf is analogous with the enlarged terminal leaflet of other leafy members of *Psorothamnus*.)

Inflorescences of few-flowered, spinescent, twiglike racemes. Flowers 12 mm; calyx 4.5–5.0 mm, the tube prominently ribbed with several large, orange-brown blister glands; petals brilliant indigo blue. Pods indehiscent, often 5.0–5.5 mm with several orange-brown blister glands and a tail-like persistent style; the single seed 4.0–4.3 mm. Spectacular, mass flowering June–July.

Large sandy washes, sand flats, and interdune troughs; Sonoran-desertscrub of the Gran Desierto of northwestern Sonora; 10–310 m (30–1015 ft). Also southwestern Arizona, southern Nevada, southeastern California, and northeastern Baja California.

This unique plant is worthy of cultivation in extremely hot, dry desert regions. It is apparently frost hardy to about −9°C (16°F). The seeds generally germinate without treatment; however, scarification leads to quicker and more uniform germination. The plants prefer well-drained soil and grow rapidly with irrigation during warm weather. Cultivated plants in the ground can grow 1 m per year and often flower in their second year. Plants in containers have flowered at 8 months of age.

Figure 172. *Psorothamnus spinosus*. (Benson and Darrow 1945). Drawn by L. B. Hamilton.

### *Robinia*—Locust

Trees and shrubs; flowers papilionoid. North America; 4 species. References: Isely and Peabody (1984), Lavin and Sousa (1995).

**Robinia neomexicana.** New Mexican locust; *garrobo* (figure 173)

Shrubs or trees to approximately 10 m, often propagating by root sprouts. Bark smooth and gray, with age becoming fissured. Nodes with paired, painfully sharp, slightly downcurved stipular spines to 1.7 cm. Leaves winter deciduous, 10–25 (36) cm, pinnate; leaflets 9–19 per leaf, 2.5–6 cm, ovate to oval, pubescent with fine, white hairs, sparse on the upper surfaces and moderately dense on the lower surfaces.

Flowers showy, pale pink to lavender, approximately 2 cm, fragrant, in densely crowded racemes shorter than the leaves. Pods 4.5–10 cm, gradually dehiscent, densely covered with conspicuous, coarse, glandular, brown hairs that stick together like Velcro. Flowers May–August.

Mountain slopes and canyons; oak and pine-oak woodland; northeastern Sonora; 1220–2590 m (4000–8500 ft). Also northern and western Chihuahua, locally in eastern California, and Arizona to Utah, Colorado, and Trans-Pecos Texas. Localized in Sonora. The southernmost Sonoran records are from the Sierra el Tigre and Sierra Bacadehuachi. Mostly a second growth species of understory shrubs or small trees in pine forest and upper oak woodland. In southern Arizona it often forms thickets on steep, eroding slopes.

Sonoran populations are var. *neomexicana*. The weakly distinguished var. *rusbyi* grows among var. *neomexicana* in portions of Arizona and New Mexico; it has glabrous or glabrate pods and generally reduced pubescence (Isely 1998).

### *Senna*

Herbs to trees; flowers caesalpinioid. Mostly New World; 260 species. Reference: Irwin and Barneby (1982).

Many *Senna* species are cultivated for their showy flowers, and others are used for medicinal purposes. The genus *Senna* is segregated from *Cassia*, to which it is closely related.

**Senna atomaria** [*Cassia emarginata*]. *Zorillo, palo zorillo, mora hedionda* (Sinaloa); *juracosi, jupachumi* (Guarijío); *jupachumi* "skunk's ass" (Mayo) (figure 174)

Erect, slender trees with few-branched, slender trunks and limbs, and a high crown sometimes reaching 10 m. Leaves drought deciduous, pinnate, velvety haired, especially when young, foul-smelling when crushed (hence the common names meaning "skunk" or "stinking"). Stipules 2–4 mm, brown, slender, and persistent. Leaves 17–21+ cm; leaflets 5–12 cm, broadly elliptic to ovate, rounded at tip, 2–5 pairs per leaf, the terminal pair largest; leafstalks without a nectary gland (many members of this genus have elongated leafstalk nectaries).

Inflorescences terminal racemes 3–8 cm. Flowers 2.5–3 cm wide; sepals green to yellow-tinged, unequal in size; petals

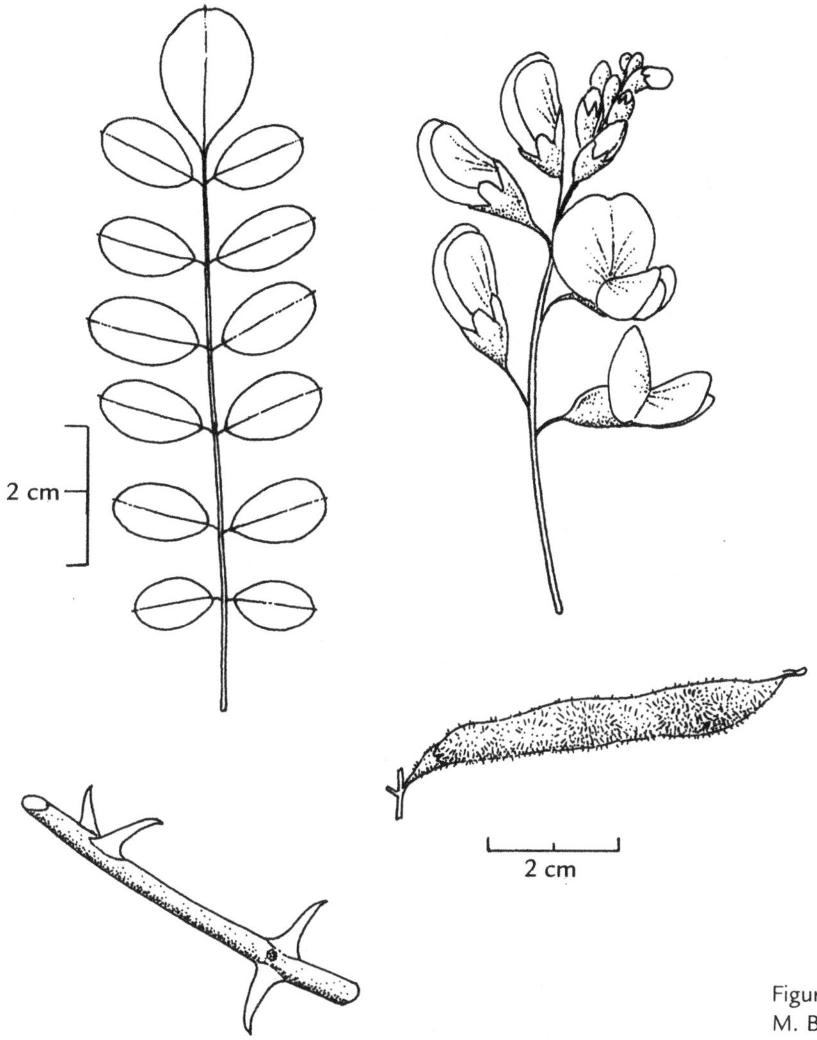


Figure 173. *Robinia neomexicana*. Drawn by M. B. Johnson.

bright mustard yellow; fertile stamens 7 with anthers 4.5–5.5 mm, sparsely hairy, opening by terminal pores (flowers buzz pollinated by bees); sterile stamens 3 with anthers 2.2–2.8 mm. Mostly flowering when leafless, February–July, the peak usually in early April. Pods 25–40 × 1 cm, straight, blackish, thick, moderately compressed, indehiscent with pulpy mesocarp, pendent, numerous and persistent; commonly hanging in great numbers from the leafless or near leafless trees during the winter–spring dry season.

Slopes, valleys, and arroyos; tropical deciduous forest and sometimes in thornscrub; southern and east-central Sonora; 30–1000 m (100–3280 ft). Also adjacent Chihuahua and Sinaloa to South America and the West Indies.

This species is one of the more abundant elements of the tropical deciduous forest of the foothills and lower mountains in southeastern Sonora. The flowering trees add yellow accents to the dry forests of spring. The pods are important dry-season forage for cattle and are harvested as fodder. In cultivation it is highly frost sensitive but could be an important drought-tolerant landscape subject and avenue tree for hot regions.

## Fagaceae—Beech Family

Trees and shrubs, mostly temperate and montane in tropical regions, Northern Hemisphere and portions of the Southern Hemisphere; 8 genera, 700 species.

This family includes beech (*Fagus*), chestnut (*Castanea*), and southern beech (*Nothofagus*, the Southern Hemisphere counterpart of the oaks). Many members of this family, especially the oaks, are important sources of dyes, edible nuts, fodder, fuel, hardwood lumber, and bark used in tanning.

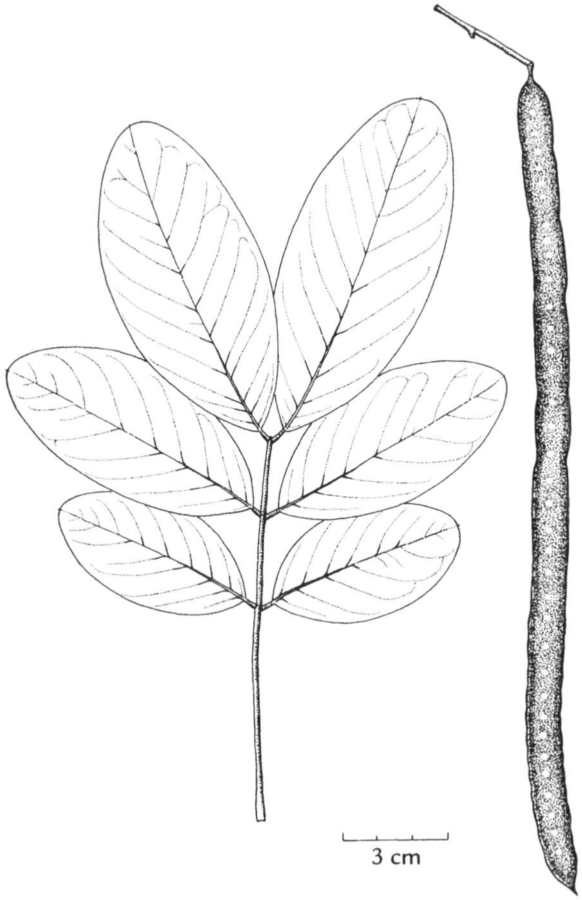


Figure 174. *Senna atomaria*. Drawn by M. B. Johnson.

## *Quercus*—Oak, Encino

Trees and shrubs with close-grained hardwood. Twigs with clusters of subterminal buds covered by bud scales, and round lenticels. Leaves alternate, simple, usually rather firm; stipules inconspicuous and deciduous. Hairs of two types, both usually present: (1) stellate hairs, whitish or yellowish, nonglandular, ascending to spreading, and (2) unbranched hairs, orange to golden, glandular, often appressed. Herbage densely woolly to glabrate.

Monoecious. Flowers wind-pollinated, small and inconspicuous, without petals; male flowers many in pendulous catkins; female flowers solitary or in small clusters, sessile or stalked. Fruit an acorn and partially surrounded by a cup. (This cup seems to represent a highly reduced system of coalesced branches and their associated reduced leaves.) Acorns 1-seeded, mostly at least 1 cm, annual (maturing within one year on the new twigs and located among the leaves of the current year) or biennial (maturing in two years and located below the leaves on the leafless part of the twig; *Q. albocincta* is an exception with annual acorns below the leaves). Cotyledons of fresh seeds white, yellow, pink, or pink-purple.

Mostly temperate and tropical montane regions. Widely distributed in the Northern Hemisphere and south to Colombia and Malaysia; 400 species, the highest diversity in montane regions of Mexico; at least 125 species in Mexico. References: González Villarreal (1986), Landrum (1994), Marshall (1957), McVaugh (1974), Nixon (1993, 1997).

The oaks are grouped into two subgenera and several sections. The New World oaks fall into three sections of subgenus *Quercus:*

Section ***Quercus*** (*Leucobalanus*), white oaks. Eastern and Western hemispheres. Bark generally light gray (lighter colored than in the black oaks) and often scaly or checkered. Acorns annual; inner shell of the hard fruit case glabrous, the fresh seeds (cotyledons) whitish, pink, or purplish. Includes *Q. arizonica*, *Q. chihuahuensis*, *Q. gambelii*, *Q. grisea*, *Q. oblongifolia*, *Q. perpallida*, *Q. rugosa*, *Q. subspathulata*, *Q. toumeyi*, and *Q. tuberculata*.

Section ***Lobatae*** (*Erythrobalanus*), red or black oaks. Western Hemisphere. Bark generally dark gray to blackish, often smooth or furrowed. Maturation time of the acorns often diagnostic and ecologically important, the acorns annual or biennial; inner shell of the fruit case woolly pubescent, the fresh cotyledons whitish to yellow. Represented in Sonora by *Q. albocincta*, *Q. coccolobifolia*, *Q. durifolia*, *Q. emoryi*, *Q. hypoleucoides*, *Q. mcvaughii*, *Q. scytophylla*, *Q. sideroxyla*, *Q. tarahumara*, and *Q. viminea*.

Section ***Protobalanus***, intermediate or golden-cup oaks. Southwestern United States and northwestern Mexico, mostly California and northern Baja California; five species. Represented in Sonora by *Q. chrysolepis*.

Although most oak species in Sonora are readily identified by general leaf characteristics, differences in pubescence and epidermal glands can be critical in distinguishing closely related taxa. For this purpose high-power magnification may be necessary. These hairs are best seen on young leaves because older leaves tend to be glabrate. (For an excellent discussion on oak trichomes, or hairs, see Hardin 1976.) Otherwise, the best time to study oaks is at the end of the growing season, in late summer or early fall, when the leaves are mature. Characteristics involving leaf pubescence discussed in the species accounts generally refer to mature leaves found from late summer or early fall until the leaves fall away in late spring. It is best to work with sun leaves because shade leaves and "water sprouts" may present unusual features.

Oaks frequently form hybrid swarms as well as intermediate populations, sometimes making identification an art. The major difficulty in oak taxonomy comes from attempts to discriminate boundaries within a many-sided continuum. Hybridization is commonplace among members of the same section.

At least 21 species of tree-sized oaks occur in Sonora, making it the largest genus of trees in the state. *Quercus gambelii* is the only Sonoran oak with a "northern" or temperate affinity, and the only one that is winter deciduous. The remaining Sonoran oaks are allied with the southern or tropical Mexican oak flora. Some, such as *Q. albocincta*, generally become leafless early in spring drought, while others shed their leaves at the end of the spring dry season before summer rains. Thus many Sonoran oaks acquire "autumn" colors during spring drought. In some cases the dry, dead leaves may remain on the trees for several months. High-elevation oaks, such as *Q. tarahumara* and the silverleaf oaks (*Q. hypoleucoides* and its relatives) tend to be evergreen. The biggest flush of new foliage occurs with the beginnning of summer rains.

The Sonoran oaks occur at elevations above the desert, although some descend into the upper elevational margins of the desert in riparian canyons. Islands of oak and pine barrens are sometimes encountered at remarkably low elevations within tropical deciduous forest and thornscrub. These woodlands, sometimes far from the continuous oak and pine forests of the sierras, owe their existence to the red, yellow, or white hydrothermally altered acidic soils on which they grow (Goldberg 1982, Martin et al. 1998).

Three oak species that occur near the border in southern Arizona are not documented for Sonora. *Quercus ajoensis* grows in several canyons in the Ajo Mountains in Organ Pipe Cactus National Monument, and the tallest ones measured 10+ m in height. *Quercus pungens* and *Q. turbinella* (figure 176E) are most often shrubs but sometimes become small trees.

Especially in southeastern Sonora, low-intensity ground-level fires during the presummer drought play a major role in maintaining the rather abrupt margin between tropical deciduous forest and the lower oak zone. Oaks readily survive the small fires, but most tropical deciduous species are not fire resistant.

The Sonoran oaks are important to many species of wildlife, livestock, and people. The leaves are eaten by deer and cattle. Acorns of *Q. emoryi* are a staple in the diet of black bears in northeastern Sonora and are commercially harvested for human consumption. Fresh or dry leaves of *Q. coccolobifolia* and *Q. tuberculata* are made into tea and drunk for the pleasant flavor. The tea is drunk with sugar and sometimes cinnamon is added. Most oaks are prized as a source of fence posts, fuelwood, and fine hardwood lumber. Many produce crown sprouts and thus may recover from cutting. In the Yécora region the oak trees are rapidly being destroyed for wood pulp. The debarked wood is trucked to the Anáhuac mill near Cuauhtémoc, Chihuahua, a huge pulping factory largely financed as a World Bank project.

Most Sonoran oaks have considerable horticultural potential. With proper greenhouse techniques, some or all can be propagated from cuttings. Oak seeds germinate immediately and generally are viable for only a short time, and when gathering acorns for planting it is best to keep them moist. However, we have grown *Q. emoryi* from acorns stored dry at room temperature for as long as one month before planting.

The upper leaf surfaces of some oaks, either shiny or dull, appear smooth, but even with 10× magnification you can see small hairs, especially along the midrib and major veins. In the key and descriptions we call such leaves "smooth" (e.g., *Q. albocincta* and *Q. durifolia*). It is difficult to construct an accurate key that is both easy to use and covers the variation occurring among the oaks. A more accurate key would utilize technical characters of the pubescence. Our key is intended to help in identification by using the usual or most common variation, and in some cases does not cover the full span of variability. The key should be used together with the illustrations. In the following text the season or month is given for ripening of acorns and does not refer to time of flowering.

1. Lower leaf surfaces uniquely white-waxy; leaves hollylike. ______ ***Q. chrysolepis***
1' Lower leaf surfaces not white-waxy; leaves generally not hollylike.
  2. Bark of twigs and smaller branches exfoliating in small flakes.
    3. Trees usually > 8 m; leaves mostly 4–12 cm, often glaucous; acorns on pedicels more than 4 cm. ***Q. perpallida***
    3' Large shrubs to small trees seldom > 6 m; leaves 1–4 (4.7) cm, green; acorns sessile. ______ ***Q. toumeyi***
  2' Bark of twigs and branches not exfoliating.
    4. Leaves at maturity smooth or nearly so, at least the lower surfaces, or only sparsely hairy.
      5. Leaves with the marginal teeth ending in a soft bristle > 5 mm. ______ ***Q. albocincta***
      5' Leaves with or without marginal teeth, the teeth without bristles or ending in a firm spine < 3 mm.
        6. Leaves with a small tuft of hair at base of lower surface of blade; acorns annual. ______ ***Q. emoryi***
        6' Leaves without a hair tuft at base of leaf blade [or if hair tufts near petiole present, then (1) acorns biennial—these being *Q. viminea* probably introgressed with *Q. emoryi* or (2) *Q. coccolobifolia*].

7. Leaves bluish glaucous, the tip blunt. ____ ***Q. oblongifolia***
7' Leaves green or at least not glaucous, the tip blunt or pointed.
8. Leaf tips pointed; leaves 5–13 cm. ____ ***Q. viminea***
8' Leaf tips blunt; leaves 6.5–22+ cm.
9. Leaves relatively thick and firm; petioles mostly 2–10 mm; above 1500 m elevation. ***Q. coccolobifolia***
9' Leaves relatively thin; petioles 7–20 mm; below 1250 m elevation. ____ ***Q. tuberculata***
4' Leaves, even at maturity, conspicuously hairy at least on the lower surface.
10. Leaf lobes large, rather evenly spaced, usually rounded, not spinose, and cleft more than halfway to midrib; winter deciduous. ____ ***Q. gambelii***
10' Leaf lobes mostly irregularly spaced, pointed or spinose tipped, cleft much less than halfway, or the leaves not lobed; evergreen or drought deciduous in spring or early summer.
11. Lower leaf surfaces conspicuously white-woolly, the upper surfaces dark green.
12. Leaf blades relatively thin, not tough, the margins entire, not thickened and not inrolled. ***Q. durifolia***
12' Leaf blades relatively thick and tough, the margins entire or spiny toothed, thickened and inrolled; silverleaf oaks.
13. Leaf margins evenly and prominently spiny toothed more than halfway from tip toward base of blade (see figure 181). ____ ***Q. scytophylla***
13' Leaf margins entire to toothed mostly with small spines above the middle of the blade.
14. Leaf blades elliptic to lanceolate or oblanceolate, entire to few toothed; north-central and eastern Sonora. ____ ***Q. hypoleucoides***
14' Leaf blades obovate, oval or oval-oblong, sometimes broadly so, with short spines toward the leaf apex; southwestern Chihuahua and perhaps adjacent Sonora. ____ ***Q. sideroxyla***
11' Lower leaf surfaces not white-woolly.
15. Leaf blades conspicuously velvety hairy on both surfaces, the tip not spinescent. ____ ***Q. chihuahuensis***
15' Leaf blades smooth or hairy above but not velvety, the tip spinescent or not.
16. Leaves 10–38 cm; acorns sessile or on short peduncles.
17. Leaves sessile or subsessile. ____ ***Q. subspathulata***
17' Leaves prominently petioled.
18. Leaves 10–18 cm, the blade nearly flat to moderately convex or cupped, the lower surfaces densely woolly with yellow-orange hairs, the branched (stellate) hairs sessile; acorns biennial. ____ ***Q. mcvaughii***
18' Leaves 13–38 cm, the blades strongly convex, the lower surfaces fuzzy and golden brown, the branched hairs stalked; acorns annual. ____ ***Q. tarahumara***
16' Leaves < 10 cm, or if 10–12 cm then acorns on long peduncles.
19. Upper leaf surfaces dark green, the blades moderately convex (cupped), the lower surfaces with netlike, raised veins. ____ ***Q. rugosa***
19' Upper leaf surfaces gray- or blue-green, the blades not cupped, the lower surfaces without netlike raised veins.
20. Upper leaf surfaces with the veins slightly sunken. ____ ***Q. arizonica***
20' Upper leaf surfaces with the veins raised. ____ ***Q. grisea***

***Quercus albocincta.*** *Encino roble, encino prieto, cusi; hachuká* (Guarijío); *kus* (Pima Bajo) (figure 175A)

Trees 5–10 (16) m, often with a single, upright trunk and spreading limbs; bark rough and dark gray. Leaves drought deciduous in late spring, 9.5–18 (27) × (1.5) 3–7 (13) cm, broadly lanceolate, rather smooth, glossy, and deep green, often moderately bicolored, each side with 4–8 short, pointed lobes tipped with a slender, flexible bristle to 1 (1.5) cm; petioles slender and prominent. Fresh new growth light green or whitish to pink. Acorns annual, sweet and edible, sessile to short stalked; acorns ripen early summer–early September.

Rocky ridges, slopes, and canyons; oak and pine-oak woodland; southeastern and east-central Sonora, (200) 490–1500 m [(655) 1605–4920 ft]. One of the northernmost populations occurs at 1500 m in the vicinity of Nacori Chico; the low elevation records are trees on islands of oak woodland on hydrothermally altered red soils within areas of tropical deciduous forest. Also adjacent Chihuahua and northeastern Sinaloa, and Baja California Sur.

This is a distinctive tree, recognized by its large, shiny leaves with bristle-tipped lobes. This attractive oak is worthy of cultivation and grows in the desert with additional irrigation. The flowers appear on the leafless twigs and subsequently the new, leafy shoots emerge from the terminal or near terminal nodes. The acorns are thus situated below the new leaves even though they are annual. Local inhabitants consider cusi to produce the best-tasting acorns. In our opinion they taste like the acorns of *Q. emoryi*.

***Quercus arizonica*** [*Q. alamosensis*]. Arizona white oak; *encino blanco*; *toa* (Pima Bajo) (figures 19, 175B and 176A)

Trees 5–18 m; crown rounded to irregular and spreading; bark fissured and pale gray. Leaves nearly evergreen to briefly deciduous in late spring, with wide variation in size and shape, 2–10 cm, relatively firm, the base cordate to truncate, the margins with several short, spinescent teeth to sometimes entire, the apex variable, rounded to obtusely pointed with a spinescent tip; petioles short. Upper leaf surfaces dull, dark green to gray- or blue-green, sparsely hairy or glabrate, with impressed veins; lower surfaces pale green, sparsely pubescent (tomentose) with fasciculate and glandular, short, orange-brown hairs, and also papillate. Cotyledons of fresh seeds pink-purple (magenta or fuchsia), or sometimes both pink and white on the same tree. Acorns sessile to stalked; ripening fall and spring.

Slopes, ridges, canyons and valleys; grassland, oak and pine-oak woodland; north-central and eastern Sonora, 915–2315 m (3000–7600 ft). Also Chihuahua, Durango, Coahuila, Arizona to Trans-Pecos Texas, and Baja California Sur.

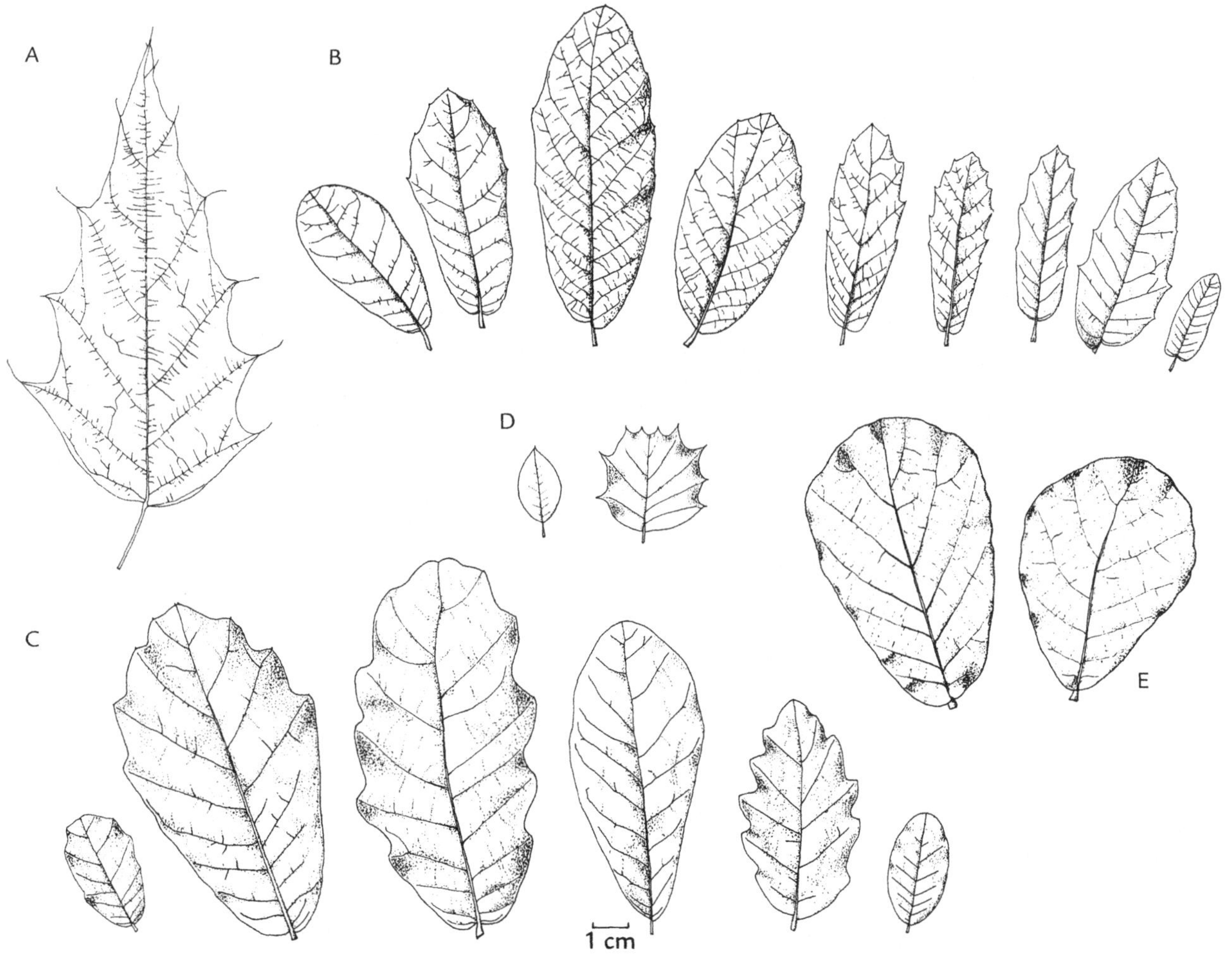


Figure 175. *Quercus*. (A) *Q. albocincta*, (B) *Q. arizonica*, (C) *Q. chihuahuensis*, (D) *Q. chrysolepis*, (E) *Q. coccolobifolia*. Drawn by M. B. Johnson.

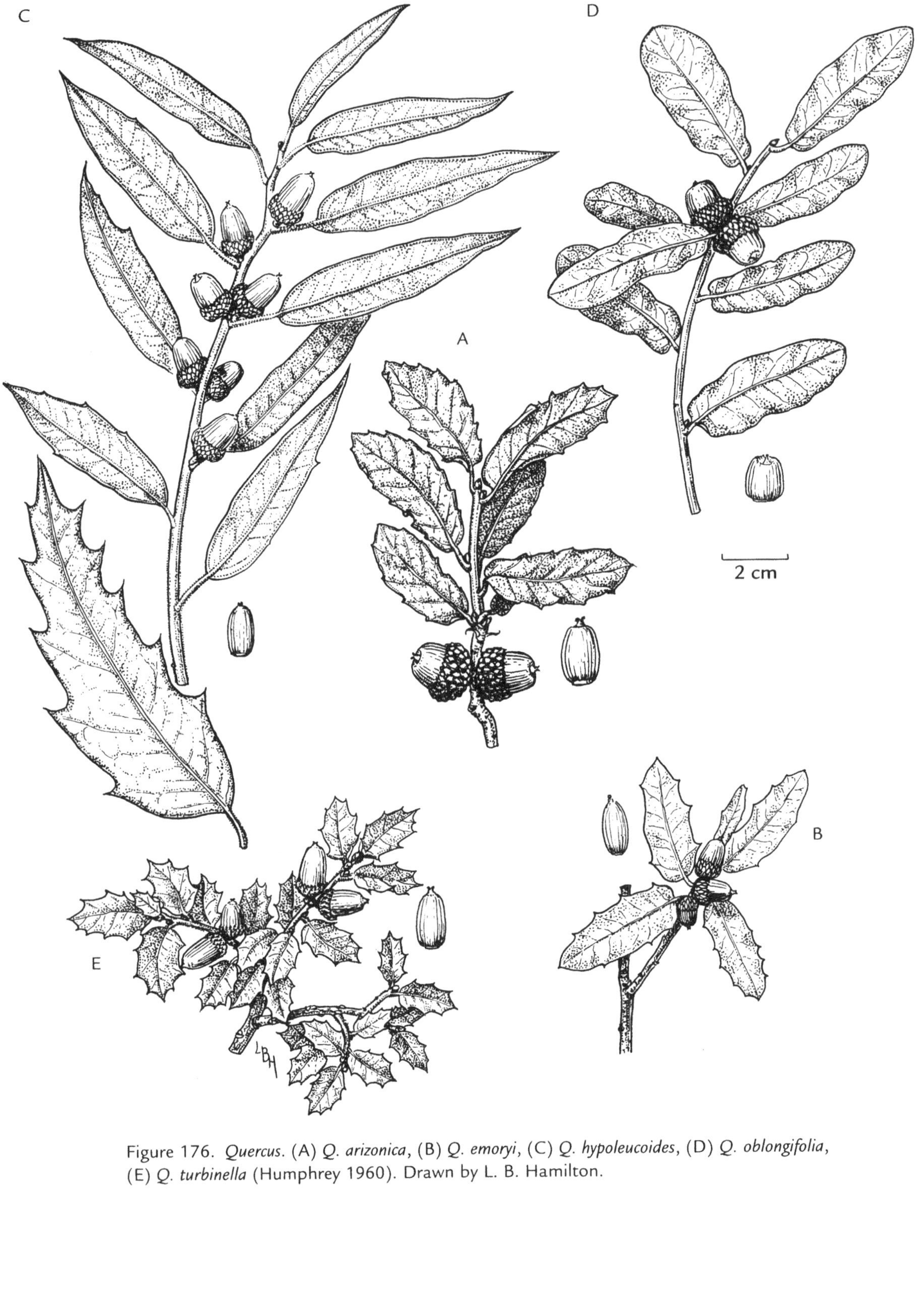


Figure 176. *Quercus*. (A) *Q. arizonica*, (B) *Q. emoryi*, (C) *Q. hypoleucoides*, (D) *Q. oblongifolia*, (E) *Q. turbinella* (Humphrey 1960). Drawn by L. B. Hamilton.

This is one of the most widespread oaks in Sonora. Southward in Mexico it merges into *Q. laeta*, which ranges from Sinaloa and Durango to Nuevo León and Michoacán. Landrum (1994) treated *Q. arizonica* as a synonym of *Q. grisea*. In New Mexico there are trees of intermediate character with both colors of acorns on the same tree (Richard Spellenberg, personal communication, 1994). At least in Sonora, however, these oaks show predictable differences in architecture, leaf color and pubescence, fresh seed color, and habitat. Nixon (1997) regarded them to be more closely related to other species than to one another.

***Quercus chihuahuensis*** [*Q. santaclarensis*]. Chihuahua oak; *encino blanco, encino chino; sajaguo* (Guarijío); *tua ush* (Pima Bajo) (figure 175C)

Small to medium-sized trees, often 5–10 (14) m, with low, spreading, irregular branches. (The largest trees, with trunks approximately 1 m in diameter, occur in the vicinity of Mulatos.) Bark deeply fissured, checkered, whitish gray, and similar to that of *Q. arizonica* and *Q. oblongifolia*. Foliage drought deciduous in spring, although dry, brown leaves sometimes persist several months. Leaves 3–9 (16) cm, the blades often oblong to obovate, soft and velvety, covered with short, whitish to pale yellow-brown hairs on both surfaces, the margins wavy (sinuate) to broadly lobed or sometimes nearly entire; sessile or short petioled, or petioles rarely 5–10 mm on larger leaves. Acorns annual, one to several per cluster, subsessile or on stalks to 8.5 cm; ripening October–December.

Slopes, ridges, canyons, and arroyos; grassland, oak and pine-oak woodland, and occasionally descending to the upper margin of Sonoran desertscrub, thornscrub, and tropical deciduous forest; eastern Sonora, 275–2075 m (900–6800 ft); northward to the vicinity of Bacoachi and the north end of Sierra el Tigre in the Río Sonora and Río Bavispe drainages. Common and widespread, often occurring as small scattered trees in oak-grassland and also forming closely spaced woodland in favorable habitats. Also Chihuahua, Sinaloa, Durango, Zacatecas, and San Luis Potosí, and locally in western Texas as putative hybrids with *Q. arizonica* and *Q. grisea*.

No other Sonoran oak has leaves so densely velvety and fuzzy on both surfaces. This oak exhibits greater variation in leaf shape than any other Sonoran oak, and apparent hybrids with several other species are commonplace. The acorns are sometimes fed to pigs.

***Quercus chrysolepis***. Canyon live oak (figure 175D)

Shrubs or shrublike trees, probably to 4+ m, with a well-formed thick trunk; bark gray and rather smooth. Nearly evergreen; leaves 1.5–3.5 cm, the blades often elliptic to ovate or orbicular, tough and hollylike, the margins thickened and wavy, entire or with large, spine-tipped teeth, the upper surfaces smooth and shiny, the lower surfaces uniquely white waxy; younger leaves with sparse white and brown hairs. Acorn cups large and saucer shaped, the acorns sessile or nearly so, reported as biennial among the golden-cup oaks, although the acorns occur among the leaves. (The leaves, at least in southeastern Arizona and New Mexico, are probably retained for two years.)

Arid slopes and canyons; probably pine-oak woodland. Probably extreme northeastern Sonora, 1830–2135 m (6000–7000 ft). Reported by Mearns (1907) from Sonora in the Sierra San Luis on the Sonora-Chihuahua border and near the New Mexico border. He mentions spiny leaved shrubs of this oak on the ridges north of Turkey Canyon (probably in Mexico) and trees 6–9 m with mostly entire leaves at lower elevations in the canyon. Southwestern New Mexico, Arizona, and the Pacific Coast from Baja California to Oregon.

In adjacent southern Arizona this oak reaches 4–5 m, with a well-formed thick trunk. Farther north in Arizona it can become a medium-sized tree with a heavy trunk, and on the Pacific coast a tree to 25 m. The Sierra San Luis population is the only report of *Protobalanus* for mainland Mexico. The nearby Arizona and New Mexico populations are most likely the result of hybridization with *Q. palmeri*, a related oak of Arizona and the Pacific Coast of North America (Tucker and Haskell 1960).

***Quercus coccolobifolia*** [*Q. endlichiana*]. *Encino manzano, encino negro, gueja; opog guiush* (Pima Bajo) (figure 175E)

Trees often 5–7 (10+) m, with straight trunks, blackish bark, and rounded crowns. Leaves 6.5–15 cm, the blades broadly obovate, conspicuously cordate at the base; upper surfaces at first brownish velvety with minute stellate hairs and golden glands, becoming smooth and light to dark green with age; lower surfaces lighter green, often golden, moderately to densely velvety with small stellate hairs and golden glands; margins entire to wavy, and sometimes with a few small, short spines; leaves subsessile or with petioles 2–10 mm. Acorns annual, sessile or nearly so, 1–3 per cluster; ripening July–August.

Slopes, ridges, canyons, and arroyos; pine-oak woodland; eastern Sonora from Sierra el Tigre southward, 1400–2135 m (4590–7000 ft). Also Chihuahua to Jalisco and eastward to Guanajuato and Nuevo León. Widespread but usually as scattered individuals.

Some leaves on individual trees turn red in late spring before falling. The distinctly cordate leaf bases and golden-colored lower leaf-surfaces distinguish this species from *Q. tuberculata*. Hybrids are reported with *Q. emoryi*; also see *Q.* ×*knoblochii*.

***Quercus durifolia.*** *Roble prieto, saucillo; cusi, shipari* (Pima Bajo) (figure 177A)

Attractive, symmetrical, upright, and characteristically large trees 15–25 m, with a high bole and a dense crown. Bark blackish, thin, fissured and checkered; twigs blackish. Leaves 3–8 cm, short petioled; blades often lanceolate, the upper surfaces dark green, smooth and glossy (sometimes with fuzzy pubescence of short, white hairs on young leaves), the lower surfaces with fine, dense white hairs; margins entire except the small terminal spine; the leaves tend to orient into flat sprays. Acorns annual, solitary, on short peduncles; ripening August–November.

Ridges, slopes, canyons, valleys, and stream banks; pine-oak woodland; east-central and southeastern Sonora, 1525–2135 m (5000–7000 ft). Also Chihuahua, Durango, Sinaloa, Coahuila, and Nuevo León; largely a high-elevation Chihuahuan species.

***Quercus emoryi.*** Emory oak, blackjack oak; *bellota; toa, doa* (Tohono O'odham) (figures 176B, 177B, and 178)

Trees often 10–20+ m, with a well-developed single trunk and spreading canopy, the branches often nearly horizontal. (A tree measured by Phil Jenkins in north-central Sonora had a spread of 27 m.) Bark blackish, rough. Leaves 5–7 (9) cm, the blades firm, the upper surfaces dark green, smooth and glossy, the lower surfaces dull green and smooth except for a conspicuous patch of fuzz at the base of the blade, the apex spine tipped; margins entire or with a few short, spine-tipped teeth. Acorns annual, nearly sessile, the cotyledons yellowish; ripening early summer.

Slopes, ridges, canyons, valleys, and arroyos; grassland, oak and pine-oak woodland; north-central and northeastern Sonora, 915–2135 m (3000–7000 ft), with a southernmost population in east-central Sonora on Cerro Verde, an

Figure 177. *Quercus.* (A) *Q. durifolia,* (B) *Q. emoryi,* (C) *Q. gambelii,* (D) *Q. grisea,* (E) *Q. hypoleucoides.* Drawn by M. B. Johnson.

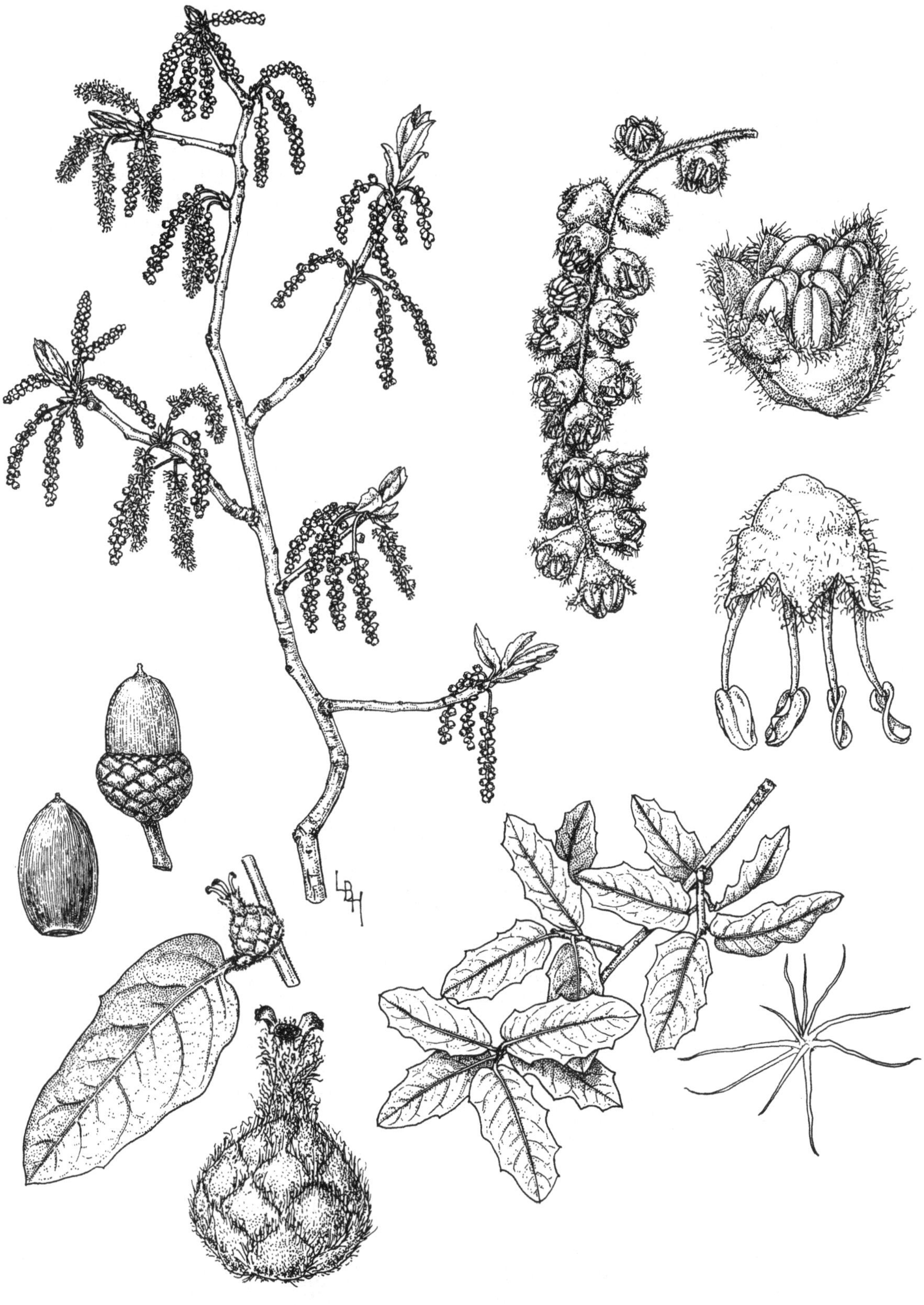

Figure 178. *Quercus emoryi*. Branch with staminate catkins and enlarged male flowers (above), female flower (lower left), stellate hair (lower right), acorn and cup, and acorn (Schmutz and Hamilton 1978). Drawn by L. B. Hamilton.

island of altered-soil southeast of San Javier. Also Chihuahua and Arizona to Trans-Pecos Texas.

This is the most abundant low-elevation oak in northern Sonora and southern Arizona, where it often grows with *Q. oblongifolia*. Emory oak is important to many species of wildlife for food and shelter. The wood is harder than most other common oaks in the region and is a preferred fuelwood. In early summer the acorns, or *bellotas*, are harvested in considerable quantity and sold in Sonoran markets and informally in southern Arizona. The acorns are eaten fresh and are often consumed in local cantinas—the floors becoming littered with the empty shells. This oak would be a desirable landscape tree at elevations above the desert. The seeds germinate even after storage for one month at room temperature. However, fresher seeds have a higher rate of germination, germinate sooner, and produce more vigorous seedlings. Some populations in northern Sonora and as far south as Cruz del Diablo northeast of Huasabas seem to be introgressed with *Q. viminea*. Emory oak resembles and is apparently closely related to *Q. devia* of Baja California Sur and *Q. peninsularis* of Baja California.

***Quercus gambelii*** [*Q. albifolia*]. Gambel oak, Rocky Mountain white oak (figure 177C and 179)

Thicket-forming shrubs to small or medium or rarely large trees, generally propagating by root sprouts. [Mearns (1907) reported a Gambel oak nearly 20 m in the Sierra San José, and trees of this size occur in neighboring Arizona.] Bark grayish and fissured. Leaves 4–14.5 × 2–6 cm, relatively soft, velvety pubescent, especially when young; petioles 0.5–1.5 cm; blades deeply lobed and not toothed; upper surfaces sometimes smooth late in the season. Acorns annual, sessile to stalked; cotyledons cream colored; ripening late summer.

Slopes, ridges, and canyons; upper elevations in pine-oak woodland and mixed conifer forest; northeastern Sonora, above 1830 m (6000 ft). Also Arizona, Utah, Colorado, New Mexico, Texas, Oklahoma, Chihuahua, and Coahuila.

This is the only winter-deciduous oak in Sonora and Arizona. It is a secondary, or successional species, and usually establishes in forested areas after a fire or where the canopy trees, usually conifers, fall or are removed. In Sonora it

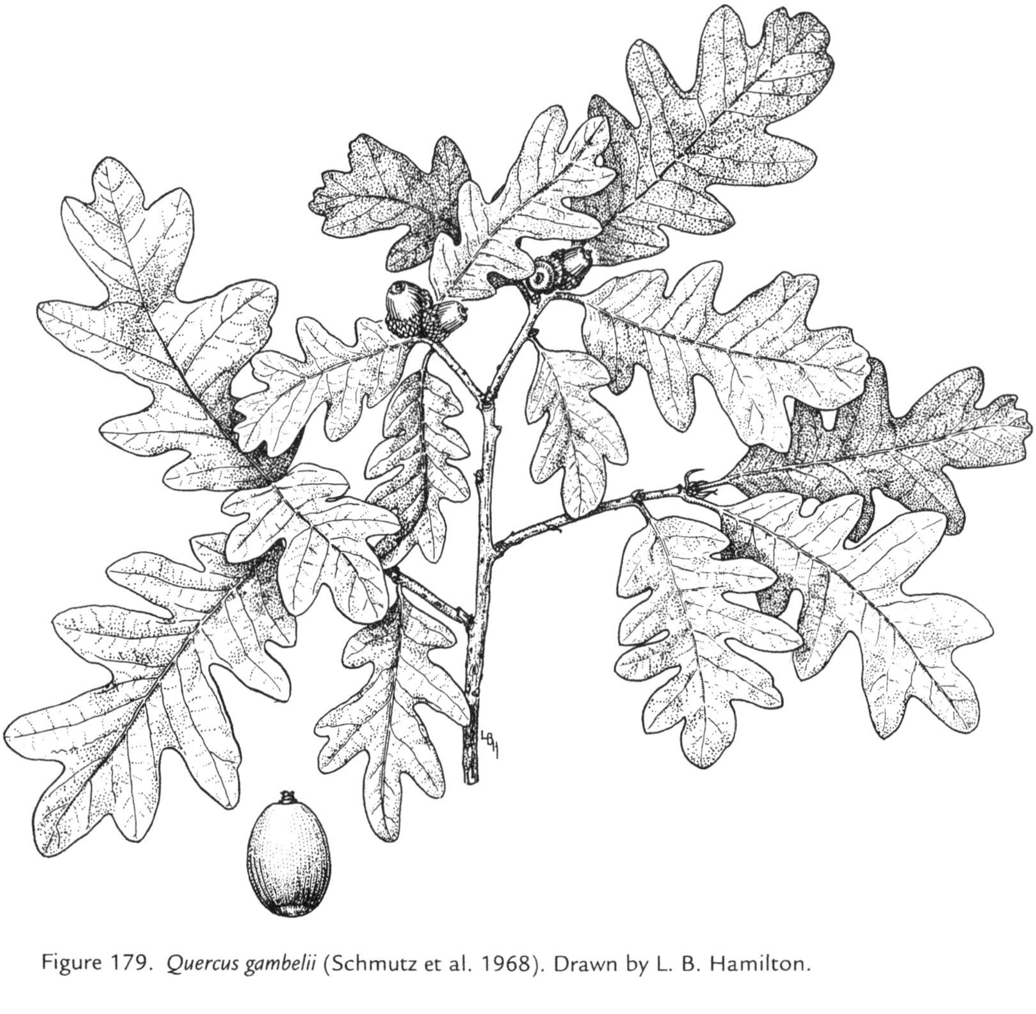

Figure 179. *Quercus gambelii* (Schmutz et al. 1968). Drawn by L. B. Hamilton.

is recorded from the Sierras de los Ajos, San José, San Luis, and el Tigre, mostly confined to limited areas at higher elevations.

***Quercus grisea.*** Gray oak (figure 177D)

Broad-crowned trees to 12 m, with low, spreading limbs; bark light colored. Foliage dense, gray-green. Twigs and leaves densely pubescent with yellow-brown hairs, the upper leaf surfaces more sparsely pubescent. Leaves rather variable, lanceolate to ovate or sometimes slightly obovate, mostly 3–7.5 × 1.2–3.7 cm, the margins entire to toothed, often irregularly so, often wavy (undulate), the tips mostly obtuse, often with a short, blunt mucro; petioles 1–5 mm. Acorns annual, often paired, on thick stalks 3–22 mm, the fresh cotyledons whitish; ripening October–November.

Valley plains and broad canyon floodplains; grassland and lower oak woodland; northeastern Sonora, 1380–1500 m (4525–4920 ft). Sonora along the lower flanks of the Sierra San Luis, the Sierra de Nacori, grassland south of Naco, and arroyo margins in grassland near Cananea. Southeastern Arizona to Trans-Pecos Texas, Coahuila, Durango, and Chihuahua. Southward in Mexico *Q. grisea* intergrades with *Q. spicata* (*Q. hartwegii*), which in turn intergrades with other Mexican white oaks.

***Quercus hypoleucoides*** [*Q. hypoleuca*]. Silverleaf oak; *encino blanco, encino colorado, cusi, cusi prieto; shipari* (Pima Bajo) (figure 176C and 177E)

Large shrubs or trees 8–10 (20+) m, often with multiple, ascending, and often straight stems, or with an erect, single, thick trunk. [Mearns (1907: 53) reported several trees from the Sierra San Luis nearly 30 m in height having trunks 1 m diameter and "one on the north side of Sierra San José 23 m tall."] Bark dark gray to blackish, smooth, fissured with age. Leaves (3.5) 6–13 (17) × 1–4 cm; blades elliptic to lanceolate or oblanceolate, leathery and rather thick, dark green and smooth above, densely fuzzy white below; margins slightly inrolled, entire or with a few small to stout teeth (see below), and a short terminal spine. New emerging leaves reddish. Acorns both annual and biennial, sessile to subsessile, often paired; ripening May–July.

Slopes, ridges, canyons, arroyos, and valleys; upper elevations in oak woodland, and pine-oak woodland; north-central and eastern Sonora, 1525–2530 m (5000–8300 ft). Often in dense stands, especially in northern Sonora. Also Durango and Chihuahua, and Arizona to Trans-Pecos Texas.

The silverleaf oaks, characterized by leathery leaves, dark green above and fuzzy white below, are taxonomically difficult. Extremes have been named, but plants and populations with intermediate characteristics are commonplace. The northern populations of *Q. hypoleucoides* tend to have smaller and entire leaves, whereas those in the south tend to have larger leaves with more numerous and larger teeth. *Quercus hypoleucoides* seems to be a northern derivative of the closely related *Q. scytophylla*, and they intergrade in canyons in western Chihuahua. In addition to its narrower leaves, *Q. hypoleucoides* usually has more abrupt leaf blade bases and is a more xerophytic species. Another closely related oak, *Q. sideroxyla*, is distinguished by its shorter and proportionally broader leaves. Sometimes a range of variation of spineless to spiny and *hypoleucoides*-like to *scytophylla*-like leaves can be found in the same population or even on the same tree.

***Quercus ×knoblochii***

Hybrid of *Q. coccolobifolia* and *Q. viminea* (Muller 1942). Slender trees to 8 m; trunk dark. Leaves thickish and appearing intermediate between the parent trees, often 7–15 cm, mostly oblanceolate, short petioled, the margins undulate or often with bristle-spinose teeth towards the tip. Produces acorns but is nonperpetuating and occurs singly or in groups of several trees. Rare in east-central and southeastern Sonora and adjacent Chihuahua where both parent species are present.

***Quercus mcvaughii.*** *Encino roble, cusi, gueja; hopakily* (Pima Bajo) (figure 180)

Spreading trees 10–24 m, the smaller trees often with dense, compact crowns, the larger trees often with irregular, open canopies. Bark dark gray and fissured. Evergreen to drought deciduous in late spring. Leaves 10–18 (29) × 2.5–12 (17) cm; blades broadly obovate or sometimes oblong, thick and often moderately convex (cupped downward), with coarse, conspicuous veins, the upper surfaces becoming smooth, the lower surfaces uniquely orange-fuzzy or pale reddish brown with dense pubescence; margins often with few shallow and coarse spinose teeth toward the leaf apex, or sometimes entire; petioles stout, 1–3 cm; new, emerging leaves commonly bright red. Acorns biennial, on short peduncles or essentially sessile, often paired; ripening late May–summer.

Slopes, mesas, ridges, and canyons; oak and pine-oak woodland; eastern Sonora, 1640–2440 m (5380–8000 ft). Also adjacent western Chihuahua, northern Durango and expected in eastern Sinaloa.

This distinctive oak is common and widespread in Sonoran mountains as far north as the Sierra el Tigre. A single large tree with many saplings in the Sierra de los Ajos appears to be introgressed with *Q. hypoleucoides*. "In its range [*Q. mcvaughii*] is the only large-leaved oak with a very pale brown tomentum of sessile hairs that nearly or completely obscures the abaxial epidermis of the leaves" (Spellenberg 1992: 1201–1202). This oak has been incor-

Figure 180. *Quercus mcvaughii*. Drawn by M. B. Johnson.

rectly called *Q. fulva*, a species known from farther south in Mexico. *Quercus mcvaughii* intergrades with *Q. crassifolia* in Durango, a species that ranges from southwestern Chihuahua to Guatemala. Where they occur together, *Q. mcvaughii* can be introgressed with *Q. hypoleucoides* and its relatives.

**Quercus oblongifolia.** Mexican blue oak; *encino azul* (figures 176D and 181A)

Small to medium-sized trees with a spreading crown; branches usually crooked and irregular; bark pale gray, fissured, and checkered. Leaves glabrous, often 3–5 (8.5) × 1–3 (4.5) cm, subsessile or petioles 1–4 mm; blades rather firm, mostly oblong, sometimes elliptic to narrowly obovate, the surfaces glaucous blue-green, the tip blunt and rounded; margins usually entire, or sometimes shallowly lobed; lobes rarely with a small tooth. Acorns annual, solitary or paired, mostly sessile or on short pedicels 3–8 mm (some trees have both sessile acorns and some on stalks reaching 1–3 cm); fresh cotyledons white with a small purplish area at the tip; ripening fall.

Slopes, canyons, valleys, and arroyos; grassland and oak woodland; north-central and eastern Sonora, 1035–1740 m (3400–5700 ft). Coahuila to Sonora, Arizona to Trans-Pecos Texas, and Baja California Sur.

Mexican blue oak and Emory oak are the most common low-elevation oaks in northern Sonora and southern Arizona. When *Q. oblongifolia* grows intermixed with another white oak, it is usually with *Q. chihuahuensis*. Apparently closely related to, and easily confused with *Q. perpallida*.

**Quercus perpallida.** *Encino cacachila.*

Trees often 8–16 m, resembling *Q. oblongifolia* and primarily differing as follows: bark on younger branches and most twigs flaking (somewhat like that of *Q. toumeyi*). Leaves often 4–12 cm, relatively thin, shallowly lobed or sinuate, or the lobes more often toothed. Acorns on slender stalks often 4.5–5.5 cm.

Slopes and canyons; oak and pine-oak woodland; east-central and southeastern Sonora, 1100–1740 m (3610–5700 ft). The northernmost record is from east of Yécora. Otherwise known only from southwestern Chihuahua and expected in northeastern Sinaloa.

The long peduncles are diagnostic but are only briefly present. This feature gives the trees a bristly or porcupine-like appearance. The flaking bark is a useful field character but not always reliable for younger twigs and herbarium specimens.

***Quercus rugosa*** [*Q. diversicolor, Q. reticulata*]. Netleaf oak (figure 181B)

Often scrubby shrubs 1–2 m, forming closely spaced thickets on dry rocky slopes, and in canyon bottoms and other favorable habitats reaching 10 m, with stout but somewhat irregular trunks, or sometimes to 25 m near the Sonora border in southwestern Chihuahua. Bark dark gray to blackish, becoming fissured with age. Leaves 3–12 × 1.5–5 (8.5) cm; petioles to 1 cm; blades broadly obovate, with strongly impressed, netlike venation especially prominent below; margins with small, spinescent teeth. Upper leaf surfaces smooth with minute hairs along the veins; lower surfaces pale green with fuzzy hairs, orange when fresh and brown when dry. Fruiting stalks 3–13 cm, pubescent, often with 2–4 acorns. Acorns annual, the fresh cotyledons pink-purple; ripening in fall.

Arid rocky slopes, canyons, and streamsides; pine-oak woodland; eastern Sonora, above 1670 m (5485 ft). Honduras northward through montane Mexico and western Texas to Arizona, and Baja California Sur.

The southern Mexico and Central America populations have substantially larger leaves than those from northern Mexico and the southwestern United States. In Sonora this distinctive oak might be confused with *Q. arizonica. Quercus rugosa* has downward cupped leaves, whereas those of *Q. arizonica* are flat or slightly wavy. Along with *Q. perpallida*, the consistently long peduncles of *Q. rugosa* are unique for the Arizona-Sonora oaks.

***Quercus scytophylla*** (figure 181C)

Forest trees 15–20 m, often with a narrow crown in dense forest and developing a stately, spreading stature in more open places. Leaves 4–18 cm; petioles 7–30 mm; blades obovate, rather tough, dark green above, densely white-fuzzy below; margins above the middle of the leaf with short-spinose teeth. Acorns subsessile, mostly biennial; ripening July–August.

Slopes, ridges, and canyons; oak and pine-oak forest; southeastern Sonora, 1200–1700 m (3935–5575 ft). In Sonora known from Arroyo Santa Bárbara and Sierra Saguaribo (Tepopa, etc.) in the southeastern part of the state, and high mesas and canyons in the Yécora region. Common in west-central and southwestern Chihuahua and southward to Oaxaca.

This silverleaf oak is similar to *Q. hypoleucoides* and might be included within the concept of that species. It differs in having leaves much broader on average, blades attenuate at the base, and margins not inrolled. There is considerable intergradation among the various silverleaf oaks along the western slope of the Sierra Madre Occidental. *Quercus epileuca*, another member of this complex, was described from Sonoran specimens. The type specimen seems to be a hybrid between *Q. scytophylla* and *Q. mcvaughii* (Richard Spellenberg, personal communication, 1996).

***Quercus sideroxyla*** [*Q. incarnata*]. *Encino prieto* (figure 181D)

Straight-trunked trees to 15 m. Bark dark gray. Leaves (4) 4.5–8.8 × 2.0–4.3 cm; blades obovate to elliptic or oval-oblong, sometimes broadly so, with cordate bases, the upper surfaces dark green and smooth, the lower surfaces densely white woolly; margins inrolled with short spines toward the leaf apex. Acorns annual.

Canyons; pine-oak woodland; east-central Sonora, 2000–2400 m (6560–7870 ft). Southwestern Chihuahua and reported from adjacent Sonora by Spellenberg (in press). Chihuahua to Nuevo León, San Luis Potosí, and Guanajuato.

This silverleaf oak, closely related to *Q. hypoleucoides* and *Q. scytophylla*, is distinguished by its shorter and proportionally broader leaves. However, these taxa appear to intergrade.

***Quercus subspathulata.*** *Encino asta, encino cochi* (figure 182A)

Robust trees often 10 (15) m, sometimes forming dense groves. Twigs glabrous, reddish, often with conspicuous raised, whitish lenticels. Leaves apparently drought deciduous, subsessile (petioles 1–2 mm), at least 9.5–20 × 5.5–10 cm, mostly obovate, tough, the margins undulate to shallowly lobed; veins of lower surfaces forming a conspicuous ladderlike pattern; apex obtuse to rounded (and probably also notched), the base narrowed, rounded to narrowly clasping. Upper leaf surfaces light green, smooth; lower surfaces of young leaves brownish (amber) with irregular, short, crinkled hairs and sessile to stalked glands. Acorns annual, 2 or 3 on short peduncles.

Slopes and ridges; pine-oak woodland; east-central and southeastern Sonora, 1200–1400 m (3935–4590 ft). In small, isolated populations in Sonora, such as the vicinity of Santa Bárbara in the Sierra Milpillas, where it occurs in a nearly pure stand, and a few localities on altered whitish or reddish soils such as in the Sierra Oscura. Common and

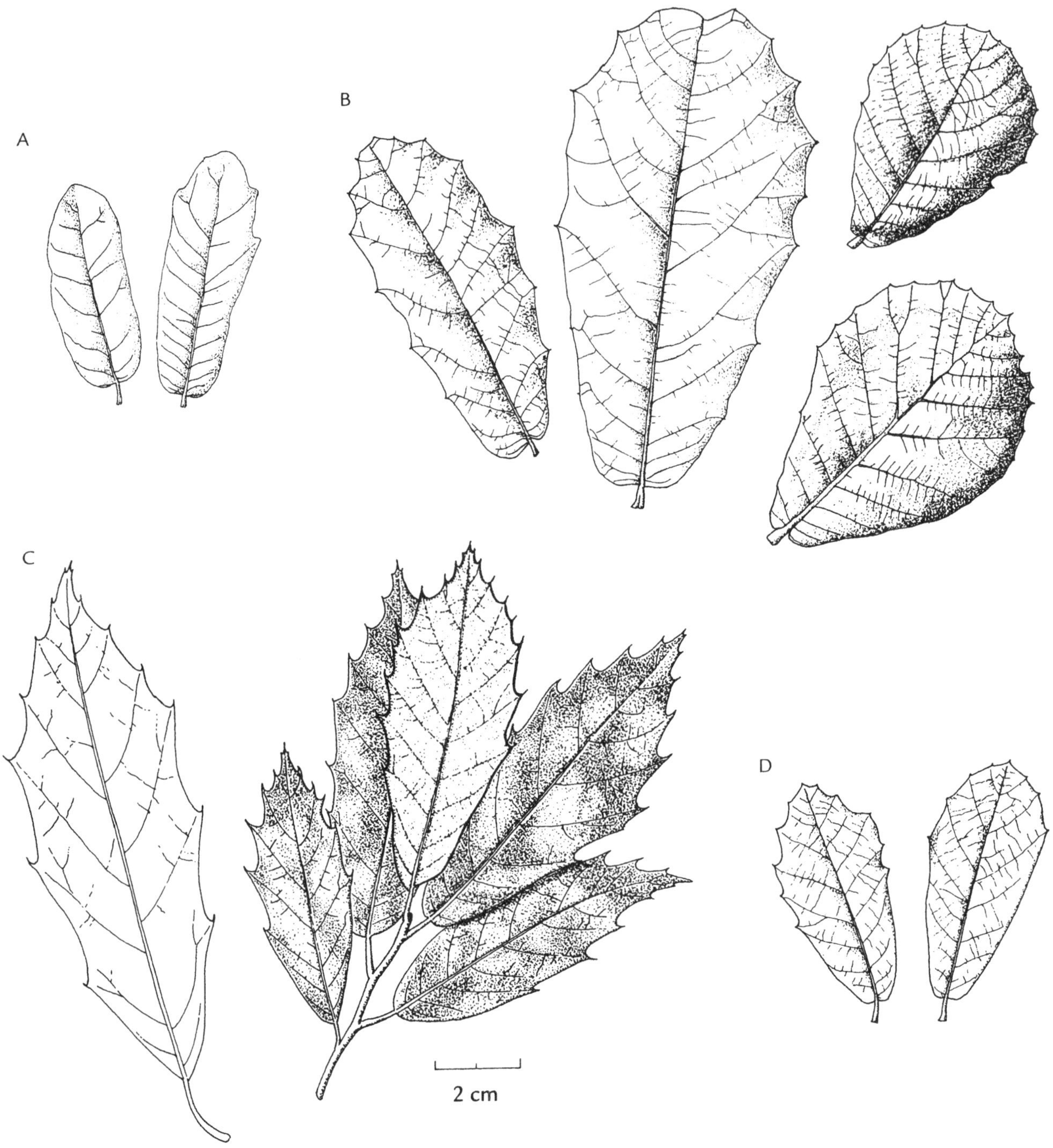


Figure 181. *Quercus*. (A) *Q. oblongifolia*, (B) *Q. rugosa*, (C) *Q. scytophylla*, (D) *Q. sideroxyla*. Drawn by M. B. Johnson.

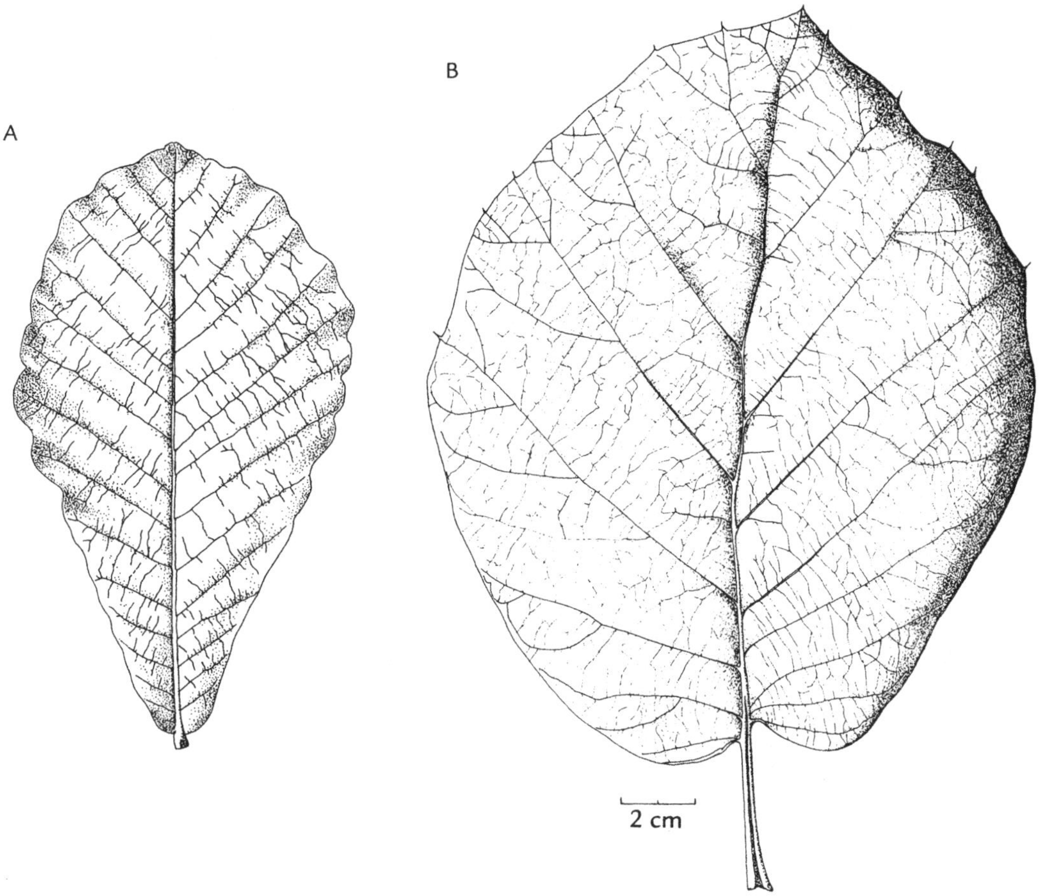


Figure 182. *Quercus*. (A) *Q. subspathulata*, (B) *Q. tarahumara*. Drawn by M. B. Johnson.

widespread farther south. Also southwestern Chihuahua, Sinaloa, Durango, and Jalisco.

This oak can be confused with the apparently closely related *Q. tuberculata* but differs in having thicker twigs, much shorter and thicker petioles, mature leaves that are lighter green, a coarser and more conspicuous ladderlike pattern of veins on the lower leaf surfaces, reddish and denser glands, and denser pubescence. Trees from the vicinity of El Chiribo on Sierra Saguaribo appear intermediate between *Q. subspathulata* and *Q. tuberculata*. Locally called *encino cochi* or *encino asta*, these trees reach 10 m and have leaves to 23 cm and petioles 5–8 mm.

***Quercus tarahumara*.** Handbasin oak; *encino cajete; gueja* (Pima Bajo) (figure 182B)

Distinctive small to large trees; trunk straight and slender to massive; bark blackish; crown dense. Twigs usually thick, with dense golden-brown hairs. Leaves enormous, 13–38 cm, tough and rigid; blades orbicular to broadly obovate, cupped downward so that a fallen, upturned leaf forms a bowl-like basin, the upper surface dull, glabrate, the lower surface with densely matted golden-brown hairs, becoming sparser with age; margins with small, sharp teeth; petioles stout and prominent. Acorns annual, sessile or nearly so, often several (1–15) crowded on racemes or spikelike clusters often 1–3 cm, the peduncles short and thick; ripening late summer and fall.

Slopes, mesas, and ridges; pine-oak woodland; eastern Sonora, (1020) 1370–1830 m [(3345) 4500–6000 ft], from the Sierra Bacadéhuachi (north of Nácori Chico) and vicinity of Mulatos and Yécora southward. Also southwestern Chihuahua, northern Sinaloa, and northwestern Durango. The low elevation record in Sonora is from hydrothermally altered soils within tropical deciduous forest.

Over much of its range in Sonora and nearby Chihuahua this oak is a small tree, often occurring on open, relatively arid slopes or mesas, or as an understory tree among pines. It often grows on whitish, yellow, or reddish hydrothermally altered soils. In the vicinity of Mulatos, where it grows on rocky, altered soils, it forms sparse but extensive forests of exceptionally large trees with thick trunks, large branches, and a broad crown. These areas are highly mineralized and scheduled to be developed into open-pit mines.

This is one of the largest leaved oaks in the world. The leaves "are tough, and endure long on the ground after they fall, where they crackle loudly under foot. They curl into basins and may catch rain water, ready for the sleepy morning face of the dry-camper" (Gentry 1942: 103). This unique oak is worthy of cultivation.

The Sonora-Chihuahua populations have been known as *Q. pennivenia*, a synonym of *Q. urbanii* (Spellenberg et al. 1995). *Quercus tarahumara* is closely related to the more southern *Q. urbanii* and *Q. radiata* from west-central Mexico, distinguished by their longer, spikelike clusters of acorns (Spellenberg and Bacon 1996).

***Quercus toumeyi*** [*Q. chuichupensis*]. Toumey oak; *encino chaparro* (figure 183A)

Dwarf, densely branched, spreading trees, often 3–5 m, rarely reaching 10 m. Bark flaking on twigs and younger branches and light gray and finely checkered on trunks and limbs. Leaves evergreen or tardily drought deciduous, 1.1–4.0 (4.7) × 0.5–1.4 (2.0) cm, short petioled; blades elliptic to oblong, rigid; margins entire or with few spine-tipped short teeth, the apex often spinescent tipped. Upper leaf surfaces glossy dark green, sparsely pubescent or glabrate, the lower surfaces dull with short hairs. Acorns annual, subsessile to prominently pedunculate, solitary or in small clusters.

Slopes, ridges, and mesas; grassland, oak and pine-oak woodland; eastern Sonora, 915–2010 m (3000–6600 ft). Also southern Arizona to westernmost Texas and western Chihuahua.

This oak is often highly localized at the lower elevational limits of oaks or may form more extensive elfin woodlands in relatively xerophytic habitats at somewhat higher elevations. Extensive woodlands of this oak occur in the southern part of the Sierra San Luis and in the vicinity of Mulatos. This is the smallest leaved oak in Arizona and Sonora. It tends to have irregularly spiralled hairs, especially among the southern populations, which distinguishes it from other white oaks in Sonora. These hairs tend to be absent from plants in the northern part of the state. *Quercus toumeyi* might be confused with the scrub oak *Q. turbinella*, but the latter is not documented for Sonora. *Quercus toumeyi* has papillae on the lower leaf surfaces that are not present on *Q. turbinella*, and *Q. turbinella* has acorns on peduncles > 10 mm with the leaves generally consistently spiny margined and often larger. *Quercus toumeyi* hybridizes with *Q. oblongifolia* in the vicinity of Maycoba, east of Yécora.

***Quercus tuberculata.*** *Encino amarillo; toché* (Guarijío) (figure 183B)

Low-branching, spreading trees, often 6–10 (15) m. Bark pale gray, checkered, and flaking in rectangular plates. Markedly drought deciduous, the leaves turning reddish in the fall or spring dry seasons. Leaves 8–22 × 3–14 cm; petioles 7–20 mm; blades broadly oblanceolate to obovate, relatively thin, the veins prominent, the upper surfaces glabrate, glossy dark green, the lower surfaces green to brownish, densely papillate, with fine soft hairs especially on the veins; margins often wavy and shallowly lobed, sometimes entire. Acorns annual, sessile, single or paired; ripening October–November.

Slopes, canyons and arroyos; upper margins of tropical deciduous forest, grassland, and oak woodland; eastern Sonora as far north as the mountains southeast of Bacadehuachi and canyons near Nacori Chico; 455–1220 m (1500–4000 ft). Also Baja California Sur, Chihuahua, Coahuila, Durango, Nuevo León, and Sinaloa.

Common at lower elevations of the oak zone in much of southeastern and east-central Sonora, and often descending into tropical deciduous forest where it grows alongside palms and burseras. Often growing with *Q. chihuahuensis* with which it hybridizes.

***Quercus viminea.*** Madrean willow oak; *saucillo, cusi saucillo* (figure 183C)

Attractive, often slender-trunked trees 12 (20) m, with an open and often high canopy, the major limbs usually sharply ascending. Bark dark gray-brown, smooth, becoming rough and furrowed with age. Leaves evergreen to tardily drought deciduous in late spring, often turning reddish as they fall, 5–13 × 0.7–2.5 (3) cm; petioles 7–14 mm; blades relatively thin, shiny yellow-green and glabrous on both surfaces, or sparsely hairy and sometimes with a patch of fuzz near the petiole (these might be introgressed with *Q. emoryi*); margins entire or with several small, widely spaced, spine-tipped teeth toward the apex, the apex spine tipped. Acorns subsessile or on short peduncles, biennial, along several nodes on twigs just below the leaves; ripening July–September.

Slopes, ridges, valleys, mesas, and canyons; oak and pine-oak woodland; eastern and north-central Sonora, 1220–2285 m (4000–7500 ft). Common and widespread in mountains as far north as the Sierra el Tigre, Sierra del Oso (ca. 25 km northeast of Imuris), and Sierra Baviso southeast of Magdalena, although the northern populations might be introgressed with *Q. emoryi*. A small population, consisting of shrubs or small trees, occurs on the south end of the Sierra de los Ajos. These plants might be limited by the occasional, extreme winter cold. A 1923 collection from Red Mountain (several kilometers south-southwest of Patagonia) in Santa Cruz County, Arizona, represents the sole specimen for the United States. Also Chihuahua, Sinaloa, Durango, Nayarit, and Jalisco.

This tall, slender tree with its narrow, shiny green, willowlike leaves is distinctive among Sonoran oaks. It would make an attractive landscape tree.

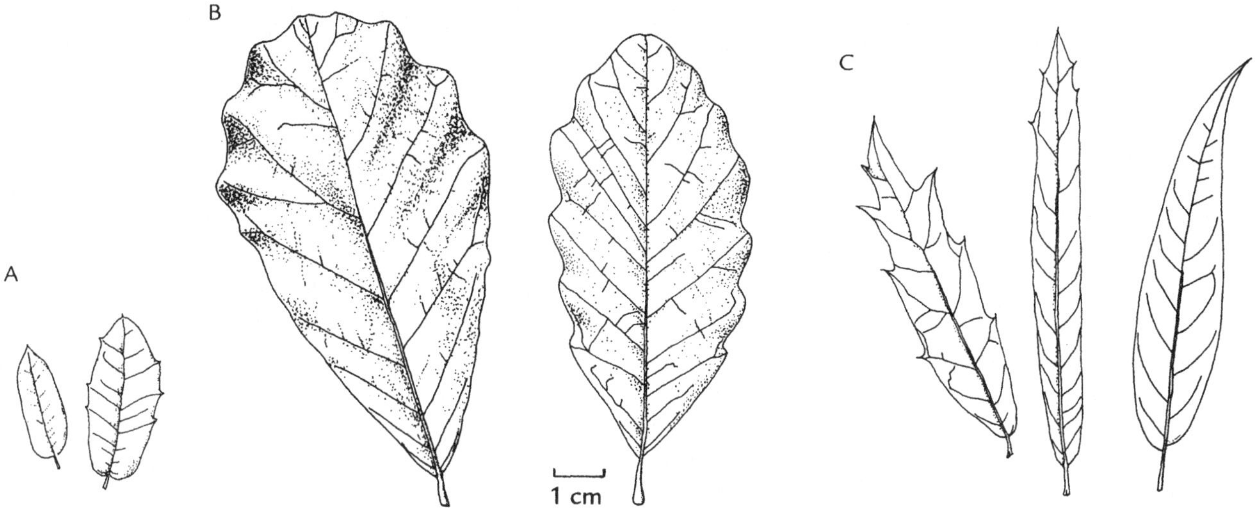


Figure 183. *Quercus.* (A) *Q. toumeyi*, (B) *Q. tuberculata*, (C) *Q. viminea*. Drawn by M. B. Johnson.

## Flacourtiaceae—Flacourtia Family

Trees and shrubs. Widespread, tropical and subtropical and some in temperate regions; 86 genera, 800+ species. Reference: Sleumer (1980).

### *Xylosma*

Trees and shrubs; Sonoran species with many branches bearing axillary spines. Leaves simple, alternate or fascicled, with conspicuous pinnate veins. Dioecious. Flowers small, in small axillary fascicles (clusters); pedicels jointed; sepals hairy (Sonoran species), persistent; petals absent; flower base with a gland-bearing disk. Fruit a berry. Tropics and subtropics in the Americas, Asia-Malaysia and the Pacific; 85 species. Two species in Sonora; not ranging farther north in western North America.

1. Leaves with petioles to about 5 mm long, the blades oval or broadly elliptic to broadly ovate or broadly lanceolate. ________ ***X. flexuosum***
1' Leaves sessile, narrowly elliptic to oblanceolate. ***X.*** sp.

***Xylosma flexuosum*** [*Flacourtia flexuosa; F. celastrina; Myroxylon flexuosum; M. celastrinum; Xylosma celastrinum*]

Mostly shrubs, occasionally small bushy trees 6-8 m with slender, often descending branchlets. Axillary spines often 1-2 cm long. Glabrous except youngest growth minutely pubescent. Leaves to about 7.5 cm long; petioles mostly less than 5 mm; blades oval to ovate or broadly elliptic to broadly lanceolate, somewhat leathery, moderately bicolored and shiny above; leaf margins serrate with small, forward-pointing teeth; stipules minute and quickly deciduous.

Flowers inconspicuous, whitish or yellowish; pedicels 4–5 mm slender, jointed near base; flowers with 8 fused glands forming a lobed disk. Sepals 5 (4–8 elsewhere), reddish brown, 2.0–2.5 mm, broadly triangular, ciliate. Male flowers with many slender stamens about 3 mm. Fruits rounded, 7–8 mm in diameter, reddish; style persistent, 2-branched (2- or 3-branched elsewhere), short, and thick. Flowering in March; fruits ripe March–April.

Shaded, riparian canyon bottoms in oak forest and lower edge of pine-oak forest, sometimes mixed with upper edge of tropical deciduous forest. Southeastern Sonora, known from Sierra Saguaribo; 1050–1350 m [3445–4430 ft]. Southern Texas and northeastern and northwestern (Sonora) Mexico, and southward to northern South America.

***Xylosma*** sp. *Güigüiri* (at Curea)

Shrubs or sometimes trees 7–8 m; branches rigid and heavily galled, the galls elongate, prominent, and woody; bearing slender spines to 3 cm, or the larger plants unarmed. Leaves fascicled in short-shoots, pubescent with small, spreading and slender hairs especially along major veins, sessile, to 6+ cm, narrowly elliptic to oblanceolate, bicolored, dark olive green above with conspicuous, lighter-colored veins; margins narrowly and inconspicuously revolute, entire or subentire to moderately crenate or crenate-serrate. Male flowers reddish. Female flowers and fruits unknown; flowering in January.

Riparian canyons in foothills thornscrub and tropical deciduous forest, 300–560 m [985–1340 ft]. Scattered localities in southeastern Sonora including near Cedros and Tesopaco in the Río Mayo region (Martin et al. 1998), and east-central Sonora including the vicinity of Curea in the Río Yaqui drainage (Tom Van Devender, personal communication, 2000).

## Fouquieriaceae—Ocotillo Family

This unique family, with no obvious close relatives, has a single genus, *Fouquieria*.

### *Fouquieria*

Glabrous, spiny shrubs or trees. Long shoots with alternate, widely spaced nodes, producing petioled leaves with a rigid spine developing from the petiole midrib. Short shoots in axils of the spines, these leaves clustered with internodes reduced to a few cells, sessile to short petioled, not developing spines. Inflorescenses of panicles or racemes. Sepals 5, separate; corollas 5-lobed, tubular below, white or red to orange-red. Stamens 10 or more, uneven in length, surpassing the corolla. Fruits of capsules; seeds wind dispersed, flat with thin, papery wings. Arid regions of Mexico, one extending into southwestern United States; 11 species. Reference: Henrickson (1972).

Fresh *Fouquieria* seeds germinate readily with warm temperature, and seedlings are easily grown in well-drained soils. Growth rates vary greatly among the species. Except for the boojum, they are easily grown from cuttings.

The red-flowered species, frequented by hummingbirds, have slender, tubular corollas and conspicuously exserted stamens and even longer styles with three slender stigma branches. The white-flowered species, such as the boojum, adapted for insect pollination, have much shorter and relatively broader flowers with relatively short and unbranched styles.

The system of branching and spine formation is unusual but not unique. For example, some *Sesamothamnus* (Pedaliaceae), native to eastern and southern Africa, have similar leaf-spine formation and are strikingly similar in appearance to certain *Fouquieria*. *Alluaudia* (Didiereaceae), endemic to Madagascar, has several species with ocotillolike growth forms, and the stems likewise are made into living fences.

1. Boojum tree, many times taller than wide; trunk usually single, erect, occasionally few-branched above, very thick and succulent; flowers cream colored. ***F. columnaris***

1' Ocotillos, usually less than 3 times as tall as wide; trunkless or with one to several well-developed thick, spreading, or ascending trunks or limbs; flowers red to red-orange.

 2. Stems long, mostly straight and wandlike, ascending to erect, usually not branched above; trunk very short, usually appearing trunkless. ___ ***F. splendens***

 2' Stems and major limbs branched above; trunk(s) thick and well developed.

  3. Inflorescences relatively compact, usually longer than wide; pedicels 2–6 (13) mm; coastal Sonora. ___ ***F. diguetii***

  3' Inflorescences usually relatively loose and open, usually as wide or wider than long, pedicels (3) 5–30 mm; widespread in lowland Sonora. ___ ***F. macdougalii***

***Fouquieria columnaris*** [*Idria columnaris*] Boojum tree; *cirio*; *cototaj* (Seri) (figures 13 and 184)

Trunk (main axis) often 5–14 m, unbranched, or sometimes with one to several major limbs from the upper part of the plant; trunks and major limbs usually upright but sometimes bending and contorted. Lateral branches slender and spreading. Short-shoot leaves 11–21 mm, mostly obovate, often slightly glaucous, entire or shallowly notched at tip.

Panicles (including peduncle) 17–65+ cm, mostly on the upper stem of the current year's growth at the top of the plant, and sometimes also from the ends of some of the lateral branches; the old dry, straw-colored inflorescences persisting for much of the year. Flowers crowded in the upper part of the inflorescencs, cream white, 10–12 mm; stigmas minutely 3-lobed, relatively short. "The flowers have a strong, sweet, honeycomb fragrance and produce a noticeably sweetened nectar" (Henrickson 1972: 534) and are visited by numerous species of bees as well as hummingbirds. Flowering July and August; seeds late October to mid-November.

Rocky granitic slopes and canyons; Sonoran desertscrub. West-central Sonora, for about 100 km in the Sierra Bacha, from about 10 km north of El Desemboque San Ignacio north to Punta Cirio south of Puerto Libertad; 10–660 m (30–2165 ft). In Sonora limited to granite substrate; especially common on north-facing slopes and toward higher elevations. They grow intermixed with the common ocotillo, *F. splendens*. Also both states in Baja California and at higher elevations on Isla Angel de la Guarda in the Gulf of California.

Looking like great upturned carrots with scraggly little secondary branches, the boojum is one of the most unusual plants in the world. In 1922, when Godfrey Sykes sighted these strange plants he announced, "Ah, a Boojum. Yes, definitely a Boojum" (Sykes 1982: 354). The name is from Lewis Carroll's *The Hunting of the Snark*. *Cirio*, the Spanish term for candle, refers to the shape of the plant.

The *cirio* is often characterized by bizarre branching and bending, especially in Baja California. The internal cylinder of wood in the stem is often quite thin with much support coming from water stored in the trunk. As water is lost during dry seasons, horizontal air cavities develop in the trunk. This weakens the stems, causing them to bend. In contrast, columnar cacti remain upright through drought because the pleated stems allow shrinkage without com-

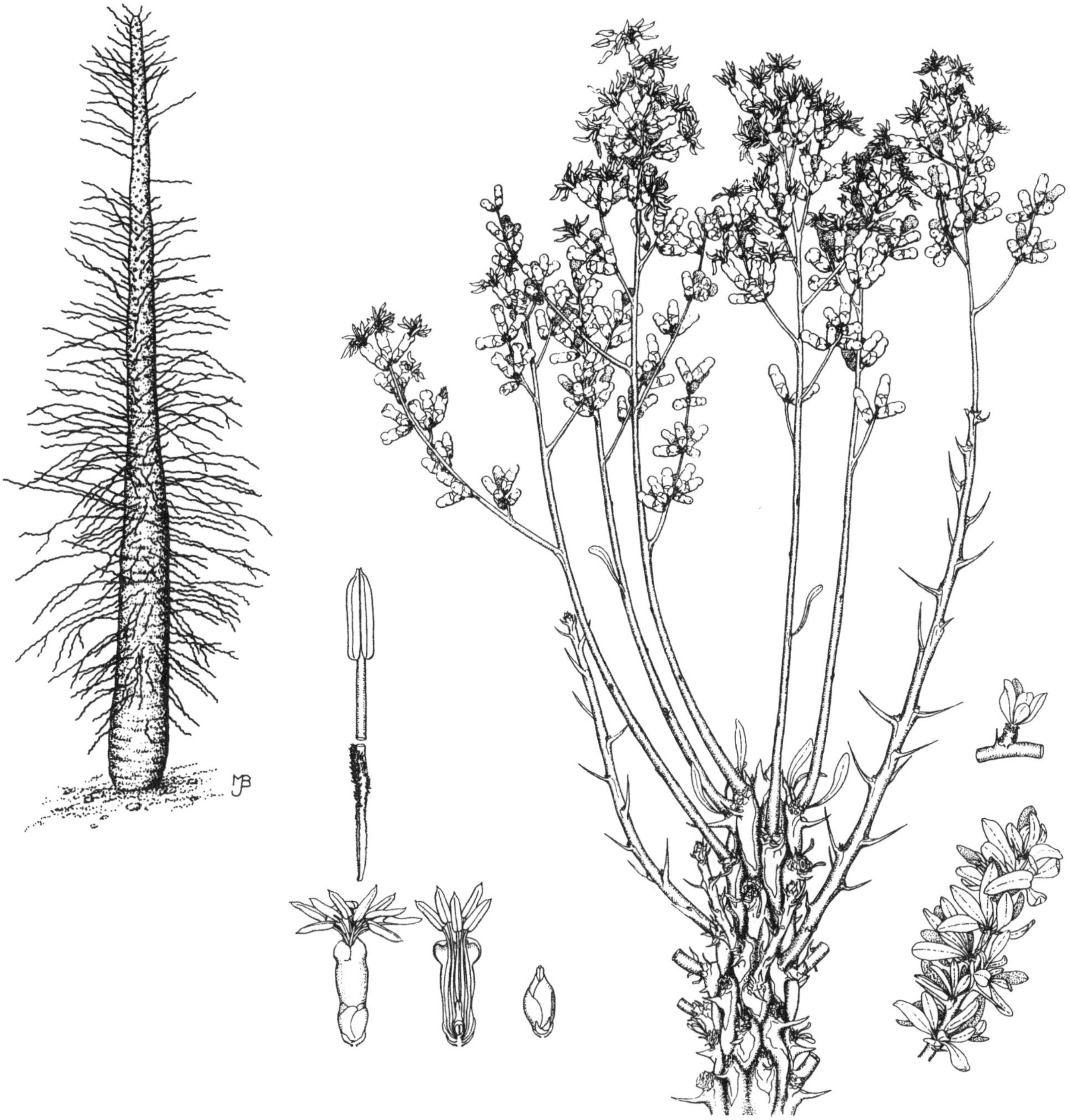


Figure 184. *Fouquieria columnaris* (Henrickson 1972), drawn by F. Runyan; plant at left drawn by M. B. Johnson.

promising rigidity. Boojums in Baja California are often taller than those in Sonora and sometimes reach 20–22 m (Humphrey 1974).

Boojums are leafless and dormant during the hot summer. Seed germination and vegetative growth occur with cool-season rains. Its distribution seems to be strongly influenced by maritime fog. The plants intercept fog—moisture condensing on the spiny tangle of twigs drips down the trunk, irrigating the root zone.

Large plants seldom survive transplanting. The boojum is readily propagated by seeds and young plants are available in the horticultural trade. In cultivation in Arizona the seedlings may grow continuously through their first year but thereafter produce stem growth only from October to April. Watering should be greatly reduced during the hot summer months to prevent rotting. For maximum growth the plants need ample root space. Older, established plants have survived overnight temperatures of −7°C (18°F) in cultivation.

***Fouquieria diguetii.*** *Palo Adán; murue* (Mayo); *xomxéziz caacöl* (Seri); *mureo* (Yaqui) (figure 185A)

Shrubs to small trees often 3–5 m, with several ascending trunks and a relatively compact crown of spiny twigs; bark of older limbs becoming greenish or bronzed and peeling in dry seasons. Leaves 20–41 mm, mostly obovate, slightly glaucous to green.

Racemes or panicles mostly near the stems tips, the flowers mostly crowded. Flowers bright red, mostly 30–43 mm (including stamens); style much elongated; stigma of 3 slender lobes; pedicels 2–6 (13) mm. Flowering primarily after rains. Tetraploid, $2n = 48$.

Sonoran desertscrub; coastal west-central and southwestern Sonora, near sea level–150 m (500 ft). Common and widespread on rocky slopes, desert plains, and alluvial plains from the vicinity of Tastiota southward along the coast to Puerto Yavaros. Also widespread in both states of Baja California. *Fouquieria diguetii* grows together with *F. splendens* in the Tastiota region and from Guaymas southward it often occurs with *F. macdougalii*. All three species occur in the vicinity of Cañón de Nacapule.

*Fouquieria diguetii* somewhat resembles the common desert ocotillo (*F. splendens*) but is more shrub- or treelike, and the flowers are darker red. *Fouquieria diguetii* and *F. macdougalii* sometimes may be difficult to differentiate. *Fouquieria diguettii* differs from the latter by its generally smaller, thicker leaves, often more compact and conical inflorescences, generally thicker twigs, and relatively thicker and shorter pedicels. *Fouquieria diguetii* reaches maximum development in desertscrub, while *F. macdougalii* shows its greatest development in tropical deciduous forest. The former is slower growing, requiring several more years to come into flower from seed than *F. macdougalii*.

***Fouquieria macdougalii.*** Tree ocotillo; *jaboncillo, ocotillo macho, torote spinosa, torote verde; culurí, chonolí* (Guarijío); *mureo, murue* (Mayo); *nüliog* (Pima Bajo); *mureo* (Yaqui) (figure 185B)

Large shrubs or trees 6–11 m, with one or more thickened, often twisted trunks, the bark waxy yellowish brown to greenish and peeling away in large, thick flakes or sheets during the dry season. (The tallest trees, in tropical deciduous forest, are slender and upright with a high bole and small, spreading crown.) Leaves 17–40 mm, mostly elliptic to oblong, green, relatively thin.

Panicles mostly axillary near the stem tips, open and often relatively few flowered, or larger with more moisture. Flowers bright red, 34–44 mm (including the long-exserted stamens); styles longer than the stamens; stigma with 3 linear branches; pedicels slender, (3) 5–30 mm. (Dry season inflorescences can be very reduced on both *F. macdougalii* and *F. diguetii*.) Flowering various seasons, especially after rains. Diploid, $2n = 24$.

Slopes, bajadas, plains, and arroyos; Sonoran desertscrub, thornscrub, and tropical deciduous forest. Most of Sonora except the northwest and northernmost areas, near sea level–915 m (3000 ft). Usually a shrub in the desert and increasingly larger in the southern thornscrub and tropical deciduous forest. Also southwestern Chihuahua, Sinaloa, and Nayarit.

This ocotillo makes an excellent container plant and can be a striking landscape subject in lowland regions of Sonora. It grows rapidly from seed, producing flowers in the second year, and can be propagated from stem cuttings. It sustains severe freeze damage below –4°C (25°C). The stems are sometimes incorporated into living fences. The bark has been used as soap for washing cloths, hence the name *jaboncillo*.

***Fouquieria splendens*** subsp. ***splendens.*** Ocotillo; *melhog* (O'odham); *xomxéziz* (Seri); *mureo* (Yaqui) (figure 186)

A unique, long-lived desert plant, often 3–7 m, with slender, few-branched, wandlike, spiny branches arising from a very short, reduced trunk. Except during the coldest weather, the leaves are produced following ground-soaking rains and quickly shed as the soil dries. Long shoots generally produced only during times of hot weather and abundant soil moisture. Flowers about 1.5–2.5 cm, red-orange, in dense panicles at the stem tips, attracting hummingbirds. Flowering February–May (June); fruits in late spring.

Slopes, ridges, canyons, bajadas, and plains; primarily in Sonoran and Chihuahuan desertscrub, also in thornscrub, grassland, sometimes in oak woodland, and rarely in pine-oak woodland; northern and central Sonora, near sea level–1625 m (5330 ft). One of the more common and widespread perennials in the drier regions of Sonora. Southern limit is on the exposed rhyolite slopes surrounding Cañón de Nacapule near San Carlos. Grows in a wide range of habitats and in many different soils.

This species ranges from Baja California and southern California to Trans-Pecos Texas and southward to Nuevo León, Durango, and Zacatecas. Subspecies *splendens* is the most wide ranging of the three subspecies. Nearly encompassing the entire range of the species, it is found throughout most of the Sonoran and Chihuahuan deserts.

The stems readily take root when planted and are made into living fences and corrals. The stems have been used in the construction of wattle and daub houses and brush shelters made by nomadic desert people such as the Seri. The usual harvesting practice is to cut all of the stems near the base of the plant. When cut too close the plants usually perish, but when cut about 1 m above ground level, or even slightly lower, the plants usually resprout. Larger plants may be more than a century old. Although the cortex below the waxy bark may contain chlorophyll, photo-

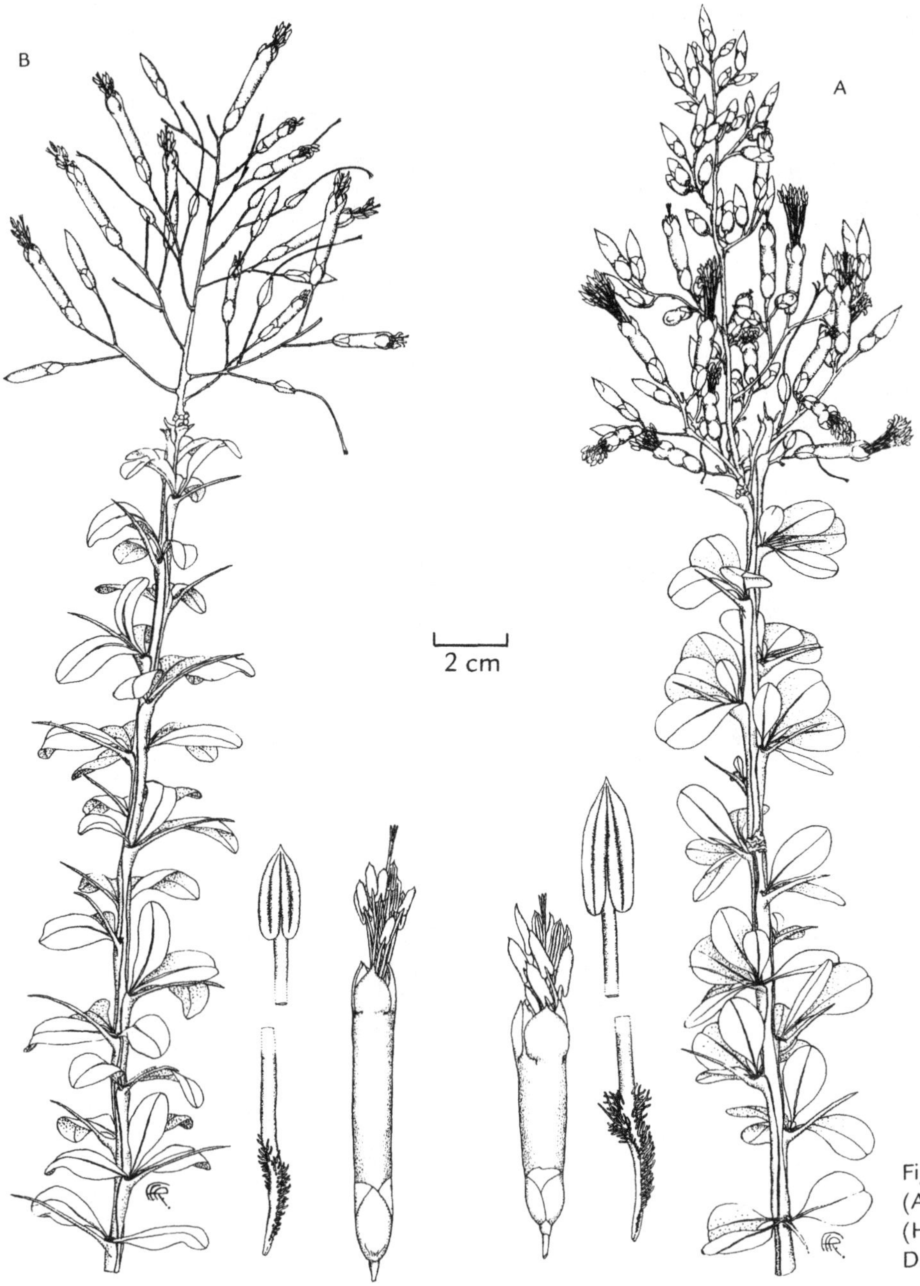


Figure 185. *Fouquieria*. (A) *F. diguetii*, (B) *F. macdougalii* (Henrickson 1972). Drawn by F. Runyan.

synthesis ceases in the drought-induced leafless condition (Mooney and Stain 1964). Ocotillos are often transplanted bare-root from the wild for landscaping. Smaller plants, usually with about six to eight stems, reestablish easily, but large plants seldom survive transplanting.

## Juglandaceae—Walnut Family

Temperate and warm regions. Northern hemisphere to South America and Malaysia; 9 genera, 60+ species. Reference: Whittemore and Stone (1997).

Pecans (*Carya illinoinensis*) are grown commercially in the Arizona–Sonora border region, mostly at elevations above the desert.

### *Juglans*—Walnut, *Nogal*

Both hemispheres; 21 species.

Walnuts are grown in temperate climates for the edible nuts, and the fine-grained wood is used in furniture and construction. The Persian or English walnut (*J. regia*) is the most important commercial species.

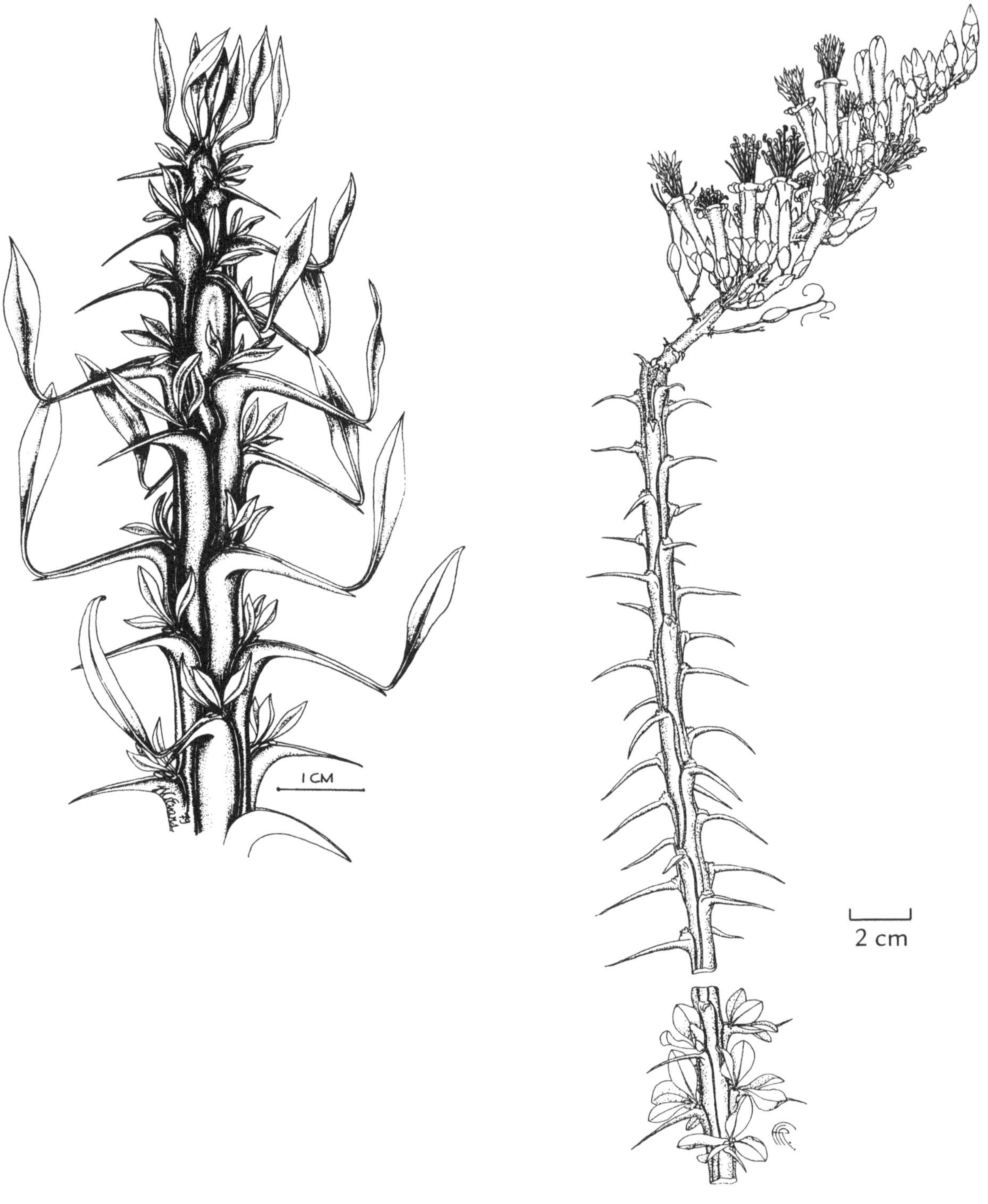


Figure 186. *Fouquieria splendens* subsp. *splendens*. Rainy-season stem showing petiole-spine development of long-shoot growth and sessile leaves of axillary short-shoots (left); leafless dry-season stem and inflorescence, and stem with short-shoot leaves (right). Drawn by F. Runyan (Henrickson 1972) and N. E. Weaver.

***Juglans major*** var. ***major*** [*J. microcarpa* var. *major*]. Arizona walnut, *nogal*; *uupio* (Pima Bajo, this term also used for "skunk") (figure 187)

Spreading, hardwood trees often 10–16 m; bark rough. Twigs and new growth densely pubescent with brown hairs; leaves with stellate and simple hairs, becoming sparsely pubescent with age; crushed leaves pungently aromatic. Leaves winter deciduous, alternate or sometimes opposite, 17–31.5 cm, odd-pinnate; petiole prominent; leaflets (7) 9–17 (19), asymmetric, mostly 5–14 cm, the margins serrated; stipules absent.

Monoecious. Flowers inconspicuous, wind pollinated; petals absent. Male flowers in drooping catkins, yellow, the stamens many; female flowers green, one to few in a cluster at stem tips, the ovary inferior, surmounted by minute calyx lobes and 2 styles. Fruit 2.5–4 cm, rounded, the single seed enclosed in a thick, indehiscent, hard shell surrounding the nut. Flowering March–May (June).

Riparian habitats: canyons, streams, riverbanks; upper margin of Sonoran desertscrub, and grassland, oak woodland, pine-oak woodland, mixed conifer forest, and rare in tropical deciduous forest (Arroyo Gochico north of Alamos); north-central and eastern Sonora, 895–1670 m (2930–5475 ft). Northern Sinaloa, Durango, Chihuahua, and eastern Coahuila to Trans-Pecos Texas, Oklahoma, New Mexico, and Arizona. Two varieties: var. *major* is replaced southward in Mexico by var. *glabrata*, distinguished by glabrous twigs and leafstalks.

This tree is relatively scarce in Sonora except in the extreme north-central and northeastern part of the state. An excellent shade tree, it is recommended for well-watered upland regions. The walnuts contain a relatively small quantity of edible meat embedded in a hard, thick shell. The binational city of Nogales received its name from "the grove of walnuts which occupied the present site of the city" (Mearns 1907: 44). The Nogales Wash is now treeless, cement-lined, strewn with garbage, and subject to violent flash flooding and criminal activities.

## Koeberliniaceae

The family has a single species.

***Koeberlinia spinosa.*** Crucifixion thorn; *corona de Cristo*; *xooml* (Seri) (figure 188)

Shrubs or occasionally small trees 3–5 (6) m with exceedingly hard wood, and thick, often twisted and contorted trunks and limbs; branches and twigs rigid and interlacing, yellow-green, slender, and thorn tipped. Twigs and inflorescences minutely pubescent. Leaves alternate, scalelike to 2.2 mm, quickly falling with drought, the plants essentially leafless.

Inflorescences of racemes to 6 cm; racemes and dried fruits semipersistent. Flowers on slender pedicels, pale yellow-green, numerous, and visited by hordes of flies. Sepals and petals each 4, separate, and deciduous; petals approximately 4.5 mm, rounded at the tip; stamens 8 and about as long as the petals. Fruits subglobose berries, 5–6.5 mm in diameter, on a short stipe, 2-chambered, green to dark purple or blackish, drying capsulelike; style persistent as a slender beak, 2–4-seeded. Flowering March–April; fruits ripen in May.

Low hills, valleys, and plains; Sonoran and Chihuahuan desertscrub, and occasional in coastal thornscrub; northeastern and western Sonora, near sea level–1525 m (5000 ft). Arborescent koeberlinias in Sonora occur only in the western part of the state, mostly on Isla Tiburón and the vicinity of Bahía Kino. A single tree was found at Bahía San Carlos, and shrub-sized plants are rare near the coast southward nearly to the Sinaloa border. Southeastern California to Texas, Baja California Sur and Hidalgo, and the Chaco of Bolivia and Paraguay.

The burning wood produces copious, oily black smoke. During the nineteenth century, Seris returning from Hermosillo fumigated their homes with koeberlinia smoke as protection against much-feared foreign diseases such as measles (Felger and Moser 1985).

## Lauraceae—Laurel Family

Hardwood shrubs to tall trees. Leaves alternate (in Sonora), evergreen, simple, petioled; blades pinnately veined with a prominent midrib, glandular punctate and aromatic, the margins entire; stipules absent. Inflorescences of axillary and terminal panicles. Flowers small, greenish to yellowish, pedicelled; perianth of 6 small sepals (tepals) in 2 whorls, those of each whorl overlapping and united at their bases; petals absent. Fertile stamens 9 in 3 whorls (whorls I–III) plus an inner whorl of 3 staminodes (whorl IV); anther of fertile stamens with 4 chambers (thecae) opening by persistent flaps. All stamens and staminodes separate. Fruits fleshy, 1-seeded.

Worldwide, mostly tropical America and Asia, a few subtropical and temperate; 50 genera depending on interpretation, 2850 species. References: Allen (1945), Kopp (1966), van der Werff (1991, 1997).

The family includes avocado, laurel or sweet bay, camphor, cinnamon, sassafras, and various timber and ornamental trees. Many lauraceous trees are prominent members of tropical forests, especially in the New World.

The two tree species in Sonora are restricted to the most tropical habitats in the state. Many members of this fam-

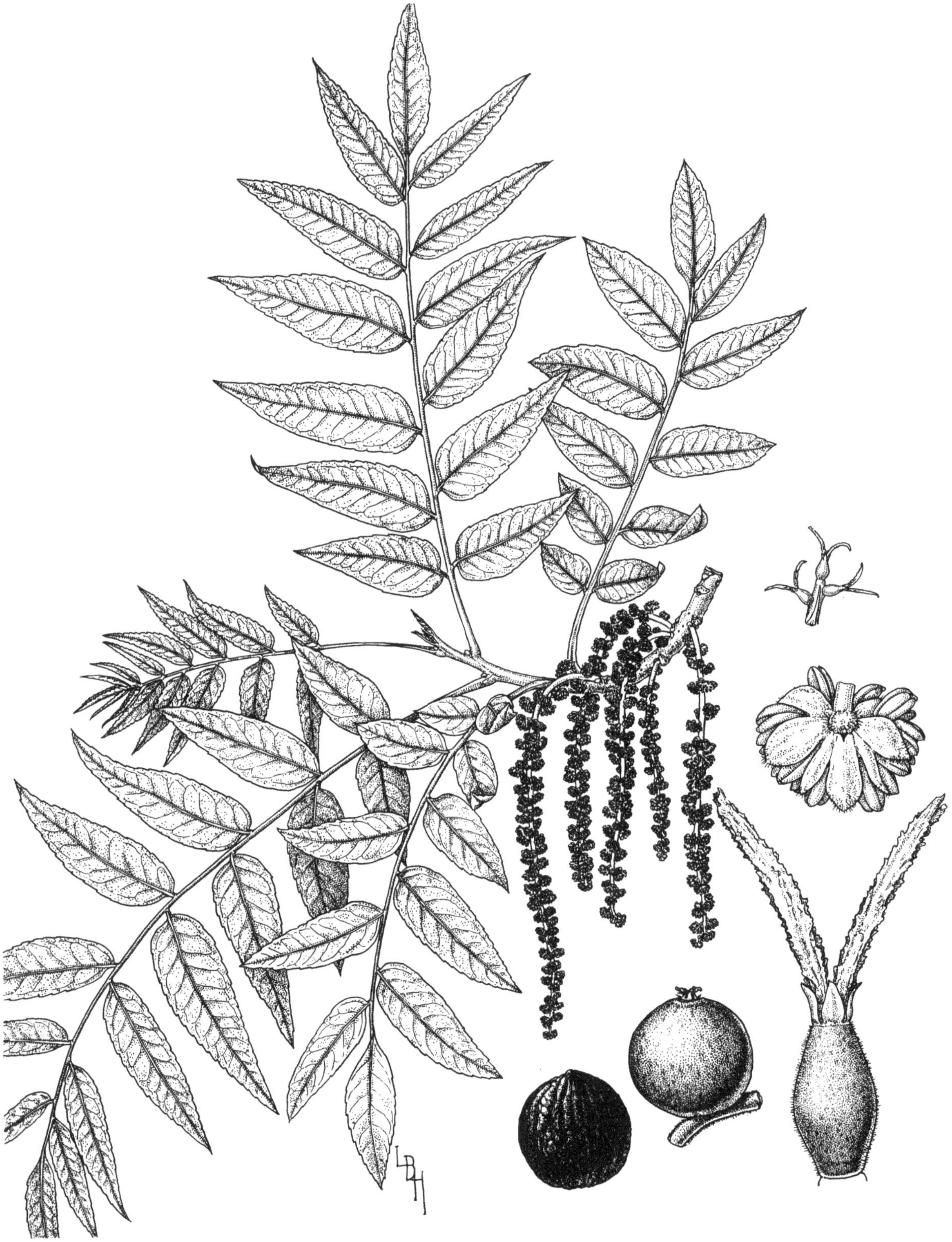

Figure 187. *Juglans major*. Drawn by L. B. Hamilton.

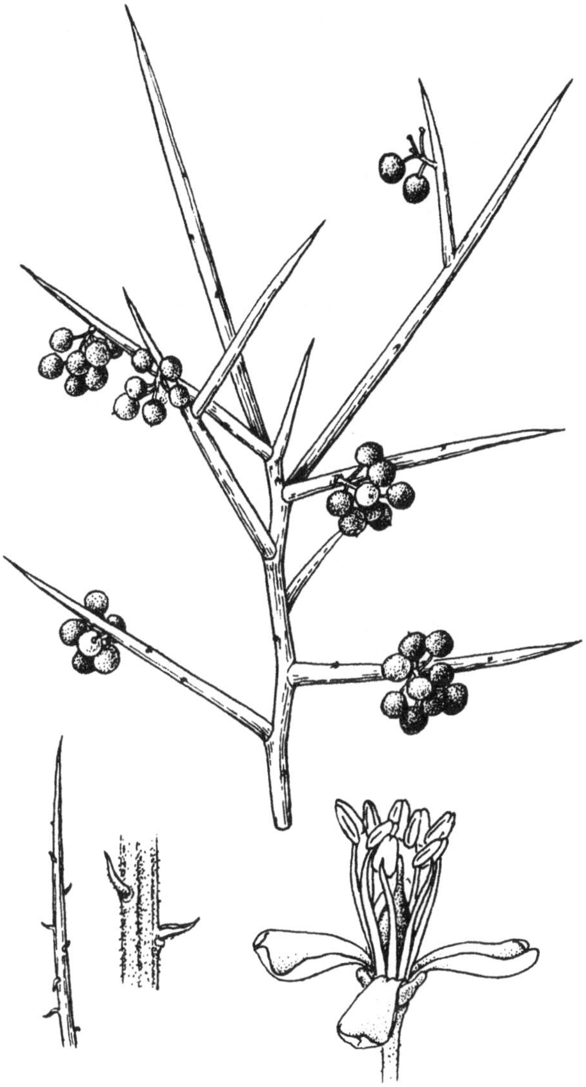

Figure 188. *Koeberlinia spinosa*. Drawn by F. Bond (courtesy James Henrickson).

ily can be difficult to distinguish except by technical characters of the very small flowers. The stamens, with their unusual anthers opening by persistent flaps, are unique in the flora of northwestern Mexico. The only other member of this family in Sonora is *Litsea glaucescens* var. *subsolitaria*, a shrub of pine-oak woodland along the Chihuahua border on Sierra Saguaribo in the upper Río Mayo drainage.

1. Leaves with 5–9 lateral veins on each side of the midrib; outer and inner sepals nearly equal in length. ___ ***Cinnamomum***

1' Leaves with 12–18 lateral veins on each side of the midrib; outer 3 sepals about half as long as the inner 3 sepals. ___ ***Persea***

### *Cinnamomum*

Evergreen trees and shrubs. Tropical and subtropical Americas, Asia, Pacific Islands, and Australia; 300+ species. Includes cinnamon and camphor trees. Reference: Kostermans (1961).

The American species were formerly included in *Phoebe.*

***Cinnamomum hartmanii*** [*Persea hartmanii*]. *Haya; chihuana cara* (Guarijío) (figure 189)

Trees to 20 (25) m, often taller than wide; with a thick trunk, sturdy and spreading limbs, dense crown, and gray bark. New growth buds minutely and densely pubescent with golden brown hairs; petioles and midribs of lower leaf-surface of younger leaves minutely and sparsely pubescent; twigs pubescent or glabrous; herbage otherwise glabrous. Evergreen; leaves mostly 9–23 × 2.2–6.0 cm, drooping (hanging down like mango foliage); petioles 1.0–2.5 cm; blades narrowly elliptic to oblanceolate or lanceolate, thickish, shiny green, moderately bicolored, the lateral veins 5–9 on each side of the midrib, the leaf tip obtuse or sometimes acute, the base cuneate, the veins pinnate and prominent below.

Panicles including peduncles 2.3–19 cm, irregularly and few-branched, glabrous and somewhat glaucous; panicle branches and pedicels slender. Flowers numerous, sweet scented, greenish cream colored; glabrous except sparsely short pilose with white hairs surrounding the stamen bases. Tepals all essentially equal in size, 3.5–4.0 mm, broadly ovate, the tips obtuse, the upper one-third abscising (breaking away). Fertile stamens with filaments about as long as the anthers, each anther 4-chambered and facing inward (introrse, the chambers or openings face inward). Stamens of the inner whorl of fertile stamens (whorl III) each flanked by a pair of sessile, rounded glands, the other stamens without glands. Staminodes (whorl IV) not more than half as long as the fertile stamens; staminodes thick, short stalked with a spear-shaped (sagittate) blade. Sepal bases and pedicels enlarging in the fruit. Fruits 1.3–1.8 cm, rounded, one to many per panicle. Pedicels often infested by a gall insect resulting in chains of fruitlike galls. Flowering March–April; fruiting June–July.

Riparian canyon bottoms in tropical deciduous forest and oak woodland; southeastern Sonora, 300–1250 m (1000–4100 ft). Also adjacent southwestern Chihuahua to Durango, Michoacán, and Guerrero.

"The fruit is sought and eaten by the natives. It is black and rather bitter" (Gentry 1942: 115). It is an impressive

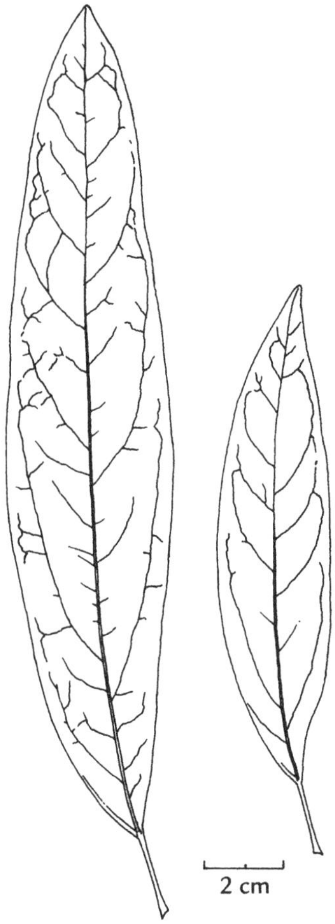

Figure 189. *Cinnamomum hartmanii.* Drawn by M. B. Johnson.

canopy tree of moist, tropical canyons, often growing with *Aphananthe, Oreopanax, Persea, Platanus,* and *Prunus zinggii.*

### *Persea*—Avocado, *Aguacate*

New World tropics and subtropics; 150 species.

Avocados (*P. americana*) are grown in gardens in the subtropical regions of southeastern Sonora, but in the desert regions the leaves burn in the hot, dry air. The genus also includes important tropical hardwood timbers including Ecuadorean mahogany (*P. theobromifolia*), now virtually extinct due to overharvesting.

***Persea podadenia*** var. ***podadenia.*** *Amolillo, laurel, laurelón; chihuana cara* (Guarijío) (figure 190)

Heavily foliaged trees 8–18 m. Bark shallowly furrowed, gray-brown. Leaf buds and new growth, inflorescences, pedicels, and inner and outer tepal surfaces densely pubescent with small, tawny brown hairs, otherwise glabrous. Probably evergreen or nearly so. Leaves 7.5–18 × 1.7–4.5 cm, often drooping; petioles short; blades narrowly elliptic to elliptic, lanceolate, or slightly oblanceolate, rather thick, dark green above, glaucous below, the lateral veins 12–18 on each side of the midrib, the leaf tip acute to sometimes obtuse or rounded.

Flowers 4–5 mm, pedicelled, inconspicuous and greenish, crowded in axillary panicles shorter than the leaves. Fertile stamens of the outer 2 whorls each with 4 chambers. Tepals persistent in fruit. Fruits rounded, approximately 1 cm. Flowering May–June; fruits September.

Canyon bottom streams; oak and pine-oak woodland; southeastern Sonora, 900–1350 m (2950–4430 ft). Also in adjacent southwestern Chihuahua to Durango, Sinaloa, and probably Jalisco. Two varieties: var. *glabriramea* occurs in the mountains of Tamaulipas, Nuevo León, and Veracruz in eastern Mexico.

This species often grows with *Alnus, Cinnamomum, Clethra, Cupressus, Ilex,* and *Magnolia.* The trees can be spotted from

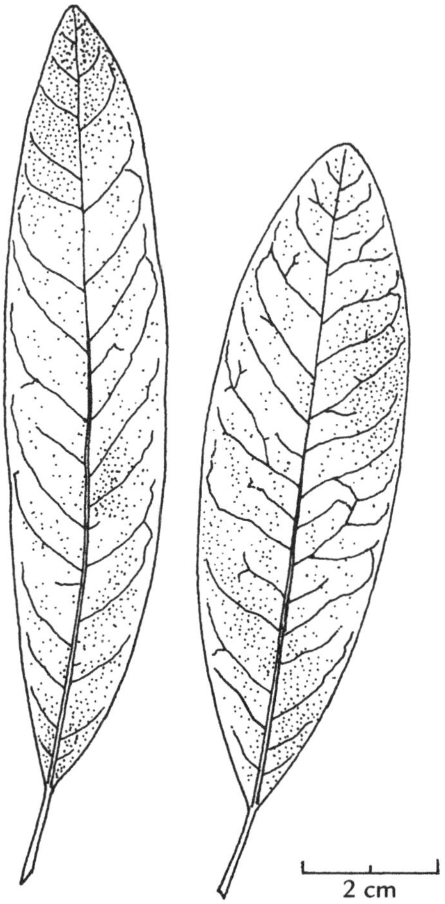

Figure 190. *Persea podadenia.* Drawn by M. B. Johnson.

a distance by the drooping leaves. The leaves, which have a flavor similar to that of sassafras, are used to season food (Standley 1922). The high elevation record is from Rancho Byerly, Sierra Charuco, in the upper Río Mayo drainage in Chihuahua near the Sonora border.

## Magnoliaceae—Magnolia Family

Trees and shrubs in humid temperate and tropical regions worldwide; 220 species in 6 genera. Reference: Meyer (1997).

### *Magnolia*

Western Hemisphere and Asia; 120 species. References: Felger (1971), Vázquez-G. (1994).

***Magnolia pacifica*** subsp. ***tarahumara***. Tarahumara magnolia (figure 191)

Elegant, evergreen trees reaching 15–20 m, the trunk sometimes to nearly 1 m diameter; bark rough, dark brown. Young herbage, especially the stems, densely hairy with pale golden silky hairs, soon glabrate. Leaves alternate, simple, often 8.5–26.5 × 3.5–8.5 cm, glabrous; petioles 1–3 cm; blades tough, slightly to moderately glaucous, elliptic to lanceolate-elliptic, both surfaces with netlike veins, the midrib and veins conspicuous below, the tip blunt to acute, the margins entire. Stipules to 5+ cm, linear, bractlike, thin, enclosing the leaf bud, separating from the petiole and deciduous as the enclosed leaf expands, leaving a thin, horizontal scar encircling the twig; similar but ovate, and densely silky-haired leafy bracts enclose the flower buds.

Flowers showy and large, varying considerably in size, often 9–15 cm wide, on a stout pedicel; floral parts spirally arranged on a stout axis. Perianth segments petal-like and white; sepals 3; petals mostly 6 in 2 series, often 5–8 cm, broadly obovate to oval; stamens numerous, often 10–12 mm. Fruits woody and conelike, 3–5 cm, falling with the peduncle. Seeds mostly 11–13 mm, bright red, generally remaining attached to the dry fruits when they fall to the ground. Flowering May (July); fruits ripen the following March–April.

Canyon streams in riparian oak and pine-oak woodland; southeastern Sonora, 940–1400 m (3085–4590 ft). Also southwestern Chihuahua, northeastern Sinaloa, and northwestern Durango. Two other subspecies southward in western Mexico to Jalisco and Zacatecas.

This species is the northernmost member of the family in western North America. In Sonora it is restricted to deep, cool, riparian barrancas or canyons on Sierra Saguaribo in the upper Río Mayo and Cuchujaqui drainages near the Chihuahua border, and there is a smaller, isolated population on Sierra de Alamos. On Sierra Saguaribo it grows in dense canyon-bottom forest of upland tropical vegetation characterized by evergreen trees and shrubs with simple and mostly large leaves (e.g., *Clethra*, *Ficus*, *Oreopanax*, *Quercus*, *Myrsine*, and *Platanus*). In Sinaloa the seeds are eaten by the chachalaca, crested guan, hooded grosbeak, masked tityra, and tufted jay. Members of the magnolia family are pollinated by beetles. The Sonoran magnolia has been successfully cultivated in southern California and deserves more extensive planting, but is not recommended for landscaping in arid regions.

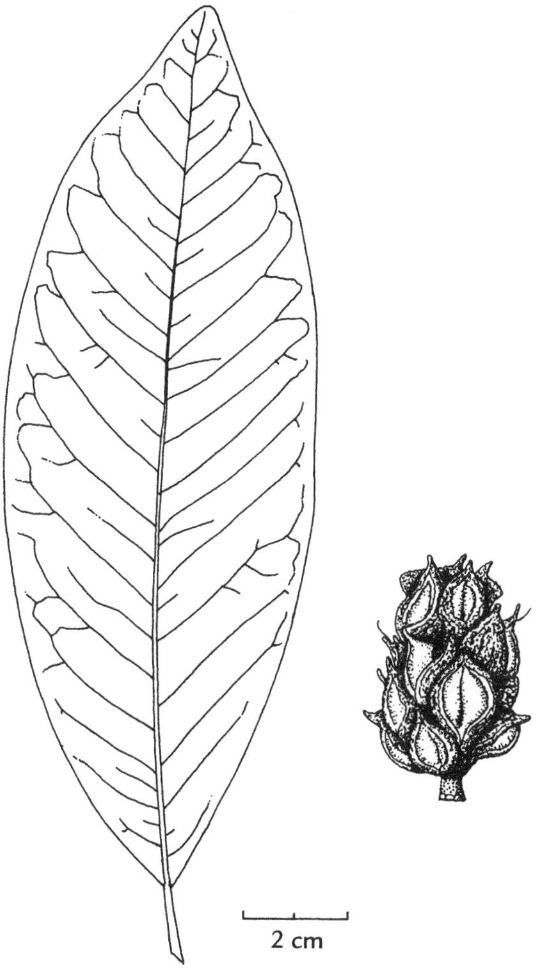


Figure 191. *Magnolia pacifica* subsp. *tarahumara*. Drawn by M. B. Johnson.

## Malpighiaceae—Malpighia Family

Woody vines, shrubs, or small trees, with double hairs (T-shaped, attached in the middle with two opposite-pointing arms) or glabrous. Leaves opposite, stipuled, simple, and entire or notched at tip. Flowers bilateral, often showy, 5-merous, on jointed pedicels with 2 bractlets. Sepals 5, usually with paired, fleshy, oil-producing glands at the base. Petals 5, stalked (clawed) with fringed or ragged-toothed blades; one petal often slightly enlarged and serving as a flag and landing pad for bees. Stamens 10 (in Sonora). Ovary of 2 or 3 (5) carpels, each with a single ovule. Fruits fleshy (drupes) among the tree species in Sonora, otherwise often dry with winged and nutlike mericarps or samaras.

Mostly tropical and subtropical, largely in the Americas; 60 genera, 1200 species. References: Buchmann (1987), Robertson (1972).

The New World species usually have sepals with oil glands, and the principal pollinators are oil-gathering anthophorid bees such as *Centris*. Several tropical South American malpighs, such as yagé (*Banisteriopsis caapi*), feature prominently in native rituals for their powerful hallucinogenic properties.

Most Sonoran malpighs are perennial vines or shrubs. *Echinopterys eglandulosa* is fairly common in tropical deciduous forest. Like many malpighs, it is a woody vine that assumes the habit of a shrub and approaches tree size when it has nothing to climb on, but the young branch tips are twining. In arid regions it remains a shrub.

1. Sepals without glands; fruits dry and bristly, separating into 3 nutlets. ____ ***Echinopterys eglandulosa*** (not treated in text)

1' Sepals bearing glands; fruits fleshy, not bristly.
   2. Herbage densely hairy, at least the twigs and petioles; flowers yellow; flowering and fruiting pedicels thick, 5–6.5 mm; fruits 2-lobed, 2-seeded. ***Bunchosia***
   
   2' Herbage sparsely hairy or glabrous; flowers pink; flowering and fruiting pedicels slender, 9–15.5 mm; fruits 3-lobed, 3-seeded. __________ ***Malpighia***

### *Bunchosia*

Trees and shrubs of tropical America; 55+ species. Many have edible fruits.

#### ***Bunchosia sonorensis.*** *Palo cenizo, garbanzo de coyote; goy carabanzo, tóroco* (Mayo); *mi'isicaochara* (Guarijío) (figure 192)

Usually shrubs, occasionally small trees 7–8 m. Branches rigid and rather thick, often branching from the base. Bark gray on younger branches, whitish and checkered on larger branches and trunks. Twigs, petioles, and pedicels densely hairy; blades densely hairy when young, with age sparsely hairy or glabrous above and moderately hairy below. Leaves drought deciduous, 2–14 cm, petioled; blades mostly broadly lanceolate to elliptic or obovate, pale green above, blue-green below, with a pair of yellowish glands on the lower surface near the base; stipules approximately 1 mm, inserted on the petioles near their bases.

Flowers in panicles, fragrant, showy, attracting bees; pedicels thick, 5.5–6.5 mm; petals yellow, approximately 1 cm; styles united into a column; stigma 3-lobed. Fruits 1.3–1.6 cm wide, pubescent, orange to red, 2-lobed, 2-seeded. Flowering at least February–April and October–November.

Mostly localized, on plains, slopes, and canyons; shrubs in thornscrub and sometimes trees in tropical deciduous forest; southeastern Sonora, 150–610 m (500–2000 ft). Also in Sinaloa.

The Guarijío name ("cat's balls") refers to the bilobed, fuzzy fruits. The largest trees were found along the canyon bottom of the Río Guajaráy in the Guarijío region.

### *Malpighia*

Trees and shrubs in tropical America, especially the Caribbean; 40 species.

The Barbados cherry (*M. glabra*) is widely grown in the tropics for the fleshy, red fruits, that are very high in vitamin C. It is occasionally encountered as a shrub in southern Sonora. Many of the species have stinging hairs.

#### ***Malpighia emarginata*** [*M. umbellata*]. *Granadilla; siré* (Guarijío), *sire* (Mayo) (figure 193)

Shrubs in thornscrub, and often small trees in tropical deciduous forest. Bark on trunk and major limbs smooth and pale gray-brown, with white markings, and flaking away in thin, irregular pieces. Twigs and leaves sparsely hairy or glabrous. Branches relatively rigid. Leaves drought deciduous, 1.5–5.5 cm; petioles short; blades elliptic to obovate with a terminal notch; stipules slender, approximately 1 mm, inserted just below the base of the petiole.

Inflorescences of short, axillary cymes. Flowers showy, attracting bees and flies; pedicels slender, 10–18 mm, elongating as the fruit matures; petals pink to sometimes white, approximately 6.5 mm; styles 3, separate; stigmas small. Fruits approximately 1 cm wide, glabrous, red, 3-lobed and 3-seeded. Flowering March–November.

Slopes, valleys, plains, and arroyos; thornscrub and tropical deciduous forest; southern and central Sonora north to Soyopa, near sea level–400 m (1310 ft). Southward to Venezuela and Ecuador.

The fruits are edible and are locally harvested; they are eagerly eaten by birds. This species is widely cultivated in the American tropics for its edible fruit, the *acerola* of commerce and dooryards. It seems to be naturalized rather easily by birds spreading the seeds.

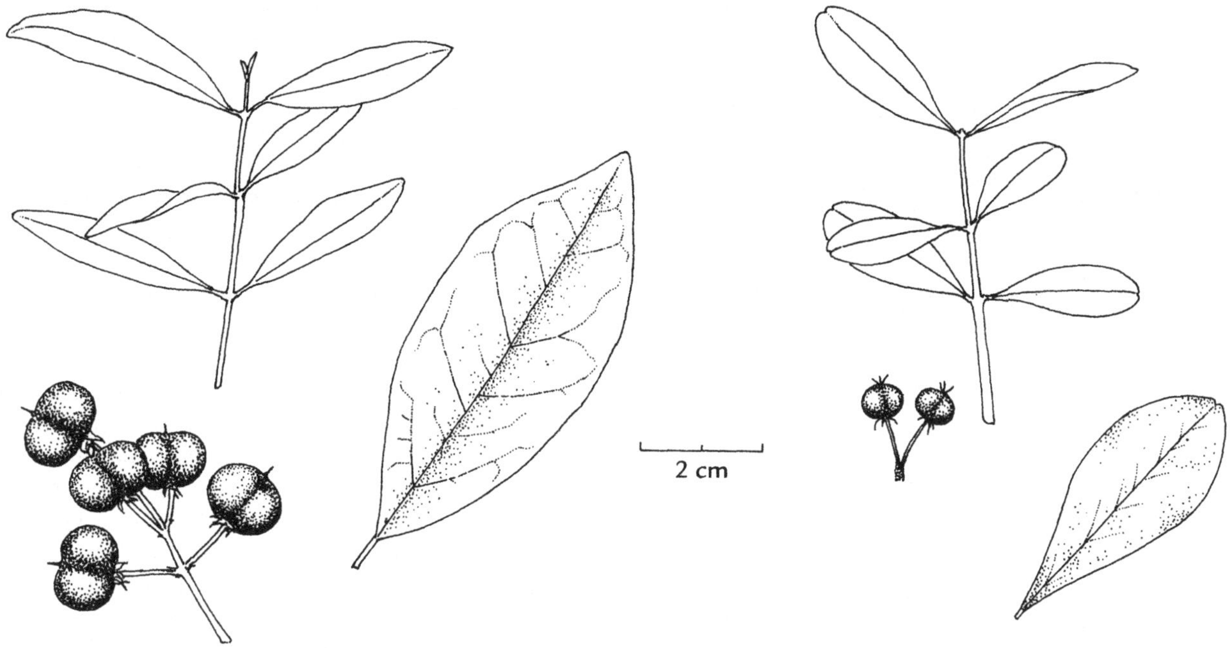


Figure 192. (*left*) *Bunchosia sonorensis*. Drawn by M. B. Johnson.

Figure 193. (*right*) *Malpighia emarginata*. Drawn by M. B. Johnson.

*Malpighia emarginata* is exceedingly variable, and the plants in Sonora and Sinaloa which have been called *M. umbellata* may be a distinct species (William Anderson, personal communication 1996).

### Meliaceae—Mahogany or Chinaberry Family

Unarmed trees and shrubs. Leaves alternate and spirally arranged, pinnately compound; leaflets stalked; stipules absent. Flowers mostly small, unisexual or bisexual, in axillary racemes or panicles. Calyx small, more or less cup shaped except in *Melia*; petals 5 (6), much longer than the calyx. Ovary superior or mostly so. Seeds two to many per fruit. Subfamily Melioideae with drupes or capsules, the seeds few, rounded (not winged) (*Melia* and *Trichilia*); subfamily Swietenioideae with 5-valved capsules, the seeds many, flattened and winged (*Cedrela* and *Swietenia*). Worldwide, tropical and some subtropical; 51 genera, 575 species. Reference: Pennington (1981).

Economically important members of this family include the neem tree (*Azadirachta indica*), native to India, and mahogany (*Swietenia*) in the New World. The neem tree would undoubtedly thrive in cultivation in the near frost-free areas of Sonora. *Swietenia humilis* (*S. cirrhata*) is known as far north as Sinaloa and should be sought in the wettest, tropical riparian canyons in southeastern Sonora. It is successfully although seldom cultivated in Hermosillo.

1. Leaves twice pinnate, the leaflet margins toothed; fruits indehiscent (drupes). ____________ ***Melia***

1' Leaves once pinnate, the leaflet margins entire; fruits dehiscent (capsules).

 2' Spindly, slender-stemmed shrubs or small trees; seeds with an aril and not winged. ________ ***Trichilia***

 2. Medium to large trees with stout, often massive trunks; seeds winged, without an aril.

  3. Leaves sparsely hairy or perhaps glabrate, the leaflet tips blunt or pointed but not extended into a drip point; capsules pendulous, opening from apex only; stamens 5, separate; seeds winged only from apex of seed body. _______ ***Cedrela***

  3' Leaves glabrous, the leaflet tips long-acuminate, extended into a slender drip point; capsules erect, opening from base and apex; stamens 10, the filaments fused into a column; seeds winged from apex and base. ____ ***Swietenia humilis*** (not treated in text)

#### *Cedrela*

Deciduous trees of the New World tropics; 8 species.

***Cedrela odorata*** [*C. dugesii; C. mexicana; C. occidentalis*]. Spanish cedar, cedrela; *cedro* (figure 194)

Trees approximately 10 m with a substantial trunk; bark regularly and evenly longitudinally furrowed or fissured.

Figure 194. *Cedrela odorata*. Drawn by M. B. Johnson.

Branchlets, bark, and immature fruits reported to usually smell of garlic when broken or crushed. Young twigs, leafstalks, and leaflet veins with small, spreading, white to golden hairs. Leaves 30–55 cm, pinnate, drought deciduous; leaflets opposite to subopposite, at least 5–8 pairs per leaf, mostly 10–17.5 cm, 4.5–5.5 cm wide; blades thin, asymmetric especially near the base, mostly oblong with rather short, acuminate tips, the margins entire. (The few Sonoran specimens seen are even-pinnate without a terminal leaflet; elsewhere this species usually has a terminal leaflet.)

Inflorescences terminal or subterminal, large and much branched, open and pendulous, and with lenticels. Flowers 5-merous, greenish white, probably with an unpleasant odor. Calyx with irregular, broad, shallow teeth, and splitting along one side; petals probably 7–8 mm. Stamens 5, separate. Capsules probably 3.5–4.5 cm, pendulous, 5-chambered with 5 woody valves opening from the apex to reveal a 5-ribbed, ovate-oblong, light-weight seed mass of potentially 50 tightly appressed, overlapping seeds that eventually breaks apart. Seeds flattened with a prominent terminal wing, probably 2–3 cm including the wing. Probably flowering with summer rains; fruits probably in early spring, about 5 or 6 months after flowering.

Riparian arroyos in tropical deciduous forest; southeastern Sonora, 400 m (1310 ft). Mexico northward along the Pacific drainage to the Sonora-Sinaloa border, and the Atlantic drainage northward to Tamaulipas, and to northern Argentina and the Caribbean.

In Sonora known only from El Cobre, where there are four trees in a large arroyo in the Agua Caliente drainage near Choquincahui, and a much larger population at Bahuiyahui (south of Alamos near the Sinaloa border). *Cedrela* is much more common in nearby northern Sinaloa, where the trees often reach 30 m. Although protected by law, some have been cut illegally. The wood is prized for furniture, doors, and door frames. It is insect resistant and is dried for several years before use. The wood is coarse and fragrant, and in other regions it has been used for cigar boxes and moth-proof chests. The wood from the Sonoran trees is lighter colored and not as desirable as the red- to deeper-colored *cedro* wood from farther south (e.g., Sinaloa). In wetter, more tropical regions it is an important timber tree, growing to 30–35 (60) m. Over much of its range it has been seriously overharvested, and larger, well-formed trees are scarce. This is one of the world's most important hardwoods and is now grown as a plantation tree.

## Melia

Trees and large shrubs; 3 species, Africa to Asia and Australia.

***Melia azedarach*.** Chinaberry tree, paradise tree; *paraiso, piocha* (figure 195)

Medium-sized trees with relatively soft wood and a rather thick trunk. Bark gray brown, rough, and fissured; branches brittle. Glabrate with simple to stellate and branched, minute, mealy hairs on youngest herbage and pedicels. Winter deciduous; leaves twice pinnate (sometimes 3-pinnate at bases of larger pinnae), often 33–58 cm. Leaflets 3.5–5.5 cm, ovate to lanceolate, asymmetric, the margins shallowly toothed.

Panicles mostly 8–22 cm. Flowers 16–21 mm wide, bisexual and staminate on the same tree, attractive, very fragrant. Calyx approximately 2 mm, of 5 (6) sepals fused basally. Petals 5 (6), spreading, whitish with violet toward tips. Stamens 10, the filaments fused into a slender tube 7–8.5 mm, violet purple and darkest toward apex. Fruits mostly 14–16 mm, rounded but slightly longer than wide,

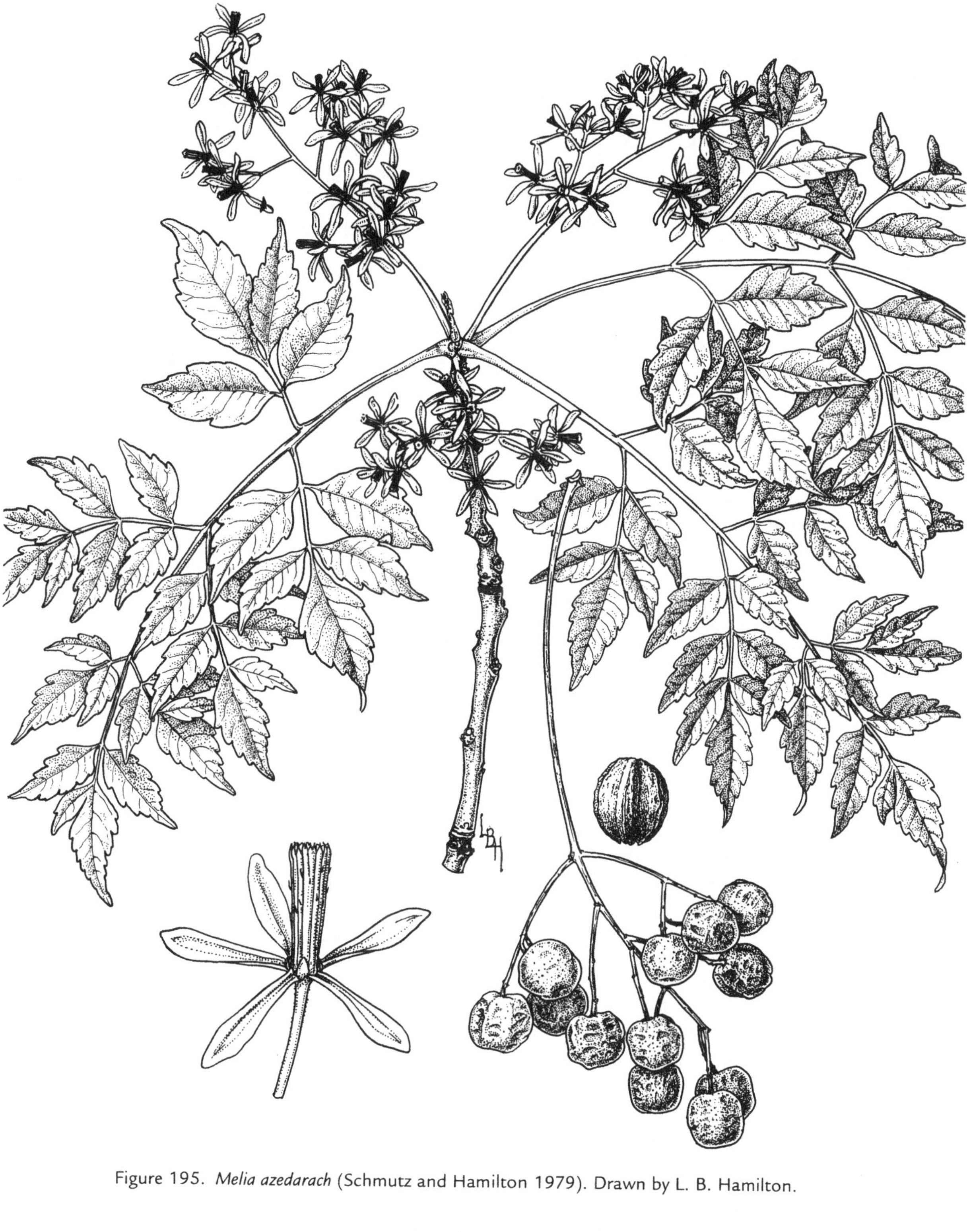

Figure 195. *Melia azedarach* (Schmutz and Hamilton 1979). Drawn by L. B. Hamilton.

hard and pale yellow-brown, many persisting about 1 year, several-seeded. Flowering March–April.

Native to Asia and Australia; cultivated and naturalized in many warm regions of the world. Cultivated through much of Sonora and Arizona; occasionally persisting at abandoned dwellings, especially in well-watered habitats in northern and eastern Sonora above the desert and in tropical deciduous forest, and less often well established and sometimes spreading in natural areas. Occasionally propagating by root-sprouts, especially if the parent tree is severely damaged, and rarely reproducing by seeds; in cultivation grown from seeds and cuttings.

The hard, rounded fruits can be dangerous for pedestrians on walkways. The fruits and bark are poisonous, although the bark, roots, and leaves have been used medicinally. The fruits are a source of flea powder and insecticide. The wood is sometimes used for cabinet making. There are several cultivated varieties, but these are apparently not grown in Sonora or Arizona.

### *Trichilia*

Spindly hardwood shrubs or small trees; bark with prominent lenticels. Foliage tardily drought deciduous, often becoming reddish before falling; leaves clustered near stem tips, odd-pinnate; leaflets alternate to subopposite, thin, the margins entire, the lateral leaflets asymmetric or not, the terminal leaflet symmetric.

Flowering branches of axillary panicles or short, raceme-like branches with small, deciduous leaves, or sometimes 1-flowered. Flowers white; probably both unisexual and bisexual, on slender, jointed pedicels, 8–10 mm wide, with a nectary at the base of the ovary; rudiments of opposite sex well developed in unisexual flowers. Calyx approximately 1 mm, usually 5-lobed. Filaments united below into a tube, hairy on the inside, each with broad lobes. Fruit a stalked capsule with 2 or 3 woody valves becoming transversely wrinkled upon opening. Seeds mostly 2–5 per capsule, partly to entirely enveloped by a thin, fleshy yellow to orange or red aril-like covering (here called the aril). The elegant trogon greedily feeds on the aril-covered seeds. Mostly New World tropics, also Africa and the Indo-Malayan Region; 86 species.

1. Bark flaking in irregular plates, otherwise smooth; calyx lobes ciliate; ovules and seeds of each ovary chamber or fruit valve 1, or 2 and superimposed (one on top of the other); capsule valves not reflexing; aril only partially covering the seed. ____________ ***T. americana***

1' Bark finely fissured or furrowed; calyx lobes not ciliate; ovules and seeds of each ovary chamber or fruit valve 1, or 2 and collateral (next to each other); capsule valves reflexing; aril enveloping the seed except for a small area at the base. ________________ ***T. hirta***

***Trichilia americana*** [*T. colimana*]. *Piocha; síquiri tájcara* (Mayo) (figure 196A)

Large shrubs or small trees; bark smooth, gray brown, peeling or scaling in large, thin, irregular plates to reveal a pale undersurface, and with conspicuous lenticels. Youngest, emerging herbage densely pubescent; older herbage glabrate or sparsely pubescent, especially on the leaflet midribs and veins. Leaves 12–42 cm; leaflets 7–13 per leaf, elliptic to oval, ovate or nearly orbicular, nearly symmetric to asymmetric especially at the base, 4–15 × 2–7 cm.

Possibly dioecious. Female inflorescences 1-flowered and 1.5 cm to multiple-flowered and 12 cm; male inflorescences paniculate, 12–18 cm. Calyx lobes ciliate; ovules and seeds 1 per chamber or valve, or 2 and superimposed (one over the other). Capsules 15–18 mm including the stipe; valves 2, the outer surfaces densely pubescent with minute nonglandular hairs, wrinkled transversely in open fruits, not reflexing or sometimes slightly reflexed, blunt tipped or often, especially on vigorous, well-watered plants, narrowed at the tip to a conspicuous point or projection to 2–3 mm. Seeds 6.5–9 mm maximum diameter, the seed coat hard; arils bright yellow to orange, only partially surrounding the seeds (covering the base, apex, and adaxial seed surface, the abaxial surface is bare). Flowering August–September; fruits at least December–March.

Canyons and slopes; tropical deciduous forest; southeastern Sonora, 250–740 m (820–2425 ft); Sierra de Alamos, the Río Cuchujaqui (Fuerte) drainage, and the Río Mayo drainage east of Alamos and San Bernardo. Usually an understory plant in shaded, moist habitats with deep, rich soil. Also southwest Chihuahua southward along the Pacific slope of tropical Mexico to Costa Rica.

***Trichilia hirta***. *Piocha, piocha cimarrón, palo amarillo; síquiri tájcara* (Mayo) (figure 196B)

Resembles *T. americana* but differs as follows: bark finely checkered (finely fissured or furrowed). Leaves (9) 15–27 cm; leaflets (9) 11–21 per leaf, 3–8.5 × 1.2–3.0 cm, elliptic to lanceolate. Probably dioecious.

Panicles or racemes 2.5–8 cm; fruiting branches sometimes with leafy bracts or small leaves, or sometimes the fruits solitary on stalks (pedicel plus the stipe) approximately 1.5 cm. Calyx lobes not ciliate (sometimes with a few scattered hairs). Ovules and seeds 1 per chamber or valve, or 2 and collateral (horizontally next to each other). Capsules 10–11 mm, the valves 3, conspicuously reflexed, blunt tipped or with only a short apiculate projection < 1 mm, the outer surfaces sparsely pubescent or glabrous, and somewhat granular and glandular. Seeds 6–7.5 mm; seed coat soft and oily; aril bright orange or red-orange, entirely surrounding the seed except sometimes for a small area of the base. Flowering mostly May–August; fruits November–April.

Moist slopes, canyon bottoms, and bajadas; tropical deciduous forest and lower elevations in oak woodland; southeastern Sonora, 305–925 m (1000–3035 ft). Locally common in moist habitats in the vincinity of Alamos including Sierra de Alamos and canyons of the Rio Guajaráy. Sonora and Tamaulipas to Paraguay and the Caribbean.

The two *Trichilia* species in Sonora are closely related and easily confused. They are best distinguished if the suite of characters is taken into account. *Trichilia americana* tends to produce fewer, broader, and larger leaflets, as well as larger capsules. However, under favorable conditions with high temperature and moist soil (at least in Sonora and Sinaloa), *T. hirta* probably produces leaves as large as those of *T. americana* but with more numerous and narrower leaflets. For example, a specimen of *T. hirta* from Sinaloa has leaves 36 cm long, and 19 and 21 leaflets reaching 12 cm long but only 3–3.5 cm wide. The appearance of the bark is a reliable field character.

## Moraceae—Mulberry Family

Trees, shrubs or perennial herbs, often with milky sap. Leaves alternate and simple; stipules deciduous, often enclosing the leaf bud like bud scales and leaving a scar. Dioecious or monoecious. Flowers small and inconspicuous; calyx 4-lobed or absent; petals absent; styles 2. Fruits 1-seeded; individual fruits often small and aggregated into a highly modified structure such as in the figs (*Ficus*); seeds usually animal dispersed. Worldwide, tropical and subtropical, some in temperate regions; 40 genera, 1100 species. Reference: Wunderlin (1997).

This large, mostly tropical family includes breadfruit and jackfruit (*Artocarpus*). *Dorstenia drakeana*, the only nonarborescent member of the family in Sonora, is an unusual herbaceous perennial in the southeastern part of the state.

1. Leaf margins serrated, or if nearly entire then some branches with sharp spines.
   2. At least some branches with sharp thorns; leaf blades glabrous or with soft hairs; extreme southeastern Sonora near the Sinaloa border at 350 m (and perhaps northern Sonora at intermediate elevations). ______ ***Maclura***
   2' Unarmed; leaf blades with rough hairs (scabrous); northern and eastern Sonora above 750 m. ***Morus***
1. Leaf margins entire or lobed (rarely serrulate in *Trophis*); unarmed.
   4. Monoecious; flowers and fruits aggregated into figs containing many seeds; relatively widespread. *Ficus*
   4' Dioecious; flowers in axillary spikes; 1-seeded fruits not enclosed in a figlike structure; narrowly restricted to riparian "tropical" canyons. ______ ***Trophis***

### ***Ficus***—Fig, *Higuera*

In Sonora mostly trees, sometimes shrubs; sap milky. Growth bud enclosed by a pair of elongated, bractlike stipules falling away as the new leaf expands, the fallen stipules leaving a ringlike scar on the twig; one stipule in the pair often partially to fully enclosing the other (see figure 197). Leaves entire, usually rather firm; petioles prominent. Flowers minute, on the inner surface of a rounded receptacle, the fig. Base of fig with a 2- or 3-lobed involucre (the lobes called scales), the fig apex with a small opening or pore (the ostiole) closed by small scales.

Worldwide, mostly in Asia and mostly tropical, a few in warm temperate regions; 750 species. References: Condit (1947), Ramirez (1969, 1970, 1986), Tomlinson (1980), Wiebes (1986).

Female flowers are of two kinds: gall flowers and seed flowers. The gall flower has a long stalk (pedicel) and short style. The seed flower has a short stalk and a long style. Figs are pollinated by tiny, delicate agaonid wasps which have a highly specialized relationship with these plants. In the usual pollination scenario, the female wasp, attracted to a young developing fig, forces her way into the ostiole. At this time the small scales around the ostiole spread apart slightly. The female wasp has barely enough room to wiggle through the passage, which mostly excludes other, larger insects. Her wings and antennae are often torn off as she squirms through the tight hole. Various wings and antennae caught on guard scales are signs that several females have entered the fig. The female wasp has a long ovipositor which she uses to penetrate the styles. The ovipositor is long enough to penetrate the short style of a gall flower but not long enough to fully penetrate the long style of a seed flower. Thus, whether a flower develops a seed or becomes a wasp nursery depends on the style length.

Within a single fig, the female flowers become functional much earlier than the male flowers. The female wasp carries pollen from her birth-fig by means of sacs on her body. She pollinates both seed and gall flowers. Once she enters a fig, she is unable to escape and the young fig becomes her tomb. The wasp larva takes the place of the seed in the gall flower and feeds on endosperm tissue and the nucellus.

The male wasp, unlike the female, is wingless and has stronger jaws. After the males develop into adults, they chew through the gall flowers containing virgin females and impregnate them. The males then chew an exit tunnel through the fig wall. The males die inside the fig or they leave through the exit tunnel, but because the males are wingless they fall to the ground and soon die. Before the newly emerged female leaves she fills her pollen sacs with pollen from the now mature male flowers, often by biting into the anther to remove the pollen. She then leaves via an exit tunnel and flies to a young fig to begin a new cycle.

Figure 196. *Trichilia*. (A) *T. americana*: (a) habit, (b) male flower, (c) female flower, (d) bisexual flower, (e) fruits.

Combinations of pore size and guard scale arrangements, flower size, phenology, and probably specific chemical cues seem to be principal factors preventing interspecies pollination and hybridization. Although fig species in Sonora often occur together, no hybrids are known.

Several varieties of common fig (*F. carica*) are widely cultivated in Sonora, and they occasionally persist at abandoned settlements. There are three subgenera of *Ficus*, two of them in the New World. Subgenus *Pharmacosycea* can be distinguished by figs borne singly in the leaf axils, three scales at the base of the fig, and male flowers with two stamens. These figs are not epiphytic and do not become "stranglers." In Sonora these figs, represented by *F. insipida* and *F. maxima*, are mostly restricted to moist canyon or arroyo bottoms and often develop into very large trees. The remaining Sonoran figs, in the subgenus *Urostigma*, are distinguished by figs borne in pairs in the leaf axils (one fig in the pair may fail to develop), two scales at the base of the fig, and male flowers with one stamen.

The presence of one or two figs per node is readily

Figure 196. *(continued)* (B) *T. hirta*: (f) habit, (g) male flower, (h) female flower, (i) male flower (Pennington 1981). Drawn by R. Wise.

determined even after the figs and leaves have fallen (see figure 198). For *Urostigma* the nodes that have borne two figs have one rounded scar to each side of the leaf scar; these appear like a pair of eyes above a mouth. However, toward the branch tips, especially at the end of the growing season, one of the pair of figs may fail to develop. The number of scales subtending the figs should be interpreted carefully because the scales of *Pharmacosycea* often split as the fig develops and thus may appear as three or four.

The Sonoran members of the subgenus *Urostigma* occur in riparian tropical canyons to desert habitats. In Sonora, where these tropical trees are mostly limited by dryness and winter freezing, they more often grow on rocks or cliffs than as hemiepiphytes on trees (strangler figs). In Sonora *F. cotinifolia* is generally the only *Ficus* that becomes a strangler fig. Occasional plants of *F. pertusa* in Sonora, especially young trees in gardens, may grow on palms. However, southward in Sinaloa the strangler fig habit becomes commonplace. The popular Sinaloan song

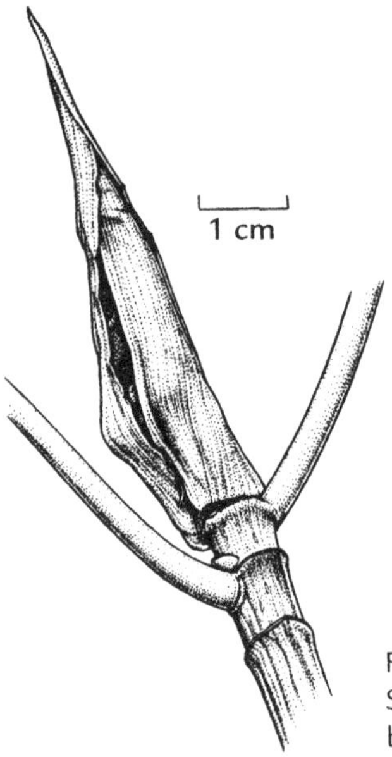


Figure 197. *Ficus petiolaris.* Stem tip with stipules. Drawn by A. K. Jakob.

"El sauce y la palma" refers to the strangler fig (*el sauce*) and the palm *Sabal rosei.*

Many of the Mexican figs

> often begin their growth upon other plants, usually palms, germinating and developing a stem from which aerial roots descend to the ground. . . . In this way the plants in their young stages are often vinelike. With age, the aerial roots increase in size and form a trunk which gradually envelops completely the host plant. The stems [actually the roots] at first are flat, broad, and thin, and as they increase in size several will unite, assuming irregular and fantastic forms. Ultimately the host plant dies but often it persists for a long time, and it is not unusual to see the fronds of a palm rising from the crown of a large fig tree. Figs of this habit of growth are known generally as *matapalo* [tree killer]. Sometimes the plants begin their growth upon cliffs, developing their thin trunks [actually roots] against the rocks. Large fig trees, too, often send down from their branches aerial roots which take hold of the soil and finally develop into trunks. In this way trees of the banyan type are formed, some of them of enormous size (Standley 1922: 205–206).
>
> Ripe fruits (figs) are fleshy, and attractive to birds which distribute the numerous small seeds, probably mostly in their droppings. [But also by wiping their beaks on tree branches and rocks.] Seeds can germinate either on the ground or in the crotches between branches of another tree or within the leaf bases of a palm crown. They are also capable of establishing within fissures of bare rock. Initially growth is dependent on sufficient humus or soil for the seedling to become well rooted. This process is assisted by swelling of the hypocotyl in the seedling, apparently a water-storage organ. The limitations of a restricted soil supply, which would inhibit further development of most trees, is overcome by the production of aerial roots which grow over the surface of the host tree or rock and can reach more distant soil resources. When the tree is established as an epiphyte the sapling continues to enlarge its crown and root system and is capable of enveloping and even killing its supporting tree. There is, however, no parasitic feeding by the *Ficus*. Popular accounts suggest that the fig "strangles" its support but this is somewhat exaggerated; death of the supporting tree is as likely to follow from shading and root competition (Tomlinson 1980: 249–250).

Six species of figs occur in southeastern Sonora, three of which reach the desert edge and one which occurs in the southern part of the Sonoran Desert. All six species occur in rather close proximity on Sierra de Alamos. These figs are recommended for landscape use but are frost sensitive and often too large for home gardens. Most of the Sonoran species have edible figs, and the figs may be produced at almost any time of year. The fruits are eaten fresh or cooked.

1. Leaf blades less than twice as long as wide, mostly oval or orbicular to ovate or cordate (heart-shaped).
   2. Leaves and figs hairy; desert regions. **F. petiolaris** subsp. **palmeri**
   2' Leaves and figs glabrous or leaves with hairs along veins of lower surface; subtropical regions.
      3. Leaves oval to orbicular, widest at about middle, the tip rounded; figs sessile. ___ **F. cotinifolia**
      3' Leaves cordate, widest below middle, the tip pointed; figs stalked. ____ **F. petiolaris** subsp. **petiolaris**

1' Most leaf blades at least twice as long as wide, not orbicular or cordate.
   4. Leaves mostly 5–10 (14) cm; blades shiny and lustrous, lanceolate to sometimes ovate (widest below the middle); mature figs ±1 cm wide. __ **F. pertusa**

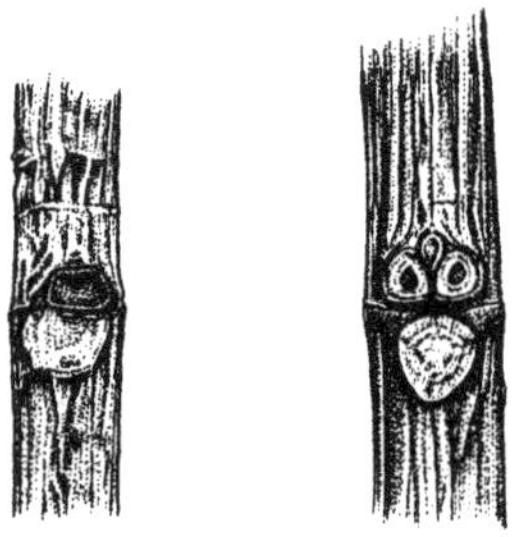

Figure 198. *Ficus.* Twigs with leaf and peduncle scars. *Urostigma, F. insipida* (left) and *Pharmacosycea, F. petiolaris* (right). Drawn by A. K. Jakob.

4' Leaves mostly 11–30 cm; blades often dull green, (mostly widest at about middle); mature figs > 1.5 cm wide.
  5. Leaves soft-pubescent when young, glabrous at maturity; figs often paired. _____ ***F. trigonata***
  5' Leaves often hairy with both or only the lower surfaces usually rough-scabrous; figs solitary.
    6. Petioles usually smooth; stipules 4–8.5 cm. ***F. insipida***
    6' Petioles exfoliating in small, thin, brown flakes; stipules < 2.5 cm. _____ ***F. maxima***

***Ficus cotinifolia***. *Nacapúl, chuna, camuchín; guaorócochi, guojtoli* (Guarijío) (figures 199A and 200)

Often reaching 15 m or occasionally larger, often with aerial roots and sometimes producing banyan roots. Bark often light colored. Partially to completely drought deciduous depending largely on habitat. Young twigs hairy; petioles sometimes sparsely pubescent. Leaves 8–20 cm, glabrous; blades mostly oval to orbicular, rather tough, the tip usually rounded; petioles usually long.

Figs edible, paired (or one fails to develop), sessile, subtended by 2 scales often splitting into 3 or 4 parts as the fig develops, globose, approximately 1 cm wide, green with reddish brown flecks, the ostiole in a sunken, circular pit.

Rocky canyon bottoms, cliffs, and rock slopes, and occasionally as epiphytes on *Taxodium* or other trees including *Pithecellobium dulce*; tropical deciduous forest; southeastern Sonora, 180–610 m (600–2000 ft). Also Baja California Sur and southwestern Chihuahua to Costa Rica. Usually a strangler fig through most of its range.

***Ficus insipida*** subsp. ***insipida*** [*F. radulina*]. *Chalate, chuna, higuera; huilocochi* (Guarijío); *tchuna* (Mayo); *kau chuuna* (Yaqui) (figures 198 and 199B)

Broad-crowned trees often 10–25 m; trunks massive and often with short buttresses, often with exposed, spreading, and gnarled roots; bark smooth and gray. Young stems often pubescent. Evergreen, except trees in the few desert-edge populations sometimes partially to almost fully defoliating during extreme drought. Leaves 12–31 cm; blades 10–22 × 3.5–10 cm, elliptic to ovate, sometimes broadly so, the lower surfaces usually scabrous to sometimes smooth, the veins especially prominent, usually 14–18 lateral pairs, the leaf base rounded, the tip usually pointed (acuminate or acute); petioles 2.5–4.5 cm, sometimes grooved when dry but usually not exfoliating; stipules linear, 4–8.5 cm.

Peduncles thick, 7.5–10.5 mm; figs solitary, subtended by 3 scales, pyriform-globose, 3.3–3.5 × 2.5–3 cm, green with yellow-green larger and smaller elliptic spots, the inside reddish, narrowed below to a stipelike base or the stipe absent, the ostiole conspicuously elevated (ca. 1 mm high).

Canyons, arroyos, and streams; tropical deciduous forest, thornscrub, and the southern margin of Sonoran desertscrub; southern Sonora, 30–900 m (100–2950 ft). Sonora and southwestern Chihuahua to Paraguay, and the Caribbean. Relatively common in canyons in tropical deciduous forest in southeastern Sonora. Reaches the desert margin in the Sierra el Aguaje north of San Carlos, especially at Cañón de Nacapule (Felger 1999). A second, allopatric subspecies in Amazonian South America.

Dense groves produce deep shade along streams in certain broad canyons in the Sierra Bacatete in the Yaqui territory where the surrounding valleys and hillsides support desert/thornscrub vegetation. The pollinator wasp is *Tetrapus costaricanus*. Fruit-eating bats (*Artibeus jamaicensis*) feed on the fruits in tropical deciduous forest. This species and *F. trigonata* develop buttressed trunks, a common characteristic of tropical forest trees. *Ficus insipida* is readily distinguished by its often rough-surfaced leaves, exceptionally long stipules, and solitary figs.

Some trees in deep, riparian canyons in southeastern Sonora have scaly, flaking twigs and petioles like those of *F. maxima*, but the long stipules (at least to 4.5 cm) are diagnostic.

***Ficus maxima***. *Chalate de burro; chuná* (Guarijío) (figure 201)

Massive, spreading trees often 15–25 m; trunk smooth and with short buttresses. Young twigs and petioles densely to sparsely hairy. Probably evergreen; leaves 10–32 cm; blades 7.5–29 × 4.5–10.5 cm, elliptic to obovate or oval, both surfaces noticeably scabrous (like sandpaper when rubbed backward), the veins especially prominent beneath, usually 9–14 lateral pairs; leaf base narrow to broad, the tip usually rather blunt, obtuse to short-acuminate; petioles 1.5–3.5 cm, the epidermis scaly, exfoliating in small, thin, brownish flakes (unusual among the Sonoran figs); stipules 1.2–2.3 cm, glabrous or hairy toward the base.

Peduncles 6–12 mm, pubescent; figs solitary, subtended by 3 scales, rounded, 1.7–2.0 cm diameter, pubescent and greenish with reddish, wartlike papillae, reddish inside, the ostiole nearly flat and at least sometimes surrounded by rather membranous green bracts.

Riparian canyons and arroyos in tropical deciduous forest; southeastern Sonora, 225–800 m (740–2625 ft), known from the Alamos region including La Aduana, Alamos, and the Sierra de Alamos, and in large riparian canyons east of San Bernardo. A common tree of lowland forests and hills in the American tropics southward to the Amazon basin and the Caribbean.

This species is readily distinguished by its rough-surfaced leaves, scaly petioles, short stipules, and solitary figs (see *F. insipida*). The leaves with their unusual, scaly petioles are

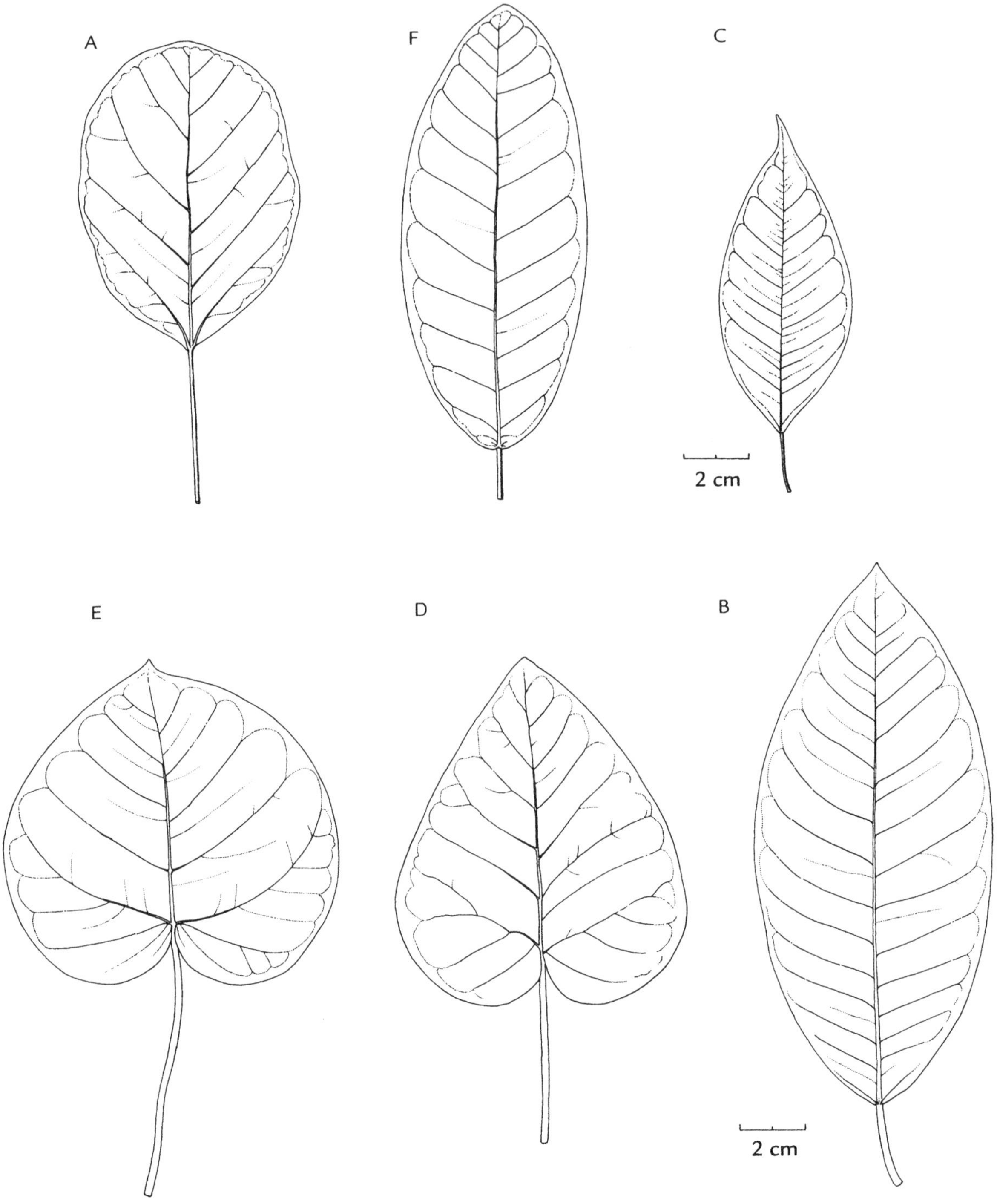


Figure 199. *Ficus*. (A) *F. cotinifolia*, (B) *F. insipida* subsp. *insipida*, (C) *F. pertusa*, (D) *F. petiolaris* subsp. *palmeri*, (E) *F. petiolaris* subsp. *petiolaris*, (F) *F. trigonata*. Drawn by M. B. Johnson.

Figure 200. *Ficus cotinifolia* on a *Taxodium*. Choquincahui, east of Alamos. Photograph by M. A. Dimmitt.

Figs usually produced in large quantities in late summer and winter (December–January), on prominent peduncles, paired (or one fails to develop or falls off), subtended by 2 scales splitting as the fig develops, edible, globose, 0.8–1.2 cm in diameter, at first green with red-brown spots or mottling, becoming darker when fully ripe, the ostiole sunken.

Rock slopes, cliffs, canyons, streams, and springs; tropical deciduous forest, thornscrub, and Sonoran desertscrub; west-central Sonora in the Sierra el Aguaje north of San Carlos and the Sierra Libre, and southeastern Sonora northward to the vicinity of Onavas and Tepoca; 10–1200 m (30–3935 ft). Also southwestern Chihuahua to Paraguay.

The trees tend to be single-trunked in desert-edge canyons such as Cañón de Nacapule north of San Carlos, whereas in more tropical regions such as near Alamos they sometimes form slender aerial roots that may develop into banyan roots supporting massive, spreading limbs. The Sonoran trees usually grow on rocks. In the more arid habitats, such as cliffs or canyon walls, the plants may remain

remarkably similar to those of another member of the same family, the *ramón* (*Brosimum alicastrum*), a large tropical tree in lowland Mexico used for cattle feed that yields an edible seed important in pre-Columbian times. The ramón ranges as far north as Sinaloa, where it is known as *apomo*. It can be distinguished from *Ficus* by its relatively large, single-seeded fruits and light-colored bark that peels up in 2 cm-thick, squarish slabs.

***Ficus pertusa*** [*F. padifolia*; *F. sonorae*]. Rock fig; *camuchín, gamuchín, nacapule*; *naka'apuli* (Yaqui) [*Nacapul* is the Cahitan (Yaqui and Mayo) term for earlobe (*naca* = ear)]; *tucuchí* (Guarijío) (figure 199C)

Medium to large trees reaching 15–20 (25) m; crown usually broad and dense; roots spreading over rock faces, sometimes producing aerial and banyan roots; bark gray, smooth. Evergreen at permanent water, or facultatively and tardily drought deciduous in drier habitats. Leaves 5.5–17 × 2.4–5.7 cm; blades lanceolate or elliptic (or sometimes ovate in well-watered and shaded habitats), shiny green, darker above than below, the tip acute to acuminate; stipules 8–22 mm, narrowly triangular-attenuate.

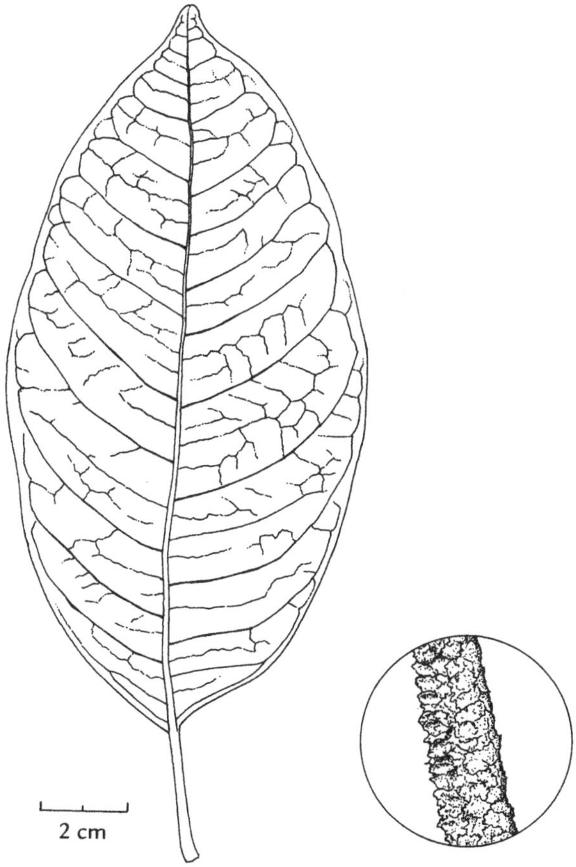


Figure 201. *Ficus maxima*. Inset shows twig with scaly bark. Drawn by M. B. Johnson.

as shrubs. The pollinator wasp is *Pegoscapus silvestrii*. This is the smallest-leaved fig in Sonora. A tree about 25 × 35 m fills a side canyon of the Arroyo Cuchujaqui near the road from Alamos to Guirocoba. An enormous *nacapule* shades the spring north of Cañón de Nacapule, but it has been severely damaged by cattle.

**Ficus petiolaris.** Cliff fig; *tescalama.*

Unusual trees that often grow on sheer canyon walls, cliffs, and mountain rocks. Seedlings germinate in crevices or cracks. Root and stem bark whitish to whitish yellow. Stipules broad in relation to other Sonoran figs. Leaves usually heart-shaped to oval with veins conspicuous and raised below; petioles well developed, often one-half to two-thirds as long as the blade.

Figs globose-obovoid, edible, paired or single when one fails to develop or falls off, subtended by 2 scales or sometimes appearing as 3 due to splitting of 1 scale as the fig develops; ostiole usually obscured by short, broad, thick, overlapping, and ascending reddish scales. Pollinated by *Pegoscapus* wasps.

The roots grasp and cascade down the surface as if melted and poured over the rock face. If the roots reach the canyon floor or moist soil the plant develops into a tree, otherwise it may remain dwarfed as a shrub.

The two subspecies are often treated as distinct species. These are the most widespread and common figs in Sonora, distinguished by their habit, broad and often cordate leaves, broad stipules, and unusual root system. Intermediate forms occur in the Guaymas region: in the more arid, exposed habitats the plants have the aspect of subsp. *palmeri*; in more favorable habitats, such as riparian canyons, they resemble subsp. *petiolaris*. Sometimes the more exposed branches have characteristics of subsp. *palmeri*, while shaded branches bear leaves like those of subsp. *petiolaris*. Plants of subsp. *palmeri* tend to be more densely pubescent, have tougher, more oval leaves with yellowish green veins, and shorter peduncles, while subsp. *petiolaris* tends to more often be glabrate or glabrous, and have thinner, cordate leaves with pink veins, and longer peduncles.

**Ficus petiolaris** subsp. **palmeri** [*F. palmeri*]. Desert rock fig; *Tescalama*; *xpaasni* (Seri) (figures 199D and 202A)

Generally large shrubs, or dwarfed shrubs on cliffs, especially sea cliffs, less often trees. Twigs, leaves, and figs densely to sparsely pubescent or glabrate. Leaves tardily drought deciduous, 6–28 cm; blades often relatively firm, broadly ovate or less often oval, the base moderately cordate or not, the veins not pink; stipules often 15–27 × 6–8 mm. Peduncles 6–20+ mm; figs 1.6–1.8 cm. Often fruiting spring and fall.

Sonoran desertscrub. Coastal west-central Sonora, near sea level–825 m (2700 ft). Endemic to the Sonoran Desert. Sonora mostly from the Guaymas region northward to the Sierra Seri and Cerros Anacoretas, Gulf of California islands including San Esteban, San Pedro Nolasco, San Pedro Martír (the type locality), and Tiburón, and the Baja California peninsula. Often on sea cliffs and extending to peak elevations on the islands and adjacent mainland mountains.

The leaves are notably large in comparison to other Sonoran Desert trees and shrubs. Apart from the riverine cottonwoods and willows, this is the largest tree in the northern Gulf of California. Occasional desert rock figs in the interior of Isla Tiburón reach 10 m and one was estimated to be 18 m tall. In a sheltered canyon on the western flank of the Sierra el Aguaje, a singular giant tree on blackened yellow volcanic tuff is about 17 m tall with a spread of about 35 m. This spectacular banyan-like tree has ten major "trunks" interconnected by horizontal limbs about 1 m or more above the substrate. Many aerial roots do not reach the soil but those that do develop into banyan trunks and become thick, somewhat irregular and sometimes moderately buttressed. Other roots are plastered against bare rocks and some have split the rock. Old, rotten parts of the largest trunks are fire-scarred. Beneath the canopy and between the rock outcrops supporting the tree is a deep litter of dry ficus leaves that crackles underfoot.

**Ficus petiolaris** subsp. **petiolaris.** Rock fig; *tescalama; guajtori* (Guarijío); *báisaguo* (Mayo); *vok ú'riduk* (Pima Bajo) (figure 199E, 202B and 203)

Trees often 10–15+ m (to 25 m in tropical canyons such as side canyons along the Río Guajaráy, or stunted as shrubs when freeze damaged at higher elevations, especially at the northern limits). Trunk and larger branches often laterally compressed. Twigs and leaves essentially glabrous, except a woolly tuft at the junction of the blade and petiole; peduncles and figs pubescent or glabrous. Foliage evergreen or sometimes drought deciduous; stipules 22–55 × 6–20 mm; leaves 12–31 cm; blades broadly heart-shaped (cordate at base), often with conspicuous pink veins. Peduncles 6–31.5 mm; figs 1.8–2.1 cm.

Rock habitats: slopes, cliffs, canyons, streams, and arroyos; at the margins of Sonoran desertscrub (riparian canyons), thornscrub, tropical deciduous forest, lower elevations in oak woodland, and occasionally shrubs on sunny cliffs in pine-oak woodland; central and eastern Sonora as far north as the Sierra el Tigre (vicinity of 30°36'N); sea level–1800 m (5900 ft). Also southwestern Chihuahua to Oaxaca.

**Ficus trigonata** [*F. goldmanii*]. *Chalate, higuera; guagüurí* (Guarijío) (figure 199F)

Trees to 25 m, with dense, spreading crowns sometimes to 30 m across; trunk massive with short buttresses at base

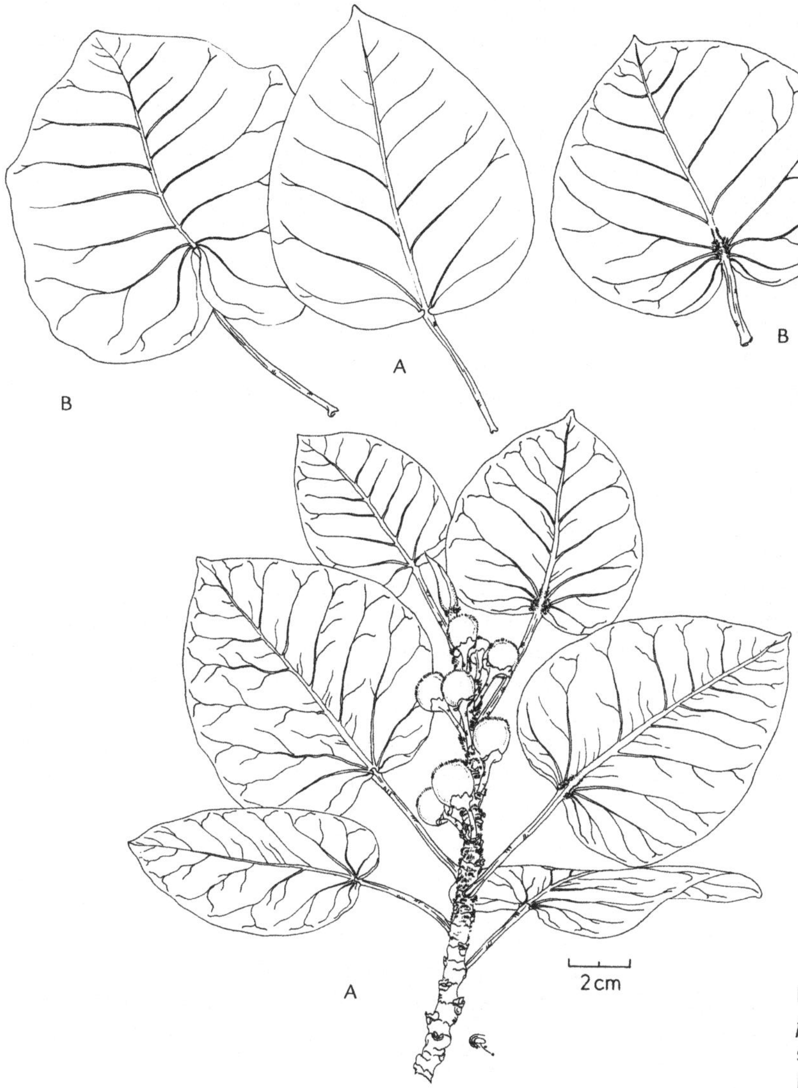


Figure 202. *Ficus petiolaris*. (A) subsp. *palmeri*, (B) subsp. *petiolaris*, upper leaf surface (left) and lower surface. (Felger and Moser 1985). Drawn by F. Runyan.

and exposed, gnarled roots; bark gray; limbs sometimes forming a few banyan roots. Twigs glabrous or glabrate. Evergreen or nearly so. Leaves 11–26 cm; blades 10.6–23 × 5–10.5 cm, with soft pubescence on both surfaces when young, shiny green above, dull green below, narrowly to broadly elliptic, ovate or oval, the tip often broadly obtuse; petioles 1.3–4.5 cm; stipules 1.5–3.5 cm, often with long, silky, deciduous hairs.

Peduncles 11–17.7 (19) mm, green with red-brown flecks, thick and expanded just below the fig. Figs mostly paired (some solitary especially toward stem tips), subtended by 2 scales with one sometimes splitting, globose, 2–3 cm diameter; ostiole slightly raised and surrounded by a thickened, yellowish ring.

Canyons and streams; tropical deciduous forest; southeastern and east-central Sonora, 220–1000 m (720–3280 ft). Sonora to northern South America and the West Indies. In Sonora known for certain only from Sierra de Alamos and the vicinity of Alamos where it is sometimes locally common, springs and canyons in the foothills between Onavas and Tepoca, and farther north on north-facing slopes of large canyons in the mountains west of El Novillo in the Río Yaqui drainage.

Gentry (1942) described and illustrated a tree at Alamos with a 30-m canopy. The figs are sweet and eaten fresh and in our opinion the fruit is the best tasting of the Sonoran figs. Although they are in different subgenera, *F. trigonata* resembles *F. insipida* in habit of growth, and in having relatively large, edible figs and large leaves. *Ficus trigonata* differs in its generally lighter colored bark, smaller stipules, soft-pubescent young leaves, glabrous older leaves often of larger maximum size, lighter colored

Figure 203. *Ficus petiolaris* subsp. *petiolaris*. Roots on canyon wall along Río Guajaráy. Photograph by D. Burckhalter.

leaf blades with blunter tips, and figs often paired rather than always solitary.

### *Maclura*

Trees, shrubs, or woody climbers, armed with spinose-tipped branchlets or thorns. New and Old World tropics and temperate North America; 11 species. References: Berg (1986), Kaastra (1973).

The Osage orange (*M. pomifera*) is native from Arkansas to Texas and introduced and naturalized elsewhere in the United States including New Mexico. It is locally established in wetland habitats in Santa Cruz County in southeastern Arizona near the Sonora border, as well as elsewhere in Arizona. It is a large shrub or small tree with sharp, stout thorns, winter-deciduous foliage, and a unique aggregated fruit or syncarp 10 cm or more in diameter, yellow-green to green with an irregular surface.

***Maclura tinctoria*** subsp. ***tinctoria*** [*Chlorophora tinctoria*]. *Mora* (figure 204)

Trees 11–20 m with spreading canopies. Trunk well developed, often forming small buttresses at the base; bark pale yellow-brown and smooth, eventually flaking in large, thick pieces, and green beneath the flaking bark, the lenticels conspicuous. Sap copious, of clear, yellowish to watery-milky latex, becoming gummy and sticky on drying. Twigs and branches near base of tree with stout, sharp, solitary or paired spines often 5–20+ mm. Youngest herbage hairy; leaf veins remaining sparsely hairy. Leaves drought deciduous; petioles 0.7–4 cm; blades ovate, 6–18 × 3–10 cm, often asymmetric, often with acuminate drip tips; veins pinnate, prominent, especially below; margins toothed to nearly entire; stipules brown, often 7–8 mm, soon deciduous.

Dioecious. Inflorescences axillary, the flowers whitish, rather inconspicuous. Male flowers densely crowded in elongate spikes, moderately fragrant; stamens "spring-loaded," inflexed in bud. Female flowers in dense, usually stalked, fleshy heads 6–8 mm wide not including the stigmas; stigmas long and slender, the head resembling a soft pin-cushion. Fruits globose, 1+ cm diameter, fleshy and edible but said to be caustic to the lips, the stigmas somewhat persistent. Reproductive during summer rains.

Tropical deciduous forest in extreme southeastern Sonora, 220–360 m (720–1180 ft), from El Zapote eastward to Rancho el Guamúchil, 3 km northwest of the Sinaloa border. Locally common with many juvenile plants along arroyo bottoms and the adjacent slopes, generally in rich soil. Common in nearby northern Sinaloa, and perhaps nearby Chihuahua. Tropical America, through much of Mexico to Argentina and the Caribbean. A second subspecies in South America.

This tropical species has a long history as a commercial dyewood. The bark has been used medicinally and for tanning, and the wood for making furniture and other items. The *moras* are well known to the local people who take pride in them, knowing these are the only ones in Sonora. The bark, resembling that of palo santo amarillo (*Ipomoea arborescens* var. *pachylutea*) is one of the most outstanding features of this tree.

### *Morus*—Mulberry

Trees or shrubs; monoecious or dioecious. Warm temperate to subtropical in the Northern Hemisphere; 10 species.

The white mulberry, *M. alba*, native to China, is extensively cultivated in Sonora and Arizona. The male trees, called "fruitless" mulberry, produce copious pollen which is a major contributor to springtime hayfever. The female trees produce fleshy, edible fruits. The leaves are a larval food of the domestic silk moth.

***Morus microphylla***. Littleleaf mulberry, *mora*; *kohi* (Mountain Pima); *gohui*, *goohi* (Tohono O'odham) (figure 205)

Mostly shrubs, occasionally small trees to 7 m; bark brown, nearly smooth, branches rather slender and pliant; lenticels sparse and conspicuous on younger stems. Sap of newest growth milky. Young growth, leaves, and inflores-

Figure 204. *Maclura tinctoria* subsp. *tinctoria*. Drawn by M. B. Johnson.

cences with white hairs. Winter buds covered by several well-developed scales. Leaves 3–15+ cm including petioles, winter deciduous, or sometimes remaining through the winter in warmer regions, highly variable in size and shape depending on soil moisture and position on the tree. Petioles 0.9–2.2 cm; blades broadly ovate, sometimes deeply 3-lobed, especially on vigorous new growth, scabrous on both sides, especially the upper surface, the tip acute to sometimes acuminate, the margins serrate-toothed (young, developing leaves with gland-tipped teeth); stipules paired, membranous, unequal in size, quickly deciduous.

Dioecious. Flowers inconspicuous, sessile in pendulous, peduncled catkins, wind pollinated, mostly flowering with new growth in spring. Male catkins white; male flowers 5–6 mm wide, with membranous sepals. Stamens 4, longer than the sepals, the filaments flattened; developing stamens inflexed (incurved), held by the cupped sepal base and elastically and suddenly bending outward like a tension-held spring to fling dry pollen in a puff. Female catkins often 1–1.5 cm; female flowers 2.5–3.0 mm wide; sepals about as long as the ovary, green with fringed margins; stigmas thick, 2-branched, yellow-green, sessile or nearly so. Fruits

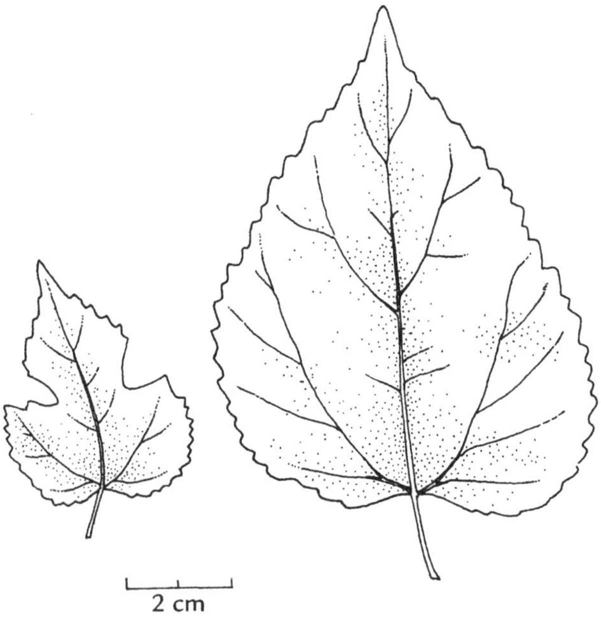


Figure 205. *Morus microphylla*. Drawn by M. B. Johnson.

red, turning blackish at maturity, fleshy and tasty, resembling the common, cultivated mulberry but much smaller. Flowering February–April.

Hills, slopes, canyons, and arroyos; Sonoran desertscrub, Chihuahuan desertscrub, grassland, oak woodland, pine-oak woodland, and occasional in tropical deciduous forest; north-central and eastern Sonora, 740–1675 m (2425–5500 ft). Arizona to Texas and Oklahoma, and southward to Durango, Chihuahua, and Sonora. In Sonora mostly in the eastern part of the state, generally at elevations above the desert, or in riparian or semiriparian habitats at upper elevations of the desert. In southeastern Sonora in pine-oak forest on Sierra Saguaribo in the upper Río Mayo drainage.

### *Trophis*

Tropical trees and shrubs from the Americas and Madagascar to the Orient and Australia; 9 species. Reference: Berg (1988).

***Trophis racemosa.*** *Granadilla*; *pu'sí* (Guarijío) (figure 206)

Trees to 20 m with a large, thick trunk; but usually much smaller; bark gray and smooth with conspicuous lenticels; sometimes flowering when only 1.5 m tall. Sap milky. Young twigs and leaves hairy. Evergreen to deciduous during severe drought in late spring. Leaves 8–17 cm; blades broadly elliptic to obovate, sometimes broadly lobed at about the middle, with a prominent acuminate tip, conspicuously net-veined below, scabrous and rough to the touch, the margins nearly entire to serrulate; stipules 1–1.6 mm, brown and scalelike.

Dioecious. Inflorescences axillary; flowers small, greenish. Male flowers in densely flowered, slender, pendent, catkinlike spikes often 2–3 cm; calyx lobes conspicuous and hairy; stamens spring-loaded. (When the stamens are released from the tepals, the anthers spring free and fling powdery white pollen into the air.) Female spikes short and few flowered; calyx 1.5–2.5 mm, densely hairy, the calyx tube pear-shaped and narrowed to a porelike opening with minute lobes; stigmas linear. Fruits densely pubescent, rounded, pink, fleshy, and berrylike, 1-seeded, the seeds approximately 1 cm. Flowering November–December; fruits ripen May–June.

Tropical deciduous forest and lower elevations in oak woodland; southeastern Sonora, 300–1050 m (985–3445 ft). Along streamways in tropical, riparian canyons in the Río Mayo and Río Fuerte drainages. Sonora and southwestern Chihuahua to South America and the West Indies. Fruits edible; the pulp is separated and cooked.

## Myrsinaceae—Myrsine Family

Mostly evergreen understory shrubs and small trees. Herbage, inflorescences, and floral parts with sunken "glands" (see *Ardisia*). Leaves alternate, petioled, simple, thickish, and entire, the midrib of the lower surface prominent; stipules absent. Flowers small, white, pink, or green, bisexual or unisexual (more than half the genera are dioecious); calyx and corolla (in Sonora) each 5-lobed and united at their base; stamens 5, attached to the corolla. Ovary superior; style 1. Fruits generally fleshy drupes, globose, 1-seeded, and with a persistent style. Tropics and subtropics of both hemispheres; 32 genera, 1000 species. Reference: Lundell (1971).

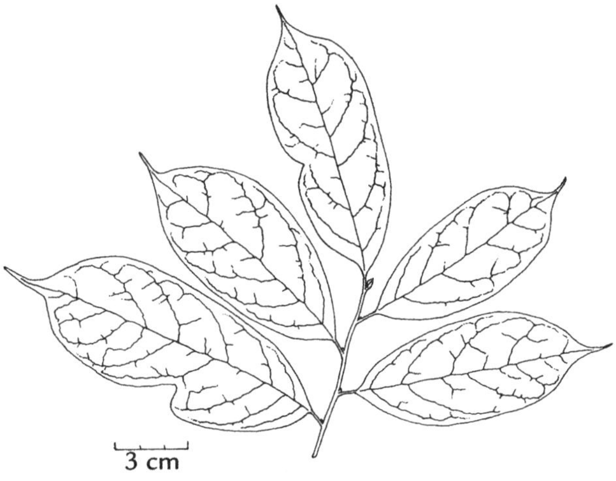


Figure 206. *Trophis racemosa*. Drawn by M. B. Johnson.

1. Leaves broadly elliptic to obovate, often 12–18 × 4–7 cm, the tip usually obtuse; flowers in open, terminal panicles; style long and slender; stigma inconspicuous. ***Ardisia***

1' Leaves elliptic, often 6–9.5 × 1.5–5 cm, the tip acute; flowers in short, dense axillary clusters; style short; stigma conspicuous, pyramid-shaped. ____ ***Myrsine***

### *Ardisia*

Trees and shrubs, mostly tropical America and Asia; 500 species. Reference: Pascarella (1997).

This is the largest genus in the family. The species do not produce nectar or other floral resources such as oils or resins and are presumed to be pollinated by small bees.

***Ardisia revoluta*** [*Icacorea revoluta*; *Ardisia scopulina*]. *Aguapepe* (Sinaloa and Durango) (figure 207)

Multiple-stemmed large shrubs or sometimes trees 8–10 m. Bark rough, fissured, and dark colored. Twigs often stout. Plants glabrous except minute red-brown glandular hairs on young shoots (especially leaf buds) and inflorescences in bud, and minute cilia on bud scales. Most of the plant conspicuously streaked or dotted with elongated, dark, often purplish "glands" especially visible on the leaf when held up to the light and on the inflorescences and virtually all floral parts except the filaments. (These "glands" are schizogenous secretory ducts or cavities and are a common feature in the family and unique in the Sonoran tree flora.) Foliage dense and bright green, the leaves often 12–18 × 4–7.5 cm; blades broadly elliptic to obovate, thickish, leathery, and shiny (looking rather like plastic), and basally decurrent onto a short, narrowly winged petiole.

Inflorescences of many-flowered terminal panicles 10–15 cm. Buds and flowers subtended by soon deciduous small scales. Flowers fragrant, milky white, long-pedicelled, subtended by soon-deciduous ovate to elliptic bracts. Sepals ovate, persistent; corolla white, 7–8 mm, deciduous, petal lobes spreading; stamens and the much longer style exserted. Anthers reddish, opening by longitudinal slits. Stigma inconspicuous, about as wide as the slender, elongated style. Fruits edible, approximately 5 mm, bright red, ripening maroon-purple, gland dotted; style longer than the fruit. Seed shaped like a mushroom cap. Apparently flowering and fruiting at various seasons.

Deep, riparian tropical canyons in tropical deciduous forest; southeastern Sonora, 480–850 m (1575–2790 ft). In Sonora known for certain only from several canyons above Bavícora, Arroyo el Cobre at Choquincahui, the canyon system near Rancho San Pedro including Arroyo Verde in the upper Río Cuchujaqui, and a disjunct, seemingly out-of-place record from Sierra San Javier southeast of Hermosillo (Alberto Búrquez, personal communication 1998). Often a locally common understory shrub or tree in deep shade of evergreen canopy trees such as *Aphananthe*, *Drypetes*, *Ficus*, and *Sideroxylon*. Probably also southwestern Chihuahua; southward to Colombia.

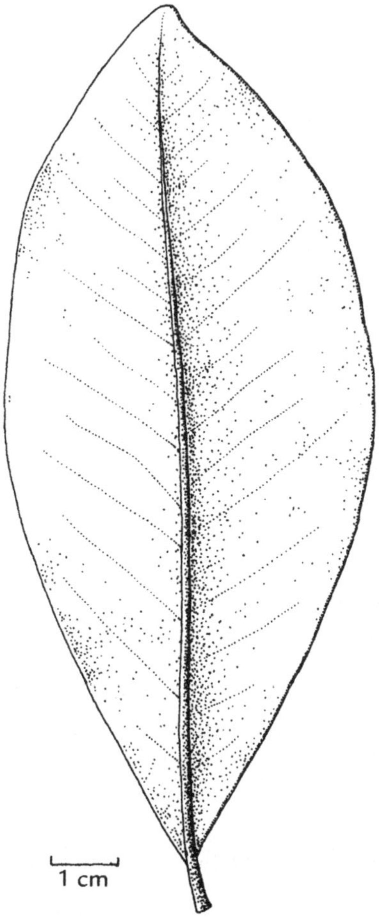


Figure 207. *Ardisia revoluta*. Drawn by M. B. Johnson.

### *Myrsine*

Tropical and subtropical trees in both hemispheres; 300 species (includes *Rapanea*), of which nearly one-quarter remain undescribed mostly in the tropics. Reference: Ricketson and Pipoly (1997).

***Myrsine coriacea*** subsp. ***coriacea*** [*Rapanea ferruginea*] (figure 208)

Slender, apparently evergreen, small- to medium-sized trees; bark mottled gray. Stem tips and developing leaves densely pubescent with red-brown hairs; older leaves sparsely hairy. Leaves 6–9.5 × 1.5–5 cm; blades mostly elliptic, thick, and leathery, gland dotted (not prominent as in *Ardisia*), the apex acute.

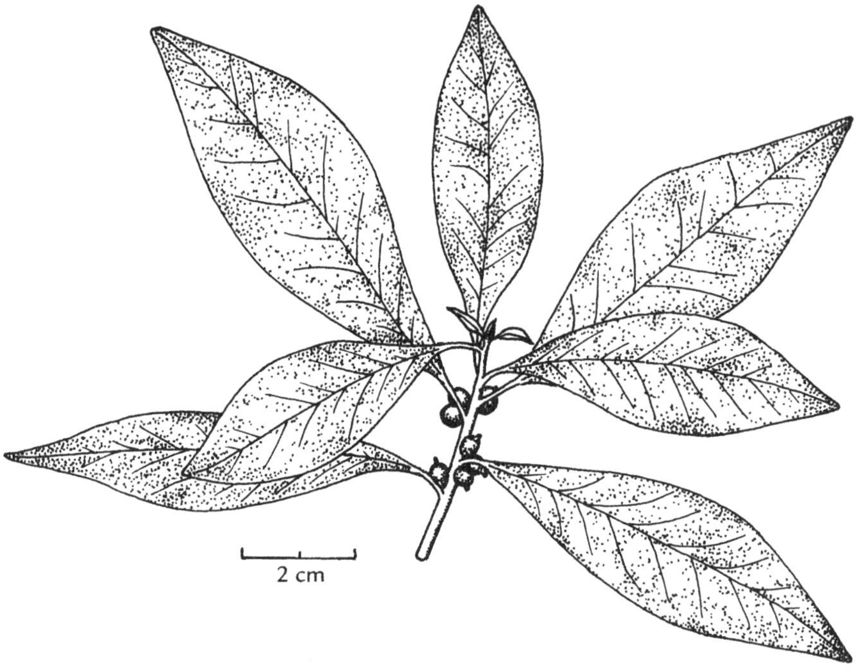


Figure 208. *Myrsine coriacea* subsp. *coriacea*. Drawn by M. B. Johnson.

Flowers apparently unisexual and the plants dioecious. Flowers densely crowded in sessile clusters (probably suppressed short shoots) at nodes below the leaves, the clusters subtended by small, glandular-fringed bracts. Flowers minute, yellow-white, the male and female flowers apparently not markedly different. Pedicels thick, 1.5–2 mm; calyx approximately 1 mm, persistent, minutely ciliate; corollas 1.6–2.0 mm; stigma subsessile, prominent, pyramid shaped, and resembling a miniature morel mushroom (the stigma is unique in the family). Fruits 3.5–4 mm, purplish, the surface roughened; stigma persistent and shorter than the fruit. Flowering at least March–April.

Riparian tropical-montane canyon bottoms in oak woodland and north-facing canyon slopes in pine-oak woodland; southeastern Sonora, 1065–1830 m (3500–6000 ft). In Sonora known only from the vicinity of Curohui on Sierra Saguaribo, where it grows with *Magnolia* and *Platanus*. Also nearby montane regions in Chihuahua and Sinaloa to South America and the Caribbean. Another subspecies occurs in wind-swept elfin forests in Central American highlands.

This "is the most widespread neotropical species and perhaps the only 'weedy' neotropical member of the Myrsinaceae" (Ricketson and Pipoly 1997: 583).

## Myrtaceae—Myrtle Family

Mostly tropical and subtropical trees and shrubs, characteristically with secretory cavities and ectotrophic mycorrhizae. Worldwide, widespread and diverse in Australia; 130 genera, 3850 species. References: McVaugh (1963), Wilson (1960).

*Eucalyptus*, with more than 500 species in Australia, and *Eugenia* with nearly as many species, mostly in tropical America, are the largest genera. Several *Eucalyptus* trees are widely cultivated in Sonora, mostly below 1200 m. The Murray red gum (*E. camaldulensis*) is the most commonly grown eucalyptus in the Sonoran Desert region and the most wide-ranging *Eucalyptus* species in Australia. *Eucalyptus* has not naturalized in Sonora or Arizona. The pineapple guava or feijoa (*Acca sellowiana*, = *Feijoa sellowiana*) is often cultivated in the region.

### *Psidium*

Evergreen trees and shrubs. Leaves opposite, simple; petioles short; blades pinnately veined and dotted with minute translucent oil glands, the midrib prominent especially below, the margins entire or nearly so; stipules absent. Peduncles axillary with 1–3 flowers. Calyx 4- or 5-lobed, the lobes united in bud and splitting irregularly between the lobes at anthesis. Petals 4 or 5, spreading, white. Stamens many, conspicuous, and showy. Ovary inferior. Fruits fleshy, globose to pear-shaped edible berries, crowned by the persistent sepals; seeds hard or bony. American tropics; 100 species.

1. Leaves often 6–15 cm, ovate to oblong-elliptic; fruits 2.5+ cm. ________ ***P. guajava***

1' Leaves often 2.5–5 cm, elliptic; fruits approximately 1.5 cm. ________ ***P. sartorianum***

***Psidium guajava*.** Guava; *guayabo; guayaguasi* (Guarijío); *gua'ibasim* (Mayo) (figure 209)

Multiple-trunked large shrubs or sometimes small, single-trunked trees to 10 m. Bark smooth, mottled greenish to coppery in color, peeling away in large, thin patches or plates. Twigs 4-angled, finely woolly. Leaves 6–15+ × 3–5 cm; blades ovate to oblong-elliptic, the upper surface finely woolly when young, soon glabrous or essentially so; margins entire.

Flowers solitary or in clusters of 2 or 3, white, approximately 3 cm wide (including the stamens). Calyx splitting vertically into lobes at or before flowering time. Fruits 2.5–3 cm, ovoid, the pulp sweet but tart, aromatic, and yellowish to pink. Seeds many. Flowering May–June.

Moist slopes and canyons; tropical deciduous forest and oak woodland; southeastern Sonora, 350–1370 m (1150–4500 ft). Naturalized in scattered places in southeastern Sonora including canyons on Sierra de Alamos; also cultivated in Sonora. Probably native in southern Mexico and elsewhere in the American tropics, widely cultivated and naturalized in tropical regions worldwide.

Guavas tolerate light frost but otherwise require tropical conditions.

***Psidium sartorianum*.** *Arrallán; choqué'i* (Gaurijio) (figure 210)

Slender evergreen trees 6–15 m, single or multiple trunked; trunk and major limbs often thick and mostly erect to ascending. Bark smooth, mottled gray, brown, and white, flaking in irregular patches or plates. Leaves 2.5–5 × 1.2–2.2 cm; blades broadly elliptic, shiny green, sparsely hairy along the midrib and petioles or glabrous or essentially so, the base cuneate, the leaf tips mostly obtuse; margins entire to slightly wavy.

Flowers 1 or 2 on long peduncles and pedicels, white, approximately 1.5 cm wide (including the stamens). Calyx limb circumscissile at flowering time, but remaining attached on one side. Fruits approximately 1.5 cm, nearly round, yellow-green with several seeds; flowering October–December.

Canyon bottoms and moist slopes; tropical deciduous forest; east-central and southeastern Sonora, 245–900 m (800–2950 ft). Irregularly scattered through the barranca country in eastern Sonora as far north as the Sierra San Javier east of Hermosillo, with extensive populations in some moist, tropical canyons. Also southwestern Chihuahua to Colombia and Venezuela and in the Caribbean Re-

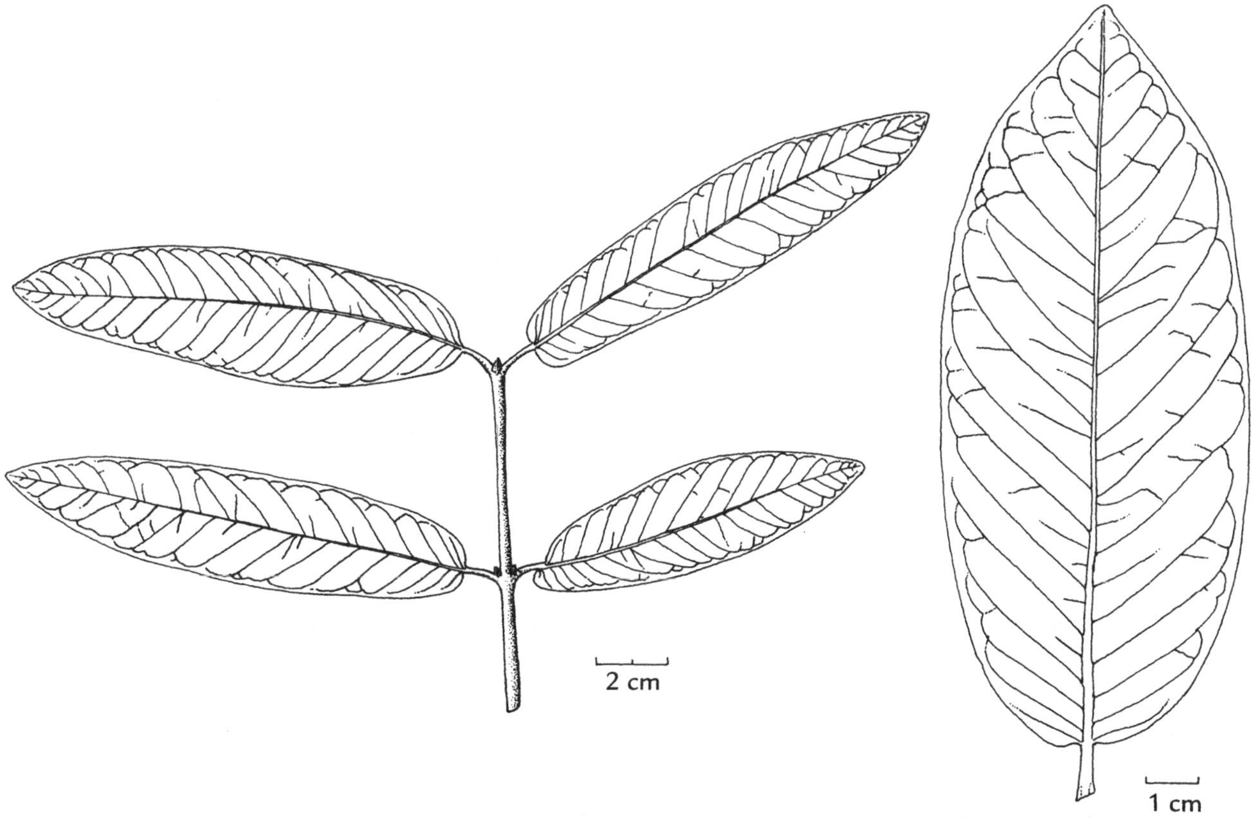


Figure 209. *Psidium guajava*. Drawn by M. B. Johnson.

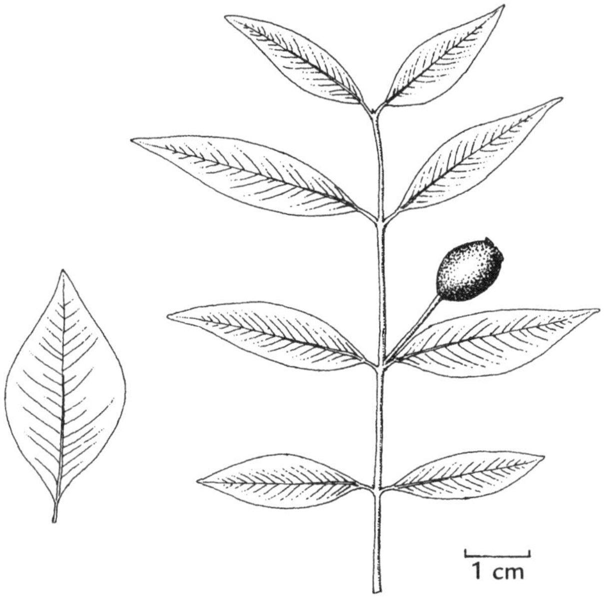


Figure 210. *Psidium sartorianum*. Drawn by M. B. Johnson.

gion. Native in southeastern Sonora such as Arroyo Verde near Santa Bárbara, but elsewhere perhaps not originally native, and sometimes seemingly persisting for many years after being planted.

The fruit, fresh or dried, is used in making refreshing drinks that are rich, spicy, and acidulous in flavor. The fruits are also used for conserves and sometimes added to *bacanora* (a local, agave-based tequilalike liquor) for flavoring. In late December people go to the hills to harvest the fruit. Substantial quantities are gathered in the hills near El Fuerte in northern Sinaloa and sold in the main market in Los Mochis. The fruit is well known in Jalisco, where it is sold in supermarkets as a sweet. The fruit pulp and seeds are rolled into round cakes. The seeds are not eaten. The bark is used for tanning, and the leaves are reported to have tonic and astringent properties.

## Nyctaginaceae—Four-O'Clock Family

Herbs, shrubs, trees, and vines. Tropical and warm temperate regions, mostly New World; 32 genera, 300 species. Many herbaceous and some shrubby members in Sonora.

Bougainvillea (*Bougainvillea* ×*buttiana*), with several varieties, is widely cultivated in Sonora for the brilliantly colored "flowers" (actually the bracts surrounding the flowers).

### *Pisonia*

Trees or shrubs, usually dioecious. New and Old World tropics; perhaps 40 species.

***Pisonia capitata.*** *Garabato, garambullo; susutí* (Guarijío); *baijuo* (Mayo); *vahewo* (Yaqui) (figure 211)

Scrambling shrubs to large or sometimes enormous scandent shrubs or trees; occasionally trees to 8 m with a single but generally irregular trunk reaching 4–5+ m and 6–55 cm in diameter; bark gray and smooth, with age often finely checkered. Branches dense and briarlike, decussate, rigid, and armed with vicious, single or paired, stout, recurved or sometimes straight thorns 1–3 cm, plus short shoots sometimes stoutly thorn tipped, or some stems unarmed. Herbage densely to sparsely hairy, becoming glabrate. Foliage drought deciduous in drier habitats or tardily drought deciduous to nearly evergreen in wetter habitats. Leaves 5–14 cm, simple, opposite, sometimes also alternate especially on new growth, and fascicled in short shoots, the opposite leaves in a pair often unequal in size; petioles prominent; blades broadly obovate to orbiculate, the base tapering onto the petiole, the margins usually entire or leaves of vegetative growth often minutely and shallowly toothed or wavy. Stipules absent.

Dioecious or rarely monoecious with unisexual flower clusters. Flowers small; sepals petal-like, 5-toothed; petals absent; ovary appearing inferior. Male flowers in compact, headlike cymose clusters (1) 1.5–2.5 cm wide, fragrant; calyx red-purple; stamens 5, cream colored. Female flowers smaller than male flowers, dark red-purple, in compact cymes expanding to become open or loose with slender branches as the fruits develop. Fruiting pedicels slender, often longer than the fruits. Fruits 7–13 mm, club shaped, 1-seeded, 5-ribbed, with sticky, thick-stalked glandular hairs along the ribs causing the fruits to adhere to feathers, fur, and clothing. Flowering January–April.

Canyons, arroyos, and floodplains; tropical deciduous forest, thornscrub, and occasional in Sonoran desertscrub; southern Sonora north to the Guaymas-San Carlos region and the Río Yaqui region near Lo de Campa, 5–760 m (15–2500 ft). Also southwestern Chihuahua (e.g., Urique, in the Barranca del Cobre) to Nayarit, and a small, disjunct population in mountains northwest of Tucson in southern Arizona. Common in densely vegetated arroyos (washes) in coastal thornscrub as far north as the vicinity of Vicam and also along arroyos and riparian canyon bottoms in the Guaymas region.

Although tree-sized *P. capitata* in Sonora are far less common than the shrubs, the larger plants are noteworthy. A collection made by Howard Scott Gentry in 1938 from Arroyo Gochico, in the Río Mayo region, is described on an herbarium label (University of Arizona) as a "tree, erect or

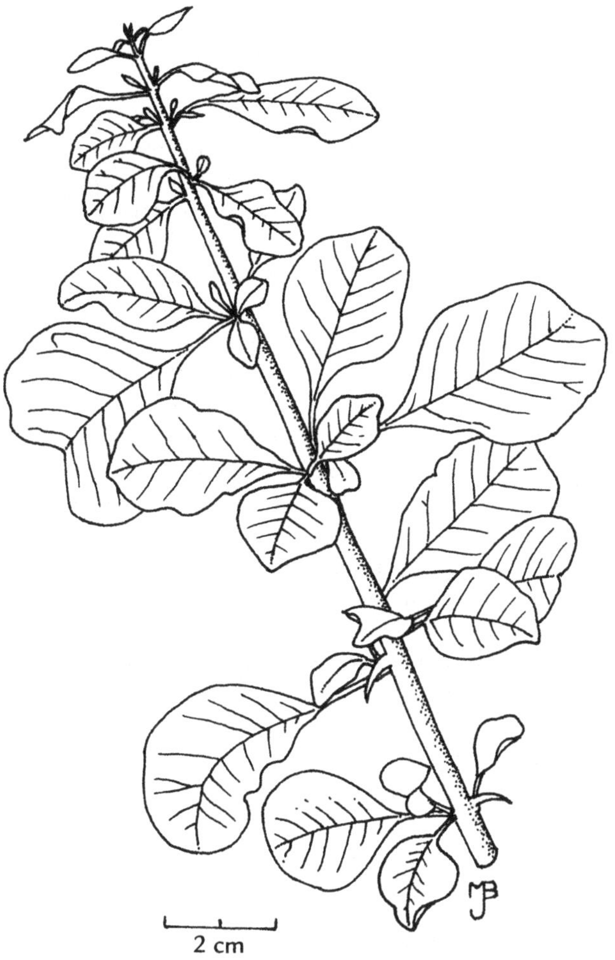


Figure 211. *Pisonia capitata*. Drawn by M. B. Johnson.

scandent with a hundred foot stem on other trees." Similar plants also occur in other riparian canyons within tropical deciduous forest in southeastern Sonora. Trees with well-formed trunks are fairly common only in certain areas of the Guarijío country north of Alamos. A few heavy-trunked trees are also found in low-lying coastal thornscrub at Camahuiroa in extreme southwest Sonora. These trees offer dense, welcome shade, but local Mayo people caution that such trees are also lairs for *pichicuates*, or cottonmouths (*Agkistrodon bilineatus*), which feed on the packrats that make their nests among the overhead briars.

Occasional yellow-flowered plants are found in southeastern Sonora. They do not appear different except in flower color, although among some the male flowers are in loose, open, paniculate clusters 3–6 cm across. These plants, with yellow flowers and sometimes larger inflorescences might be identified as *P. aculeata*, but it is doubtful that more than one species occurs in Sonora.

## Olacaceae—Olax Family

Tropical regions of the world, few extending into subtropical regions, and fewer in arid regions; 27 genera, 180 species. Reference: Sleumer (1984).

*Schoepfia* and *Ximenia* represent the northernmost limits of the family in the New World—in Sonora and the Baja California peninsula on the Pacific side, and Florida on the Atlantic side. Apart from *Schoepfia*, the spinescent shrub *Ximenia parviflora* var. *glauca* is the only other member of this family in Sonora. The only record for *Ximenia* in Sonora is the Sierra Libre south of Hermosillo.

Most or perhaps all members of the family are semiparasitic, obtaining water and minerals from host plants by means of root haustoria. The biology of the arid-adapted members of the family has scarcely been investigated. The family is allied with the Opiliaceae (see below), Santalaceae (sandalwoods), mistletoes (Loranthaceae and Viscaceae), and others in the order Santalales in which various modes of parasitism are almost universal.

### *Schoepfia*

Glabrous shrubs or trees, unarmed or with weakly spinescent twigs; most or all are root parasites (e.g., documented for *S. schreberi*). Leaves drought deciduous (in Sonora), alternate, simple, entire, somewhat succulent or thick, and with pinnate veins; stipules absent.

Flowers sessile, one to several at the top of slender peduncles, these reduced spikes one to several in leaf axils and shorter than the leaves. Flowers small, yellow, pink, or red-orange, and sweet scented. Each flower subtended by a cuplike epicalyx with three shallow teeth fringed at their tips with minute ciliate hairs. (The epicalyx is easily mistaken for the calyx. The epicalyx, composed of the fusion of one bract and two bractlets, is persistent on the fruit.) Calyx practically absent, reduced to a narrow green disk at the base of the corolla. Corollas with a short tube, 4 (perhaps sometimes 5) short, reflexed lobes, and tufts of small, bristly hairs inside the tube behind each anther. (The hairs project over each anther, and may serve to protect the anthers and "guard" the entrance to the corolla tube.) Stamens the same number as, and opposite the corolla lobes; filaments completely fused to the corolla tube, the anthers near the mouth of the corolla tube. Ovary inferior; style 1; stigma capitate. Fruit a 1-seeded, fleshy drupe; calyx or floral axis tissue becoming brightly colored, fleshy, and growing over most of the fruit.

New and Old World tropics; 23 species, 19 in the Americas. Three species in northwestern Mexico, one restricted to the Baja California peninsula, and two in Sonora.

All three species may have heterostylous flowers. In the short-styled flowers the style and stigma almost reach the base of the anthers, the stigma is reduced in size, and the anthers contain pollen. In the long-styled flowers the style and stigma extend slightly above the anthers, the stigma is somewhat enlarged, and the anthers are without pollen. Probably various intermediate conditions are present, but the actual condition in the field is poorly understood. Sleumer (1984) indicates the heterostyly may not be related to a strictly dioecious condition because both forms may bear fruits.

1. Leaves 14–22 mm wide, widest at or below middle; flowers yellow to pink or red-orange. ______ ***S. schreberi***
1' Leaves 4–8 mm wide, widest above middle; flowers red-orange. ______ ***S. shreveana***

***Schoepfia schreberi*** [*S. parvifolia*]. *Palo cachorra, pimentilla; cuta béjori* (Mayo) (figure 212A)

Slender trees 6–8 m, the trunk usually well formed; bark corky, thick, deeply furrowed, scaly, tan colored on larger limbs and trunks, and checkered on smaller branches; twigs brittle, soon becoming corky. Roots of this species reported to have white haustoria to 3 mm wide. Leaves 25–45 × 14–22 mm, rather firm, moderately thick, cracking when bent; blades elliptic to oval or ovate, tapering at the base and decurrent on the short petiole or subsessile, the tip obtuse to rounded, often somewhat crinkled. (Leaf margins often insect chewed.)

Peduncles 6–13 mm, 1–4 per node, axillary among the leaves or from nodes below the leaves. Flowers yellow to red-orange, in pairs (like twins, or sometimes 1 or 3), 7–8 mm, fleshy: calyx 4-toothed, 1.5–2 mm, green, the corollas yellow to red-orange, the tube more than twice as long as the calyx. "Fruits" 10 mm, red to purple. Flowering December–March; fruit March–May.

Slopes, valley plains, canyons, and arroyos; tropical deciduous forest; southeastern Sonora, 245–600 m (800–1970 ft). Sonora to northern South America and the Caribbean including southern Florida.

The Sonoran populations have been known as *S. parvifolia*, originally described from southern Mexico and characterized by relatively small leaves. Sleumer's (1984) broad interpretation of *S. schreberi* includes these smaller-leaved plants as well as populations in wetter, more tropical regions with much larger leaves. Although there is considerable variation in leaf size, there is relatively little variation in the flowers and fruits.

***Schoepfia shreveana***. *Palo verde; juia tóluco, júchica* (Mayo) (figure 212B)

Slender shrubs or trees 4–5 (6) m. Bark corky, deeply fissured and checkered on lower limbs and trunks, smooth on smaller branches, tan colored. Twigs brittle, sometimes moderately spinescent at tips. Leaves evergreen to tardily drought deciduous, 20–45 × 4–8 mm, linear-cuneate to oblanceolate, glaucous, gradually tapering to a short petiole or subsessile.

Flowers bright red-orange, 8 mm, single or paired on peduncles 5–6 mm. "Fruits" 10 mm, usually solitary on peduncles reaching 10 mm, red to purple. Flowering and fruiting various seasons except late spring–summer drought.

Dense Sonoran desertscrub and thornscrub; low areas with fine-textured, often alkaline, alluvial soils. Known only from western Sonora but expected in northwestern Sinaloa: north of Guaymas (Caleta Venecia at 28°08' N, Bahía Algodones, and San Carlos), often just inland from the zone of mangroves and *Maytenus*, and the coastal plain from the vicinity of Empalme southward nearly to the Sinaloa border in the vicinity of Camahuiroa; below about 20 m elevation.

The limited distribution of this species places it at risk: much of its habitat near the shore north of Guaymas has been developed for resorts, and from Guaymas southward the coastal plain has been or is threatened with urbanization and conversion to agriculture.

## Oleaceae—Olive Family

Temperate and tropical regions of both hemispheres, greatest diversity in the Orient; 30 genera, 600 species.

The olive (*Olea europaea*) thrives under cultivation in most of Sonora, especially in the northern part of the state, and there are commercial groves near Caborca. Lilac (*Syringa*) is grown in the northern part of the state above the desert, such as at Nogales, Cananea, and Agua Prieta, and privet (*Ligustrum*) is a common ornamental in the nontropical regions of the state.

Devilwood (*Osmanthus americanus*) is known in Sonora from several arroyos in the Yécora area and Arroyo Verde near Santa Bárbara, where it is a shrub to 3.5 m. Elsewhere in Mexico it is often a tree to 15 m in humid, semitropical to tropical regions. An apparently undescribed species of *Forestiera* from Sierra de Alamos in southeastern Sonora sometimes approaches tree size; it flowers in September. Several other *Forestiera* species in Sonora are woody shrubs that occasionally approach tree size. Two species of *Menodora*, both subshrubs, are the only other Sonoran members of the family.

1. Leaves compound; fruits dry (samaras). ___ ***Fraxinus***
1' Leaves simple; fruits fleshy (drupes).
   2. Flowers without petals; fruits longer than wide. ***Forestiera*** spp. (not treated in text)
   2' Flowers with petals, the corolla tubular; fruits rounded. ______ ***Osmanthus americanus*** (not treated in text)

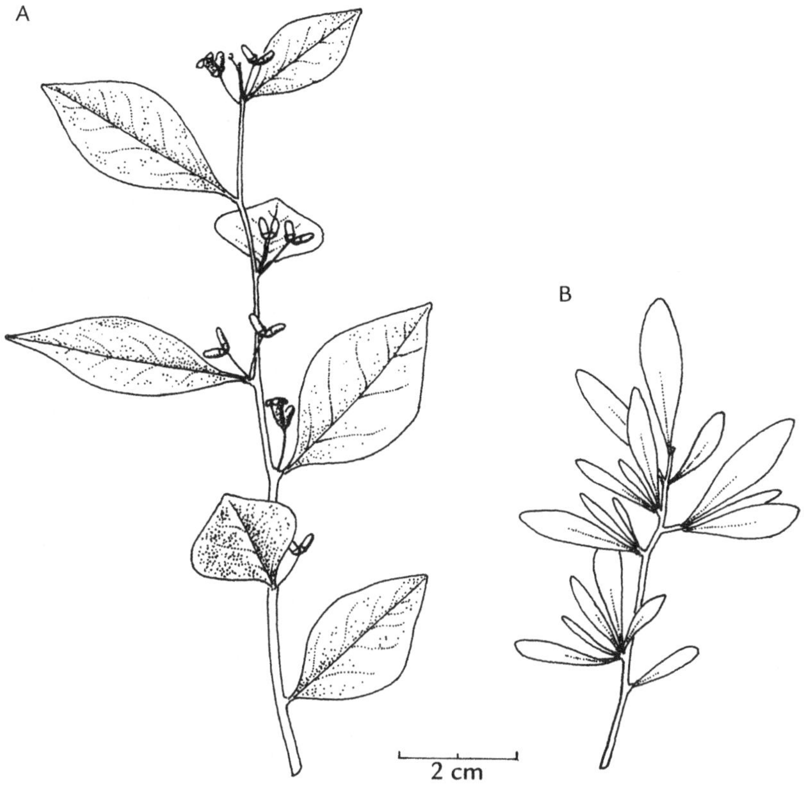


Figure 212. *Schoepfia*. (A) *S. schreberi*, (B) *S. shreveana*. Drawn by M. B. Johnson.

### *Fraxinus*—Ash, *Fresno*

Trees, occasionally shrubs, with winter or dormant buds enclosed by 1 or 2 pairs of scales. Branches with conspicuous lenticels. Leaves and twigs opposite and decussate; leaves odd-pinnate with well-developed petioles; stipules absent.

Usually dioecious, or some flowers bisexual. Flowers small, wind pollinated, in dense to somewhat loose clusters; calyx cup-shaped, green, with 4 or more somewhat irregular lobes; petals absent (in Sonora). Male flowers with 2 stamens bearing large, yellow anthers protruding well beyond the calyx cup; ovary vestigial. Female flowers green, with an elongated ovary and style and 2 short-branched stigmas, without stamens. Fruits of single samaras in dense, pendulous clusters, each with an elongated, flat, firm, terminal wing.

North temperate America, Eurasia, and tropical Asia; 65 species. References: Little (1952), Miller (1955).

1. Shrubs or small trees with small, slender trunks; leaves < 10 cm, the larger lateral leaflets mostly < 3.5 cm; stamens with well-developed filaments; anthers 1.5–2.0 mm; samaras 12–16 mm. ___________ ***F. gooddingii***

1' Small to large trees with well developed trunks; leaves usually > 10 cm, the larger lateral leaflets more than 4 cm; stamens sessile or subsessile; anthers approximately 3.5 mm; samaras > 25 mm.

2. Leaflets whitish-glaucous below. ___ ***F. papillosa***

2' Leaflets green below. _______________ ***F. velutina***

***Fraxinus gooddingii.*** Goodding ash, littleleaf ash; *fresnillo* (figure 213)

Usually shrubby, often a small, multiple-trunked tree 7–8 m, with an upright growth habit and slender twigs. Bark smooth and gray, becoming checkered on the trunks. New growth and inflorescences woolly with short, often golden brown to white stellate, glandular hairs; herbage glabrate with age. Often evergreen, or the leaves yellowing and falling in early spring at flowering time, and new leaves appearing with the flowers or later in spring, or sometimes winter deciduous during cold weather. Leaves 4.5–9.5 cm, the rachis sometimes narrowly winged; leaflets 5–9, the margins entire to serrated, the tips acute to obtuse; larger lateral leaflets 1.5–3.2 cm, broadly lanceolate to oval, slightly thickened, the terminal leaflet longer than the lateral leaflets.

Flowers inconspicuous, sometimes bisexual or appearing bisexual. Calyx approximately 1 mm with 4 or more ragged lobes; male and bisexual flowers with well-developed filaments, the anthers 1.5–2.0 mm. Samaras 12–16 mm. Flowering February–May, fruits May–June.

Rocky slopes and canyons; upper elevations in Sonoran desertscrub, grassland, and oak woodland. Widely scattered in northeastern and north-central Sonora, 855–1550 m (2800–5085 ft). Usually on dry, rocky, rhyolite slopes, and sometimes also along dry arroyos and canyons. Also adjacent southeastern Arizona and Chihuahua.

Unlike other ash trees in the region, it is not a riparian tree. It is recommended for cultivation as a small tree for patios and gardens; the moderately open canopy provides light shade. At lower elevations it would undoubtedly require supplemental water. The similar and closely related *F. greggii* occurs in west Texas, Coahuila, and Nuevo León.

***Fraxinus papillosa.*** Chihuahua ash; *fresno*

Trees 5–20 m. Resembling *F. velutina* but distinguished by

> the papillose, glaucous or whitish under surface of the leaflets. . . . These minute papillae, though not shown by a hand lens, are revealed by a high power (40×) dissecting microscope or by low power (50× to 100×) of a compound microscope. . . . Under high magnification the lower surface appears as a solid mass of whitish beads or sleetlike particles, the minute opaque papillae of each epidermal cell. . . . In *F. velutina* the greenish lower surface varies from hairy to glabrous and also may bear many minute peltate scales (characteristic of the family Oleaceae) and often numerous irregular or scurfy papillae not in a solid sleetlike mass. Specimens of *F. papillosa* generally can be recognized by the whitish, mostly glabrous, under surface of the leaflets (Little 1952: 377).

However, in specimens from Sonora the lower leaf surface is hairy rather than glabrous.

Slopes and arroyos; grassland, oak woodland, pine-oak woodland, and mixed conifer forest; northeastern Sonora, 1310–1980 m (4300–6500 ft).

In northeastern Sonora, such as in the Sierra de los Ajos and the Sierra San Luis including Cajón Bonito, trees that might be assigned to *F. papillosa* occur at higher elevations than does *F. velutina*. The *F. papillosa*-like trees grow in riparian canyon-bottom forests, and at higher elevations such as in mixed conifer forest, often on north- and east-mountain slopes. These *papillosa*-like plants tend to have smaller leaves and are often small trees about 5 m, although occasional canyon-bottom trees reach 15–20 m in height. *Fraxinus papillosa* is reported from Sonora and southeastern Arizona to Chihuahua, southwestern New Mexico, and west Texas. This distribution is wholly within the range of *F. velutina*. Although most authors treat them as two distinct species, Standley (1924: 1136) expressed the opinion that they "would have to be united ultimately" and Powell (1988) likewise casts doubt on their distinctiveness. We find the distinctions confusing and likewise questionable. *Fraxinus papillosa* is said to be the western counterpart of *F. americana*, and *F. velutina* the western counterpart of *F. pennsylvanica*.

Figure 213. *Fraxinus gooddingii*. Drawn by M. B. Johnson.

***Fraxinus velutina*** [*F. pennsylvanica* subsp. *velutina*]. Velvet ash, *fresno*; *bitoi* (Tohono O'odham) (figure 214)

Trees (5) 10–25 m, often broad- and high-crowned. Bark gray, with age deeply furrowed into many forking ridges. Winter buds and youngest shoots in spring with two pairs of glandular-pubescent scales, the outer pair brown, the inner pair larger and green. Young herbage densely hairy, the leaves sometimes becoming glabrate with age. Leaves winter deciduous, turning golden yellow in fall, mostly

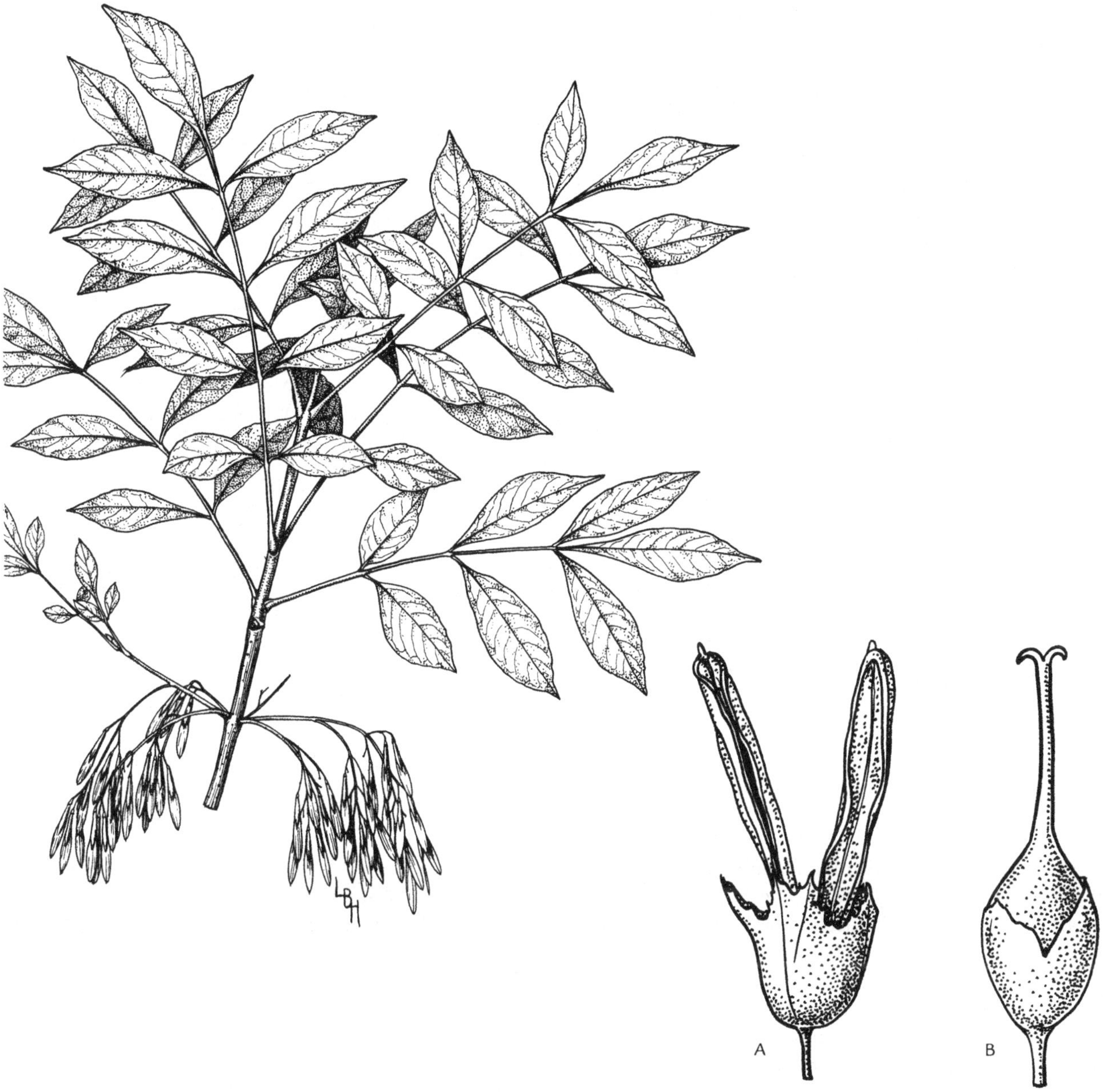


Figure 214. *Fraxinus velutina*. (A) Male flower with exserted anthers, (B) female flower (Benson and Darrow 1945). Drawn by L. B. Hamilton.

10.5–21 cm, the rachis occasionally narrowly winged; leaflets (3) 5–9, the tips mostly acute, the margins minutely toothed to nearly entire; larger leaflets 4–10 cm.

Dioecious. Mass flowering before leaves emerge, and sometimes sporadically with the leaves during summer and early fall. Flowers inconspicuous. Male flowers with the calyx ≤ 0.5 mm, radial; stamens 2 or 3; anthers subsessile, approximately 3.5 mm. Female inflorescence buds covered by 2 pairs of bud scales, the outer pair very reduced; female flowers with calyx approximately 1 mm, bilaterally symmetrical with 4 ragged-edged lobes; ovary green; styles and stigmas yellow-green. Samaras 26–41 mm. Flowering February–April; fruits summer and fall.

Canyon bottoms and streams; Sonoran desertscrub, Chihuahuan desertscrub, grassland, oak woodland, pine-oak woodland, and pine forest; north-central, northeastern, east-central and perhaps southeastern Sonora, 760–2340 m (2500–7680 ft). Common riparian tree in northeastern Sonora and much of Arizona; usually along streams with running water or watercourses with a very shallow water table. Widely cultivated in Sonora and Arizona as a large, spreading shade tree. Widespread in southwestern United States and northern Mexico.

Velvet ash shows considerable variation, especially in leaf characters such as presence or absence of hairs, leaflet thickness, and fine structure of the lower leaflet surfaces, which has led to the description of a number of varieties. The general opinion is that the varieties do not seem to be worthy of taxonomic recognition. Velvet ash also has been treated as a subspecies of the widespread green ash (*F. pennsylvanica*) of temperate North America.

## Opiliaceae—Opilia Family

Trees and shrubs; semiparasitic, attaching to host plants by means of root haustoria. Tropical regions worldwide; 10 genera, 32 species. Reference: Standley (1920).

### *Agonandra*

American tropics; 10 species.

This is the only New World genus in the family. The one species in Sonora represents the northern limit of the family in the New World.

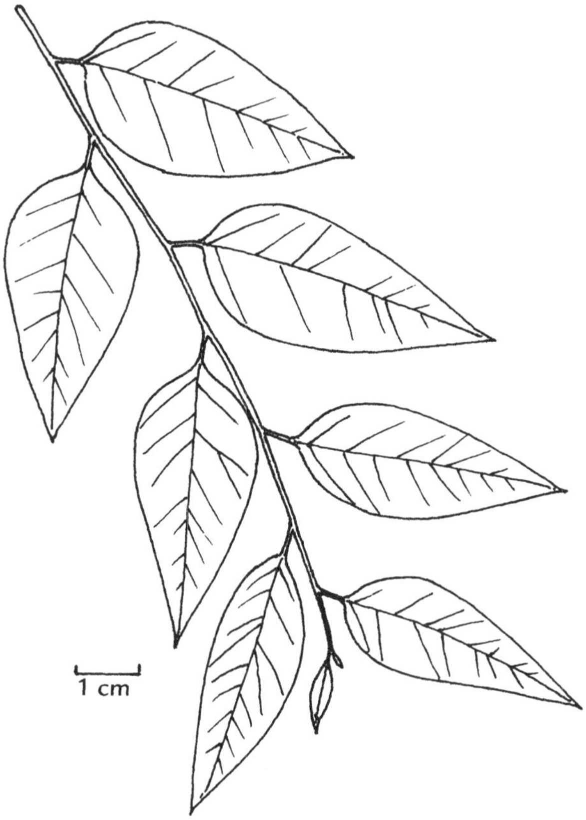


Figure 215. *Agonandra racemosa*. Drawn by M. B. Johnson.

***Agonandra racemosa.*** *Palo verde, mata chamaco; murió* (Guarijío); *úsim yuera* (Mayo) (figure 215)

Glabrous shrubs or slender trees to 12 m; trunks often 14–20 cm in diameter, the trunk and limbs erect-ascending, the branches slender, brittle, drooping, leafy, and lacy; bark dark brown with warty lenticels, otherwise mostly smooth or with age sometimes checkered. Leaves alternate, simple, 4–9 cm, slightly succulent, firm, shiny green to often slightly glaucous, petioled; blades narrowly to broadly elliptic, lanceolate, ovate, or rounded, the lateral veins few, irregular, and poorly defined, the apex acute to rounded with an apiculate tip, the margins entire; stipules absent.

Dioecious. Inflorescences of axillary and terminal racemes or sometimes panicles, shorter or longer than the leaves, the buds covered by small bracts falling before the flowers open; flowers pedicelled, yellow-green. Female flowers 1.5 mm, yellow-green except for the cream-colored stigma, the sepals irregular, appressed, 0.3–0.7 mm; stigma sessile, without petals or stamens. Male flowers approximately 5 mm wide; calyx reduced to a minute green disk; petals 4 or 5, spreading, 2.0–2.5 mm; stamens 4 or 5, exserted, alternating with 4 or 5 prominent, fleshy, glandlike scales. Fruits fleshy drupes 10.0–13.5 mm, rounded, greenish white. Flowering December–May.

Slopes and canyons in foothills and mountains; coastal thornscrub and tropical deciduous forest; southern and east-central Sonora, 15–915 m (50–3000 ft). Northward to the vicinity of San Javier, and coastal thornscrub in the vicinity of Camahuiroa in extreme southwestern Sonora. The wood is used to make corrals. Magpie jays greedily feed on the ripe fruits. Sonora and southwestern Chihuahua to Central America.

## Platanaceae—Sycamore or Plane Tree Family

Single genus; 8 species of trees. North temperate regions of the world. Reference: Kaul (1997).

The London plane is a hybrid tree widely planted along avenues in temperate regions of the world.

***Platanus wrightii*** [*P. racemosa* var. *wrightii*]. Arizona sycamore; *aliso* (figure 216)

Trees often 10–20 m, perhaps occasionally taller; older trees often spreading; trunk often massive, gnarled, and hollow, sometimes several-trunked, usually irregular and often leaning. Bark whitish or greenish beneath the outer bark, which flakes off in irregular, thin plates; bark of older portion of trunk becoming darker, fissured, and not flak-

Figure 216. *Platanus wrightii.* (A) Stamen, (B) pistil (Benson and Darrow 1945). Drawn by L. B. Hamilton.

ing. Herbage and flower stalks with branched (dendritic) hairs, generally dense and woolly, becoming sparsely pubescent or glabrous with age, especially the upper leaf surfaces. Leaves alternate, 24–38 × 18–36 cm, simple, palmately lobed and veined with (3) 5 (7) lobes cleft more than halfway and pointed at the tips, the margins irregularly toothed to entire; petioles long, their bases enlarged and covering the axillary buds. Stipules conspicuous, on rapid-growing twigs often united at the base and clasping the stem, the margins coarsely toothed to sometimes entire.

Monoecious. Inflorescences axillary, unisexual, flowers small, wind pollinated, in dense, many-flowered globose heads, the flower bases with bristly tufts of irritating, golden brown hairs; sepals and petals mostly inconspicuous or reduced, or soon deciduous. Male inflorescences shorter than female inflorescences and soon deciduous; sepals several; petals greatly reduced; anthers several, forming the conspicuous part of the flower, approximately 2.5 mm, subsessile, the tips with a terminal glandular tuft. Female inflorescences with a main axis 8–30 cm, pendulous, with 2–8 racemosely arranged heads, or some or all of the heads sometimes sessile especially in southeastern Sonora and northern Sinaloa, the flowering heads 14–20 mm in diameter; flowers with 3 or 4 sepals, no petals, 3 or 4 staminodes; stigmas elongated to 3+ mm, red-purple, and forming the conspicuous part of the flower. Fruiting heads 2.5 cm in diameter, the individual fruits of achenes 6.0–7.5 mm with persistent stigma bases, 4-angled, 1-seeded, surrounded by and falling with a ringlike tuft of bristly hairs; fruiting heads break apart at maturity and are wind dispersed. Flowering spring; fruits ripe in the same season and through the summer.

Riparian canyon bottoms; Chihuahuan desertscrub, grassland, oak woodland, pine-oak woodland and upper elevations in Sonoran desertscrub; north-central and eastern Sonora, 335–1830 m (1100–6000 ft). Central and southeastern Arizona, southwestern New Mexico to Chihuahua and Sinaloa.

The Sinaloa trees often have sessile flower heads, a character of *P. racemosa*, but the leaf segments are relatively long and slender like those of *P. wrightii* in Arizona and Sonora rather than broader and blunter like those of *P. racemosa* of California and the Baja California peninsula.

The autumn leaves turn golden to coppery colored, contrasting with the white bark and blue sky. The large, dry, crackly leaves often carpet boulder-strewn streambeds. Honeybees and various birds such as the elegant trogon nest in the hollow cavities in the trunks and larger limbs. Arizona sycamore sometimes occurs in pure stands but is more commonly associated with other riparian trees, forming gallery forests near permanent streams. In the northern part of the state it often grows with cottonwood

(*Populus*), willow (*Salix*), and ash (*Fraxinus*). In the Sierra Madre Occidental of southeastern Sonora, it commonly grows with *Cinnamomum*, *Magnolia*, *Salix bonplandiana*, and other trees.

## Polygonaceae—Buckwheat Family

Annual and perennial herbs, vines, and shrubs to large tropical trees. Leaves simple, alternate (in Sonoran trees), petioled; blade with secondary venation. Stems often swollen at nodes with fused stipules enclosing (sheathing) the stem.

Plants sometimes dioecious. Flowers mostly small and bisexual, or sometimes unisexual. Calyx in 1 or 2 series, often petal-like; petals absent. Fruits 1-seeded, an achene. The stipule, unique among the trees in Sonora and called an ochrea, is a tubular sheath that forms a cap enclosing the developing bud. On its inner surface, especially toward the base, are numerous minute hairs that secrete mucilage which may protect the bud from drying out (Tomlinson 1980). When the ochrea falls it leaves a scar encircling the stem. Similar, tiny ochrea occur below the bractlets subtending the individual flower clusters on the inflorescence rachis.

Mostly North Temperate regions in both hemispheres, also extending into the tropics, and well developed in many desert or semiarid temperate regions; 46 genera, 1100 species.

This family includes many herbs and some shrubs in Sonora including *Eriogonum*, *Polygonum*, and *Rumex*. Queen's wreath or *San Miguelito* (*Antigonon leptopus*) is a native vine widely cultivated in Sonora. Polygon trees, represented by about seven genera, are almost entirely confined to the New World tropics. Two genera of polygon trees reach their northernmost limits in Sonora, and the species are rare in the state.

1. Leaves more or less orbicular, about as wide as long; fruits rounded. ______ ***Coccoloba***

1' Leaves more or less elliptic, clearly longer than wide; fruits 3-angled, longer than wide. ______ ***Ruprechtia***

### *Coccoloba*—Sea Grape

Dioecious trees, shrubs, and vines. Tropical and subtropical America; 125 species. Reference: Howard (1959).

Only a few species of Mexico and Central America have heavily coriaceous (tough) leaves like those of *C. goldmanii*.

#### ***Coccoloba goldmanii***. Sonoran sea-grape (figure 217)

Usually multiple-stemmed woody shrubs, occasionally small trees 3–5+ m. Herbage and inflorescences pilose with short hairs, the stems becoming glabrous, the upper leaf blade surfaces glabrous or nearly so with age, the lower surfaces persistently pubescent especially along the major veins. Essentially evergreen, or sometimes nearly leafless during extreme drought. Leaves mostly 6–19 cm long; petioles 0.7–2.2 cm long; blades mostly 5.5–17 cm wide, orbicular, often slightly broader than long, less often with some broadly elliptic or oval leaves, relatively firm and tough, the lower surfaces with a conspicuous network of veins, the margins entire although usually slightly wavy.

Flowers many on flowerstalks (peduncle and raceme) 7–30.5 cm, slender, pendent, and appearing terminal (actually in the axil of the forthcoming leaf). Flowers small, white, and star shaped, subtended by small bracts, the pedicels often 2 mm; sepals 5. Male flowers in clusters of usually three, 4–5 mm wide, with a prominent disk (hypanthium); calyx greenish white; lobes translucent, at first cup shaped and then reflexed after the pollen is shed; stamens (including anthers and pollen) and pistils (presumably sterile) white; stamens (7) 8, 2.5 mm, conspicuously exserted, with glistening nectar droplets produced on the disk between the filament bases and the ovary (at least during the day). Female flowers solitary, presumably with the stamens not exserted and not producing pollen; styles 3. Fruits 6–7 mm, globose (the apex not narrowed) and hard. Flowering June–July, fruiting October–December.

Canyons; riparian Sonoran desertscrub, thornscrub, and tropical deciduous forest; west-central and southeastern Sonora, 10–400 m (30–1310 ft). In Sonora occurs in several riparian canyons north of Bahía San Carlos (e.g., Cañón de Nacapule and Cañón las Barajitas) and is known from a collection in mountains southeast of Ciudad Obregón, and another from Arroyo Gochico east of San Bernardo. Otherwise known from the type collection in southwestern Chihuahua and a number of widely scattered localities in Sinaloa.

This species seems most closely related to *C. barbadensis*, a widespread species of Central America and Mexico as far north as Culiacán, Sinaloa. It differs from *C. goldmanii* by its narrower, elliptic leaf blades, nonpersistent pubescence, and larger and more elongated fruits with the tips narrowed into a broad, tapering beak.

### *Ruprechtia*

Dioecious shrubs and trees; American tropics, 20 species, mostly in South America. References: Cocucci (1961, 1997).

The largest polygon trees are species of *Ruprechtia* and the closely related *Triplaris*, which reach 20+ m in Central and South America.

#### ***Ruprechtia fusca*** [*R. occidentalis*] (figure 218)

Trees 7–9 m, usually with several trunks; bark light colored. Young herbage sparsely pubescent, soon glabrous;

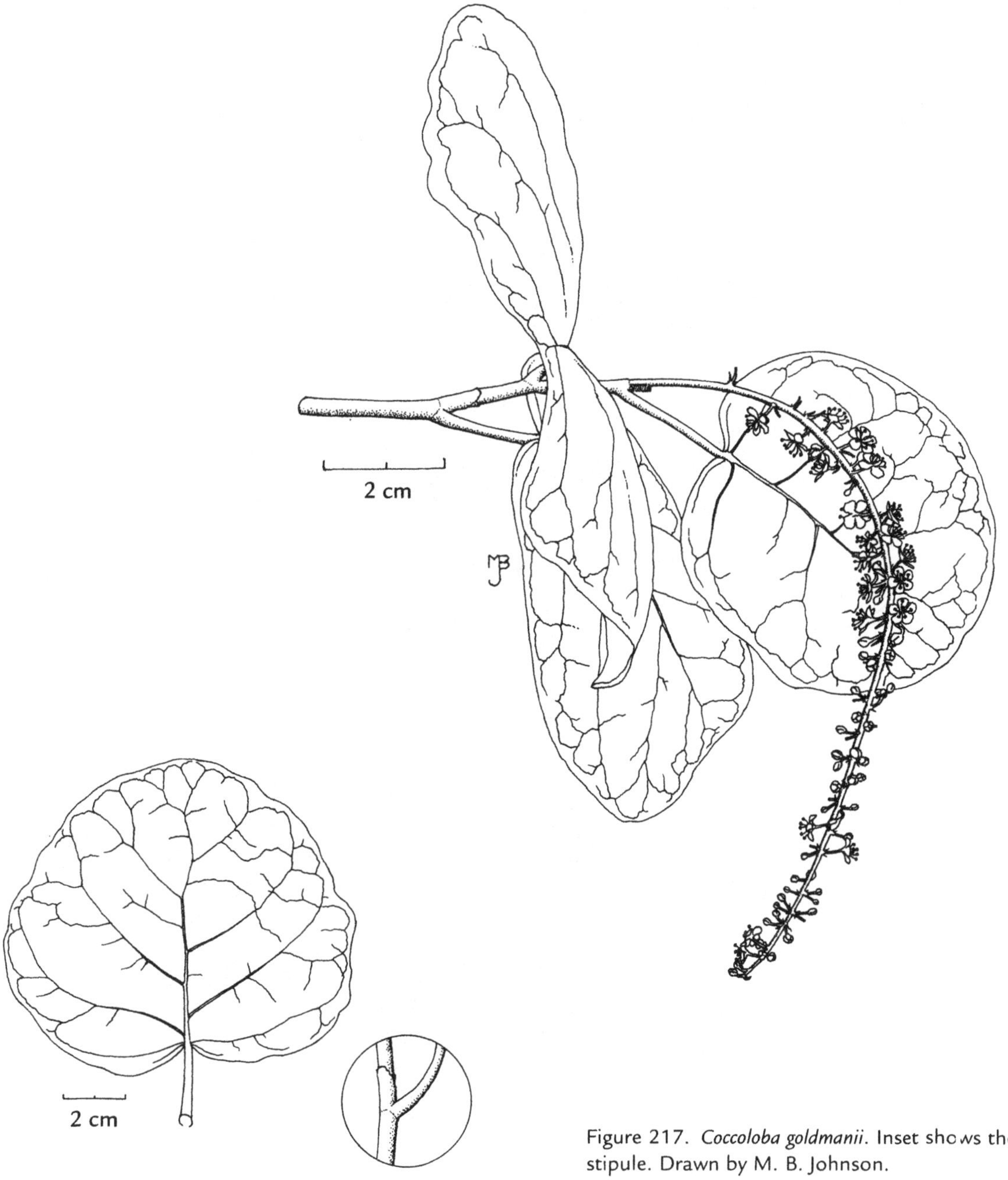


Figure 217. *Coccoloba goldmanii*. Inset shows the clasping stipule. Drawn by M. B. Johnson.

lower leaf surfaces, especially along the midrib, generally glabrate or sparsely pubescent. Drought deciduous. Leaves 4.8–9.5 × 1.3–4 cm, narrowly to sometimes broadly elliptic or ovate-elliptic, the margins entire and slightly wavy, the tip mostly obtuse; petioles short, to approximately 5 mm; blades somewhat firm, pinnate-veined with 10–14 pairs of lateral veins, the veins conspicuously raised below with a reticulate pattern, the midrib prominent below.

Inflorescences axillary, the axes slender, generally shorter than the leaves, 1–9 cm, often several at nodes, pubescent. Flowers pedicellate. Male inflorescences commonly in clusters of several racemes or paniculate with few, short branches; flowers rather inconspicuous, 1.5–2.5 mm; stamens exserted. Female inflorescences showy, generally paniculate and densely flowered; pedicels 3–7 mm, slender, appearing jointed about midway; calyx papery, with an

Figure 218. *Ruprechtia fusca*. (A) Branch with fruits, (B) female flower, (C) male flowers. (Pennington and Sarukhán 1998). Drawn by E. Esparza.

inner and outer whorl each with 3 segments fused below, the outer whorl pink, moderately pubescent especially near the base, enlarging in fruit to 20–28 mm, the segments straight to moderately spreading, narrowly oblanceolate, 4–6 mm wide, the tips rounded, the inner calyx segments inconspicuous, 5–7 mm. Achenes to 12 mm, 3-angled, protected by the persistent calyx; styles and stigmas 3 and persistent. Flowering midwinter.

Rare in Sonora; along arroyos in coastal thornscrub near Agiabampo in extreme southwestern Sonora at about 10 m elevation. More common in Sinaloa; southward in western Mexico to Oaxaca.

## Rhamnaceae—Buckthorn Family

Mostly hardwood trees and shrubs, some woody vines and rarely herbaceous perennials. Leaves alternate or opposite, simple, pinnately veined or several main veins from base, or leaves sometimes reduced; stipules usually small or modified into spines. Inflorescences of small axillary clusters shorter than the leaves. Flowers pedicelled, small, greenish to yellow or white, usually with a short floral cup (hypanthium) resembling a calyx. Sepals united at base, the lobes 4 or 5, usually triangular and quickly deciduous. Petals 4 or 5, or sometimes absent. Fruits drupes or capsules. Mostly tropical and subtropical but extending into desert and temperate regions; 55 genera, 900 species. Sonoran genera are mostly of tropical origin. References: Brizicky (1964b), Fernández Nava (1996).

1. Leaves mostly opposite or subopposite and often with some alternate ones, not fascicled; veins on underside of leaves dotted and dashed with blackish glands. ______ ***Karwinskia***
1' Leaves alternate or fascicled; veins uniformly colored, not glandular as above.
   2. Leaves < 2.5 cm, widest well above middle; petals none; fruits 3–4.5 mm. ______ ***Condalia***
   2' Leaves > 3 cm, widest at or below middle; petals present but often deciduous; fruits ≥ 5 mm in length.
      3. Leaf veins conspicuously pinnate; oak and pine-oak woodland. ______ ***Rhamnus***
      3' Leaves with 3 prominent veins arising at base of blade; desert, thornscrub, and tropical deciduous forest.
         4. Ovary partially inferior; fruits 5–6 mm, dry and dehiscent, 3-seeded. ______ ***Colubrina***
         4' Ovary superior; fruits 8–10 mm, fleshy and indehiscent, 1-seeded. ______ ***Ziziphus***

### *Colubrina*—Snakewood

Warm parts of the world; 31 species. Three species in Sonora. Reference: Johnston (1971).

*Colubrina californica* and *C. viridis* are hardwood shrubs in western Sonora or the dryer parts of the tropical deciduous forest and rarely approach tree size.

1. Leaves often > 5 cm, the margins serrated; plants unarmed. ______ ***C. triflora***
1. Leaves mostly < 5 cm, the margins entire; twigs often spinescent tipped.
   2. Leaves dull gray-green, conspicuously pubescent. ***C. californica*** (not treated in text)
   2' Leaves bright green, glabrous or nearly so to sometimes sparsely pubescent. ______ ***C. viridis*** (not treated in text)

***Colubrina triflora*** [*C. glomerata*]. *Palo cachorra, manzanita del monte* (Sinaloa); *cuta guícori* (Mayo) (figure 219)

Shrubs or trees sometimes 8–12 m, with one to several trunks often curved and leaning, the upper branches sprawling across the canopies of forest trees; unarmed. Bark rough, dark colored, and finely checkered. Branches

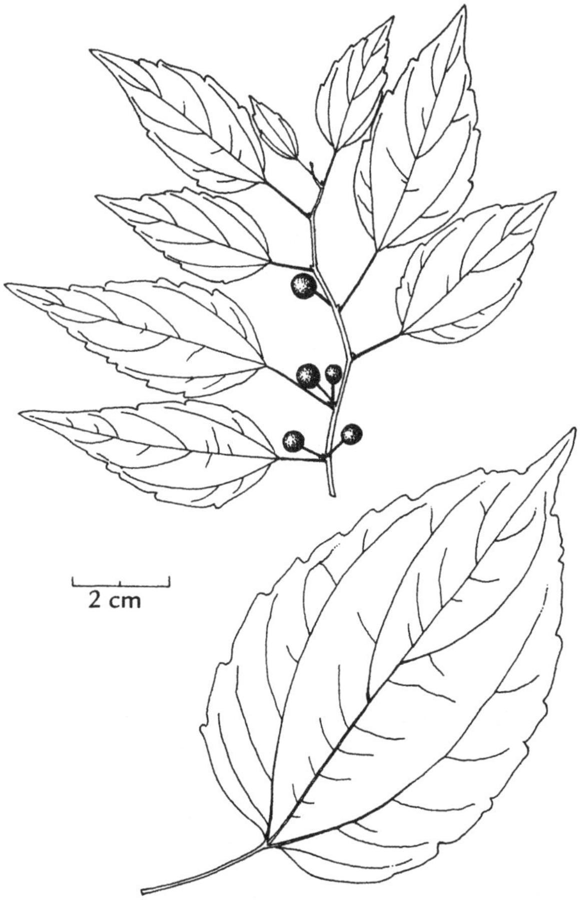


Figure 219. *Colubrina triflora*. Drawn by M. B. Johnson.

and stems slender, the ultimate stems long and leafy. Herbage glabrous or sparsely pubescent on young growth, especially in arid habitats. Leaves alternate, tardily drought deciduous or perhaps evergreen in riparian habitats, 6–18.5 × 3.5–8.5 cm; petioles prominent; blades ovate to broadly lanceolate, thin, shiny green, with 3 prominent veins from the base, acute to acuminate at the apex, the margins serrate toothed; stipules small and quickly deciduous.

Inflorescences one to several-flowered, the pedicels elongating in fruit. Flowers 4–5 mm across, greenish, with a floral cup nearly filled with a thick, nectar-producing disk; rim of the floral cup bearing the sepals, petals, and stamens, each 5 in number; sepals and petals deciduous; ovary partly inferior. Capsules 5–6 mm wide, globose, green, explosively dehiscent into three 1-seeded segments. Flowering at least August–November.

Tree-size plants in moist canyons and shaded slopes in tropical deciduous forest in southeastern Sonora. In these places they seem to have sprung up in gaps in the canopy caused by hurricane damage (Phil Jenkins, personal communication 1998). Shrub-sized plants occasional on dry hillsides in tropical deciduous forest and canyons or arroyos in thornscrub of southwestern Sonora, and one record from pine-oak woodland near Santa Bárbara in the upper Río Cuchujaqui drainage; 20–1200 m (65–3940 ft). Also the Cape Region of Baja California Sur, southwestern Chihuahua and Sinaloa to Costa Rica.

### *Condalia*

Woody shrubs, rarely small trees. Southwestern United States to South America and the West Indies; 18 species. Reference: Johnston (1962b).

**_Condalia globosa._** Bitter condalia; *bachata; jutuqui, ju'upa quecara* (Mayo); *kawk koawul* (O'odham) (figure 220)

Densely branched, large shrubs or rarely small trees 2.5–4 (6+) m, more or less symmetrical, with hard wood and thick, sometimes massive trunks; branches rigid. Long shoots of primary branches bearing secondary, thorn-tipped twigs, both having alternate leaves from relatively widely spaced nodes, the secondary twigs with short axillary short shoots (spur branches) bearing closely set, fascicled leaves. Larger leaves (on long shoots) petioled and quickly drought deciduous; smaller leaves (on short shoots) subsessile, gradually drought deciduous. Leaves 3.3–22 × 1.5–4 mm, spatulate, the margins entire; stipules small, not spinose. Lower leaf surfaces with the veins convex and conspicuous, the upper surface with the midrib concave. Lower surfaces of fresh, young, but mature leaves often yellow with minute, glistening golden glands.

Flowers in small clusters on short shoots, yellow-green, approximately 3 mm wide; calyx tube shallow, with a flat disk and 5 lobes, the disk at anthesis awash with sticky, glistening nectar; petals absent. Fruit a drupe 3–4.5 mm, globose, dark brown or black, with a 2-mm wide basal disk, usually very bitter, 1-seeded. Flowering various seasons, probably in response to soil moisture.

Valleys, plains, and arroyos; Sonoran desertscrub and coastal thornscrub; western Sonora, 10–815 m (30–2675 ft); southeastern California and southwestern Arizona to northwestern Sinaloa and both Baja California states.

Apparently only var. *pubescens* occurs in the northernmost part of the range and only var. *globosa* in the southernmost part (e.g., south of Guaymas). However, both varieties occur intermixed over the majority of the distribution. Variety *pubescens* apparently differs from var. *globosa* only in having hairy leaves and sometimes develops into small trees, whereas var. *globosa* does not seem to become arborescent. Many of the larger individuals now found in the desert have regrown from large, axe-cut stumps, and the largest plants are in remote areas where there has been little or no woodcutting. The hard wood is highly desirable as firewood. In the remote interior of Isla Tiburón some individuals are > 6 m with massive woody trunks. The label on a University of Arizona herbarium specimen collected by Robert Peebles in 1939 states that it is "common in the vicinity of Sonoita, Sonora, and there attains a height of about 20 feet." Plants of this size are no longer found near Sonoyta. The larvae of the saturniid moth *Agapema anona* feed on the leaves and often form large, spongy masses of cocoons among the thorny branches.

Kearney and Peebles (1960: 530) and Benson and Darrow (1954: 262) claim that the bitterness of the fruit is of taxonomic value, but Johnston (1962: 351) says this quality is not a dependable character. *Condalia globosa* is a member of the *C. spathulata* complex that includes four closely related allopatric species from central Mexico to southwestern United States. *Condalia globosa* occurs in the most arid and westernmost area of any member of the complex.

### *Karwinskia*

Dry to moist tropical regions, Texas and Mexico to Bolivia and the Caribbean; 12 species. Includes timber trees in Central America. Reference: Johnston (1966).

**_Karwinskia humboldtiana_** [*K. parvifolia; K. pubescens*]. Dot and dash plant; *cacachila; guasalaco, jimuari* (Guarijío); *aroyoguo* (Mayo); *kaio u'us* (Pima Bajo) (figure 221)

Mostly shrubs, sometimes trees 5–10 m with a well-formed trunk 6–38 cm in diameter and a dense, spreading crown; wood heavy and hard; bark light or dark, checkered and flaking in small, vertical rectangles. Herbage glabrous (or

Figure 220. *Condalia globosa* var. *pubescens* (Abrams 1951, above); (Little 1950, Hunt/USDA, below). Drawn by L. Hughey.

shrub populations sometimes minutely pubescent); inflorescences and flowers dotted and streaked with blackish glands that are especially conspicuous on the lower leaf surfaces. Foliage light bluish green; often evergreen, or tardily drought deciduous. Leaves mostly opposite or subopposite and often with some alternate, not fascicled, mostly 2.5–8.5 cm; blades mostly elliptic to ovate-elliptic, entire, pinnately veined, the tips acute to obtuse; petioles 4–8 mm.

Flowers yellowish with dark spots, approximately 4 mm wide, in small, umbellate clusters about 1 cm. Fruits rounded, approximately 1 cm; pericarp blackish when ripe; seeds often 1 (to probably 4). Flowering with rains, especially during summer; fruits often ripe in fall and winter.

Widespread, plains to mountains in many habitats, generally nonriparian; Sonoran desertscrub, thornscrub, and tropical deciduous forest. Widespread, mostly in nondesert areas in the eastern three-quarters of Sonora northward to about 30° N, generally absent from desert areas north of Guaymas and in lowland, coastal plains only in extreme southwestern Sonora; near sea level–1200 m (3940 ft). Texas, Baja California Sur, and most of mainland Mexico.

Tree-sized *Karwinskia* occur on foothill and mountain slopes, ridges, and valley plains in tropical deciduous for-

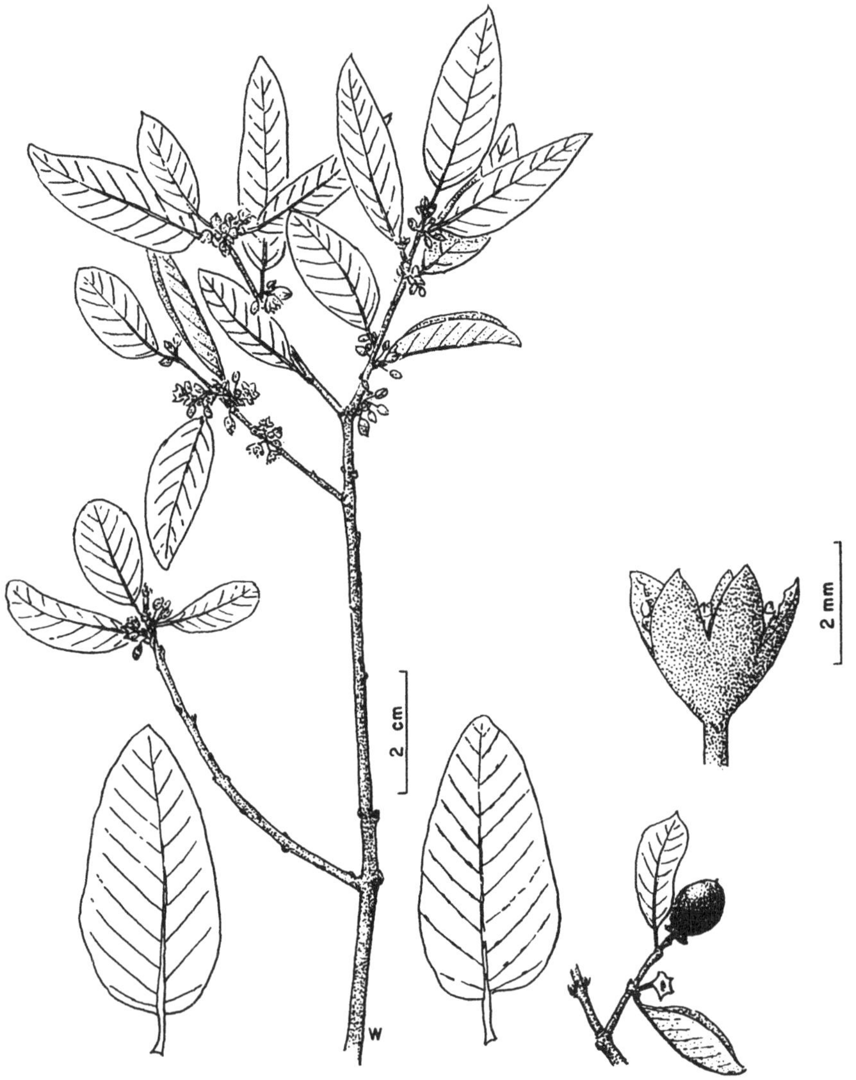


Figure 221. *Karwinskia humboldtiana* (Wiggins 1980). Drawn by I. L. Wiggins.

est in southeastern Sonora, 240–700 m (790–2300 ft). Trees 5–10 m are fairly common only in the Guarijío region north of Mesa Colorado and are sparsely scattered among mostly shrub-sized plants elsewhere in southeastern Sonora.

*Karwinskia parvifolia*, distinguished primarily by its smaller leaves, and *K. pubescens*, distinguished by being pubescent, seem to be morphologically continuous with *K. humboldtiana*, and there is a recent trend to treat them as synonyms (e.g., Turner et al. 1995). Although larger leaved populations generally occur in better watered areas, the tree-sized plants and populations have the smaller leaves of *K. parvifolia*. The smaller leaved form is also the only one in the more arid regions.

The light colored, evergreen foliage and lacy appearance of the branching pattern would make a handsome cultivated tree, and as shrubs they should be excellent as hedges. It is expected to be frost sensitive but quite drought resistant.

The fleshy pericarp is sweet and edible, but the seeds are highly toxic, paralyzing the motor nerves. When people ate the fruits they would spit out the seeds. Gentry (1942: 174) stated that "the fruit is sometimes eaten, and is reported to make young boys weak and to produce trembling." Gentry observed young coati mundis (*Nasua nasua*) "eating the fruits with great eagerness." It is an important medicinal plant. The leaves are boiled in water and the tea drunk for fevers, or made into a decoction as a wash for sores. For chest pain, the ground (powdered) stem is rubbed on the chest. *Karwinskia* is also used as an abortive (see *Montanoa*, Asteraceae). The Guarijío often use *cacachila* for roofing cross-beams, and the branches and twigs are bound into a broom for outdoor use.

### *Rhamnus*

Mostly in temperate regions, both hemispheres; 125 species.

Several shrub species occur in the mountains of northern and eastern Sonora including *R. betulifolia*, *R. californica* and *R. crocea*. In southern Arizona these species sometimes become arborescent.

1. Leaves about as wide as long, the margins spinose toothed. ___________ ***R. crocea*** (not treated in text)

1' Leaves mostly at least twice as long as wide, the margins entire to minutely toothed but not spinose.

  2. Leaves uniformly green on both surfaces, usually > 3 cm wide. ___ ***R. betulifolia*** (not treated in text)

  2' Leaves bicolored, conspicuously lighter colored below, usually < 3 cm wide.

    3. Lower leaf surfaces densely pubescent with whitish hairs; flower clusters (umbels) on a common stalk or peduncle. ______ ***R. californica*** (not treated in text)

    3' Lower leaf surfaces relatively sparsely pubescent with pale yellowish hairs; flowers in sessile clusters, without a common stalk. __***R. scopulorum***

**Rhamnus scopulorum** [*R. serrata* var. *scopulorum*]. *Ojo de venado; shiik wupil* (Pima Bajo)

Evergreen shrubs or rarely small trees 6–8 m. Twigs below leaves glabrous, moderately glaucous, becoming red-brown; leafy portion of twigs with short, often whitish hairs; newest herbage densely pubescent with short, pale golden hairs. Leaves alternate, 5.5–9 × 1–3 cm, petioled; blades lanceolate to nearly elliptic, upper surface darker green, the lower surface lighter colored and more densely pubescent, the margins minutely serrate, the teeth often gland-tipped. Stipules to approximately 3 mm, dark red-brown, subulate or linear-filiform.

Flowers in sessile, few-flowered clusters shorter than the petioles; pedicels 3–4 mm; pedicels and flowers densely pubescent with short, pale golden hairs. Flowers white, 3–4 mm; sepals 5-lobed; petals 5, cupping the stamens; styles short and persistent. Fruits 7–8 mm wide and about as long, fleshy, blackish purple at maturity, 3-lobed and 3-seeded; fruiting pedicels approximately 1 cm. Flowering June–July.

Canyons, mostly along streams, in oak and pine-oak woodland, and mixed conifer forest. Mountains in eastern Sonora from the Río Bavispe region and Sierra Nacori southward, 1050–2040 m (3445–6690 ft). Also western Chihuahua and Sinaloa to Jalisco.

Tree-sized plants in Sonora have been found only in Arroyo Santa Bárbara in the upper Río Cuchujaqui drainage in southeastern Sonora, in riparian, tropical evergreen vegetation at 1050 m. The surrounding mountains support an ecotone of tropical deciduous forest and oak woodland.

### *Ziziphus*

Shrubs and trees; fruits usually edible. Mostly tropical to subtropical regions and extending into many of the deserts of the world; 100 species, about two-thirds in the Old World, 7 species in North America. The New World and Old World groups are distinctive. Reference: Johnston (1963).

The only other species in Sonora, *Z. obtusifolia* var. *canescens*, is a large, thorny shrub that sometimes attains treelike stature. This untidy, irregularly branched shrub has paired, thorn-tipped lateral branches that superficially look like stipular spines. The leaves are quickly drought deciduous and it is often leafless.

Jujube (*Z. jujuba*) in China and *ber* (*Z. mauritiana*) in India are important fruit trees. Jujube is sometimes grown in Sonora and Arizona, but the better, grafted varieties are seldom seen. Superior *ber* varieties developed in the arid zones of India would be worthwhile introductions into arid regions of Mexico, especially Sonora. Both species thrive under cultivation in deserts. Jujube is frost hardy, but *ber* is frost sensitive.

**Ziziphus amole** [*Z. sonorensis*]. *Amole, saituna; bahi cápore* (Mayo) (figure 222)

Large shrubs or trees 5–10 (13) m, often with a well-developed trunk, or sometimes forming briarlike, wind-swept shrubs along the coast. Bark flaking in large, thick pieces. Branches and twigs rigid, woody and spinescent with thick, stout thorns. Glabrate, especially older twigs and leaf blades, or with short, fine hairs. Leaves alternate, facultatively and tardily drought deciduous, bright green, often 3.5–7 × 2–5 cm; blades broadly ovate to broadly elliptic or oval, with 3 prominent veins especially when seen from below; petioles short; nodes often with stout, stipular spines.

Flowers in small subumbellate clusters, inconspicuous, yellowish, approximately 3 mm wide, very fragrant, attracting large numbers of bees. Fruits fleshy, edible drupes, approximately 1 cm, red-orange when ripe, 1-seeded. Flowering July–September; fruits in fall and sometimes February–March.

Valleys, arroyos, rocky slopes, and fine-textured soils of low places and saline flats; Sonoran desertscrub, thornscrub, and tropical deciduous forest; southern Sonora from the Guaymas region and the vicinity of Soyopa southward, near sea level–305 m (1000 ft). Seldom common except in the extreme southwestern part of the state. Often near the shore, sometimes just behind rocky beaches or on slopes or alluvium near the beach. Largest trees grow along washes

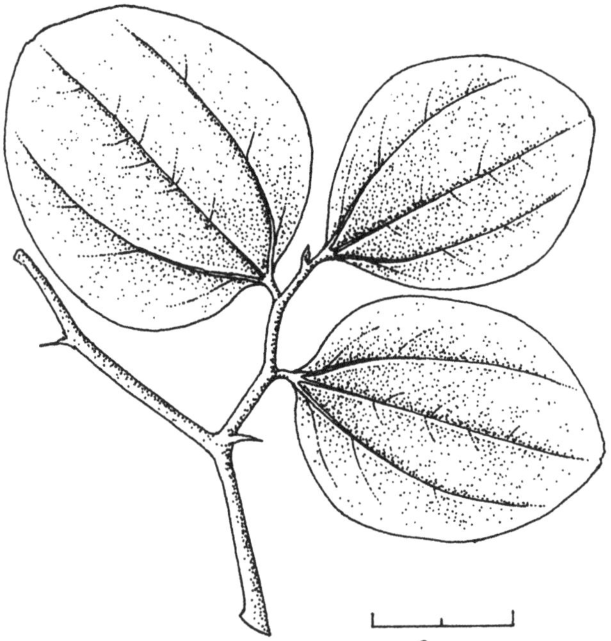

Figure 222. *Ziziphus amole*. Drawn by M. B. Johnson.

in the coastal plain near the Sinaloa border. Also Sinaloa to Oaxaca and Tamaulipas.

The fruit is eaten fresh, and the Mayo of southern Sonora brewed the roots into tea to expel amoebas and cleanse the intestines.

## Rhizophoraceae—Red Mangrove Family

Trees and shrubs of tropical regions, mostly Old World; 16 genera, 130 species. The 4 genera and 17 species of the tribe Rhizophoreae are mangroves.

### *Rhizophora*—Red Mangrove

Tropical shores around the world; 6 closely related species. References: Tomlinson (1980, 1986).

Some *Rhizophora* species contain sufficient tannin in the bark to be of commercial value.

***Rhizophora mangle*.** Red mangrove; *mangle rojo; canaro* (Mayo); *pnaacoj-xnazolcam* (Seri); *nahnawa'ara* (Yaqui) (figures 4 and 223)

Shrubs or small trees with hard wood and high tannin content, mostly 3–5 m (in Sonora); unique stilt roots beginning as arching aerial roots from stems and branches; glabrous except the petals. Leaves opposite, simple, 7–15 cm, leathery, semisucculent, ovate to obovate or broadly elliptic, the midrib prominent and raised below, the tips blunt, moderately recurved, the margins entire; petioles 8–20 mm. Leaves opposite, not quite decussate because successive "pairs of leaves diverge from each other at an angle of less than 90°. This arrangement has important architectural consequences, since mutual shading is reduced and branches diverge at diverse angles" (Tomlinson 1986: 319).

Each pair of developing leaves in the terminal bud enclosed by two interlocking stipules 3.5–6.2 cm that fall away as the enclosed leaves unfold. Each leaf pair has an associated pair of stipules above the leaf insertion so that the stipules do not enclose the associated leaves but rather the next younger leaf pair and its associated organs. Two-ranked palisade of glands (colleters) at the base and inner surface of each stipule secrete mucilage that coats the enclosed leaf pair, probably protecting them from desiccation and salt damage. (This unusual, stipule-covered, slender, conical leafbud is superficially similar to the leaf buds in *Coccoloba* and *Ficus*. Mucilage-producing colleters are also found in the Apocynaceae and Rubiaceae.) Exudate somewhat milky and possibly edible to birds or ants. Stipules produce a ring-shaped scar on the stem; petioles produce an elliptic scar below the associated stipular scar.

Inflorescences axillary, among and much shorter than the leaves, cymose, usually 2- to 5-flowered; peduncles and pedicels thick; the first node with 2 (3) branches or flowers, each node and flower with 2 short, broad bracts fused at their bases. Flower buds yellow, 4-angled, bluntly pointed at the tip. Flowers white and yellow, bisexual, soon turning downward (pendulous), wind pollinated. Calyx lobes 4, united basally, 10–12 mm, yellow to green, thick and leathery. Petals 4, cream colored, densely white-hairy along the margins of the inside surface, moderately recurved, soon deciduous. Stamens 8; anthers 4–6 mm, sessile, thick, and honeycomb-like. Ovary partially inferior. Style thick and fleshy, 2-lobed; stigma receptive after the stamens and petals fall, promoting out-crossing. Flowering during warmer months.

Muddy-sandy, intertidal shallow sea water in quiet bays, *esteros*, or inlets, and river deltas; in slightly deeper water, or the seaward side from the other two Sonoran mangroves. North end of the Canal del Infiernillo at Estero Sargento and Punta Perla on Isla Tiburón southward, and on the Gulf Coast of the Baja California peninsula. Tropical shores of the New World on both the Pacific and Atlantic sides and in west Africa.

Red mangrove is highly frost sensitive. The mangroves at Estero Sargento suffer severe freeze damage or some are killed during the occasional years of freezing temperatures. The sizes of the plants and their parts, especially the leaves, their stipules, and the inflorescences seem generally smaller among the Gulf of California populations than those from more tropical regions. In the tropics red mangrove trees sometimes reach 25 m in height.

Figure 223. *Rhizophora mangle* (Felger and Moser 1985). Drawn by L. B. Hamilton.

The anthers open in the bud before the flower opens, so that the pollen is trapped among the numerous hairs of the petals. The pollen is drawn out with the petals and disperses as the petals expand and recurve. The petals fall within a day. The wind-pollinated flowers of *Rhizophora* have evolved from an insect-pollinated ancestor. Nectar secretion is low. The flowers have a high pollen/ovule ratio (the ratio of the number of pollen grains to the number of ovules, or potential seeds, per flower). This is a common trait of wind-pollinated plants. The ratio for *Rhizophora* is an order of magnitude higher than among related, insect-pollinated genera. However, the transition to wind pollination may not be complete. The style is not typical of a wind-pollinated plant, and the flowers are visited by bees that probably are collecting pollen. Wind pollination allows *Rhizophora* to escape the competition for animal (insect) pollination that is otherwise commonplace in mangrove communities.

Flowering and fruiting occurs throughout the warmer months. The fruits are viviparous, the seedlings germinate while still attached to the parent plant, becoming green, elongated and club-shaped, and are often 20–30 cm long before falling. Most of the seedling is made up of hypocotyl, the true radicle (seedling or embryo root) is very short, and the plumule (shoot) is also short and exposed only after the seedling falls. The time from flowering to seedling development and detachment is reported from other regions as slightly more than one year; in Sonora and Baja California it should be less than one year (spanning the warmer months). The cotyledons form a tube that surrounds the plumule but remains behind on the parent tree.

After the seedling falls into the water, it can be moved about by the tide, often drifting far from its parent. The seedling can establish within hours or remain viable for months because the hypocotyl contains abundant food reserves. The Seri Indians roasted the seedlings in hot coals and ate the pithy hypocotyl (Felger and Moser 1985). After drifting to a suitable place, the roots grow rapidly, taking hold in sandy mud and pulling the leafless seedling into an upright position. The shoot then rapidly develops. Unlike the other mangroves, *Rhizophora* seedlings can grow while fully submerged.

Aerial roots can develop from any branch, including branches in the crown, and grow rapidly, perhaps as much as 1 cm per day. These roots are 5.8–19 mm in diameter near the tip and for the first season or year have thin brown bark but the inner bark, beneath the "skin" is bright green. Young growing tips of aerial roots are frequently damaged and killed, probably by isopods, resulting in several branches at that point and producing the familiar multibranched aspect. Barnacles (*Balanus*) and oysters (*Ostrea columbiensis*) sometimes form thick colonies on the stilt roots. Oysters, however, have been seriously depleted or eliminated by overharvesting except in inaccessible areas such as Isla Tiburón (Felger and Moser 1985).

Red mangrove is distinguished from the other Gulf of California mangroves (*Avicennia* and *Laguncularia*) by its stilt roots, shiny green semisucculent leaves, large sheathing stipules, four-merous flowers, hairy petals, large and sessile anthers, and unique fruits.

## Rosaceae—Rose Family

Trees, shrubs, or herbs. Leaves alternate, simple (Sonora tree species) or compound, stipules usually present. Flowers radial, the ovary superior (Sonora tree species) to inferior. Sepals and petals each 5. Stamens usually numerous, arising from near rim of floral cup. Fruits variable. Worldwide, especially temperate and warm regions of the Northern Hemisphere; 100 genera, 3100 species.

This family includes many economic and ornamental plants including almonds, apples, blackberries, pears, peaches, plums, raspberries, roses, and strawberries. Although a number of species are listed for the Sonoran Desert, they only enter the very uppermost margins of the desert proper or the ecotone with the desert.

1. Leaves mostly less than twice as long as wide; flowers in racemes; fruits fleshy, 1-seeded. ________ ***Prunus***
1' Leaves more than twice as long as wide; flowers in flat-topped panicles; fruits dry, with 5 segments and 10 seeds. ________ ***Vauquelinia***

### *Prunus*

Trees or shrubs. Leaves often with conspicuous glands near the base of the blade. Inflorescences of racemes (in Sonora), mostly on short branchlets. Flowers (in Sonora) white or cream-white about 1 cm wide; petals deciduous; calyx persistent or deciduous. Pistil 1. Fruit a drupe, mostly rounded to ovoid, with a fleshy mesocarp surrounding a hard pit (endocarp) containing a single seed. Mostly temperate regions of the Northern Hemisphere; 150 species. Reference: McVaugh (1951).

Many cultivated trees and shrubs (e.g., almonds, apricots, cherries, peaches, plums, and prunes) belong to this genus. Peach trees (*P. persica*), growing from discarded pits, are occasionally seen along roadsides at higher elevations in north-central and eastern Sonora.

1. Racemes at ends of short, leafy branches: *P. serotina*.
   2. Leaves 6–11 cm, the apex mostly abruptly acuminate. ________ ***P. serotina*** subsp. ***capuli***
   2' Leaves mostly 3.5–9 cm, the apex acute to obtuse (not abruptly so and not acuminate). ________ ***P. serotina*** subsp. ***virens***
1' Racemes axillary and leafless.
   3. Mature leaf blades relatively thin; stipules 1.2–4 mm; calyx persistent on the fruit; seeds 7.5 mm. ***P. gentryi***
   3' Mature leaf blades relatively thick and oaklike; stipules 2.5–10 mm; calyx deciduous; seeds 15–17 mm. ________ ***P. zinggii***

***Prunus gentryi.*** *Aigre; guasiqui* (Guarijío); *mo'oshkam* (Pima Bajo) (figure 224A)

Shrubs or trees to 12 m. Especially the young twigs, petioles, and lower leaf surfaces fuzzy brown pubescent, becoming glabrous or sometimes with small tufts along the midrib and veins. Leaves probably evergreen, 5–12.5 cm; blades leathery, shiny green above, paler below, lanceolate to elliptic or narrowly oblong, often with one to several oval glands near the base, the margins essentially entire; petioles 5–20 mm; stipules 1.2–4 mm, linear-triangular, quickly deciduous.

Racemes axillary, leafless, 3–5 cm. Flowers white. Calyx persistent on the fruits. Fruits edible, approximately 1 cm in diameter; seeds approximately 7.5 mm. Flowering May–June.

Canyons and streams; pine-oak woodland. Canyons leading off Mesa el Campanero and high mesas and mountains east of Yécora in east-central Sonora, and mountains along the Chihuahua border in southeastern Sonora; 1000–1600 m (3280–5250 ft). Often with *Cupressus*, *Quercus*, and *Pinus*. Known otherwise only from southwest Chihuahua, where it is more widespread and extends to 1800 m.

***Prunus serotina*** subsp. ***capuli*** [*P. capuli*]. Mexican chokecherry; *capulín* (figure 224B)

Trees often to 10 m; young leaves pubescent, otherwise glabrate. Bark of young branches smooth, shiny, red-brown, with horizonal streaks of lenticels. Leaves winter deciduous, mostly 6–11 cm; blades shiny green above, paler below, mostly broadly elliptic, sometimes lanceolate to ovate, relatively thin, the apex mostly abruptly acuminate, the margins finely serrated with forward-pointing and mostly gland-tipped teeth, often with glands near the junction of the petiole and leaf blade; petioles 6–20 mm; stipules minute or apparently absent.

Flowers white. Racemes at ends of short leafy branchlets; branchlet and its raceme 9–18 cm. Calyx persistent on the fruits. Seeds 7.5–9.0 mm. Flowering March.

Streams and arroyo margins; oak woodland, pine-oak woodland; eastern Sonora, 1100–1980 m (3610–6500 ft). The northernmost populations are on Sierra el Tigre. Mexico to South America, and Baja California Sur.

Subspecies *capuli* is a southern, more tropical and weakly distinguished relative of subsp. *virens*. They intergrade in east-central and southeast Sonora, and even elsewhere they may be difficult to distinguish. The fruits are edible and are made into alcoholic and nonalcoholic beverages.

***Prunus serotina*** subsp. ***virens*** [*P. virens*]. Southwestern chokecherry; *capulín*; *aguasique* (Pima Bajo) (figures 224C and 225)

Similar to subsp. *capuli*, differing by the smaller and blunter leaves and shorter racemes. Leaves winter deciduous, mostly 3.5–8.7 cm; blades elliptic to sometimes moderately

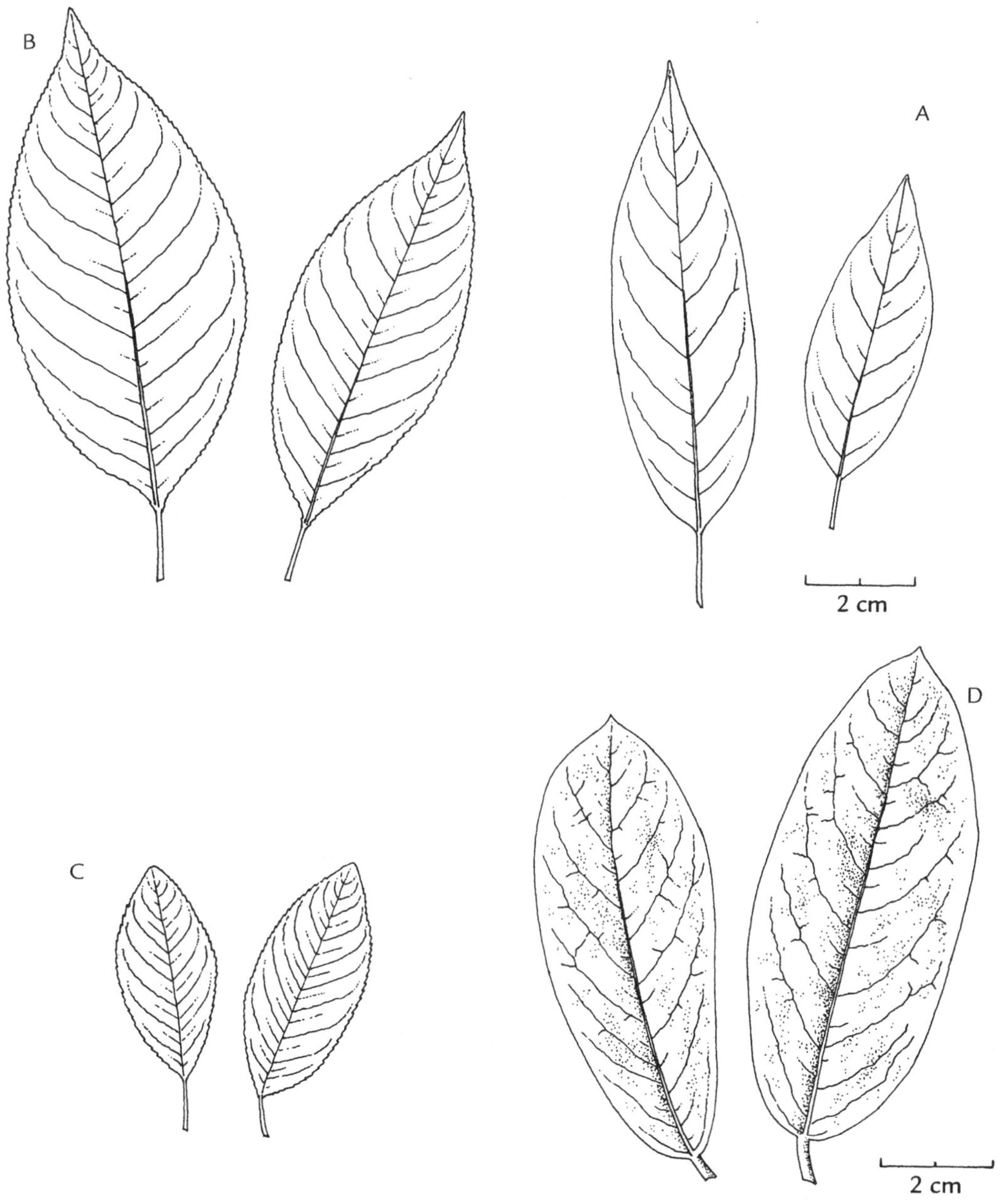


Figure 224. *Prunus*. (A) *P. gentryi*, (B) *P. serotina* subsp. *capuli*, (C) *P. serotina* subsp. *virens*, (D) *P. zinggii*. Drawn by M. B. Johnson.

Figure 225. *Prunus serotina* subsp. *virens.* Drawn by L. B. Hamilton.

ovate or obovate, the apex acute to obtuse (not abruptly so and not acuminate); petioles 5–15 mm. Racemes at ends of leafy branchlets; branchlet and its raceme 3–12 cm. Seeds 6.5–7.8 mm. Flowering March.

Canyons and streams; oak woodland, pine-oak woodland; eastern Sonora, 1065–2285 m (3500–7500 ft); southern Arizona to Trans-Pecos Texas and the northern states of Mexico.

***Prunus zinggii.*** *Jeco guasiqui* (Guarijío) (figure 224D)

Sometimes large canopy trees to 12 m or more with a thick, well-formed trunk and dense crown. Young twigs, petioles, and stipules with coarse, often curved, brown, velvety hairs. Leaves evergreen, 7–15 cm; blades broadly elliptic to ovate-elliptic or obovate, rather firm, often oaklike with conspicuous veins, the upper surface bright green, glabrous, the lower surfaces lighter in color, pubescent with tawny brown hairs becoming darker and sparser with age, with 2 rounded, sessile, maroon glands 1–2 mm across at the base of the blade, the margins entire; petioles 5–8 mm; stipules 2.5–10 mm, linear, brown.

Racemes, pedicels, and floral cups hairy. Racemes axillary and leafless, 3–10 cm. Flowers cream colored, approximately 1 cm wide; calyx deciduous. Seeds 15–17 mm. Flowering March.

Canyon bottoms and streams; tropical deciduous forest and oak woodland; east-central and southeastern Sonora, 510–1300 m (1670–4265 ft); the northernmost record is from El Reparo, near Santa Ana de Yécora. Otherwise known only from southwestern Chihuahua.

Its closest relative seems to be *P. erythroxylon*, with glabrous leaves, known from Colima and Michoacán. *Prunus zinggii* might be confused with *Sideroxylon persimile* (Sapotaceae) if the twigs of the latter lack spines. In the absence of flowers, the glands on the lower leaf surface and presence of stipules in *P. zinggii* are diagnostic, and the hairs are distinctive. However, the stipules are sometimes deciduous, and in herbarium specimens the glands may be obscured if the midrib is folded over the gland while drying. The Guarijío made tortillas and atole from the ground seeds. They also eat the fresh fruits.

## *Vauquelinia*—Rosewood

Large shrubs and small trees. Arid and semiarid habitats, often above or marginal to deserts, Mexico and southwestern United States; 3 species. Reference: Hess and Henrickson (1987).

### ***Vauquelinia californica***—Rosewood

Evergreen, hardwood, multiple-stemmed large shrubs or small trees. Herbage with short crinkled to matted hairs, sometimes glabrate with age. Evergreen, the leaves firm (sclerophyllous); petioles short; blades mostly narrow, leathery; margins finely crenulate (wavy) with small, gland-tipped teeth and glands often between the teeth.

Inflorescences of terminal, compound corymbs; pedicels, floral cups, calyx lobes, and fruits with small, woolly hairs. Flowers approximately 1 cm wide; petals white. Fruits of capsules, 4.5–6.5 mm, firm and somewhat woody at maturity, with 5 carpels united below and separate above, each with 2 seeds; seeds with a terminal wing.

*Vauquelinia californica* comprises four subspecies of large shrubs or small trees ranging from central Arizona to both Baja California states, and southwestern New Mexico to Durango and Coahuila; it does not occur in California, the specific name refers to Baja California.

Subspecies *californica* is cultivated as a landscape plant in southern Arizona and is useful as a large hedge or dense screen. Likewise, the other subspecies would be interesting landscape subjects for nontropical regions. These plants tolerate considerable cold and drought.

There are few records for *V. californica* in Sonora, which seems unusual given its widespread occurrence and diversity (three subspecies) in adjacent southern Arizona. Perhaps the paucity of Sonoran records is due in part to a lack of botanical exploration in remote and now often dangerous borderland mountains.

1. Leaves green or yellow-green on both surfaces. subsp. ***pauciflora***

1' Leaves bicolored with whitish, usually woolly undersides.

  2. Leaves lanceolate to oblong-lanceolate or narrowly elliptic, mostly 10–20 mm wide. ________ subsp. ***californica***

  2' Leaves linear to linear-lanceolate, mostly 7–10 mm wide. ________ subsp. ***sonorensis***

### ***Vauquelinia californica*** subsp. ***californica***

Similar to subsp. *pauciflora* but with bicolored leaves. Both states of Baja California and southern Arizona including the Baboquivari Mountains. It should be sought in nearby north-central Sonora. An observation of rosewood in the Sierra del Viejo near Caborca (Turner et al. 1995) may be this subspecies or subsp. *sonorensis*.

### ***Vauquelinia californica*** subsp. ***pauciflora.*** Arizona rosewood (figure 226A)

Large shrubs to small trees 3.5–4 m; larger stems 10–16 cm in diameter. Herbage with short crinkled to matted hairs, glabrate with age. Leaves (3) 4.0–9.6 × 0.6–1.8 cm, mostly glabrous or with few hairs mostly along midribs, the blades mostly narrow, lanceolate to oblong or elliptic, green to yellow-green on both surfaces; margins finely crenulate (wavy) to mostly small-toothed, the teeth gland-tipped and glands often between the teeth; petioles 2–

15 mm. Inflorescences approximately 4 cm wide; pedicels, floral cups, calyx lobes, and fruits with small, woolly hairs. Flowering May–August.

Northeastern Sonora in canyons and slopes at the north end of the Sierra el Tigre, at 1140 m, where it is apparently rare. Documented for extreme southeastern Arizona and southwestern New Mexico and reported from the adjacent corners of Chihuahua and Sonora by Mearns (1907: 57). Also southern Chihuahua, Coahuila, and Durango.

The substrate at Sierra el Tigre is rhyolite although limestone intrusions occur nearby. Here *Vauquelinia* grows at the upper margin of Sonoran desertscrub, merging with thornscrub, and just below the oak zone (*Quercus oblongifolia*). Associated plants include *Coursetia glandulosa, Dodonaea viscosa, Fouquieria splendens, Fraxinus gooddingii, Hechtia montana, Juglans major, Prosopis velutina, Rhus microphylla,* and *Yucca arizonica*.

Elsewhere this subspecies is known from limestone substrate in ecotone of Chihuahuan desertscrub and oak woodland, and lower elevations in oak woodland.

***Vauquelinia californica*** subsp. ***sonorensis***. Sonoran rosewood (figure 226B)

Shrubs at least 2.5 m, probably becoming small trees. Young stems, lower leaf surfaces, inflorescences, and calyx tubes densely white-woolly with tightly crinkled hairs. Leaf blades linear to linear-lanceolate or narrowly oblong, (3) 5–12 cm × 5–11 mm wide, markedly bicolored, white-woolly below and dull green and generally glabrate above, the margins thickened, subentire to mostly finely toothed; petioles 3–8 mm.

A population in the Sierra Cubabi in northwestern Sonora is the only documented record for the state.

This subspecies grows in protected areas, such as north-facing drainages on Sierra Cubabi, a granitic mountain in Sonoran desertscrub, with *Dodonaea viscosa, Eriogonum fasciculatum, Fouquieria splendens, Galium stellatum, Simmondsia chinensis, Solanum hindsianum,* and *Stenocereus thurberi*. A much larger and well-known population occurs in nearby southern Arizona in large canyons in the Ajo Mountains in Organ Pipe Cactus National Monument. Sierra Cubabi, about 30 km directly south-southwest, is the nearest range. Subsp. *sonorensis* has also been found farther north in Arizona in the Sand Tank Mountains. It is otherwise disjunct in northern Baja California on the eastern slopes of the Sierra Juarez. An observation of *Vauquelinia* in the Sierra del Viejo near Caborca may be this subspecies (see subsp. *californica*).

Hess and Henrickson (1987) give a maximum size of 10 m for any *Vauquelinia*, 8 m for any of the four subspecies of *V. californica*, and 7 m for subsp. *sonorensis*. A tree in Alamo Canyon, Ajo Mountains, measured by Zahner and associates (National Register of Big Trees 1996) is 14.3 m in height with an average crown spread of 12.2 m, and a trunk 2.0 m in girth at 1.4 m above ground level (original measurements in English units). Thus the most xeromorphic taxon (Hess and Henrickson 1987) in the genus contains the largest-sized individual.

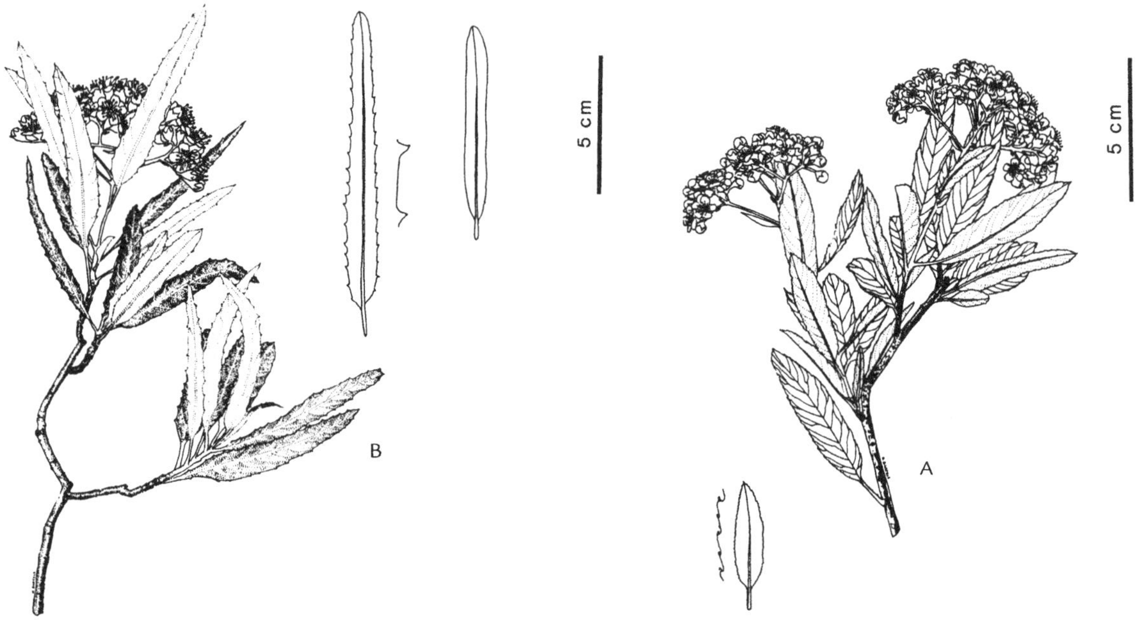


Figure 226. *Vauquelinia californica*. (A) subsp. *pauciflora*, (B) subsp. *sonorensis* (Hess and Henrickson 1987). Drawn by N. Bartels.

## Rubiaceae—Madder Family

In Sonora, herbs, shrubs, and a few tree species. Leaves simple, mostly entire, often opposite and decussate or whorled. Stipules (Sonoran tree species) often small, usually united at their bases, interpetiolar (between the opposite petiole bases; if the stipule has fallen then look for the stipular scar connecting the petiole bases), bearing colleters on the inner surface that produce mucilage, apparently protecting the growing tip of the shoot (see *Randia* and also Rhizophoraceae).

Flowers sometimes unisexual, usually radial, sometimes bilateral; calyx and corolla each united below, 4- to 6-lobed. Corollas deciduous. Stamens fused to the corolla tube and as many as the corolla lobes. Ovary inferior. Fruits highly variable, fleshy or dry; seeds one to many.

Mostly tropical and subtropical but many in temperate and even arctic and antarctic regions, on every continent; 630 genera, 10,200+ species.

This family includes quinine (*Cinchona*), coffee (*Coffea*), and *Gardenia*. Madder (*Rubia tinctorum*) was the source of the red dye alizarin, now produced synthetically.

The Rubiaceae is the fourth largest plant family on earth, ranking close to the Poaceae (grasses) in number of species. It is the largest plant family in the wet tropics of the world. Most tropical members are woody, whereas the temperate and cold climate species are usually herbaceous. About 55 species in 18 genera occur in Sonora, mostly in the wetter or more tropical areas.

1. Flowers approximately 1 cm; fruits 7–8 mm, in dense heads on long peduncles, with small, rounded, woody receptacles; peduncles and receptacles usually semipersistent; riparian habitats with permanent water. ***Cephalanthus***

1' Flowers 0.6–9 cm; fruits ≥ 8 mm, solitary, in few-flowered clusters, or in open, loose inflorescences, the receptacles not especially woody; peduncles and receptacles not persistent; not strictly riparian.

  2. Leaves opposite and mostly in whorls of 3; corollas bilateral, yellow; fruit a fleshy berry with a soft, relatively thin rind. ____ ***Hamelia***

  2' Leaves opposite; corollas radial, white or yellow; fruits dry or with a hard or leathery rind.

    3. Twigs with rigid spines. ____ ***Randia*** (in part)

    3' Plants unarmed.

      4. Calyx and corolla 6(7)-lobed; fruits ≤ 3 cm in length, dry capsules; seeds winged. ***Hintonia***

      4' Calyx and corolla 5-lobed; fruits > 5 cm, indehiscent with fleshy, black pulp; seeds not winged. ____ ***Randia laevigata***

### *Cephalanthus*

Shrubs and some trees. Temperate to mostly tropical North and Central America, Asia, and southern Africa; 17 species. Reference: Ridsdale (1976).

***Cephalanthus salicifolius*** [*C. occidentalis* var. *salicifolius*]. Button bush; *mimbre, cuaje; guatájori* (Guarijío); *mimbro* (Mayo) (figure 227)

Large shrubs or trees to 10 m; trunk sometimes thick, along rivers often gnarled from flood damage, or in the more protected canyons the bark thick and deeply furrowed, the crown commonly broad and spreading; roots often in water. Herbage glabrous or finely pubescent. Leaves fully to sometimes partially winter deciduous, opposite or whorled (often on the same branch) with 3 leaves per node, mostly 4.5–12 × 0.9–2.8 cm, moderately bicolored, narrowly to broadly elliptic to oblanceolate, the midrib prominent; petioles 2.0–6.6 mm; stipules 1.3–3.0 mm, broadly triangular, red-brown, thin toward the margins and with two to several dark-purple, terminal, deciduous glands.

Inflorescences of densely flowered heads 20–25 mm wide exclusive of the exserted styles, terminal and axillary from upper nodes; peduncles 2–8 cm with one or more leafless, bracteate nodes. Flowers white, approximately 1 cm, each subtended by slender, blunt-tipped, tiny bracts. Calyx 0.5 mm, reddish, pubescent, 4-lobed; corollas 8–10 × 3–4 mm, with a slender tube and 4 short, spreading lobes. Stamens 4; anthers not extending beyond the corolla tube. Styles elongating 5–10 mm beyond the corolla; stigmas club-shaped. Flowers with sweet fragrance strongest at night ("in the evening it throws out a sweet rangey odor," note by Howard Gentry on a University of Arizona herbarium specimen). Individual fruits 7–8 mm, obpyramidal, 1- or 2-seeded, aggregated into a semiwoody ball. Dry, dead receptacles globose, woody, and semipersistent on the dry peduncles. Flowering April–September.

Riverbanks and streams in canyons and valley bottoms in mountainous areas, usually with permanent, flowing water; riparian thornscrub and riparian tropical deciduous forest; east-central and southeastern Sonora, 275–800 m (900–2625 ft); generally rather rare or sometimes locally common in widely scattered populations. The northernmost trees are known from the Sierra Aconchi, and the greatest concentration of thick-trunked, larger trees is in remote regions of the Guarijío territory in the Río Mayo drainage. Also Chihuahua and Texas to Honduras.

The roots are often submerged, and the trunk may be under water in flood season. Within any single flower head the flowers open simultaneously. The anthers mature first, apparently within the bud. The stigma

Figure 227. *Cephalanthus salicifolius.* Cipriano Buitamea Romero next to a tree 8 m tall with a trunk 2.5 m in circumference in a riparian canyon near Rancho San Pedro, mountains about 10 km south of Guajaray, Río Mayo region. Photograph by R. S. Felger.

elongates after the flower opens and subsequently becomes receptive. The flowers might be both insect and wind pollinated.

The closely related *C. occidentalis* is a riparian shrub in southern Arizona, and elsewhere sometimes a tree. It is widespread in temperate to tropical North America and Cuba, and may occur in northeastern Sonora. The calyx tube is densely hairy in *C. salicifolius* but glabrous in *C. occidentalis*.

### *Hamelia*

Shrubs or small trees, often with showy flowers. Tropical America; 16 species. Reference: Elias (1976).

Scarlet bush (*H. patens*) is cultivated in the tropics and subtropics for its red flowers.

***Hamelia xorullensis*** (figure 228)

Small trees 6–9 m, with a well-formed trunk and spreading crown. Twigs and petioles as well as inflorescences and fruits, especially when young, with sparse to dense pubescence of short hairs; leaf blades sparsely and finely pubescent beneath, especially in the angles of the major veins, often glabrous with age. Nearly evergreen or partially drought deciduous before the summer rains; leaves opposite and in whorls of 3, often 6–15 cm, petioled; blades elliptic or lanceolate to oval or ovate, tapering at both ends,

Figure 228. *Hamelia xorullenis*. Drawn by M. B. Johnson.

thin, slightly darker above, pinnately veined, the veins prominent below; stipules small, deciduous.

Flowers 3 or more in small, terminal cymes; pedicels with a conspicuous, raised nectary at about midpoint. Calyx with 5 slender, persistent teeth 2–3 mm. Corollas approximately 3.5 cm, 5-lobed, bilaterally symmetrical, yellow, and showy. Fruit a firm, semifleshy berry approximately 1 cm, oval, with low ribs, the surface more or less smooth between the ribs except sometimes with a few wartlike lenticels and a prominent ring near the tip representing the scar of the floral parts; green, turning blackish, littering the ground under the trees in May. Seeds small, numerous. Flowering October and perhaps much of the rest of the year.

Upper Río Cuchujaqui drainage in southeastern Sonora at 480–900 m (1575–2950 ft); fairly common locally in riparian canyon bottoms among tropical evergreen vegetation surrounded by tropical deciduous forest. Known for Sonora from several records including Arroyo Verde, Arroyo Santa Bárbara, and nearby in the Cajón of the Río Cuchujaqui at San Pedro. Southward to Michoacán and Colima, and expected in southwestern Chihuahua.

### *Hintonia*

Shrubs and trees. Mexico to South America; 4 species.

*Hintonia latiflora* [*Coutarea latiflora; C. pterosperma*]. *Copalquín, amargo; jutitió* (Guarijío); *tapichogua* (Mayo) (figure 229)

Slender, upright trees or shrubs, 4–8 (10+) m, often with a single straight, slender trunk; bark on trunk of larger trees corky, deeply furrowed, and checkered, the bark of smaller branches or stems pale gray with conspicuous lenticels. The largest trees, in the Guarijío region above Mesa Colorado (north of Alamos), have dense, spreading crowns. Petioles, leaf veins, pedicels, and calyx often with small white hairs, otherwise glabrous. Leaves produced with summer–fall rains and gradually drought deciduous, opposite, 5–11 cm; blades thin, broadly elliptic or oval, the veins prominent, the leaf tips mostly acute and often acuminate; stipules 2–3 mm, broadly triangular above the fused bases.

Flowers white, fragrant, especially in the evening with a strong jasminelike perfume, solitary in leaf axils, often with several flowers near ends of twigs, pedicelled. Calyx and corolla lobes each 6 (7); calyx lobes slender, semipersistent, 5–33 mm; corollas 6–9 cm, campanulate funnel shaped, showy, the tube often greenish, otherwise white, often falling in the morning; buds yellow-green. Anthers 2 cm. Capsules woody, (1.5) 2–3 cm, oval, 2-chambered, with 6–11 low ribs and conspicuous, wartlike lenticels on the ribs and scattered on the surface between the ribs, this surface rough (rugulose); capsules often persistent for many months, separating partially to fully into two longitudinal halves to reveal the seeds, the calyx lobes eventually wearing away, their bases forming scars at the top of the capsule; seeds numerous, flat, papery winged. Flowering mostly with summer rains, (May) July–September (November–February); fruits mostly in early fall.

Slopes, valley plains, canyons and arroyos; Sonoran desertscrub, thornscrub, tropical deciduous forest, and lower elevations in oak woodland; central and southern Sonora, near sea level–1065 m (3500 ft). Bahía San Pedro, north of San Carlos, is the northernmost coastal locality. Also southwestern Chihuahua to Guatemala.

The bark is used as a febrifuge and antimalarial remedy in many parts of Mexico; the trunks, especially if near a road, are invariably scarred from bark collecting. The bark is harvested from the Alamos region, made into capsules in Navojoa and sold commercially, and it is likewise harvested in many other parts of Mexico. Known as *amargo* because of the bitter flavor, the tea is drunk as a purgative for intestinal parasites, as an energy tonic, and to "restore the blood," and reduce fevers. This tea is often used when the seasons change from hot to cool weather. It should

Figure 229. *Hintonia latiflora*. Drawn by M. B. Johnson.

be brought into cultivation in near frost-free regions, both as an ornamental landscape plant perhaps and as an agronomic crop for the bark. It is one of the few attractive flowering trees blooming during the summer rainy season.

### *Randia*

Woody shrubs and sometimes trees. Leaves opposite. Stipules small. Flowers bisexual or unisexual, showy, white or yellow, and fragrant, with spreading lobes from a slender tube. Fruits indehiscent, globose to oval, with a thick leathery to hard rind (pericarp); seeds multiple, flattened, embedded in blackish, sweet and edible, fleshy mesocarp.

Tropical regions worldwide; 200 species, mostly shrubs; 7 species in Sonora.

At least some of the species are dioecious or mostly so. *Randia sonorensis* is found in scattered riparian canyons in the Sierra El Aguaje north of San Carlos and eastern Sonora as far north as the Sierra el Tigre, and occasionally approaches tree size. In early June, the leafless shrubs are often snow white in full bloom, spectacular against a dull landscape. The poorly known *R. mollifolia* is a spindly shrub that apparently does not reach tree size in Sonora. It is somewhat similar to *R. laevigata* but differs in having softly pubescent herbage; both have yellow flowers.

1. Plants unarmed; leaves 15–30 cm, glabrate and glossy green; flowers yellow. ________ ***R. laevigata***

1' Plants armed with sharp spines; leaves 1–11 cm, often pubescent and not glossy; flowers white.
   - 2. Twigs with spines in pairs; leaves glabrous to pubescent; fruits ≤ 1 cm in diameter. ___ ***R. obcordata***
   - 2' Twigs with spines in clusters of 2–4; leaves pubescent; fruits ≥ 3 cm in diameter.
     - 3. Leaf tips mostly acute; fruits > 5 cm in diameter, persistent, armed with conspicuous blunt spines, the rind hard, yellowish brown; flowers ≥ 3 cm in width. ________ ***R. echinocarpa***
     - 3' Leaf tips mostly obtuse; fruits approximately 3 cm in diameter, falling soon after ripening, unarmed; the rind soft, leathery, and green; flowers ≤ 2 cm in width. ___ ***R. sonorensis*** (not treated in text)

***Randia echinocarpa*.** Spiny-fruited randia; *papache, papache borracho*; *josocura* (Guarijío); *jósoina* (Mayo); *piisi* (Yaqui) (figure 230)

Usually large shrubs, and sometimes trees in riparian tropical deciduous forest in southeastern Sonora, especially in the lower Guarijío region, reaching 8 (12) m, with a well-developed but rather slender trunk; larger limbs often supported by other kinds of trees. Bark smooth, gray, or on older limbs or trunks somewhat irregular and flaking. Branches and twigs or short shoots stout, opposite, and decussate; leaves clustered at the ends of short shoots above 2–4 usually stout, woody thorns often 9–20 mm. Foliage drought deciduous, the dormant buds covered by stipule-like scales; inner scales green, their inner surfaces slimy with mucilage from colleter glands. Leaves 4–11 cm, obovate, minutely and often densely hairy at first, especially below, to often glabrous with age, tapered at the base and without a distinct petiole, or the blades narrowed to form a short, winged petiole, or the larger leaves with petioles 5–6 mm, the veins prominent and raised on the lower surface. Stipules often 5–6 mm, broadly triangular-ovate, at first thick, green, and firm, enclosing the next pair of leaves, and with maggot-shaped colleters often 0.25–0.5 mm on the inner surface near the base.

Flowers with gardenialike fragrance, especially strong at night, on short shoots, both bisexual and unisexual flowers on the same plant, or the plants apparently at least sometimes dioecious; female flower buds and flowers with a markedly tuberculate-knobby ovary; male flowers with partly exserted anthers and a reduced ovary. Corolla tube and especially the calyx, ovary, and developing fruit densely hairy. Calyx tube 7–9 mm, the lobes slender, 1.5–10 mm. Corollas white, 3.2–5.8 cm wide; tube 1.5–4.2 cm,

Figure 230. *Randia echinocarpa*. Drawn by M. B. Johnson.

pale green, densely pubescent with appressed hairs outside and longer, spreading hairs inside, these densest near the mouth and surrounding the anthers; lobes white, 4 or 5 (6), spreading, oval, 1.5–3.5 cm. Mature fruits yellow, persisting on the branches long after drying and becoming brown or black, somewhat rounded or slightly longer than wide, the body often 5–8 cm maximum diameter, with a hard, thick shell or rind and stout, irregular, rather blunt and broadly flattened or winged irregular tubercles or spines often to 2–3 cm. Seeds many, 8–10 mm maximum diameter. Flowering mostly June–July and sporadically December–April.

Slopes, valleys, plains, canyons, and washes; thornscrub and tropical deciduous forest; central, southeastern, and extreme southwestern Sonora, 30–800 m (100–2625 ft). Sonora and adjacent Chihuahua to Guerrero and Veracruz.

As with other randias, the gooey, blackish pulp is sweet and edible. However, it is said that eating too much of it causes one to become *borracho* (drunk), hence the common name. The fruits are widely sold for medicinal purposes in markets in Sonora and elsewhere in Mexico. The Guarijío boil the fruits and roots to make eyedrops used when eyes get "clouded." The fruits are almost invariably pecked or chewed open by birds or mammals.

***Randia laevigata.*** *Sapuchi; guacabé* (Guarijío).

Unarmed shrubs or sometimes trees 3–6 m (mostly in southeastern Sonora), unsymmetrical with few, irregular branches. Glabrous or nearly so, or the young herbage densely to sparsely pubescent with short, thick, and often curved or hooked white to tawny hairs, especially along the major veins and vein axils. Leaves 15–30 × 8–15 cm, broadly ovate to oval or elliptic, glossy green, narrowed to a short, winged petiole; stipules 6–13 mm, triangular.

Calyx 4–7 mm including slender lobes 1–4 mm. Corollas yellow, often 7 × 5 cm with a slender tube 4 cm. Fruits on a stipe approximately 3 cm, the body 6–7 cm, rounded to oval, with a smooth, firm, leathery rind. Seeds approximately 1 cm diameter, light brown. Flowering June.

Slopes and canyons, mostly in small, scattered colonies, sporadically in tropical deciduous forest and foothills thornscrub, especially just below the oaks and lower oak woodland; eastern Sonora; 200–1050 m (655–3450 ft). The northernmost record is from the Río Bavispe region (see *Stemmadenia tomentosa*). Also southwestern Chihuahua to Durango and Nayarit.

***Randia obcordata.*** *Papache borracho; cabijósocura* (Guarijío); *pisi, namuquia* (Mayo)

Shrubs or rarely slender trees to 8 m with several trunks, the largest 12 cm in diameter, somewhat gnarled, with brown-gray bark flaking away in large pieces. Leaves glabrous or sparsely pubescent with short, white hairs. Long-shoot branches rigid and divaricate, with paired, often stout thorns to 1+ cm. Short shoots or spur branches knoblike, with fascicled leaves 1–3 (4) cm; blades often broadly obovate with a shallowly notched apex and narrowed to a winged petiole or nearly sessile.

Flowers white, approximately 6 mm; calyx approximately 3 mm including lobes 1 mm. Fruits globose, 8–10 mm in diameter, blackish, smooth and glabrous. Flowering mostly June–July.

Tree-sized plants known from central Sonora in the vicinity of Mazatán and San Javier, at the ecotone of Sonoran desertscrub and thornscrub. Plains, valleys, slopes and arroyos; Sonoran desertscrub, thornscrub, the dry margin of tropical deciduous forest; southern and central Sonora; near sea level–600 m (1970 ft). Also in northwestern Sinaloa.

This plant is called *papache borracho* because eating too much of the fruit is said to make one drunk. It can be confused with the more common and widespread shrub *R. thurberi*, distinguished by larger flowers and much larger green-and-white mottled fruits.

## Rutaceae—Citrus or Rue Family

Mostly shrubs or trees, armed or unarmed, deciduous or evergreen. Leaves and other organs gland dotted, producing essential oils and often aromatic. Leaves alternate or opposite, simple or compound; without stipules. Flowers in panicles or racemes, unisexual or mostly bisexual, and with a disk; sepals, petals, and stamens each 4 or 5. Ovary superior. Fruits various, fleshy or dry.

Nearly worldwide, mostly tropical and subtropical; 150 genera, 1500 species. Reference: Brizicky (1962).

Many kinds of citrus (*Citrus*) are cultivated in Sonora, and feral, shrub-sized plants are occasionally encountered in extreme southwestern Sonora.

1. Twigs with sharp, stout spines.
    2. Leaves simple. ______ ***Citrus*** (not treated in text)
    2' Leaves pinnately compound. ____ ***Zanthoxylum***

1' Plants unarmed.
    3. Leaves simple. __________________ ***Esenbeckia***
    3' Leaves compound.
        4. Leaves opposite. __________________ ***Amyris***
        4' Leaves alternate.
            5. Leaflets 5, 7–25 cm; fruits fleshy and globose; riparian tropical deciduous forest in southeastern Sonora. _______ ***Casimiroa***
            5' Leaflets 3, 3.5–10 cm; fruits papery and disk shaped; elevations above desert and thornscrub in eastern Sonora. _________ ***Ptelea***

### *Amyris*—Torchwood

Trees and shrubs. Tropical America northward to southern Florida, southern Texas, and Sonora; 30 species.

***Amyris balsamifera.*** *Limoncillo, palo limón; usim yuera* (Mayo)

Unarmed; multiple-stemmed shrubs or trees 3.5–6 m with a well-formed trunk. Bark gray-brown, smooth. Young herbage, especially stems, leafstalks, leaflet midribs, inflorescences, and calyces pubescent, with minute, mostly curved white hairs. Leaves evergreen (or perhaps sometimes tardily drought deciduous), opposite, shiny green, 3–15.5 cm, odd-pinnate with a prominent petiole; leaflets 5–7, stalked, often 2–5+ cm, mostly ovate to oval or elliptic, bright shiny green above, duller below, the veins pinnate, the midrib especially prominent on the lower surface, the margins entire or crenate.

Panicles axillary, shorter than the larger leaves. Flowers probably bisexual (or perhaps unisexual?), small; pedicels 0.5–2.0 mm, with 2 small bracts. Calyx approximately 1 mm, cuplike, 4-lobed, persistent, the lobes thickish, broadly triangular. Petals 4, white, deciduous, and probably more than twice as long as the calyx. Stamens 8. Ovary short-stalked (stipitate); stigma capitate, sessile (style absent). Fruits fleshy drupes, 1+ cm, blackish, mostly elliptic to oval, with conspicuous, rounded glands embedded in the pericarp; 1-seeded. Flowering at least February–March.

Southwestern and west-central Sonora, known only from arroyos in coastal thornscrub in the vicinity of Camahuiroa near the Sinaloa border and a small, isolated population of shrubs farther north in riparian Sonoran desertscrub in the canyon bottom at Cañón las Barajitas north of San Carlos; 20–70 m (65–230 ft). Tropical America including southern Florida and the Caribbean to South America and western Mexico.

### *Casimiroa*

Tropical highlands of Mexico and Central America; 6 species.

***Casimiroa edulis.*** White sapote; *zapote blanco, chapote, zapote dormilón* (Sinaloa) (figure 231) *ajpé, jijira* (Guarijío); *ja'apa ahuim* (Mayo)

Small to medium-sized leafy trees forming dense shade; trunk thick, sometimes as much as 1 m in diameter; branches and twigs sometimes knotted and conspicuously swollen. Bark with conspicuous lenticels. Herbage at first finely pubescent, the twigs soon glabrous, the leaves glabrate except the petioles. Leaves facultatively drought deciduous, alternate (a few occasionally subopposite), long-petioled, palmately compound; leaflets (3) 5, stalked (or sometimes sessile if stressed), elliptic with blunt to pointed tips, dotted with small glands; larger leaflets 7–25 cm, 3–8 cm wide, the margins essentially entire to minutely and shallowly toothed with broad, flat-topped teeth; fresh, crushed leaves sometimes with a vague citruslike aroma or foul smelling, or sometimes nearly odorless.

Panicles axillary or terminal. Flowers greenish white, bisexual (reportedly sometimes with some unisexual flowers), fragrant, approximately 6 mm wide, 5-merous; stigmas sessile, star shaped. Fruits 8–10 cm in diameter, the skin thin and yellow, the pulp juicy, of delicate texture, soft and cream colored, with a delicious sweet flavor. Seeds 5, approximately 2.5 cm, enclosed by a tough parchmentlike membrane. Flowering winter and spring; fruits in summer.

Riparian canyons in tropical deciduous forest; southeastern Sonora, 250–1060 m (820–3475 ft). Probably not native to Sonora except for an isolated population in Arroyo Verde on the upper Río Cuchujaqui and perhaps in certain other similar canyons. Often persisting at abandoned

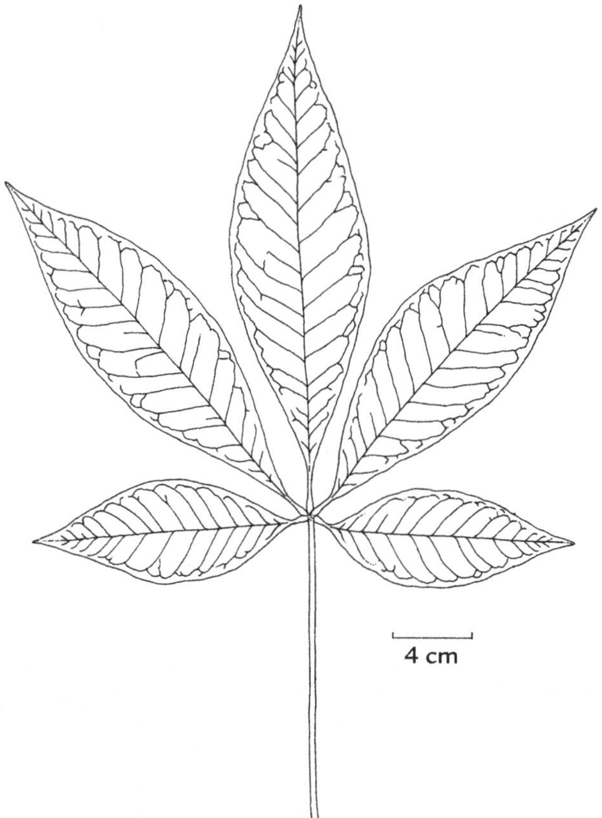


Figure 231. *Casimiroa edulis.* Drawn by M. B. Johnson.

habitations and weakly naturalized along certain canyons (e.g., the Río Chico several kilometers north of Nuri; tributaries of Arroyo San Bernardo; the north side of Sierra de Alamos, above La Aduana; and the vicinity of La Vinatería).

This tree is native from Sinaloa to Central America. It is widely cultivated in tropical and subtropical regions of Mexico and Central America. The trees withstand several degrees of frost and are cultivated in California and occasionally in southern Arizona. The delicious fruit is delicate and does not tolerate shipping. There are many cultivated varieties.

During the 1980s many of the Sonoran trees were cut down, apparently because the fruit is host to various fruit flies which are major agricultural pests. The tree has been much used medicinally in Mexico, especially the seeds (e.g., Martínez 1969, Standley 1923). In Sinaloa and elsewhere it is called *zapote dormilón* because of its use as a remedy for insomnia, either by eating the fruit or placing the leaves in pillows. The bark, and especially the seeds have been used as a sedative, although they may also be toxic. The leaves have been used as a remedy for diarrhea. In Sonora an infusion of the leaves has been used to lower blood pressure (Gentry, 1954, note on University of Arizona herbarium specimen). In Gentry's (1942) Río Mayo Plants this tree is misidentified as *Manilkara zapota* (*Achras zapota*, Sapotaceae).

### *Esenbeckia*

Subtropical and tropical America, from the southern border of Texas and northern Mexico to Chile and Argentina; 24 species.

Two species occur in the Sonoran Desert region. *Esenbeckia flava*, endemic to Baja California Sur, has fruits much larger than those of *E. hartmanii*. The bark of *E. febrifuga* from South America has been used as a substitute for quinine in treating fevers.

***Esenbeckia hartmanii.*** *Palo amarillo; toshcura* (Guarijío); *jójona* (Mayo) (figure 232)

Mostly shrubs 2.5–4 m, sometimes trees 5–7 (10) m, the wood hard, heavy, and strong. Bark gray, smooth, or exfoliating in small, thin, irregular plates on the trunk and larger limbs, the inner bark yellow (hence *palo amarillo*). Twigs and lower leaf blade surfaces with short, white hairs or sparsely pubescent with age. Leaves gradually drought deciduous, alternate, simple, 2.5–8 × 1–3 cm; petioles 1–10 mm; blades elliptic to obovate, glabrous above, tough and dry to the touch, dark green, the tip rounded to obtuse or often shallowly notched, the margins otherwise entire or very shallowly wavy.

Panicles often 3–5.5 cm, terminal, often on short shoots. Flowers white, bisexual, 10–12 mm wide, 5-merous. Petals and sepals conspicuously gland dotted; sepals one-quarter as long as the petals. Style 2.0–2.5 mm; stigma rounded; ovary base surrounded by a yellow to orange disk. Fruits solitary or several per panicle, of hard, woody capsules 2–2.5 cm wide, covered with short, blunt spines and projections; carpels often 5, each with a single seed, although some often aborting. Seeds 7–8 mm, rounded, the surface patterned like a minute woodblock print. Flowering June–September, apparently mostly with summer rains, and sometimes in March.

Slopes, valleys, plains, canyons, and coastal dunes; tropical deciduous forest, thornscrub, and margins of the Sonoran Desert; northeastern to southern Sonora, sea level–760 m (2500 ft). Along the eastern margin of the Sonoran Desert from the Río Bavispe region southward, approaching the coast in the Sierra el Aguaje north of Guaymas, and coastal in southwestern Sonora. Also in Sinaloa.

Tree-sized esenbeckias mostly occur in tropical deciduous forest in southeastern Sonora, especially the Guarijío region north of Mesa Colorado.

### *Ptelea*

Shrubs or sometimes small trees. United States and Mexico; 3 species. Reference: Bailey (1962).

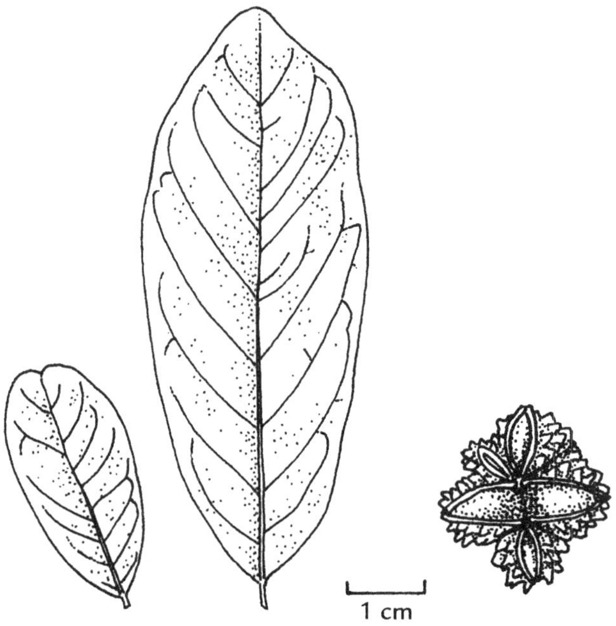


Figure 232. *Esenbeckia hartmanii*. Drawn by M. B. Johnson.

***Ptelea trifoliata*** subsp. ***angustifolia*** [*P. angustifolia*]. Hoptree (figure 233)

Shrubs and occasionally small, slender trees; herbage aromatic. Leaves winter deciduous, alternate, glabrous or glabrate; leaflets 3, 3.5–10 × 1.5–5 cm, lanceolate to obovate, thin; lateral leaflets unequal, the margins minutely toothed to entire.

Panicles terminal. Flowers greenish white, unisexual (female) and bisexual, 11–15 mm wide, 4- or 5-merous; sepals 1–2 mm; petals 4–5 mm; style short; stigmas 2. Fruits flattened, papery, disk-shaped samaras 1–2.3 cm wide, with a conspicuously veined, broad wing encircling the fruit. Flowering July.

Slopes and arroyos; grassland, oak and pine-oak woodland; eastern Sonora, 915–2285 m (3000–7500 ft). Only a few collections known from Sonora.

Subspecies *angustifolia* is widespread in Mexico, and also found in Texas, New Mexico, and Arizona. This highly polymorphic species occurs in the eastern and southern United States to Arizona and southern Mexico. The leaves may cause dermatitis in some people. The fruits have occasionally been used as a substitute for hops in brewing beer.

### *Zanthoxylum*

Trees and shrubs. Nearly worldwide, mostly tropical. Two species in Sonora, *Z. fagara* in subgenus *Fagara* with about 200 species, and *Z. mazatlanum* in subgenus *Zanthoxylum* with about 15 species primarily in temperate eastern Asia and North America and at least 2 species in Mexico and Central America.

*Zanthoxylum mazatlanum* is shrub 2–3 m tall, known in Sonora from widely scattered localities including canyons in the Sierra el Aguaje north of San Carlos, and a few records in thornscrub and tropical deciduous forest in southern Sonora, Durango, and Sinaloa (Felger 1999 and unpublished notes).

1. Leaf rachis winged; fruits with a stipe. ___ ***Z. fagara***
1' Leaf rachis not winged; fruits sessile. ___ ***Z. mazatlanum*** (not treated in text)

***Zanthoxylum fagara*** [*Z. sonorense*; *Fagara pterota*]. Lime pricklyash; *limoncillo; o'ouse sutu* (Mayo) (figure 234)

Spinescent hardwood shrubs or rarely small trees to 5+ m, with single to multiple trunks 10+ cm in diameter; bark gray and smooth except conspicuous lenticels; bark and leaves aromatic. Branches rigid, with sharp, slightly curved, paired spines in the position of stipules to 1 cm on long shoots, and with unarmed short shoots. Herbage, inflorescences, and calyces pubescent with short, spreading white hairs, or glabrate or glabrous with age. Leaves alternate, pinnately compound, with citruslike fragrance when crushed, 3.5–10.5 cm, the rachis winged; leaflets (3) 5–9, ovate to obovate, elliptic or oval, 8–30 mm, with a pair of calluslike glands near the base on the lower surface, the margins crenulate.

Probably dioecious. Flowers in small, dense, axillary racemes, minute, dull white, on pedicels approximately 1 mm, fragrant and visited by bees; sepals, petals, and stamens 4. Sepals minute, persistent, overlapping, the petals to approximately 2 mm, probably smaller on female flowers, deciduous. Female flowers with 2 separate carpels mounted together on a common, disklike stalk. Fruits capsular (follicles), each carpel (follicle) with a stipe 1–3.5 mm, the body 3.5–4 mm, rounded, gland dotted, splitting open to reveal the single seed. Seeds (2.5) 3.0–3.3 mm, rounded to broadly ellipsoid, shiny blackish, remaining attached to the gaping fruit and partially protruding. Flowering with warm to hot weather, especially following rains in summer; fruiting primarily late summer and fall.

Canyons, arroyos, open slopes, and valley plains; widely scattered in the southern and eastern margin of the Sonoran desert usually in riparian habitats, thornscrub, and tropical deciduous forest; western Sonora northward to the Sierra el Aguaje and eastern Sonora northward to the vicinity of Arizpe; near sea level–950 m (3115 ft). Also Baja California Sur, Sinaloa, Texas, and Florida to South America and the Caribbean.

In Sinaloa the wood and bark are chewed as a remedy for toothache.

## Salicaceae—Willow Family

Trees and shrubs, the wood often soft and the trees not especially long lived in the lowlands of Sonora. Leaves usually winter deciduous, alternate, simple; stipules usually present and soon deciduous; petioles usually present. Dioecious. Inflorescences of catkins, the flowers each subtended by a scalelike bract, mostly wind pollinated. Sepals vestigial; petals absent. Fruits small capsules. Seeds minute, each with a tuft of long silky hairs for wind dispersal.

Nearly worldwide, mostly temperate regions of the Northern Hemisphere; 2 genera, 340 species. Even the Sonoran Desert and subtropical Mexican species are winter deciduous, indicating their north temperate affinity.

Great gallery forests of willow and cottonwood lined the rivers of Sonora before construction of large dams during the twentieth century. In the hot Sonoran climate their growth was fast and biomass enormous, especially in the lower reaches of the Mayo, Yaqui, and Colorado rivers. Today only remnants of these riverine forests remain along the Mayo and Yaqui rivers. The lower Colorado River forests disappeared before the mid-twentieth century. The

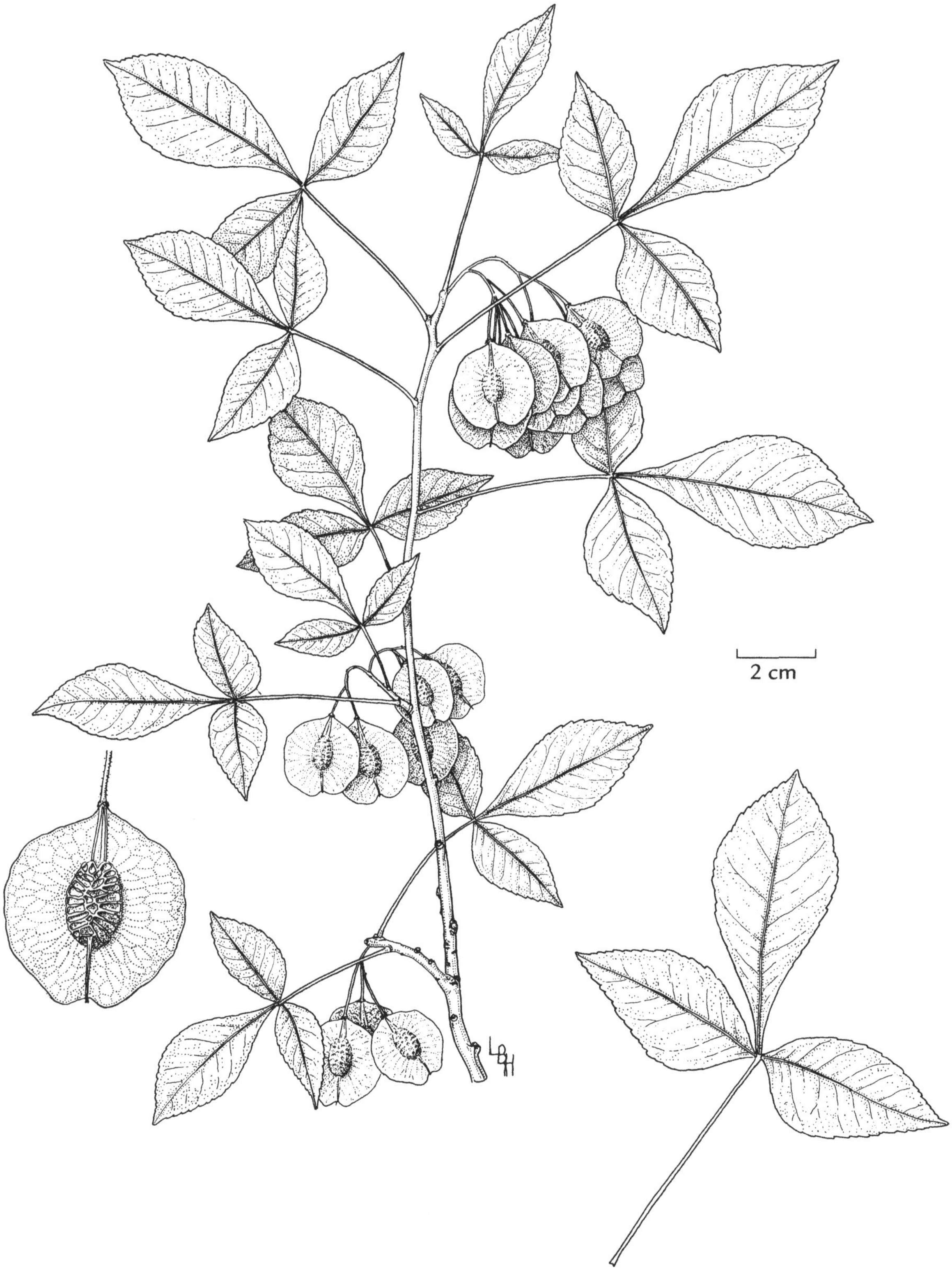


Figure 233. *Ptelea trifoliata*. Drawn by L. B. Hamilton.

Figure 234. *Zanthoxylum fagara* (Sargent 1890). Drawn by C. E. Faxon.

plants form extensive networks of shallow roots, which are extremely important in slowing soil erosion and often harbor ectomycorrhizal fungi. The destruction of cottonwood and willow forests undoubtedly has contributed to the periodic severe river flooding in Sonora and Arizona.

Cottonwoods and willows are often protected or planted as shade trees and for various utilitarian purposes. They readily grow from cuttings of any size. The best time to obtain cuttings is in winter before the buds leaf out. Along river valleys, such as the Río San Miguel north of Hermosillo, cottonwood and willow cuttings, placed along streamside fields, are woven together and develop into living fences to protect adjacent fields (Nabhan and Sheridan 1977). The leaves and young twigs of *Populus* and *Salix* are eagerly eaten by cattle and deer, and seedlings and saplings can be eliminated by cattle. For the most part the wood is soft and of little value for fuel and construction but was much used because it was so readily available. Wood from these trees fired the boilers of steamboats that carried freight and passengers on the lower Río Colorado during the latter half of the nineteenth century.

Along the southern Sonoran rivers, in an essentially frost-free climate, the new leaves often emerge at the end of the short winter even as the old leaves are still falling, and the trees are seldom, or only briefly, leafless. Along the northern rivers and streamways the leaves turn yellow in winter—the cottonwoods more brilliantly than the willows—and new leaves burst forth in late winter or early spring. The blaze of yellow and flush of green along the Río Colorado must have been spectacular.

1. Leaf blades often broad, lanceolate to triangular or rounded, less than twice as long as wide, or if more than twice as long as wide, then with resinous buds (*Populus angustifolia*) or with narrow leaves only on juvenile plants and some lower branches (*P. mexicana*); vegetative and floral buds resinous or not, with several scales; catkins drooping; stamens eight to many. ________ ***Populus***

1' Leaf blades linear to lanceolate, more than twice as long as wide; vegetative and floral buds not resinous, with a single scale; catkins mostly upright; stamens 1–8. ________ ***Salix***

## *Populus*—Cottonwood, Aspen, *Alamo*

Trees. Early-season leaves often different from late-season leaves; stipules often similar to the bud scales; petioles laterally compressed, especially near the blade, or dorso-ventrally compressed and channeled above. Vegetative and floral buds with several scales, resinous or not. Catkins stalked and drooping, appearing in spring before the leaves; female flowers yellow-green, the male flowers yellow.

Northern Hemisphere; 40 species, 8 in Mexico. References: Eckenwalder (1977), Rzedowski (1985).

The southern limit for the genus is in Chiapas. The name "cottonwood" refers to the copious cottony hair that surrounds the seeds and aids in wind dispersal. Female trees can produce enormous quantities of cottony seed drifts that are often seen floating high in the spring skies. The Lombardy poplar (*P. nigra* var. *italica*) is sometimes cultivated in the cooler, higher elevation regions of northeastern Sonora.

1. Leaf blades more than twice as long as wide; petioles less than one-third as long as the blade.
   2. Leaves not dimorphic; petioles dorso-ventrally compressed. ________ ***P. angustifolia***
   
   2' Leaves dimorphic; petioles laterally compressed (flattened perpendicular to the plane of the blade). ________ ***P. mexicana***

1' Leaf blades less than twice as long as wide, often as wide or wider than long; petioles more than half as long as the blade and laterally compressed.
   2. Bark smooth and whitish; leaf blades mostly widest at about middle.
      3. Leaf blades green above, glaucous or lighter below, mostly ovate, the margins coarsely toothed; below 1100 m. ________ ***P. brandegeei***
      
      3' Leaf blades green on both surfaces, broadly ovate to nearly orbicular, the margins finely toothed; above 2000 m. ________ ***P. tremuloides***
   
   2' Bark rough or fissured, brown to gray; leaf blades widest below the middle (near base).
      4. Leaves not dimorphic, all long petioled; blades broadly triangular; north of Guaymas including the upper Río Yaqui drainage. ________ ***P. fremontii***
      
      4' Leaves dimorphic; juvenile-growth leaves short petioled with lanceolate blades; adult-growth leaves long petioled with broadly triangular blades; south of Guaymas in the lower Fuerte, Mayo, and Yaqui river drainages. ***P. mexicana***

***Populus angustifolia***. Narrowleaf cottonwood (figure 235)

Trees to 15+ m; bark light colored, often whitish and rather smooth and almost invariably dotted with regularly spaced pits made by woodpeckers. Vegetative buds elongate-conical, extremely resinous. Young herbage, scales, and petioles sparsely pubescent with minute hairs; leaf blades becoming glabrous or essentially so. Leaves 4–11 (14) × 1.7–3 cm, willowlike, more than twice as long as wide; petioles 4–18 mm, appearing dorso-ventrally compressed (broadly channeled above and rounded below); blades darker above, linear-elliptic to linear-lanceolate, usually larger and broader (lanceolate) on vigorous shoots, the margins serrated with forward pointing, usually gland-tipped teeth.

Canyon bottoms; riparian pine-oak woodland; northeastern Sonora; 1625–1800+ m (5330–5905+ ft). In

Sonora known for certain only from the Sierra San Luis in the upper reaches of Cajón Bonito, and expected elsewhere in the same mountain system (also see *P.* ×*hinkeleyana*). Western North America: Canada to northernmost Mexico including northwest Coahuila and northwest Chihuahua, and the Chiricahua and Santa Catalina mountains in southern Arizona.

Narrowleaf cottonwood is one of the balsam poplars (section *Tacamahaca*). The report of this species from the Río Bavispe region by White (1948) is based on a misidentified specimen of *P. brandegeei*.

***Populus brandegeei*** [*P. monticola*]. *Güérigo* (figure 236C)

Trees to 15–22 m, sometimes with 2 or 3 major trunks and branching rather low or sometimes rather slender with a high, clean bole; trunk often to 1 (1.8) m in diameter. Bark rough and gray at base of larger trees, otherwise smooth and whitish like an aspen, scarring blackish. Leaf and floral buds not resinous; buds and young herbage with whitish, somewhat woolly hairs, otherwise sparsely pubescent or glabrous. Petioles 2–8 cm, laterally compressed; leaf blades 4.5–10.5 cm, mostly broadly ovate, sometimes lanceolate to elliptic or nearly orbicular, the surfaces dull, usually dark green above and somewhat glaucous below, the margins crenate to bluntly toothed.

Canyon bottom streams; oak woodland, and rare in upper elevations in Sonoran desertscrub, thornscrub, and tropical deciduous forest; eastern Sonora; 825–1190 m (2700–3900 ft). Also adjacent Chihuahua, and higher mountains in Baja California Sur. The northernmost records are from the Río Bavispe region near Colonia Morelos and near Arizpe, and Saracachi, northeast of Cucurpe about 85 km south of the Arizona border.

*Güérigo* is a white poplar, in section *Populus* (*Leuce*) which includes the aspens. The red and somewhat hard wood has been used for making furniture and for beams in Spanish colonial mission churches. Occasionally cultivated in northern Sonora and southern Arizona, it is a spectacular landscape tree suitable for situations with ample water.

***Populus fremontii*** subsp. ***fremontii*** [*P. arizonica; P. fremontii* var. *pubescens; P. macdougalii*]. Western cottonwood; *álamo; 'auppa* (Hia-ced and Tohono O'odham); *oopa* (Pima Bajo) (figures 236A and 237)

Large trees with soft wood, capable of very fast growth (along the lower Río Colorado and perhaps elsewhere these trees reached 25–30 m with trunks 1–2 m in diameter). Bark on trunk and older branches gray-brown and deeply fissured. Leaf and floral buds resinous, elongated, sharp

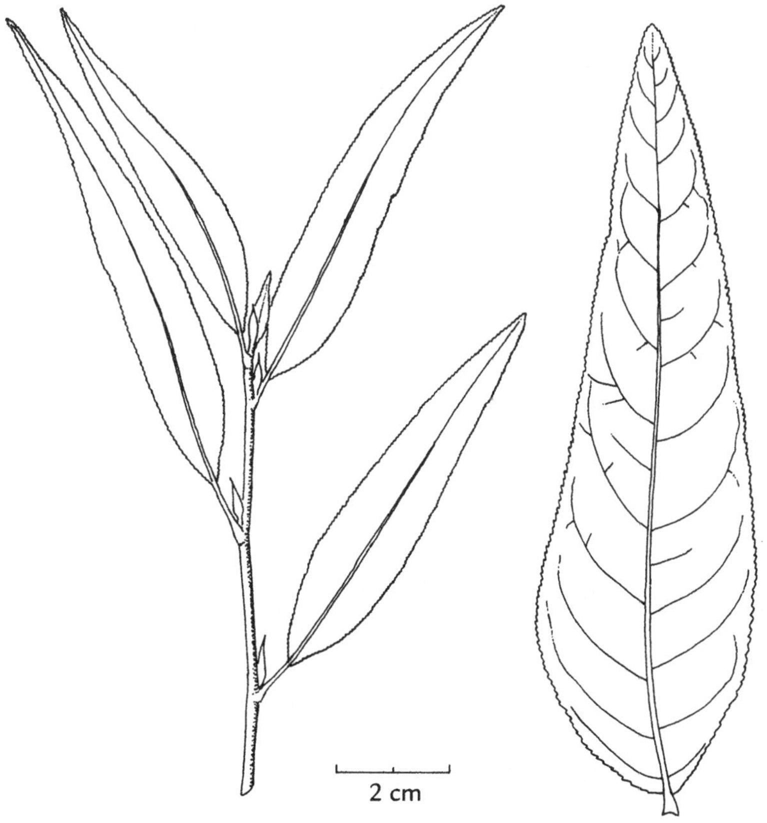


Figure 235. *Populus angustifolia*. Drawn by M. B. Johnson.

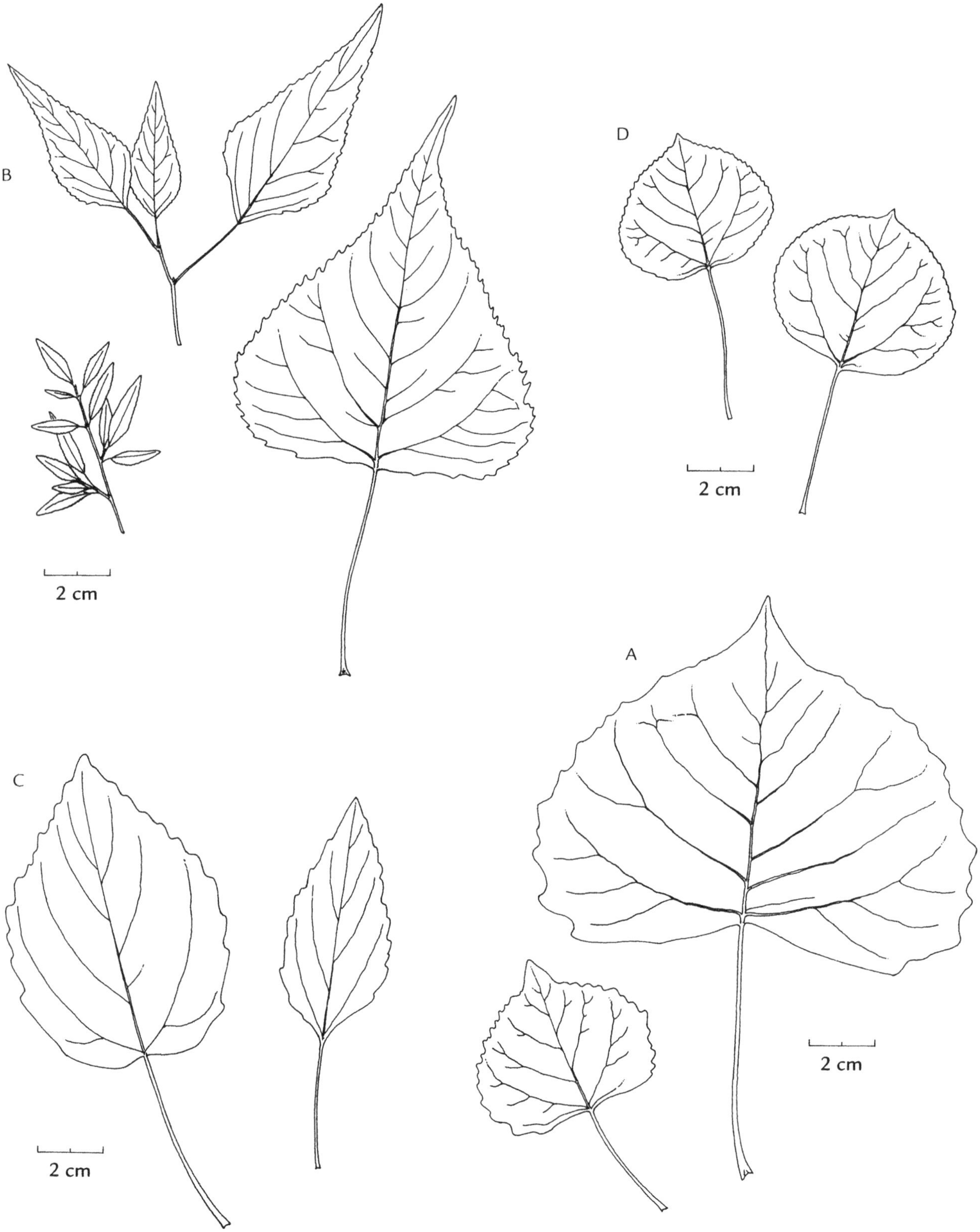


Figure 236. *Populus*. (A) *P. fremontii* subsp. *fremontii*, (B) *P. mexicana* subsp. *dimorpha*, (C) *P. brandegeei*, (D) *P. tremuloides*. Drawn by M. B. Johnson.

Figure 237. *Populus fremontii* subsp. *fremontii* (Benson and Darrow 1945). Drawn by L. B. Hamilton.

pointed, tan, and often pubescent. New shoots and leaves pubescent, becoming glabrate. Petioles 3–9 cm, laterally compressed; leaf blades shiny green on both surfaces, more or less triangular, 4.5–17 cm long and about as wide, the base truncate to cordate, the apex short acuminate to acute. Early-season leaves often with larger marginal teeth than late-season leaves. Flowering in February, the leaves enlarging as the flowers fade; fruits ripen late March–early April.

Canyon bottoms, rivers and streams, larger arroyos, irrigation canals, and fields; Sonoran desertscrub, thornscrub, grassland, and oak woodland; northern, central and east-central Sonora; near sea level–1830 m (6000 ft).

This is the largest native tree in the Sonoran Desert. It was abundant along the Río Colorado and its delta channels and is now cultivated or semicultivated in agricultural areas of the delta (Felger 2000). In addition, some thousands of young trees to about 10 m tall occur in scattered groves along the river channels above the confluence of the Río Hardy and below the limits of agriculture. The river water here has a salinity of 1000–2000 ppm (Ed Glenn personal communication 1997). It is also a common riparian and gallery forest tree in many other parts of Sonora. This tree is much appreciated and widely planted. Mearns (1907: 48), writing about the Mexico–United States border, reported that "No tree is more common, more beautiful, nor more valuable as a shade tree than the cottonwood. It grows naturally on almost every stream along the Boundary, and is planted around the houses and along the irrigation 'acequias' of nearly every ranch." Approaching Sonoyta in late February 1910, Lumholtz (1912: 287) wrote, "Already at a distance the attractive light green color of the cotton-wood trees and willows were evident, the new leaves being half-grown."

There are two subspecies. In Eckenwalder's (1977) view, subsp. *fremontii* occurs west of the continental divide, primarily California and Utah to Baja California and Sonora. It is replaced by subsp. *mesetae* in southern New Mexico and western Texas southward to the Valley of Mexico. Subspecies *fremontii* is distinguished primarily by its relatively broader and shorter leaf blades with relatively shorter tips and truncate to cordate leaf bases. Subspecies *mesetae* has leaf blades longer than broad with relatively longer leaf tips and the bases often not so blunt (cuneate). Some trees from the Río Bavispe region in northeastern Sonora are intermediate between the two subspecies. This species is a member of section *Aigeiros*, which includes the black poplars and other cottonwoods.

***P. ×hinkeleyana*** [*P. acuminata* var. *rehderi*]

Trees to 20 m with a trunk 1 m in diameter; bark whitish. First-season twigs, bud scales, and sometimes petioles sparsely and minutely pubescent with small, white hairs. Buds elongate-conical. Petioles 1.5–3 cm; leaf blades 4–8 cm, broadly to narrowly ovate to elliptic, the tip acuminate (pointed), the margins serrulate.

The only record for Sonora is a 1941 herbarium collection by Stephen S. White in the Río Bavispe region: Cañón de la Mescalera, Sierra de la Caballera, part of the Sierra el Tigre system, north of Colonia Oaxaca, 1510 m (4950 ft) (also see White [1948]). Utah to Chihuahua.

These trees are hybrids between *P. angustifolia* and *P. fremontii*, species in different sections of the genus. The presence of this hybrid in Sonora is problematic because there are no verifiable records of *P. angustifolia* in the Sierra el Tigre.

***Populus mexicana*** subsp. ***dimorpha*** [*P. dimorpha*]. Yaqui cottonwood; *álamo; aba'aso* (Mayo); *abaso* (Yaqui) (figure 236B)

Trees to 30 m, with soft wood, broad, spreading crowns, and often with long, drooping, leafy branches. Leaf and floral buds stout, blunt, slightly to mostly not resinous (dry). Herbage minutely pubescent or glabrate or glabrous. New leaves appearing at flowering time while the old leaves still present or falling. Leaves dimorphic: leaves of mature growth with petioles 4–8.5 cm and laterally compressed; blades 5.5–12 cm, shiny green on both surfaces, triangular, about as broad as long. Leaves of the juvenile phase (seedlings, saplings, and slender, smaller and lower branches of larger trees) willowlike: much smaller, narrower and shorter petioled than those of the adult phase. Flowering January–February.

River banks and large riparian washes and arroyos, irrigation ditches, margins of irrigated farmland, and occasionally in roadside ditches; thornscrub and tropical deciduous forest; southern Sonora; near sea level–395 m (1300 ft).

This cottonwood once formed extensive forests of immense trees along the lower Río Mayo and Río Yaqui and the lower Río Fuerte and other Sinaloan rivers. Although large numbers of individual trees and smaller groves remain, the few remnants of the once great riverine forests are doomed due to damming and diversions of the rivers. Seedlings are abundant along the lower Río Mayo and Yaqui, but saplings are relatively scarce. These riverine trees are often host to large clusters of *Psittacanthus calculatus*, a mistletoe bearing showy, bright orange flowers. Tent caterpillars (*Malacosoma*) often defoliate the trees.

The Mayo and Yaqui people have long appreciated the trees, and many are planted or protected around homes and along the larger irrigation ditches and drainage canals. *Abaso* root is one of the preferred woods used by Mayo and Yaqui men for making Pascola masks, and the green boughs and leafy branches are incorporated into religious ceremonies and fiestas. The Mayo cook the bark with salt and apply the liquid to wounds and bruises. *Abaso* is capable of rapid growth.

There are two widely disjunct subspecies; subsp. *mexicana* occurs in eastern and southern Mexico. This species is the only extant member of section *Abaso*, although there are fossil species from Texas and the northwestern United States. *Populus mexicana* occurs along the lower Río Yaqui, while *P. fremontii* occupies the upper portions of the river drainage.

***Populus tremuloides.*** Quaking aspen; *álamo, temblón* (figures 236D and 238)

Slender trees often to 10 m, or sometimes to 20 m (Sierra de los Ajos), propagating by root suckers, and usually forming groves. Bark smooth, often white with blackish scars. Herbage glabrous; buds conical, moderately resinous. Leaves turning brilliant yellow in fall; petioles (1.5) 2–6 cm, laterally compressed; blades 2.5–7 cm, often slightly wider than long, broadly ovate to suborbicular, the margins shallowly crenulate.

Slopes and canyons; pine and mixed-conifer forest; northeastern Sonora above 2075 m (6800 ft): Sierra de los Ajos, Sierra la Mariquita near Cananea, Sierra el Tigre, Sierra San José, and Sierra San Luis on the Chihuahua border. Alaska to Labrador southward to Baja California and the Valley of Mexico.

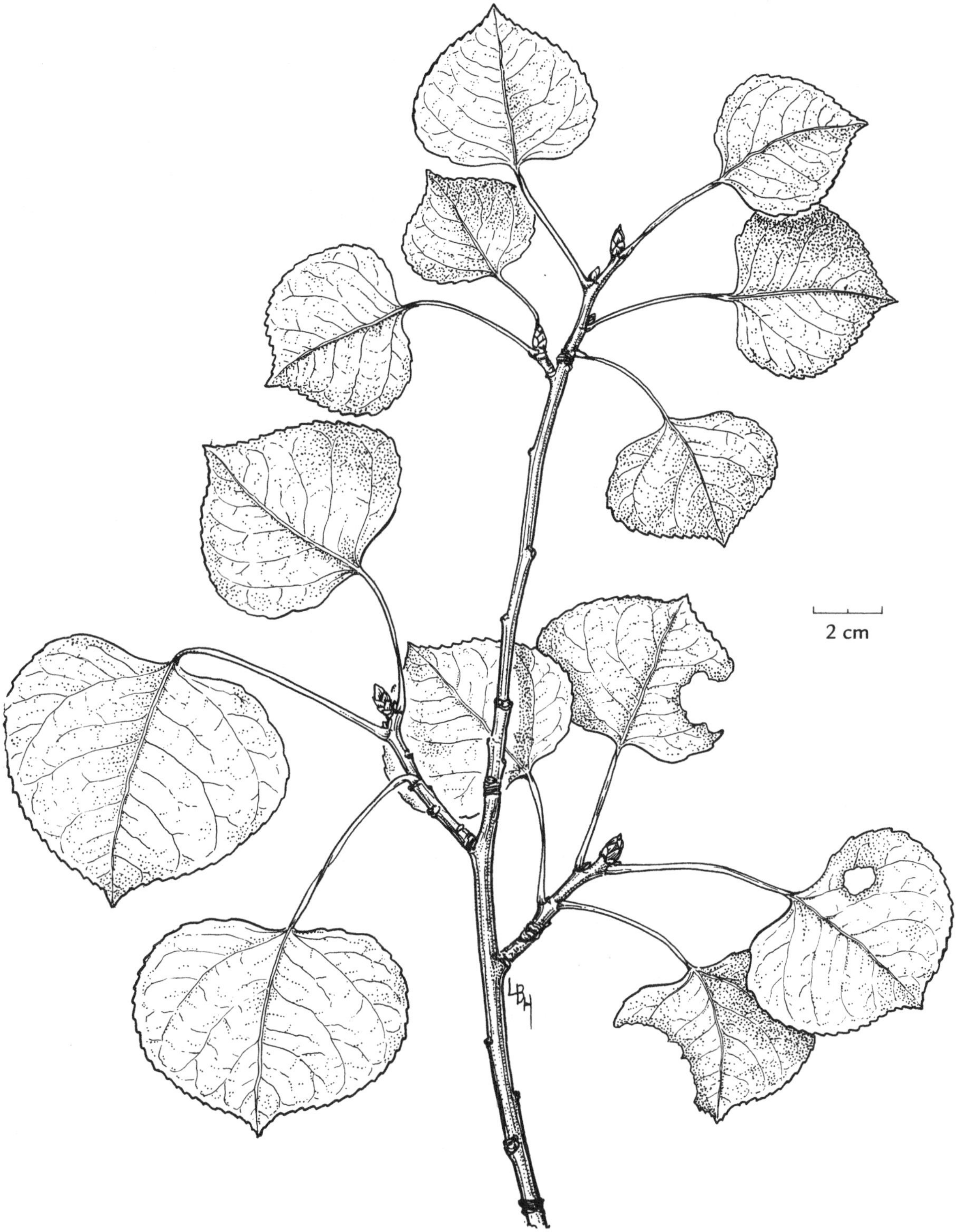


Figure 238. *Populus tremuloides* (Humphrey 1960). Drawn by L. B. Hamilton.

Quaking aspen is closely related to the Eurasian aspen, *P. tremula*, and they are sometimes considered to be conspecific. Aspens are in section *Populus* (*Leuce*).

## *Salix*—Willow, *Sauce, Sauz*

Shrubs to trees. Vegetative and floral buds not resinous, covered by a single, often tubular or caplike scale. Leaves often lanceolate (in Sonora); stipules (in Sonora) thin (scarious) and brown, variously deciduous. Catkins erect, appearing before or with the leaves.

Northern Hemisphere, mostly temperate and arctic regions, and a few in the Southern Hemisphere; 400 species. References: Argus (1986, 1995), Dorn (1976, 1977, 1998).

Several shrub species occur at higher elevations in Sonora, and one at low elevation (see notes on *S. exigua* under *S. exilifolia*). Crosses made of leafy willow twigs are often seen on the walls of homes in southern Sonora, especially in Guarijío, Mayo, and Yaqui settlements.

1. Young twigs and leaves silvery with silky hairs; leaves mostly spreading to ascending (upright), 1–3.5 × 0.2–0.4 cm; stamens 2. ________________ ***S. exilifolia***

1' Young twigs and leaves green (or twigs yellowish or reddish), glabrous or pubescent but the hairs not silky; leaves pendulous, 5.5–20 × 0.8–2.0 cm; stamens 4 or more.
   2. Leaves bicolored, green above, bluish glaucous or at least paler below. ___________ ***S. bonplandiana***
   2' Leaves similarly green on both surfaces. ***S. gooddingii***

***Salix bonplandiana*** var. ***bonplandiana***. Bonpland willow; *sauce, sauz; huatosí* (Guarijío); *huata* (Mayo) (figures 239 and 240A)

Common trees 10–15 (20) m with a well-formed trunk; bark gray-brown, checkered to fissured; twigs reddish. Leafy stems and leaves mostly drooping. Young twigs and leaves glabrous or hairy and becoming glabrate. Stipules ovate with slightly wavy margins. Leaves 7–20 × 1–2 cm; petioles 2.5–10 mm; blades narrowly lanceolate, green above, bluish to whitish-gray glaucous below; margins minutely serrated.

Flowering catkins 2.2–6.3 cm. Female flowers and fruits on a pedicel-like stipe 1.0–2.0 mm. Male flowers with 4–5 (7) stamens. Flowering February–October.

Riparian habitats: canyon bottoms, streams, rivers, and springs; Sonoran desertscrub, thornscrub, tropical deciduous forest, grassland, and lower elevations in oak woodland. Wide ranging in eastern Sonora but absent from the lowland, western part of the state; 150–1680 m (500–5510 ft). Also Baja California, southern Arizona, and extreme southwest New Mexico to Guatemala.

Figure 239. *Salix bonplandiana*. Río Guajaray north of Alamos. Photograph by G. M. Ferguson.

Exceptionally tall trees grow along the Río Guajaráy upstream from the village of Guajaráy in the Río Mayo region. These trees are evergreen, showing no signs of being winter deciduous, although elsewhere in Sonora the trees are winter deciduous.

The red willow, *S. bonplandiana* var. *laevigata*, has also been treated as a synonym of *S. bonplandiana* or a closely related but separate species. It is distinguished in part by relatively broader leaves and shorter floriferous branchlets. The red willow is more or less a northern, temperate segregate and occurs mostly in central and northern Arizona, Nevada, Utah, and northern Baja California but not in the Sonoran Desert. *Salix bonplandiana* and *S. gooddingii* are the most common willow trees in Sonora. They generally occupy different geographic ranges but occasionally occur together.

***Salix exilifolia***. Silver willow; *taraiz* (figures 240B and 241)

Large shrubs or sometimes trees 8+ m with a compact crown and short trunk. Bark on trunks and limbs grayish and fissured; bark on twigs dark reddish and sometimes

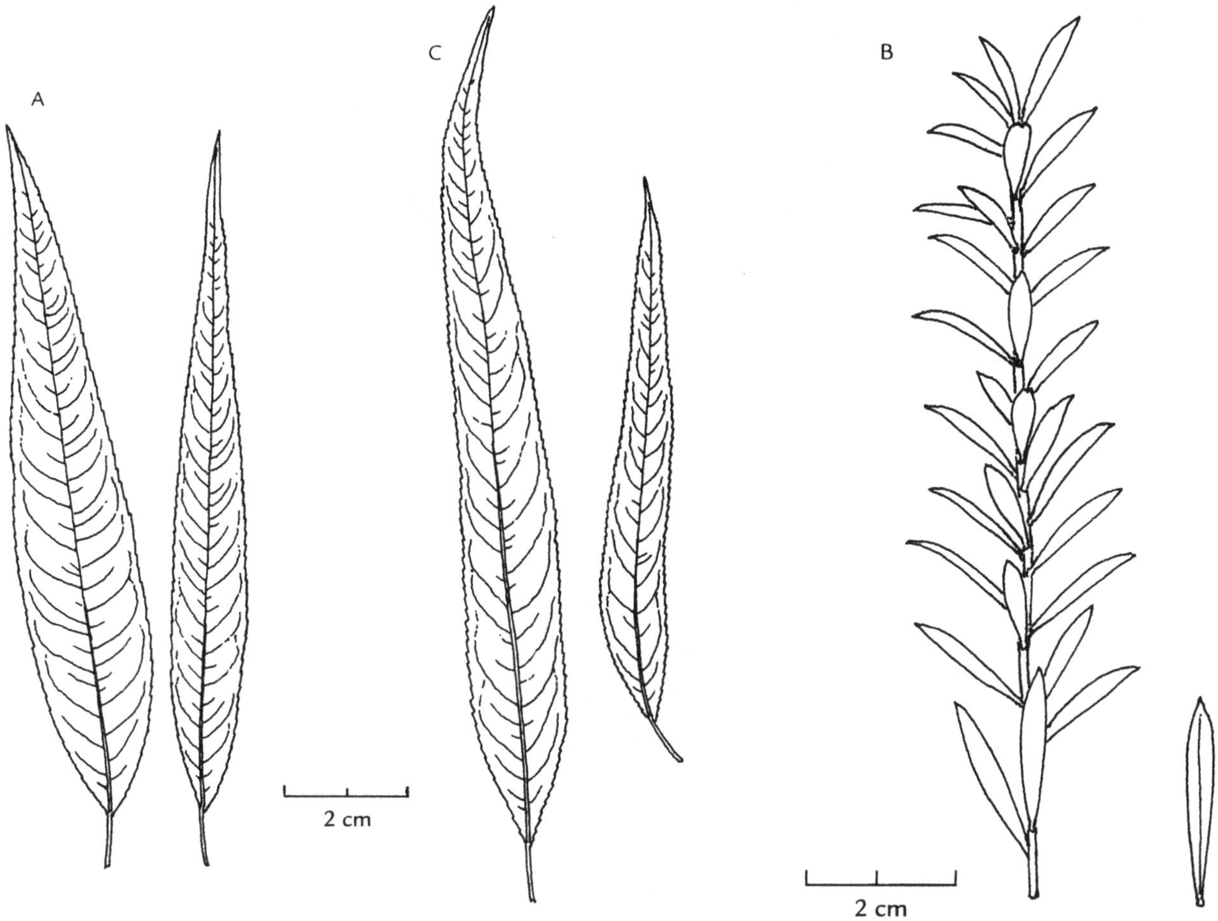


Figure 240. *Salix*. (A) *S. bonplandiana*, (B) *S. exilifolia*, (C) *S. gooddingii*. Drawn by M. B. Johnson.

flaking; sometimes forming conelike galls to 2.5 cm. Leaves and twigs densely silvery pubescent with fine, silky hairs when young; older leaves dull olive to gray-green. Leaves held upright, 12–35 × 2–4 mm, linear, same color on both surfaces, sessile or subsessile (petioles < 1 mm), the margins entire or sometimes with a few, minute, gland-tipped teeth; stipules minute, soon deciduous. Catkins 0.8–15 mm. Male flowers with 2 stamens.

Streams and arroyos; grassland, oak woodland, and occasional in Chihuahuan desertscrub. Relatively widespread in scattered localities in northeastern Sonora southward to the Río Bavispe region and isolated populations in east-central Sonora in the Maycoba-Yécora region; 945–1830 m (3100–6000 ft). An isolated stand of fewer than 8 shrubs of this willow occurs together with *Taxodium* along the streambed of the Río Cuchujaqui near Sabinito Sur in southeastern Sonora at 350 m in tropical deciduous forest. Also both Baja California states (not in desert habitats), southeast Arizona to west Texas, Chihuahua, Durango, and Coahuila.

This species often extends into drier habitats than do other willows. Little (1950) points out that it is a good soil binder and an excellent browse for livestock. This unique tree cannot be confused with any other willow in the region. It would make an attractive ornamental tree or hedge for temperate or montane-tropical regions.

This willow has long been known as *S. taxifolia*. However, Dorn (1998) considers *S. taxifolia* (*S. microphylla*) to be a closely related species which occurs farther south in Mexico and Guatemala (also see Argus and McJannet 1992). These willows belong to section Longifoliae, which includes five species in North America south to Guatemala. Also included in this section are the sandbar willows, *S. exigua*, with four infraspecific taxa across North America as far north as Alaska and Canada, and southward to limited areas of northern Mexico. *Salix exigua* occurs as a shrub in Sonora in the delta region of the Río Colorado, at a waterhole on Isla Tiburón and in the delta region of the Río Fuerte in northwestern Sinaloa.

Figure 241. *Salix exilifolia.* (1) Branch with male catkins, (2) male flower and scale, (3) branch with female catkins, (4) female flower and scale, (5) branch with fruits, (6) capsule. (Sargent 1890). Drawn by C. E. Faxon.

***Salix gooddingii*** [*S. gooddingii* var. *variabilis*]. Goodding willow, western black willow; *sauce, sauz; huatasí* (Guarijío); *huata* (Mayo); *ce'ul* (O'odham); *tewer, te'evur* (Pima Bajo); *paij* (Seri); *wata* (Yaqui) (figures 240C and 242)

Large shrubs or trees to 12 (20+) m; the trunk well formed; bark on trunk and limbs gray-brown and fissured; twigs often yellow-brown. Young twigs and leaves pubescent; mature herbage glabrate or glabrous. Leafy stems and leaves mostly drooping. Leaves 5.5–14 × 0.8–2.0 cm; petioles 1–9 mm; blades narrowly lanceolate, green on both surfaces, the margins minutely serrated; stipules ovate-cordate or sometimes orbicular-cordate, the margins often glandular-serrated.

Flowering catkins 2–7 cm. Female flowers and fruits with pedicel-like stipes (0.8) 1.0–1.3 mm. Male flowers with 4–9 stamens. Peak flowering as new leaves unfold in spring and sporadically through late summer.

Figure 242. *Salix gooddingii.* (1) Habit, (2) branch with male catkins, (3) male flower with scale, (4) branch with female catkins, (5) fruiting catkin, (6) female flower and scale. (Benson and Darrow 1981). Drawn by L. B. Hamilton.

Canyon bottoms, streams, rivers, springs, and agricultural ditches; riparian and wetland habitats in tropical deciduous forest, coastal thornscrub, Sonoran desertscrub, Chihuahuan desertscrub, grassland, oak woodland, and pine-oak woodland. Statewide; near sea level–1625 m (5330 ft). Southwest United States and northern Mexico including Baja California, Sonora, northwest Sinaloa, Chihuahua, and Coahuila.

This is the most widespread tree in Sonora, extending to each corner of the state in wetland habitats. Photos from the early twentieth century show willows along the lower Colorado, Mayo, and Yaqui rivers reaching perhaps 20 m. By the 1980s large willow trees were scarce along the channels of the vanquished rivers of southern Sonora. Shrubs and small trees of this species are now common along irrigation and drainage ditches in the agricultural lands of the these three rivers. Scattered groves of trees about 20 m tall can be found along the lower Río Guajaray north of San Bernardo.

At the end of the nineteenth century, Mearns (1907: 45) recorded "This is the common willow everywhere in the Southwest. . . . I do not remember a stream of New Mexico or Arizona that lacks it." *Salix gooddingii* is replaced by the closely related *S. nigra*, the black willow, in the eastern two-thirds of the United States and northeastern Mexico.

## Sapindaceae—Soapberry Family

Trees, shrubs, or vines; rarely herbaceous. Leaves alternate; stipules absent (in Sonora). Often more or less dioecious but with some bisexual flowers. Flowers in panicles or small axillary clusters, small, with a nectar disk; sepals 4 or 5; petals mostly 4 or 5, often with interior appendages, or absent; stamens 5–10. Fruits highly variable, dry to fleshy. Worldwide, tropical and subtropical, few in deserts and very few in warm temperate regions; greatest diversity in Asia and America; 140 genera, 1500 species. Reference: Brizicky (1963).

1. Leaves simple; petals absent. ______ ***Dodonaea***
1' Leaves compound; petals present.
   2. Leaflets 5–19 per leaf, the margins entire; fruits rounded and leathery. ______ ***Sapindus***
   2' Leaflets 3, the margins serrated; fruits flattened, maplelike samaras. ______ ***Thouinia***

### *Dodonaea*

Shrubs and occasionally small trees. Mostly Australia, several from Madagascar, Java, and Hawaii, 1 in the Americas; 50 species. Reference: West (1984).

***Dodonaea viscosa*** [*D. viscosa* var. *angustifolia; D. viscosa* subsp. *angustifolia; D. angustifolia*]. Hop bush; *jarilla, tarachiqui* (the usual name in Sonora); *guachomó* (Guarijío) (figure 243)

Shrubs or rarely trees 5–8 m with a slender trunk to 8 cm in diameter; branches mostly erect-ascending. Bark brown to light gray, shredding in narrow, braided ridges. Herbage, inflorescences, sepals, and ovaries densely resinous-viscous, especially when young; at first with somewhat sparse, minute hairs. Leaves evergreen or nearly so, simple, those of tree-sized plants 6–15.5 × 0.8–1.8 cm, narrowly elliptic to oblanceolate, narrowed at the base and essentially sessile, the midrib prominent, the margins entire.

Mostly dioecious but often with some bisexual flowers. Flowers in small clusters among the leaves, pedicelled, small, and yellow-green. Sepals often 4, 2 mm; petals absent; male flowers with large, subsessile anthers. Fruits papery capsules 1–2 cm wide, 3-winged, drying straw color, often with massive fruit crops in spring; capsules commonly splitting to release seeds while the fruit still on the plant; each single-winged segment 1-chambered with 1 (2?) seeds. Seeds 3.0–3.6 mm in diameter, dull blackish, thickly lens shaped with a low crest.

Mostly on rocky slopes; widespread, from low elevations in Sonoran desertscrub to oak woodland, and higher elevations in thornscrub and tropical deciduous forest; 10–1800 m (30–5900 ft). Most numerous in ecotone areas with oak; absent from extreme northwestern Sonora and seldom common in lowlands.

Tree-sized *Dodonaea* generally occur among shrub-sized plants, mostly in southeastern Sonora near the lower oak zones, and often on rather barren, exposed mountain slopes or mesas of gray, indurated ashy rock soils. *Dodonaea* trees were also found in relatively mesic areas on Sierra de Alamos at 675 m among tropical deciduous forest and the lower oak zone with *Quercus chihuahuensis*. Elsewhere in Sonora it is almost always a shrub. Cultivated plants from local stock in southern Arizona occasionally develop into small trees.

This species occurs on every vegetated continent of the world. Several infraspecific taxa have been named, including a purplish-leaved horticultural variety. Some authors treat the narrow-leaved forms, such as those found in Arizona and northern Sonora, as *D. angustifolia* or *D. viscosa* subsp. (or var.) *angustifolia*.

### *Sapindus*

Trees and shrubs with dense wood; bark smooth and gray, becoming rough and somewhat fissured on larger trunks. Stems and leaves with soft, white to golden colored, curved hairs. Leaves pinnate; leaflets moderately asymmetric with entire margins.

Figure 243. *Dodonaea viscosa* (Benson and Darrow 1945). Drawn by L. B. Hamilton.

Often dioecious or monoecious, but with some bisexual flowers. Flowers white, in dense, often large terminal panicles; sepals and petals each 5. Male flowers with 8–10 stamens, the filaments with conspicuous white hairs, and a reduced pistil. Female flowers with a 3-chambered ovary and reduced stamens. Fruits (in Sonora) mostly 1- or sometimes 2-lobed by abortion (or rarely 3-lobed); each lobe 1-seeded, somewhat drupelike, rounded, smooth, with leathery, yellowish, and translucent pulp, soon becoming hardened, wrinkled, and yellow-brown; seeds hard and rounded. (*Sapindus drummondii* seems to generally produce 1-seeded fruits, and among *S. saponaria* the tendency to produce fruits with more than 1 lobe is stronger in moister seasons or habitats and in more tropical areas.) New World, Asia, and Oceania; 13 species, 2 in Mexico.

1. Leaflets 10–19 per leaf, 1–2 cm wide; petals longer than wide. ______ ***S. drummondii***
1' Leaflets 5–10 per leaf, 2–6 cm wide; petals as wide as long. ______ ***S. saponaria***

**Sapindus drummondii** [*S. saponaria* var. *drummondii*]. Western soapberry; *amolillo, cirioni, jaboncillo* (figures 244 and 245A)

In Sonora and Arizona mostly slender-stemmed, spindly shrubs, often propagating by root sprouts, but sometimes trees to 10+ m with a well-formed trunk. Leaves winter deciduous, often 15–30 cm; leaflets 10–19, lanceolate, tapering to a slender point, the larger leaflets 4.0–8.5 × 1–2 cm; leafstalks, especially the larger ones, narrowly winged.

Flowers 4–5 mm wide; resembling those of *S. saponaria* but often more densely pubescent especially at the petal bases, and petals more or less elliptic to broadly obovate, 2.5–2.8 × 1–2 mm. Fruits (or each lobe) 12–15 mm in diameter. Flowering June.

Canyons and arroyos, and sometimes on dry slopes and valley floors; Sonoran desertscrub, Chihuahuan desertscrub, grassland, and oak woodland; northeastern and north-central Sonora; 760–1525 m (2500–5000 ft). Essentially a temperate, frost-hardy species; Sonora to northeastern Mexico and Arizona to Missouri, Kansas, and Louisiana.

**Sapindus saponaria.** Soapberry; *amolillo, jaboncillo; jutuhuí* (Guarijío); *tupchi* (Mayo) (figure 245B)

Trees 5–10 (15) m, often taller than wide, with a well-developed trunk 20–35+ cm in diameter; bark gray, smooth to granular. Leaves evergreen to tardily drought deciduous, often (12) 20–45 cm; leaflets (5) 6–8 (10), elliptic-lanceolate; larger leaflets 5.5–20+ × 2–6 cm; leafstalks often conspicuously winged. (The first few leaves on a twig may be reduced, sometimes with a single, large leaflet.)

Flowers 4–5 mm wide, with long, white hairs, especially near the bases of the sepals, petals, and filaments, and fringing the sepals and petals; sepals 1.5–1.8 mm; petals 2.2–2.5 mm, cupped, obovate to orbicular, about as wide as long, attracting numerous flies and medium-sized wasps; filaments, anthers, and petals pure white; sepals and pistil at anthesis white with green tinges. Fruits (or each lobe) 12.5–18.5 mm in diameter. Flowering November–February.

Mostly in riparian habitats, canyons and streams; Sonoran desertscrub, thornscrub, and tropical deciduous forest; southern Sonora from near sea level–915 m (3000 ft). Northward in western Sonora to the Guaymas region (Sierra el Aguaje) and inland, its range corresponds to the distribution of thornscrub and tropical deciduous forest. Widespread in tropical America, Sonora to northern Argentina and in Florida. Introduced in tropical Asia and Africa.

The wood is hard, heavy, and strong but the working properties are rather poor and the durability low (Record and Hess 1943), yet the wood has served for lumber in southeastern Sonora. The fruits have been used for soap in many parts of the world and the generic name *Sapindus* means "soap of the Indians." The fruits contain as much as 37% saponin; crushed in water they produce suds, which in Sonora have been used for washing clothes and shampoo. The seeds have been used as a source of oil (Uphof 1968), as fish poison (Turner et al. 1995), as rosary beads, for making buttons, and to treat arthritis, fevers, rheumatism, and kidney disease.

In addition to floral differences, *S. saponaria* differs from *S. drummondii* in nearly always being a tree with a well-developed trunk, having much larger and drought-deciduous (rather than winter-deciduous) leaves with conspicuously winged leafstalks, slightly larger fruits, and being frost sensitive. Their distributions in Sonora are allopatric, and there is no indication of intermediate forms. The northern range of *S. saponaria* is undoubtedly limited by winter cold; in cultivation it perishes with moderate freezing.

### *Thouinia*

Trees or shrubs. Leaves with 3 leaflets, moderately bicolored, the margins serrated; terminal leaflet long stalked and decurrent at its base; lateral leaflets slightly smaller, on short stalks, and often asymmetric at their bases.

Inflorescences of axillary panicles. Sepals and petals each 4; stamens 8. Fruits of 2 or 3 maplelike, 1-seeded samaras. Mexico to Costa Rica and the Greater Antilles; 27 species. The Sonoran populations are rather poorly known.

1. Terminal leaflets to 6 × 2 cm. ______ ***T. acuminata***
1' Terminal leaflets > 7 × 4 cm. ______ ***T. villosa***

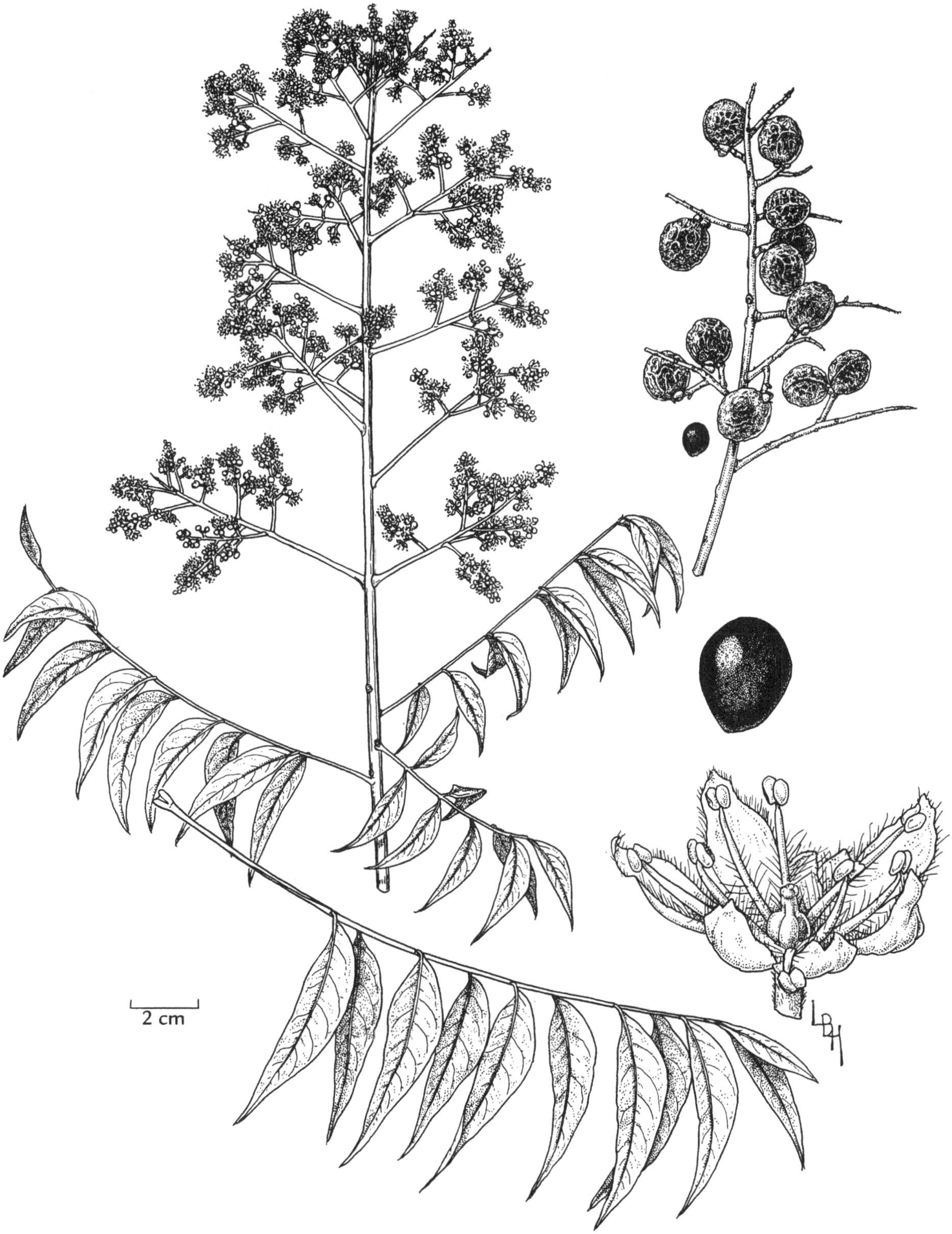


Figure 244. *Sapindus drummondii* (Schmutz and Hamilton 1979). Drawn by L. B. Hamilton.

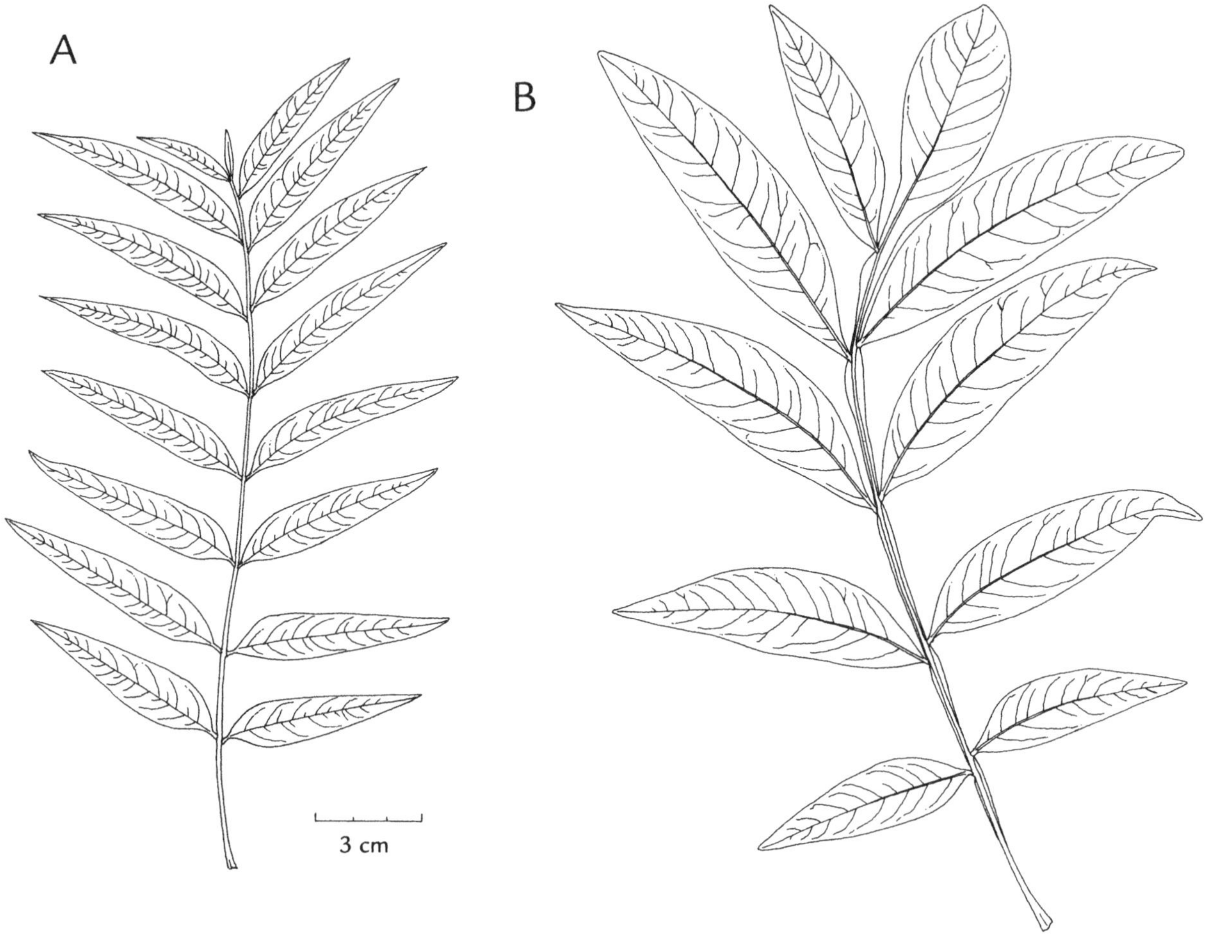


Figure 245. *Sapindus*. (A) *S. drummondii*, (B) *S. saponaria*. Drawn by M. B. Johnson.

***Thouinia acuminata*** (figure 246)

Large shrubs or trees to 5 m. Branches and twigs slender. Herbage and inflorescences pubescent with minute hairs. Leaves 5.2–12 cm; petioles 2.0–4.5 cm; leaflets lanceolate, cuneate at the base, pointed at the tip, the terminal leaflet 3.4–6.0 × 1.2–2.0 cm; lateral leaflets with asymmetric bases. Flowers cream colored, in slender axillary racemose panicles often 7–12 cm. Samaras 2, each approximately 1 cm. Flowering at least August–September.

Canyon bottom with riparian tropical deciduous forest; southeastern Sonora; 380 m (1245 ft). This species is known from the Cajón de Caujara of the Río Mayo, about 5 km below Carimechi near the Sonora-Chihuahua border, where Phil Jenkins found it in sand and rocks near the stream (personal communication 1997). Southwestern Chihuahua to Oaxaca. It is distinguished from *T. villosa* by its smaller stature, sparser and smaller hairs, smaller leaves with narrower leaflets, more strongly asymmetric lateral leaflets, and smaller-toothed margins.

***Thouinia villosa***. *Guasimilla roja* (figure 247)

Trees to 10 m, sometimes semiscandent; trunk to 20 cm in diameter; bark gray or light colored, narrowly furrowed; branches relatively slender and weak, often somewhat weeping. Herbage, inflorescences, and flowers densely hairy, often with tawny or golden brown hairs; upper leaflet surfaces not as densely hairy as lower surfaces. Leaves apparently drought deciduous, 7.5–20 cm; petioles 2.8–9 cm; terminal leaflets 7–12.5 × 4.3–6.2 cm, rhombic (ovate-diamond shaped) to oval, the tips pointed to blunt; lateral leaflets scarcely to moderately asymmetric at the base.

Panicles shorter than the leaves. Flowers crowded, greenish white, 3–4 mm wide; filaments with spreading white hairs. Samaras unknown for Sonoran specimens, probably 2 or 3 and approximately 1.5 cm. Flowering at least February–March and May.

Slopes and riparian canyon bottoms; upper elevations of tropical deciduous forest and ecotone with oak and pine-

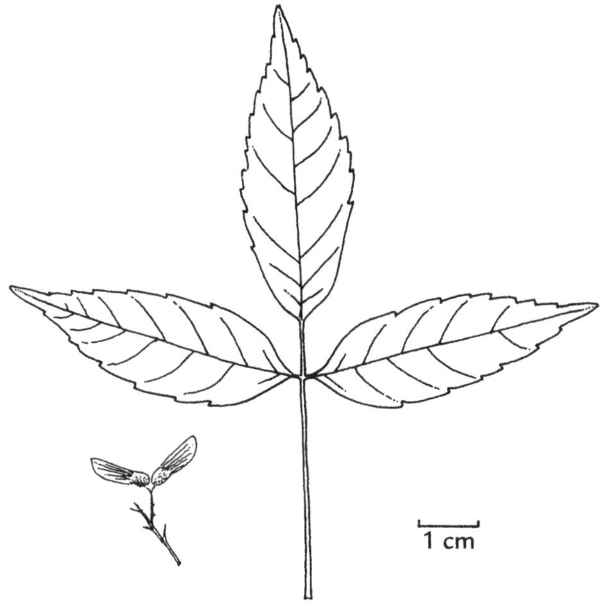


Figure 246. *Thouinia acuminata*. Drawn by M. B. Johnson.

oak woodland; southeastern Sonora; 450–1500 m (1475–4920 ft). Upper Río Mayo drainage in Sonora and southwestern Chihuahua and southward to Guatemala.

## Sapotaceae—Sapote Family

Trees or shrubs. Worldwide, tropical to subtropical and rarely in warm temperate regions; 70 genera, 1100 species. Reference: Pennington (1990).

This family includes many tropical trees with edible fruits such as chicle (*Manilkara zapota*), formerly the source of chewing gum. Only one genus occurs in Sonora.

### *Sideroxylon*

Hardwood trees and shrubs, spinose or unarmed, with latex in trunks, limbs, and fruits; hairs 2-branched (T-shaped or malpighian hairs—a characteristic of the family), mostly appressed, with one long and one short branch. Leaves alternate in a spiral arrangement, or fascicled on axillary short shoots, occasionally opposite, simple, petioled, entire (rarely slightly undulate); stipules absent.

Inflorescences of short axillary clusters or the flowers sometimes solitary. Flowers bisexual or unisexual, small, mostly 5-merous. Sepals persistent, separate, usually 5. Corollas short-tubular, with (4) 5 (6–7) lobes, each divided into 3 segments or undivided. Stamens and/or staminodes 5–10: fertile stamens exserted, usually 5, attached at the top of the corolla tube, the staminodes well developed, mostly 5 (sometimes 10 in unisexual female flowers). Style 1, often persistent in fruit. Fruits fleshy, with an edible pericarp, 1(2)-seeded.

Mostly tropical and subtropical; 49 species in the Americas, and 20 in Africa, Madagascar, and Mascarene Islands. Reference: Landrum (1992).

*Sideroxylon leucophyllum* is a large-leaved shrub or small tree on the Baja California peninsula and occurs at higher elevations on Islas San Esteban and Tiburón where it often reaches 3–4 m.

1. Herbage, especially the leaves, densely white woolly. ***S. leucophyllum*** (not treated in text)

1' Herbage not especially white woolly.
  2. Leaves 1–4.5 cm, mostly fascicled in axillary short shoots; not in southeastern Sonora.
    3. Twigs often zig-zag; leaves with raised, reticulate veins; styles 1–1.5 mm. ____ ***S. lanuginosum***
    3' Twigs straight; leaf veins obscure, not raised or reticulate; styles 4.5–5.2 mm. ____ ***S. occidentale***
  2' Leaves 5–20 cm, not fascicled; southeastern and east-central Sonora.
    3. Twigs and branches often spinescent; leaves 5–13.5 cm; corolla lobes each with lateral segments; fruits approximately 1.5 cm. ***S. persimile***
    3' Unarmed; leaves 10–23 cm; corolla lobes without lateral segments; fruits approximately 2.5–4 cm. ____ ***S. tepicense***

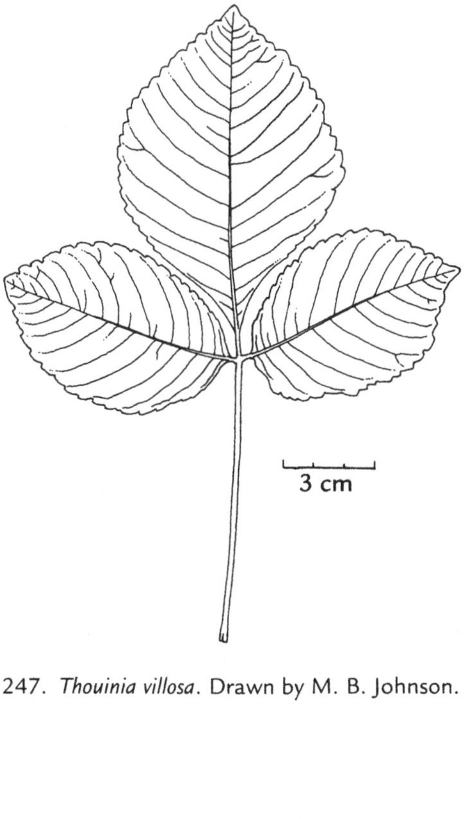


Figure 247. *Thouinia villosa*. Drawn by M. B. Johnson.

***Sideroxylon lanuginosum*** subsp. ***rigidum*** [*Bumelia lanuginosa* var. *rigida; B. lanuginosa* subsp. *rigida*]. Gum bumelia, chittam wood (figure 248)

Mostly shrubs and sometimes trees to 7 m (to 10 m in southeastern Arizona). Bark dark brown, almost blackish, granulated and somewhat powdery, rubbing off when handled, becoming rough and scaly with age. Twigs often zig-zag, often thorn tipped, woolly when young, glabrous with age, and generally with straight or curved lateral thorns (short shoots). Leaves (1.2) 1.6–5.5 cm; petioles short; midrib prominent below, lanceolate, elliptic, obovate or oblanceolate, pubescence of white to golden-brown matted hairs, especially the lower surfaces, becoming more sparsly pubescent or sometimes glabrate, especially on the upper surfaces, the veins raised and reticulate; long-shoot leaves generally larger and broader; short-shoot leaves fascicled. Flowers bisexual, 2–3 mm, in umbel-like clusters; pedicels 1–3 mm; corollas approximately 3 mm, cream colored, with 5 lobes, each with 3 coarsely fringed segments, the middle segment broadly ovate with a rounded apex, the lateral segments lanceolate; stamens 5; styles 1–1.5 mm. Fruits approximately 1 cm, bluish black.

Riparian habitats among mesquite-grassland and oak woodland; 1180–1725 m (3825–5600 ft); northeastern Sonora. Along washes in valley bottoms and canyons, often forming thickets; localized and widely scattered, mostly near the Arizona border from near the Baboquivari Mountains eastward. Much more common on the Arizona side of the border.

Subspecies *rigidum* in northern Mexico from Sonora to Tamaulipas and the southwestern United States from southeastern Arizona to Texas and Oklahoma. Two other subspecies occur in the midwestern and southeastern United States.

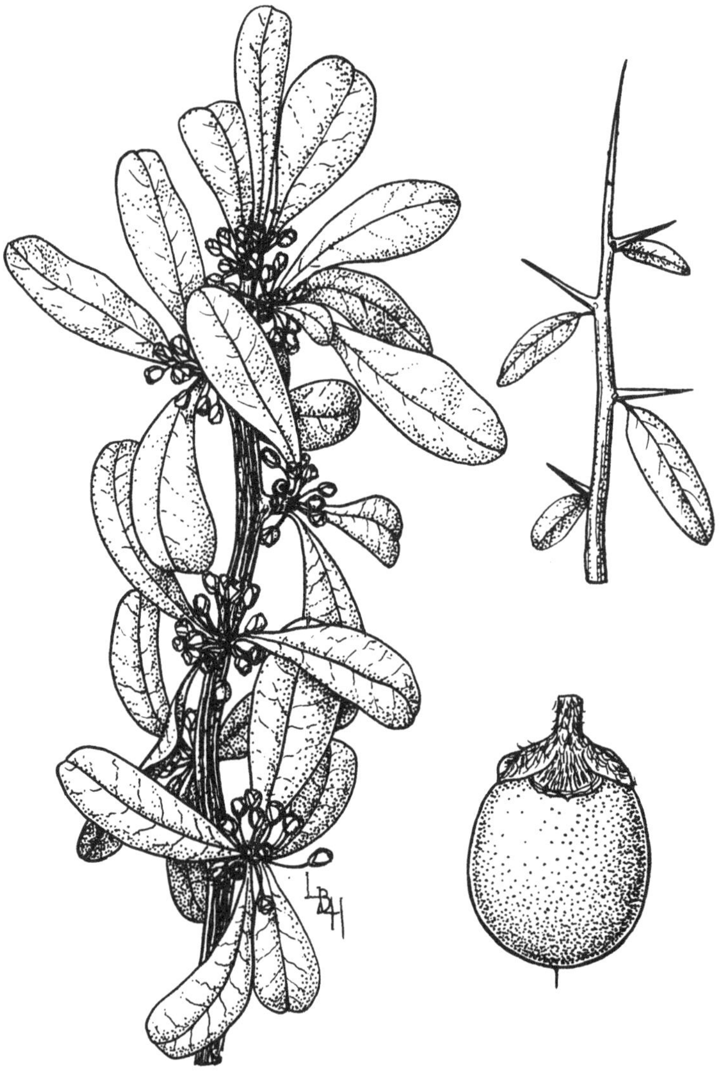

Figure 248. *Sideroxylon lanuginosum* subsp. *rigidum* (Benson and Darrow 1945). Drawn by L. B. Hamilton.

This species is the only member of the genus that ranges into temperate North America. The wood of gum bumelia and perhaps that of *S. occidentale* is host to larvae of the spectacular, diurnal, metallic green or blue cerambycid beetle, *Plinthocoelium suaveolens*.

***Sideroxylon occidentale*** [*Bumelia occidentalis*]. *Bebelama; júchica* (Mayo); *paaza* (Seri); *vapsa* (Yaqui) (figure 249)

Large shrubs to spreading trees 6–16 m with heavy trunks. Bark checkered, light gray. Young stems, leaves, and flowers moderately to densely pubescent with short, mostly white hairs, or glabrous with age. Twigs often thorn tipped and often with axillary spines (subtending the leaf fascicles). Leaves nearly evergreen, or tardily and partially or sometimes entirely drought deciduous, on long shoots and mostly fascicled in short shoots, 1–3 cm, 0.2–1 cm wide, obovate to spatulate, the surfaces mottled, tapered below to an inconspicuous, short petiole.

Flowers several in fascicles or solitary, white, prolific, fragrant, attracting bees, reportedly bisexual or unisexual (dioecious); pedicels longer than the flower; pedicels and flowers 1–2 cm. Corollas approximately 5 mm, the lobes each with 2 large lateral segments with fringed margins. Fertile stamens 5; staminodes 5 (10 in female flowers). Fruits approximately 12 mm, 1-seeded, with white latex, ripening dark purple and prunelike, oblong, the pericarp sweet and edible. Flowering October–February (June–July).

Slopes, valleys, plains, canyons, and arroyos; Sonoran desertscrub, thornscrub, and grassland. Widely scattered in Sonora except the northwestern, northeastern, and eastern edges of the state; near sea level–1065 (1370) m [3500 (4500) ft]. From the vicinities of Masiaca in the lower Río Mayo region in southern Sonora northward to Saric, Magdalena, and Arizpe, and on Isla Tiburón. Also Baja California Sur.

This species seems most common at its northern limits. Larger trees, either scattered or in groves, occur sporadically throughout the geographic range. It is probably slow

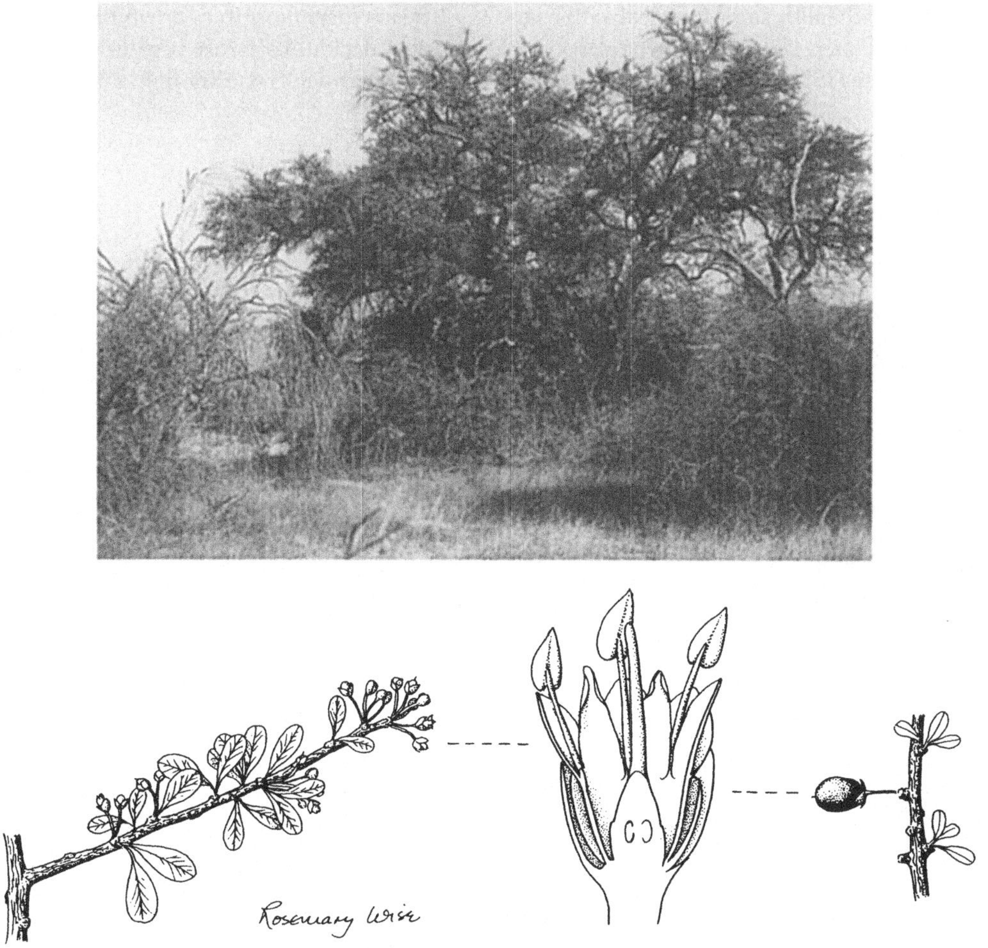


Figure 249. *Sideroxylon occidentale.* (Top) Large isolated tree in the interior of Isla Tiburón (Felger and Moser 1985), photograph by R. S. Felger; (Bottom) twig, bisexual flower, and fruit (Pennington 1990), drawn by R. Wise.

growing, and the scarcity of larger trees may be due to destruction by people. At its northern limits, in the general vicinity of Saric, it is common and widespread as a shrub on dry slopes and less common as a tree to 16 m along riverine floodplains. Although less than 15 km from the Arizona border, it is not known from the United States. The northern limit seems to be determined by freezing weather (Turner et al. 1995). One tree on Isla Tiburón measured 12 m with a trunk nearly 1 m in diameter (figure 249A). The female flowers with stamens converted into large staminodes is an unusual character, known elsewhere in the genus only among Old World species (Pennington 1990).

### *Sideroxylon persimile* subsp. *subsessiliflorum*

[*Bumelia subsessiliflora; B. persimilis* subsp. *subsessiliflora; B. arborescens*]. *Bebelama; jachojcá* (Guarijío) (figure 250)

Trees 8–15 (30) m. Bark gray, minutely checkered on younger branches, the older branches and trunks checkered and often dark with algae; wood beneath the bark orange-brown and oozing milky sap when wounded; stems and green fruits also with milky sap. Many leaf axils bearing stout, sharp thorns often 1–2.5 cm; thorns sometimes forming spinescent twigs to approximately 5 cm bearing nodes and several small leaves. Young twigs, spines, herbage, petioles, and lower leaf surfaces densely pubescent with coarse, velvety, brown hairs, with age sparsely pubescent or glabrate. Foliage dense, usually briefly drought deciduous before leafing out in mid-June just before the summer rains. Leaves spirally arranged, not clustered, (4) 5–13.5 cm; petioles 4–10 mm; blades elliptic to ovate or sometimes obovate, the upper surface often bright green, glabrous, and shiny, the lower surfaces especially hairy along the conspicuous midrib, the leaf tips blunt (rounded to obtuse).

Flower clusters usually shorter than the thorns; pedicels often 3–5 mm. Flowers bisexual, white; perianth 4–5 mm; styles and stamens exserted; corolla lobes with well-developed lateral segments; stamens 5; staminodes 5, fringed to laciniate. Fruits 15+ mm, 1-seeded, ellipsoid-oblong, bluish black, smooth, ripe in late spring. Flowering November–January.

Rocky canyons; tropical deciduous forest and oak woodland; southeastern and east-central Sonora; 305–1100 m (1000–3610 ft).

The northernmost trees are in riparian canyons within tropical deciduous forest vegetation west of Tepoca on Mexico Highway 16, although it is much more common

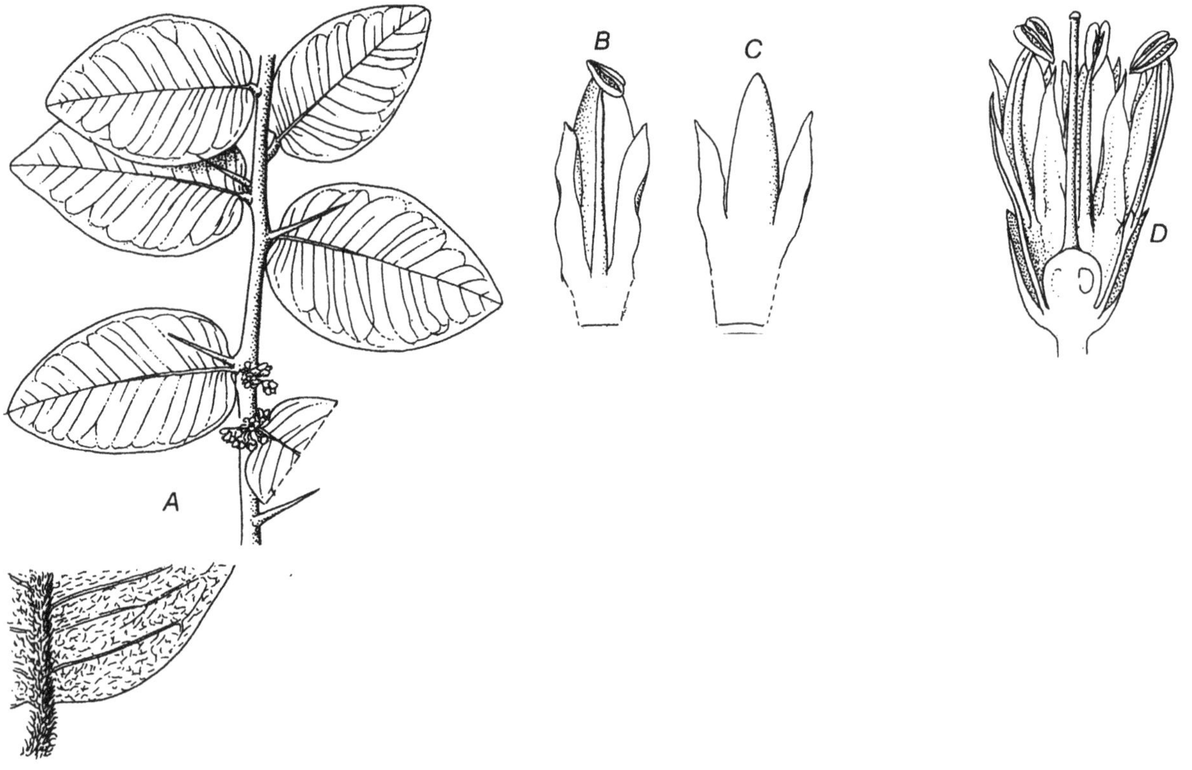


Figure 250. *Sideroxylon persimile* subsp. *subsessiliflorum.* (A) Habit and Leaf, (B) stamen and staminodes, (C) corolla lobes, (D) half flower. (Pennington 1990). Drawn by R. Wise.

farther south in the Río Mayo region. Subspecies *subsessiliflorum* ranges from Sonora and southwestern Chihuahua to Jalisco and Colima; it is replaced southward by subsp. *persimile*, which extends to Venezuela.

Deer feed on the fallen flowers. The ripe, blackish, and fleshy part of the fruit is sweet and eaten fresh or cooked, often in hot coals. It is said to burn the mouth if not cooked. The hard seed is discarded. Leafy, thornless branches might be confused with those of *Prunus zinggii*.

***Sideroxylon tepicense*** [*S. angustifolium*]. *Tempisque; cajé* (Guarijío); *ca'ja* (Mayo) (figure 251)

Spreading trees often 10–20 m; unarmed, with white latex sap. Bark gray, exfoliating from the trunk in plates. Young twigs and the leaves, especially the lower surfaces, velvety with crinkled, often reddish to golden-brown T-shaped hairs (best seen with high magnification); upper surfaces of older leaves darker green and glabrous or essentially so. Foliage briefly deciduous, usually in May, the new leaves emerging just before the summer rains. Leaves spirally arranged, clustered at the stem tips but with conspicuous internodes; leaves 10–23 cm; petioles 1.5–4 cm; blades rather tough, elliptic to elliptic-obovate or oblanceolate, the midrib and veins prominent and raised on the lower surfaces, the tips mostly acute, the margins shallowly undulate.

Flowers in dense clusters on several or more nodes on stems below the leaves; pedicels 8–10 mm; pedicels and sepals densely hairy. Flowers bisexual, inconspicuous, greenish white, 5 mm including the exserted style and stigma; corolla lobes elliptic-oblong, without lateral segments; stamens 5; staminodes 5, lanceolate, glabrous. Fruits approximately 2.5 × 1.5 cm, yellow, 1-seeded. Flowering July–August, fruits ripe in June.

Valleys, canyons and arroyos; tropical deciduous forest; southeastern and east-central Sonora, 285–455 m (935–1490 ft), northward to the vicinity of Tepoca. Widespread but usually infrequent, mostly arroyo margins in shaded montane canyons. Also adjacent southwestern Chihuahua and southward along the Pacific drainage of Mexico to El Salvador.

The fruits, sold in the marketplace in Alamos, are eaten fresh, cooked with sugar, or made into jam. The fleshy portion of the fruit is sweet and has a granular texture. The surface of the fruit is rough, and said to be caustic if too many fruits are eaten. The bark is used to curdle milk in making cheese.

Trees 25–30 m tall in riparian tropical evergreen forest in Arroyo Verde at 850 m (2790 ft) in southeastern Sonora may be *S. tepicense* or perhaps *S. capiri* subsp. *tempisque* (*Mastichodendron capiri* var. *tempisque*), known from Jalisco to Central America and the Caribbean islands.

## Simaroubaceae—Simarouba Family

Trees and shrubs, often with bitter-tasting bark. Leaves alternate, simple or odd-pinnate, without stipules; or essentially leafless. Dioecious (in Sonora). Flowers small. Fruit a drupe, berry, capsule, or samara with 1–5 carpels, each 1-seeded. Warmer regions of the world, few in deserts; 20 genera, 120 species.

*Ailanthus altissima* [*A. glandulosa*], known in New York as the "garbage tree," and in the Orient as the "tree of heaven," is native to eastern Asia. It is widely cultivated and sometimes grown in Sonora, especially in nondesert regions in the northern part of the state. It has a slender, erect habit, rather thick twigs, winter-deciduous leaves that have an acrid, unpleasant odor, and is essentially dioecious. This tree can be an indicator of prior settlement because it often persists around abandoned dwellings and may propagate by root shoots.

1. Leafless or with occasional scale leaves; twigs thorn tipped. ____ ***Castela***
1. Leafy with pinnately compound leaves; unarmed.
   2. Leaves usually > 25 cm; leaflets with 1 or 2 pairs of gland-tipped basal teeth. ____ ***Ailanthus altissima*** (not treated in text)
   2' Leaves usually < 25 cm; leaflets entire. ***Alvaradoa***

### *Alvaradoa*

Shrubs or trees. Mexico and the West Indies to Argentina; 5 species. Reference: Cronquist (1944).

***Alvaradoa amorphoides.*** *Palo torcido, peinecillo; ma'ácita* (Guarijío); *guaji, paenepa* (Mayo) (figure 252)

Shrubs often 2–5 m, sometimes trees to 7 (10) m, usually slender and upright with a single trunk. Bark gray, nearly smooth or with slightly rough areas. Densely pubescent, the leaf hairs appressed. Leaves 9–20 cm, pinnate with 11–20 pairs of leaflets and sometimes also with a terminal leaflet; foliage often with a slight bluish cast; leaflets 11–25 mm, elliptic to oblong, darker green above, the margins entire.

Flowers white, many, and crowded on spikes. Fruiting spikes on short peduncles, pendent, 3.5–11 cm; fruits 3-winged (or crested) samaras, pinkish red to red-brown or yellow-red, with long, dense, silky hairs, 1-seeded. Flowering September–October.

Slopes and arroyos; thornscrub, tropical deciduous forest, and grassland; east-central and southeastern Sonora; 305–1065 m (1000–3500 ft). Sonora to Central America, West Indies, and Florida. The common name *peinecillo* refers to the comblike fruiting spikes.

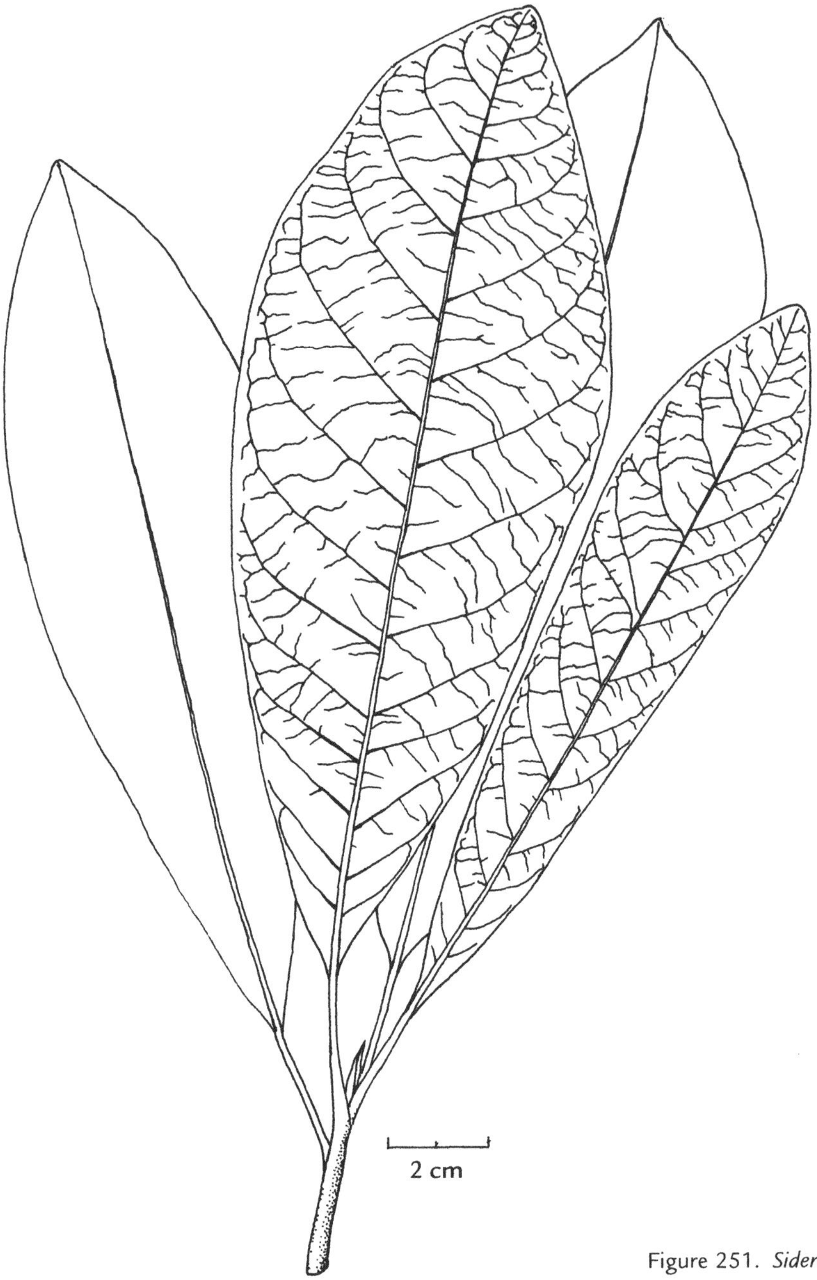


Figure 251. *Sideroxylon tepicense*. Drawn by M. B. Johnson.

## *Castela*

Mostly leafy, drought-deciduous shrubs. Semiarid to arid subtropical regions of the Americas; 15 species.

*Castela polyandra* is a low, sprawling, nearly leafless shrub occurring in Baja California and along the east side of Isla Tiburón. Reference: Moran and Felger (1968).

***Castela emoryi*** [*Holacantha emoryi*]. Crucifixion thorn; *corona de Cristo*; *zazjc caacöl* (Seri) (figure 253)

Usually shrubby, sometimes small trees to 5 m, often with several well-developed trunks; wood very hard and twisted. Branches rigid, thick, thorn-tipped, and moderately to densely pubescent with short, white hairs, becoming gla-

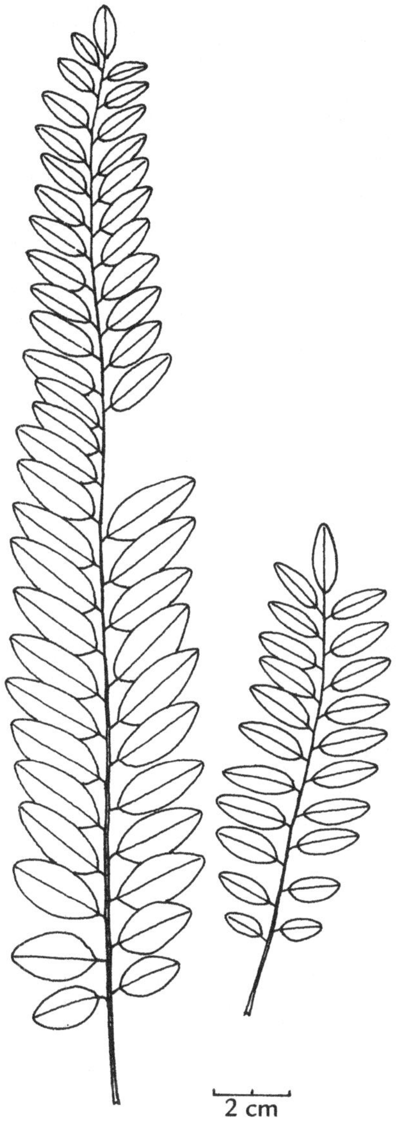


Figure 252. *Alvaradoa amorphoides*. Drawn by M. B. Johnson.

brous with age. Mature plants leafless or with few scale leaves; stems of seedlings and young plants slender, moderately leafy with leaves to approximately 1 cm and quickly drought deciduous.

Panicles mostly < 5 cm, with rather thick branches, pubescent and pinkish. Flowers 8–9 mm wide, cream yellow to greenish or rose-pink; petals often 7; often tended by black ants. Fruits of separate woody carpels in a star-shaped arrangement, persisting in dense clusters for several years or more in the outer branches. Flowering late April–June.

Desert plains; Sonoran desertscrub; northwestern Sonora; 140–195 m (460–640 ft). Open desert habitats with *Larrea divaricata* on silty-sandy soils. In Sonora rare and known from only several widely scattered localities (Felger 2000; Turner et al. 1995). Also southwestern Arizona, southeastern California, and extreme northern Baja California.

The rareness of this plant in Sonora suggests these plants may be relictual or that their occurrence may be due to chance introductions, probably by birds or mammals. It is fairly common in certain areas of southern Arizona near the Sonora border including parts of Cabeza Prieta National Wildlife Refuge in Pima County.

## Solanaceae—Nightshade Family

Herbs, shrubs, trees, and vines. Leaves alternate (often fascicled on short shoots), simple (in Sonora), without stipules. Flowers usually 5-merous, often showy. Calyx 5-lobed, persistent. Corolla united throughout or 5-lobed, usually plicate (plaited or folded) in bud, often rotate (wheel or plate shaped, the tube essentially absent, the limb flat and circular) or tubular. Stamens usually 5; filament bases fused to the corolla. Ovary superior. Fruit usually a berry or capsule.

Mostly tropical and temperate, greatest diversity in Central and South America; 94 genera, 2950 species.

This diverse family includes perhaps 100 species in Sonora but only 3 treelike species, and these only marginally qualify as trees. Among the many economically important members are chili peppers, eggplant, potato, tomato, tobacco, garden plants such as petunia, and many medicinal and poisonous plants.

1. Plants glabrous; fruit a capsule. ________ ***Nicotiana***

1' Plants conspicuously hairy (at least the new shoots); fruit a berry.

  2. Leaf margins entire; upper leaf surfaces glabrate or glabrous with age; flowers tubular; anthers small, not opening by pores; fruits few seeded. _____ ***Cestrum***

  2' Leaves entire or lobed, densely pubescent on both surfaces; flowers rotate; anthers large, opening by pores; fruits many seeded. __________ ***Solanum***

### *Cestrum*

Trees and shrubs. Tropical America; 150 species. Reference: Benítez de Rojas and D'Arcy (1998).

Night jessamine (*C. nocturnum*) is cultivated in tropical and subtropical regions for its sweet-scented flowers and has been used medicinally, especially for epilepsy.

### ***Cestrum tomentosum*** [*C. lanatum*]

Erect, slender shrubs or occasionally small trees to 6 m. Herbage, inflorescences, and calyx densely pubescent with feltlike, dull white to yellow-brown dendritic and some stellate hairs, the upper leaf surfaces glabrate with age.

Figure 253. *Castela emoryi* (Benson and Darrow 1945). Drawn by L. B. Hamilton.

Petioles 6.5–20 mm; leaf blades mostly lanceolate to ovate, sometimes nearly elliptic to oval, mostly 10–23 cm, pointed at both ends or sometimes the tip rounded or blunt, the margins mostly entire.

Inflorescences densely to loosely flowered axillary and terminal cymes, approximately 1 cm and few flowered, to approximately 10 cm and many flowered; peduncles slender. Flowers nocturnal. Calyx 5–7 mm long, densely woolly with stellate hairs, with 5 broad lobes. Corollas greenish white, deciduous, with a narrow tube 12–15 mm and lobes 4–6 mm. Fruits fleshy berries approximately 1 cm, ellipsoid, purple-black, with 2–4 seeds. Flowering at least February–June and November; fruits March–July.

Riparian canyon bottoms, mostly in dense shade in the understory of broadleaf evergreen trees within tropical deciduous forest, oak woodland, and pine-oak woodland; southeastern Sonora, 570–1400 m (1870–4590 ft). Also southwestern Chihuahua to Peru.

The wood and leaves have been used for medicinal purposes, including treatment of fevers and to repel insect pests

(Uphof 1968). Foul-smelling foliage and minor leaves reported for this species in other regions have not been confirmed for Sonoran plants. The "normal 'major' leaf [may be] accompanied by a much smaller, often sessile 'minor' leaf, a condition sometimes suggestive of stipules" (Benítez de Rojas and D'Arcy 1998: 274). The minor leaf is generally soon deciduous.

### *Nicotiana*—Tobacco

Tropical and warm temperate regions, mostly in the Americas; 70 species.

This genus includes commercial tobacco (*N. tabacum*). There are several native species in Sonora, all of which are herbaceous and glandular hairy.

***Nicotiana glauca*.** Tree tobacco; *cornetón, juan loco, palo loco; toripipa* (Guarijío); *sanwán* (Pima Bajo) (figure 254)

Spindly, sparsely branched, glabrous shrubs or small trees often 3–6 m. Leaves deciduous with freezing weather and gradually shed and reduced in size in drought; petioles 3–7.5 cm; leaf blades more or less ovate, 4–16 cm, glaucous, smooth, somewhat thickish. Flowers tubular, frequented by hummingbirds; calyx green, 7–8 mm, the corolla yellow, 28–40 mm. Capsules approximately 1.5 cm with numerous minute seeds. Nonseasonal.

Mostly in disturbed habitats including roadsides and urban and agricultural areas, and sometimes in natural areas including floodplains and riverbanks. Nearly throughout the state; near sea level–1065+ m (3500 ft); desertscrub, thornscrub, and tropical deciduous forest to oak woodland. Native to northwestern Argentina and southern Bolivia, now adventive in warm regions of the world.

### *Solanum*

Mostly tropical; 1700 species, one of the larger plant genera in the world, and the largest in the family.

This genus includes potato, eggplant, numerous minor crop plants, and various medicinal, ornamental, weedy, and/or highly toxic plants. *Solanum ferrugineum* [*S. madrense*] is a shrub in southeastern Sonora that occasionally approaches tree size. There are many other herbaceous and some shrubby species in Sonora.

1. Unarmed; pubescence of whitish hairs; leaf margins entire to slightly wavy. ______________ ***S. erianthum***

1' Herbage often armed with spines or prickles; pubescence of yellow-brown or rusty-colored hairs; leaf blades mostly broadly toothed or lobed. __________ ***S. ferrugineum*** (not treated in text)

***Solanum erianthum*** [*S. verbascifolium*, misapplied]. Tree nightshade; *cornetón del monte, lengua de buey; huatauhui* (Guarijío); *buey nini, mambia* (Mayo) (figure 255)

Unarmed; in Sonora mostly large shrubs, occasionally small trees reaching 5–6+ m, with a trunk to 20 (30) cm in diameter, and smooth, light-colored bark. Herbage, inflorescences, and calyces densely pubescent with minute, long-stalked, stellately arranged, branched hairs. (These hairs, best seen at magnifications of 20× or higher, are unique among the tree species of any family in Sonora.) Leaves 15–30 (40) cm; petioles 2.5–6.5 cm; blades lanceolate to ovate, the margins entire to slightly wavy.

Inflorescences at first appearing terminal, becoming lateral as the vegetative stem grows. Flowers crowded on dichotomously branched corymbs shorter than the leaves. Corollas 15–17 mm wide, white, sometimes with greenish veins, the lobes broad and spreading. Anthers bright yellow and conspicuous, usually connivent (close together but not united) in a cone around the style, each anther chamber opening by a terminal pore (the pores opened by vibrations from pollen-collecting bees—buzz pollination). Fruits rounded, approximately 15 mm in diameter, fuzzy-pubescent when immature, dark colored and glabrous when ripe. Flowering and fruiting any time of year with sufficient moisture.

Canyon and riverine gallery forests, rocky slopes and disturbed areas; tropical deciduous forest and thornscrub; east-central and southern Sonora; 10–915 m (30–3000 ft). Along the Río Yaqui from the vicinity of Potam to Soyopa and Onavas, and southward and more common in southeastern Sonora. Tropical America including the Caribbean and northward to northern Mexico including southwestern Chihuahua, and southern Florida and Texas. The foliage is foul smelling even upon slight contact. This species is highly frost sensitive.

The fruits are said to be eaten uncooked and have a buttery flavor with a slight orangelike taste.

## Sterculiaceae—Sterculia Family

Mostly trees and shrubs. Worldwide, mainly tropical and subtropical; 65 genera, 1000 species. Seven genera in Sonora; the others are shrubs and perennial herbs.

This family includes cacao (*Theobroma cacao*) the source of cocoa and chocolate, and cola nut (*Cola nitida*).

### *Guazuma*

Tropical America; 4 species. Reference: Freytag (1951).

Figure 254. *Nicotiana glauca* (Benson and Darrow 1981). Drawn by L. B. Hamilton.

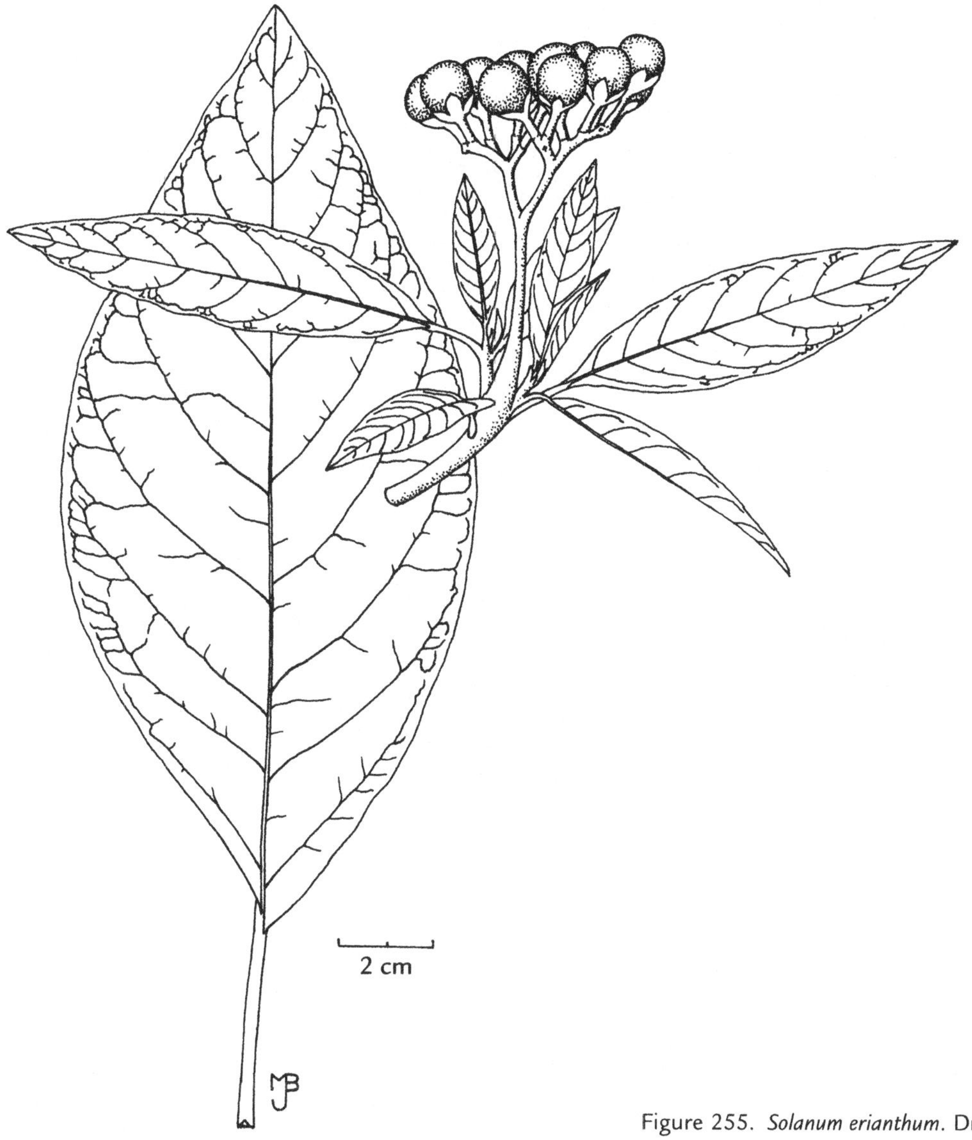


Figure 255. *Solanum erianthum*. Drawn by M. B. Johnson.

***Guazuma ulmifolia.*** Picklenut; *guásima*; *agiyá* (Guarijío); *agia* (Mayo); *aiya* (Yaqui) (figure 256)

Spreading trees often 8–10+ m, with a stout, well-developed trunk and soft wood. Bark on trunks checkered, finely furrowed or fissured, and dark gray, becoming checkered on older trunks, the younger branches with conspicuous lenticels. Twigs, leaves, flowering branches, and calyces velvety with stellate hairs. Foliage dense, nearly evergreen to often partially or sometimes fully drought deciduous in mid- to late spring. Leaves simple, alternate, 4–12+ cm; petioles prominent; blades mostly ovate, slightly to moderately asymmetric, especially toward the base, dull green above, paler below, the margins finely toothed; stipules often 3 mm.

Flowers yellow-green, approximately 6 mm wide, crowded in short, branched clusters (cymes) in leaf axils. Calyx deeply 3-lobed, approximately 4 mm; petals 5, the main body of each petal 4.5–5.0 mm, cupped or hooded over the stamens, with stringlike appendages about as long or longer than the body. Stamens united into a tubular column with 5 lobes or staminodes; anthers in groups of 3 between each lobe. Fruits 2.3–4.5 cm, oval, dark brown, conelike, hard, and woody, semipersistent, covered with short, blunt prickles. Flowering May–September.

Valleys, canyons, drainageways and arroyos; tropical deciduous forest, thornscrub, and Sonoran desertscrub; southern and central Sonora; near sea level–915 m (3000 ft). Northward to arroyos at the southern margin of the Sonoran Desert in the vicinity of Guaymas and farther

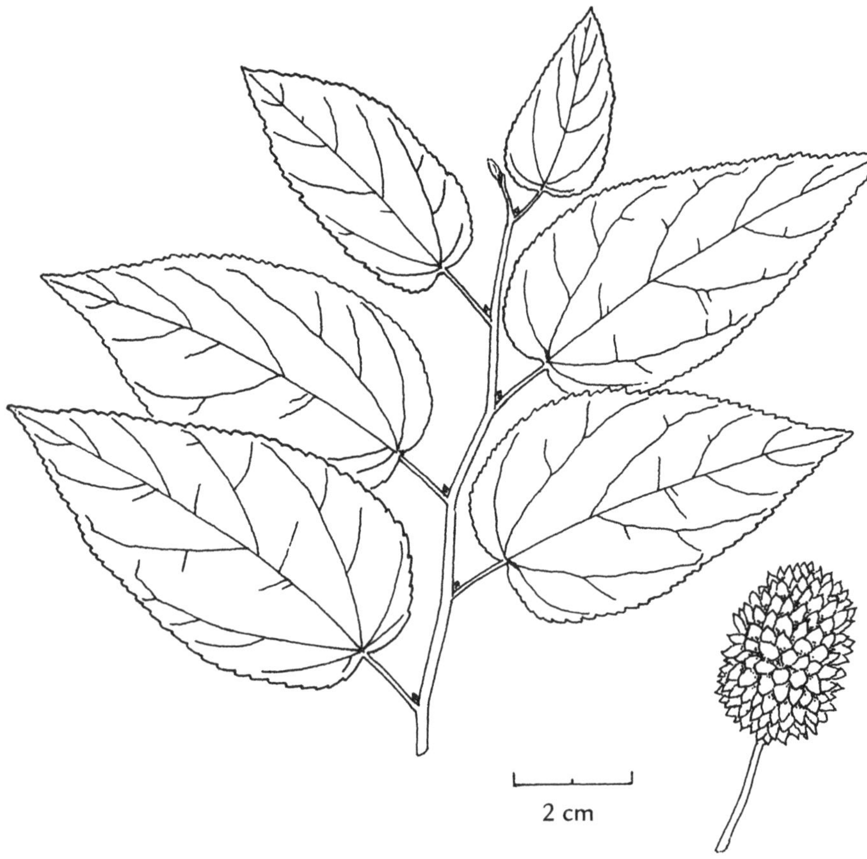


Figure 256. *Guazuma ulmifolia*. Drawn by M. B. Johnson.

inland northward in thornscrub to the vicinity of Soyopa and east of Mazatán. Lowlands of Mexico from Sonora and Tamaulipas to Central America, and widely distributed in tropical South America.

The fresh, tender green fruits become mucilaginous when chewed and can relieve thirst. The flavor is bland and not sweet. The mature fruit and seeds are ground and used to make tortillas, atole, and pinole, and the seeds are roasted and ground as a coffee substitute. The wood is light, soft, pliable, and strong, and has been used to make furniture, spoons, the frame for the Yaqui *tampaleo* drum, and violins. Women in Aduana, west of Alamos, weave baskets from strips of the bark. The Guarijío use the bark in a mixture to clean wounds (see *Bursera lancifolia*), the Mayo chew the bark to aid in digestion, in the vicinity of Tepoca the bark is used as kidney medicine, and it has been much used medicinally elsewhere in tropical America. This tree would be worthy of cultivation in hot and dry lowland tropical regions for its dense shade and rapid growth. It is frost tender and probably limited to the north by cold temperatures (Turner et al. 1995).

The fruits are eaten by coati mundi and other animals, and the foliage is grazed by horses and cattle. Horses relish the fruits, and the seeds can pass through the animals intact. The horse, or a horselike animal, may have been the primary and original agent of dispersal (Janzen 1983).

## Tamaricaceae—Tamarisk Family

Trees and shrubs, mostly xerophytes and halophytes in arid and semiarid regions, native to the Old World; 5 genera, 78 species.

### *Tamarix*

Trees and shrubs. Leaves alternate, sessile, scalelike, entire, with salt-excreting glands; stipules absent. Flowers small, subtended by bracts, 5-merous (the two species in Sonora) and in dense, branched spikes or racemes (panicles). Fruits of capsules; seeds many, minute, with feathery hairs. Fifty or fewer species.

1. Trees usually with massive limbs and trunk; leaf completely encircling the stem; blade reduced to a cusp or tooth < 0.5 mm; flowers white, sessile. __ ***T. aphylla***
1' Shrubs, rarely small trees; leaf not completely encircling the stem; blade scalelike but evident, 0.7–3 mm, triangular-ovate; flowers pink or white fading to pink, on pedicels 0.5–0.8 mm. ________ ***T. ramosissima***

***Tamarix aphylla**.* Tamarisk, salt cedar; *pino salado; hocö hapéc* (Seri) (figure 257)

Trees often 7–15 m; trunk and major limbs well developed and often massive; wood soft. Bark deeply furrowed, light brown to gray on trunks and major limbs, or red-brown on smaller branches. Branchlets not deciduous. Leaf bases encircling the stem (amplexicaul), dotted with salt-excreting glands; blade greatly reduced.

Flowers sessile, white except the rose-purple anthers; perianth 2.5–3 mm; petals longer than the sepals. Capsules 3.3–4.0 mm. Peak flowering mostly August–October.

Often planted as a shade tree and persisting in lowland regions, especially near the shore and at settlements in western Sonora; primarily Sonoran desertscrub and thornscrub.

This tree grows readily from cuttings of almost any size. During the cooler months dew condenses on the salt-encrusted twigs and drips onto the soil. Although flowers are produced in profusion, no seedlings are known from Sonora or Arizona, and the trees in Sonora and Arizona are probably derived from a single clone (Benson and Darrow 1981). Salt cedar trees are slowly spreading along the bed of the Santa Cruz River in Tucson, but these trees seem to result from branches buried by flood waters. This tree is native to the Mediterranean region.

***Tamarix ramosissima**.* Tamarisk, salt cedar; *pino salado; onk 'u'us, vepegi: 'u'us* (O'odham) (figure 258)

Mostly shrubs 2.5–5 m, occasionally small trees. Bark gray to reddish, mostly smooth, with conspicuous lenticels on small branches. Branchlets (short shoots or ultimate twigs), winter and tardily drought deciduous, their internodes shorter than, and obscured by overlapping scale leaves. Long shoots perennial. Scale leaves of short shoots 0.7–1.4+ mm, those of the long shoots mostly 2.3–3.0 mm.

Pedicels 0.5–0.8 mm. Flowers white or pink, 1.5–2.0 mm; floral disk glands or nectary maroon, producing glistening nectar droplets. Capsules 3.7–4.5 mm. Seeds 0.5 mm, yellowish. Flowering profusely during warm to hot weather, dormant with drought and cold weather, especially when temperatures are below freezing.

Common and naturalized nearly statewide below about 1065 m (3500 ft), below the oak zones in many riparian and semiriparian habitats, especially disturbed places and on alkaline or semisaline soils: riverbeds, streams, floodplains, arroyos, playas, irrigation canals, farmlands, waterholes, and roadside ditches. This noxious weed has even spread to remote, wholly natural areas such as waterholes on Isla Tiburón and the bottom of the Elegante Crater in the Pinacate region (Felger 2000, Felger and Moser 1985). Now widespread, weedy and invasive in many of the warmer, drier parts of the world.

The regularly spaced, alkali- or salt-excreting glands are readily seen on the leaves and young stems with a 10× or higher power lens. When fresh and first developing, these glands are globular and glistening golden in color, but they soon become white with buildup of alkali or salt.

## Theophrastaceae—Theophrasta Family

Trees and shrubs. Tropical America; 5 genera, 90 species. Reference: Stahl (1989).

### *Jacquinia*

Mexico to South America and the Caribbean Region; 35 species.

***Jacquinia macrocarpa*** subsp. ***pungens*** [*J. pungens*]. *San Juanico; tásiro* (Mayo); *cof* (Seri); *tahsi'o, tassio* (Yaqui) (figure 259)

Trees often 5–8 m, with a dense, somewhat rounded, dark green crown. Trunk thick and often single, the bark smooth and gray. Twigs with short, coarse hairs. Leaves simple, alternate, and spirally arranged, evergreen, 2.5–7.0 cm, elliptic to lanceolate, mostly narrowly so, rather thick, firm, and tough, sessile or subsessile, both surfaces dotted with pores, the midvein prominent below and extending into a sharp terminal spine, the margins inrolled; stipules absent. Leaves of juvenile plants linear and conspicuously narrower than those of adult trees.

Flowers in small terminal racemes. Calyx 3.5–4.5 mm, green, with 5 rounded lobes with thin margins; fresh corollas 10–14 mm wide, with a short tube and 5 spreading, overlapping lobes; fertile stamens 5; staminodes 5, petal-like and smaller than the corolla lobes; corollas and staminodes bright orange and somewhat fleshy, the petals and the attached staminodes and stamens readily detaching and falling intact. Fruits 2.5–2.8 cm, ovoid, hard-shelled and indehiscent, tipped with a hard, spinose, persistent style; fruit pulp fleshy-gelatinous. Seeds several per fruit, 5–7 mm, red-brown, lens shaped. Flowering and fruiting at various seasons; peak flowering late spring–early summer.

Plains, valleys, slopes, canyons and arroyos; Sonoran desertscrub, thornscrub, and tropical deciduous forest; cen-

Figure 257. *Tamarix aphylla* (Benson and Darrow 1981). Drawn by L. B. Hamilton.

Figure 258. *Tamarix ramosissima* (Parker 1972). Drawn by L. B. Hamilton.

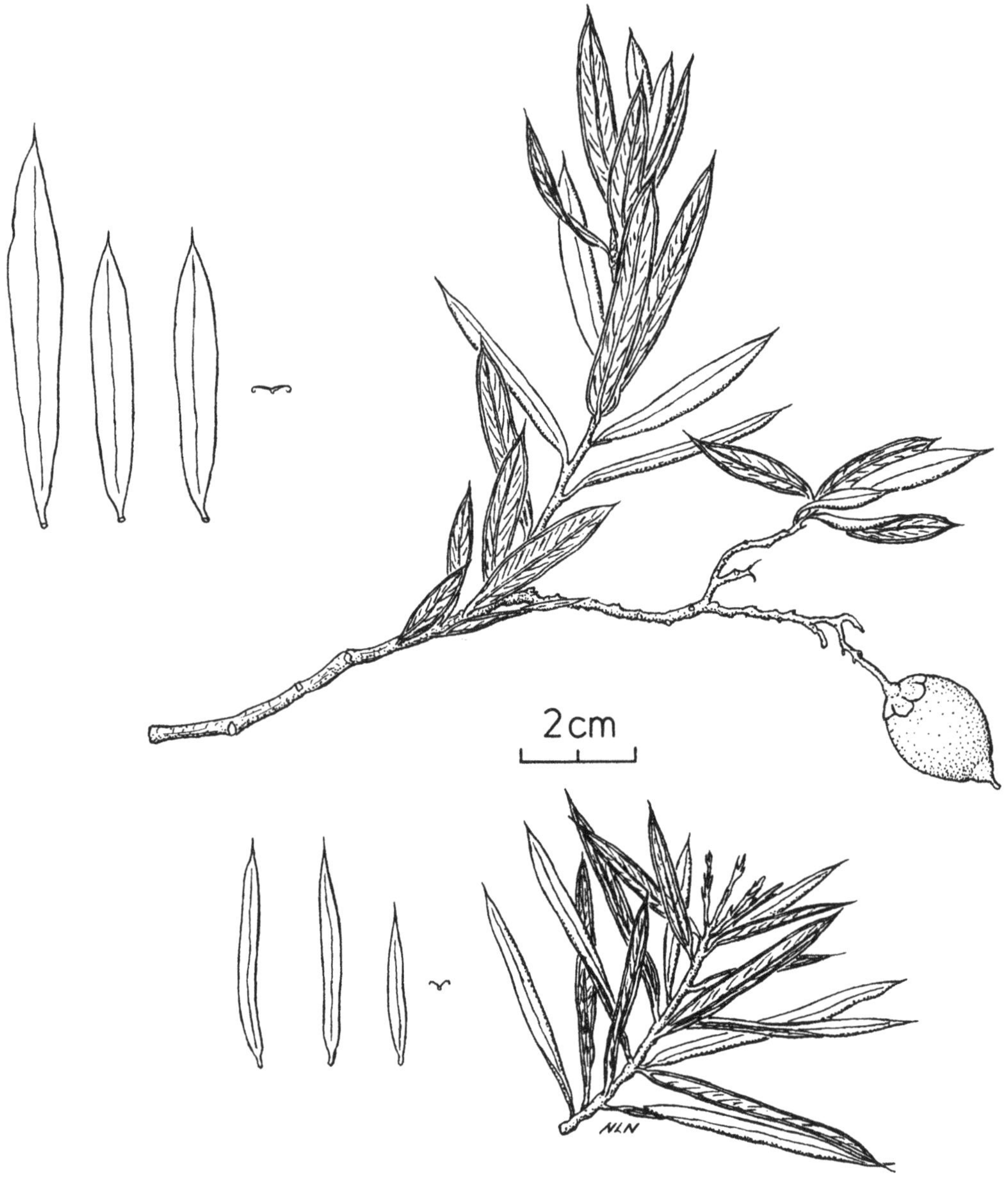


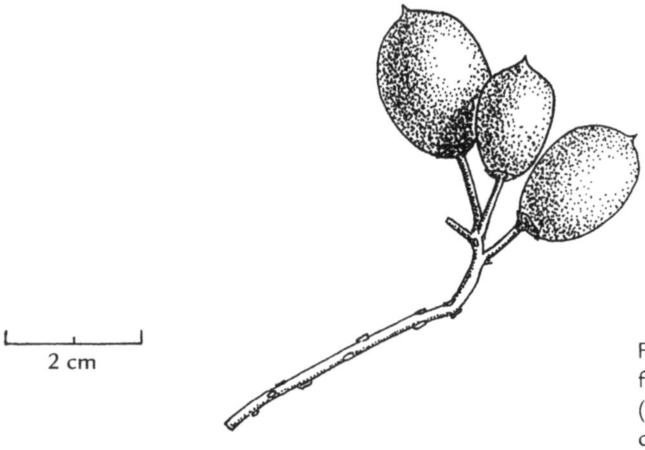


Figure 259. *Jacquinia macrocarpa* subsp. *pungens*. Larger leaves from adult tree, the narrow leaves from a juvenile plant (Felger and Moser 1985); drawn by C. M. Marlett. Fruit cluster drawn by M. B. Johnson.

tral and southern Sonora, near sea level–760 m (2500 ft). Often common on sandy desert and thornscrub plains and extending into low foothills in the vicinity of Alamos and Navojoa. This species ranges from Panama to Sonora; subsp. *pungens* occurs along the west coast of Mexico from Sonora to Guerrero, Jalisco, and Michoacán.

The Seri Indians, of coastal Sonora, ate the sweet fruit pulp, although "if one ate too much it was said to cause dizziness and possible intestinal disorder" (Felger and Moser 1985: 373). The leaves and fruits have been used to stupefy fish (Turner et al. 1995). Seri girls and women string the fresh flowers into attractive necklaces, and the Mayo and Yaqui made similar use of the flowers. The dried flowers reopen when put in water, taking on a fresh and newly-picked appearance, and may be revived repeatedly even after more than 100 years of storage. In earlier times the fruit commonly was used for soap, especially in the Alamos region. This tree would make an interesting landscape subject and barrier hedge. It is frost sensitive and damaged below about –4°C (25°F).

## Tiliaceae—Linden Family

Mostly trees and shrubs, often with stellate hairs. Bark often fibrous and mucilaginous. Stipules small and deciduous. Leaves alternate, simple; petioles well developed; blades rather thin with prominent veins, the margins serrated (in Sonora) with small, forward-pointing, blunt teeth, at first gland tipped. Flowers in cymes or panicles, not very showy (in Sonora), pedicellate; mostly bisexual. Sepals and petals each 5 and separate (in Sonora). Stamens 15–60.

Nearly worldwide, mostly tropical and especially diverse in the Southern Hemisphere; 48 genera, 725 species. Four genera occur in Sonora including *Corchorus* and *Triumfetta*.

Jute, or gunny, is made from *Corchorus capsularis*, an herbaceous species.

1. Shrubs to small trees, widespread; herbage densely stellate-pubescent; leafy bracts not present; fruits disk-shaped with marginal feathery bristles. ___ ***Heliocarpus***
1' Medium-sized to large trees, highly localized in the Yécora region; herbage glabrate; lower half of peduncle attached to a large leafy bract; fruits globose nuts without bristles. ___ ***Tilia***

### *Heliocarpus*—Sun Fruit

Shrubs or trees with stellate hairs. Leaves drought deciduous; petioles shorter to about as long as the blades; blades with palmate veins; stipules deciduous. Flowers in panicles, bisexual (or some unisexual flowers on the same plant are reported for the genus), small and not showy, pedicellate. Sepals greenish, the male flowers with 5 (in Sonora) separate, thin, narrow, white petals shorter than the sepals; petals not seen on female flowers. Stamens and style exserted. Fruits disk shaped with a double row of many raylike feathery bristles forming a sunburst pattern (hence the generic name); seeds probably 2, probably dehiscent while still on the tree, each valve may be carried away separately with the seed still attached.

Tropical America; 10+ species, mostly in Mexico and Central America. References: Lay (1949), Watson (1923).

The fruits are unique among the Sonoran flora. The flowers are probably pollinated by indiscriminate pollen-collecting insects.

1. Leaves 5.5–12.5 cm; stipules 1–4 mm, of brown bristles; female flowers without petals; stamens about 15; fruits 10–13 mm, the body completely obscured by long, tangled hairs. ___ ***H. attenuatus***
1' Leaves 11–35 cm; stipules 9–12 mm, green, flattened, and linear; female flowers with short, white petals; stamens 40–58; fruits 7–10 mm, the body densely hairy, but most of the arms of the longer stellate hairs not overlapping and not tangled, the surface often not entirely obscured. ___ ***H. palmeri***

***Heliocarpus attenuatus.*** *Samo, samo blanco, samo baboso; sanarí* (Guarijío); *sa'amo* (Mayo) (figure 260A)

Mostly shrubs in the desert and often small trees 2–6 m in tropical deciduous forest; bark gray. Herbage stellate pubescent. Leaves 5.5–12.5 × 2.3–5 cm; blades ovate, the lower surface densely pubescent, the upper surface sparsely pubescent, the marginal teeth sometimes glandular, especially at the base of the blade, the leaf tips long-acuminate (sometimes very long and slender); stipules 1–4 mm, soon becoming brown and bristlelike.

Flowers greenish yellow; sepals 3.5–4 mm, greenish, the outer surface densely stellate pubescent; stamens about 15 (reported to be as many as 20). Fruits 10–13 mm including the bristles, the body oval, high-domed, the hairs stellate with relatively long arms, very dense and obscuring the surface. Flowers visited by wasps and other insects. Flowering August–October.

Canyons and slopes; Sonoran desertscrub, thornscrub, tropical deciduous forest, and lower elevations in oak woodland; southern and east-central Sonora, as far north as the Sierra Libre south of Hermosillo and the Sierra San Javier; 150–1350 m (492–4428 ft). Also southwestern Chihuahua and northeastern Sinaloa.

This is a very common large shrub or small tree in the Alamos region. Beneath the bark the stems are highly mucilaginous, as indicated by the name *samo baboso* (*baboso* = slimy, drooling). This mucilage is used as a plaster additive or to make a whitewash for walls. The bark is stripped away, the wood soaked in water (usually in a barrel), and the next day lime (*cal*) and salt are added to produce whitewash. This whitewash, used extensively in southeastern Sonora, is also made from tree cholla (*Opuntia thurberi*) or prickly pear (e.g., *O. ficus-indica*), although *samo baboso* is preferred. The bark is used for twine. *Samo* is the local folk genus for various shrubs including the sun fruits and the legume shrub *Coursetia*.

***Heliocarpus palmeri*** [*H. polyandrus*]. *Samo prieto; samo ochóname* (Guarijío) (figure 260B)

Often spreading shrubs or sometimes small trees to approximately 5 m with a well-formed trunk and hemispherical crown; bark papery, peeling, and reddish bronze or burgundy, especially in young plants. Twigs, petioles, and leaf blade bases with small brownish stellate hairs. Stipules 9–12 mm, flattened, linear-attenuate. Leaves 11–37 × 4.5–16 cm; blades ovate, dark green, thin, at first sparsely pubescent and with age glabrate or glabrous except at the base, the marginal teeth gland-tipped (the glands maroon and prominent, especially at the leaf blade base and on young, enlarging leaves), the leaf tips mostly long-acuminate; stipules 9–12 mm, flattened, linear-attenuate.

Panicles 5–16 cm. Sepals 5 mm, green, with small stellate hairs on outer surfaces; stamens white, mostly 40–58. Fruits 7–10 mm including the bristles, the body oval, with minute stellate hairs, the hairs dense but not wholly obscuring the surface. Flowering August–October.

Canyon bottoms and slopes, mostly in mountains; tropical deciduous forest and lower elevations in oak woodland; southeastern Sonora, 170–1100 m (550–3610 ft). Also southwestern Chihuahua and northern Sinaloa.

The flowers are visited by bees and cetoniid beetles (flower scarabs). The Guarijío used the wood to make violins.

This species differs from *H. attenuatus* by its sparser stellate hairs with much shorter arms, larger, thinner, and darker green leaves, more sparsely pubescent or glabrous leaf blades, prominent glands on leaf margins, larger flowers, and smaller fruits with the body not as prominently domed. In addition, *H. palmeri* is more often treelike and generally occupies more mesic habitats than *H. attenuata*. Although their geographic ranges broadly overlap, we have not seen them growing together.

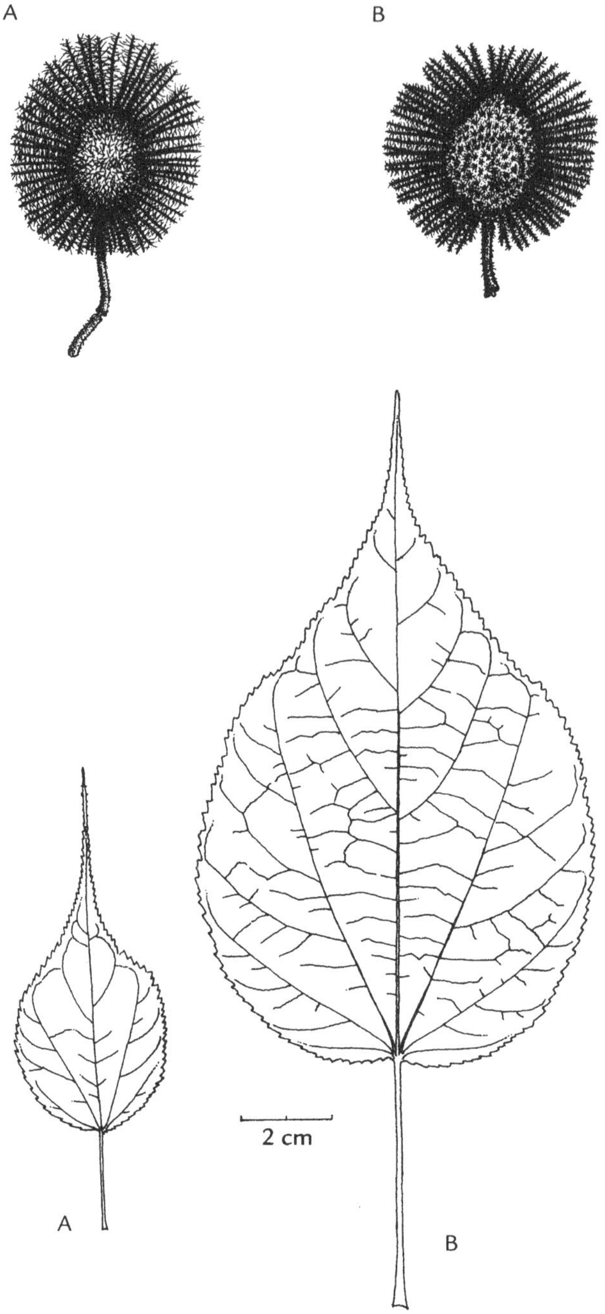


Figure 260. *Heliocarpus*. (A) *H. attenuatus*, (B) *H. palmeri*. Leaves drawn by M. B. Johnson; fruits drawn by A. K. Jakob.

## *Tilia*—Basswood, Lime, Linden

Trees of mesic and moist north temperate regions, mostly in Asia, with one species in eastern North America and mountains in Mexico; 45 species. Reference: Hardin (1990).

These trees generally do not thrive in dry climates and require moist soil. A few basswood trees, however, are successfully being grown in Bisbee, at 1630 m, in southeastern Arizona. They are important street trees in temperate regions, and in Europe are commonly known as lime or linden. The wood is soft and light but tough, and has been used for excelsior, cheap furniture, and many other uses. The fibrous inner bark has been made into mats and cordage. The sweet-scented flowers are an excellent source of honey (Uphof 1968).

***Tilia americana*** var. ***mexicana*** [*T. mexicana*]. Mexican linden (figure 261)

Trees to approximately 10 m in Sonora, with a spreading crown; bark relatively smooth and grayish. Pubescence mostly of slender-rayed stellate hairs; winter buds, new growth, and flower buds densely pubescent. Leaves winter deciduous; petioles 2.5–4.5 cm; blades 8–25 × 6–20.5 cm, moderately bicolored, becoming glabrate, broadly ovate to sometimes nearly orbicular, asymmetric toward the base, the base cordate to truncate, the tip acute to short- or mostly long-acuminate. Major leaf blade veins raised below, their axils with small tufts of hairs that apparently serve as domatia (habitats for predatory mites).

Inflorescences of small axillary or terminal cymes; long peduncle attached below to a large, leafy bract unique among the trees of northwestern Mexico; bract sessile or stalked, 6–15.5 × 1.3–4.0 cm, linear-oblanceolate to sometimes linear-oblong, the margin entire, the tip rounded to obtuse. Cymes solitary in each bract, pendulous from the free portion of the peduncle, generally few flowered. Flowers probably sweet scented, approximately 1.5 cm wide; petals yellow. Stamens numerous, those opposite the petals modified as staminodes. Fruit indehiscent, small, globose, and nutlike, probably 1- or 2-seeded. Flowering June.

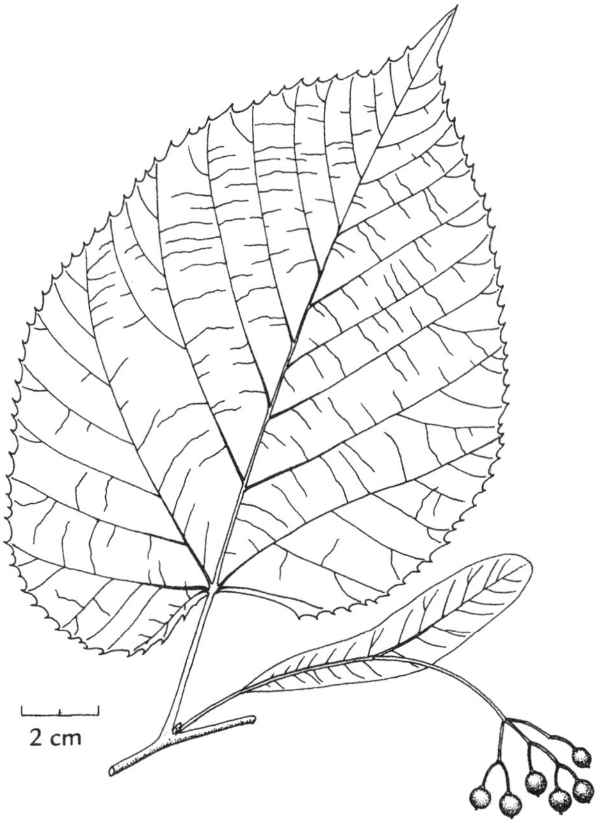


Figure 261. *Tilia americana* var. *mexicana*. Drawn by M. B. Johnson

Localized in moist canyons in pine-oak forest in east-central Sonora and southwestern Chihuahua, mainly 1690–1900 (5545–6230 ft) m (Spellenberg et al. 1996). This tree is rare in Sonora and known only from a narrow gorge with a flowing stream at Trigo Moreno, southeast of Yécora, at 1470 m. The gorge has walls draped with *Hydrangea seemanii;* associated trees are *Pinus maximinoi, Alnus oblongifolia, Clethra mexicana,* and *Cupressus lusitanica*.

This species occurs in disjunct distribution in eastern North America from Canada to Florida, Texas, and mountains in Mexico. Among the four allopatric and often rather weakly distinguished varieties, var. *mexicana* is the only one in Mexico and occurs in 17 states from Sonora, Chihuahua, and Coahuila to Oaxaca.

## Ulmaceae—Elm Family

Trees or shrubs. Stipules shed as the leaves unfold, each pair enclosing and protecting the next younger leaf bud. Leaves almost always alternate, simple; blade often asymmetric at the base, the margins often serrate. Flowers in small axillary clusters shorter than the leaves, wind or sometimes self-pollinated, small, unisexual or bisexual; sepals green to nearly transparent (membranous), united to separate or nearly so; petals absent; styles 2-branched. Fruits 1-seeded. Two subfamilies (sometimes treated as separate families): Ulmoideae, with dry, wind-dispersed fruits, commonly of samaras (includes *Ulmus*), and Celtoideae, with fleshy, animal-dispersed drupaceous fruits (includes *Aphananthe* and *Celtis*).

Tropical and north temperate regions worldwide; 18 genera, 150 species. References: Elias (1970), Sherman-Broyles et al. (1997).

1. Leaves obviously scabrous, palmately 3-veined from base, pinnately veined above with ≤ 4 (5 or 6) pairs of lateral veins. ________ ***Celtis***
1. Leaves not scabrous (or perhaps slightly so), pinnately veined with ≥ 10 pairs of lateral veins.
    2. Flowers unisexual, flowering on the new growth; the fruits rounded, fleshy; southeastern Sonora. ***Aphananthe***
    2' Flowers bisexual, flowering on the previous year's leafless branches; fruits flat, winged, dry; northeastern Sonora. ________ ***Ulmus***

### *Aphananthe*

Trees; 5 species: 1 in the New World, 1 in Madagascar, and 3 dispersed from India to Japan and Australia.

***Aphananthe monoica*** [*Aphananthe mexicana; Celtis monoica; Mirandaceltis monoica*]. *Guasimilla* "little *guasima* (*Guazuma ulmifolia*)"; *mohuarí* (Guarijío) (figure 262)

Trees commonly 20–25+ m; trunk solitary. Herbage mostly glabrate. Evergreen; leaves elmlike, 5–16 cm; petioles short; blades lanceolate to narrowly ovate, shiny green, dark green above, lighter green below and sparsely pubescent with small, appressed hairs (not scabrous or only slightly so), the veins pinnate, prominent, with 11–18+ pairs of lateral veins, the margins with forward-pointing, sharp teeth of varying sizes, the leaf tip attenuate.

Flowers inconspicuous; male flowers in small, dense clusters; female flowers solitary. Fruits rounded, fleshy, approximately 1 cm in diameter, orange at maturity and pleasantly tart in taste. Flowering late May–summer or later; fruiting summer–fall.

Riparian canyon bottoms; tropical deciduous forest and lower oak woodland; southeastern Sonora; 300–1100 m (985–3610 ft). Central America to Sonora and San Luis Potosí.

In Sonora these trees grow in a number of usually narrow canyons with intermittent or permanent water. These are the most tropical habitats in the state, and *Aphananthe* is locally common in evergreen tropical forest with a canopy often 16–25+ m high that may include *Cinnamomum hartmannii, Bursera simaruba, Drypetes gentryi, Ficus* spp., *Platanus wrightii, Sideroxylon persimile, S. tepicense,* and *Trophis racemosa*. Guasimilla lumber is prized for housebuilding but is seldom used because it occurs in remote places.

### *Celtis*—Hackberry

Trees or shrubs, spinescent or not. Leaves petioled; blades often asymmetric at the base, with 3 main veins from the

Figure 262. *Aphananthe monoica*. Drawn by M. B. Johnson.

base and few lateral veins and scabrous. Flowers usually unisexual, both sexes often on the same branch, with some bisexual flowers. Flowers solitary or in small clusters in axils of new growth; pedicelled, small, inconspicuous, green to yellow-green. Sepals mostly 5. Female flowers with a large, sessile or nearly sessile 2-branched, feathery stigma. Stamens opposite the sepals; filaments at first turgid and oval in cross-section, soon drying and becoming flattened (at least in *C. reticulata*), held by the cupped sepals, expanding like tension-held springs as they mature, straightening rapidly and flinging the dry, powdery pollen from the anthers. Fruit a drupe; seeds primarily bird dispersed.

Worldwide, tropical, temperate, and some in deserts; 60 species, 3 in Sonora.

1. Stems unarmed; fruiting pedicels as long as or longer than the fruits. ____ ***C. reticulata***

1' Stems usually armed; fruiting pedicels shorter than the fruits.

 2. Spines and thorns mostly curved; leaves mostly > 2.5 cm wide; southeastern Sonora, not in the desert. ____ ***C. iguanea***

 2' Thorns mostly straight; leaves mostly < 2.5 cm wide; deserts and thornscrub. ____ ***C. pallida***

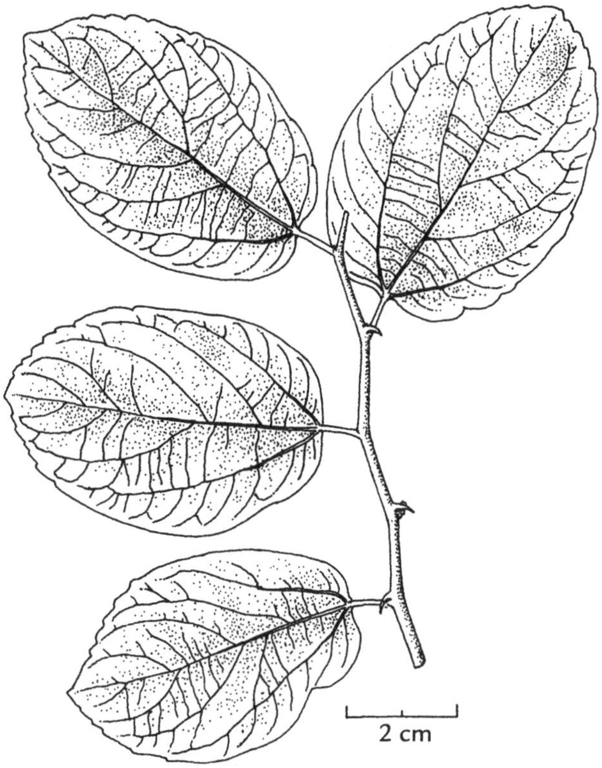


Figure 263. *Celtis iguanea*. Drawn by M. B. Johnson.

***Celtis iguanea.*** *Cumbro, garabato, vainoro; carocá'a* (Guarijío); *cumbro* (Mayo) (figure 263)

Large, sprawling tangled shrubs or lianas, sometimes forming enormous, impenetrable woody briars to 20+ m across and reaching 5–13 m in height, the main stems to 20 cm in diameter; sometimes treelike when young. Many branches armed with wicked, rigid, stout, usually recurved axillary spines and also often with short-shoot thorns. Herbage pubescent, especially when young; leaf blades minutely scabrous or the upper surfaces sometimes strongly scabrous (not nearly as rough-scabrous as *C. reticulata*). Mostly evergreen or tardily drought deciduous. Leaves 4.5–11.5 × 2.5–7 cm; blades dark green, broadly ovate to oval, the base mostly or nearly symmetrical, often moderately cordate, the margins serrated, especially toward the tip to nearly entire.

Fruits globose, sparsely pubescent or glabrous, fleshy, pale to bright orange at maturity, 9.5–11 mm in diameter, slightly sweet and insipid (often not as tasty as those of *C. pallida*). Flowering at least March–July, fruiting at least late summer–fall.

Banks of streams and rivers, arroyos and canyon bottoms; thornscrub and tropical deciduous forest; southeastern Sonora; 200–350+ m (660–1150+ ft). Northern Mexico, from Sonora and Tamaulipas, to Argentina and the Caribbean.

*Celtis iguanea* resembles *Pisonia capitata* in habit but is readily distinguished by the alternate rather than opposite leaves. Unlike the other *Celtis* species in Sonora, it is frost tender (Tom Van Devender, personal communication 1998).

***Celtis pallida*** subsp. ***pallida*** [*C. tala* var. *pallida*]. Desert hackberry; *cúmero, garambullo; cumbro* (Mayo and southern Sonora); *koom* (O'odham); *kuwavul* (Pima Bajo); *ptaacal* (Seri); *cungo* (Yaqui) (figure 264)

Briarlike, thorny shrubs or occasionally trees to 6 m with a trunk to 20 cm in diameter in tropical deciduous forest in the vicinity of Alamos. Wood hard and flexible. Young twigs, thorns, and inflorescences with stiff, appressed white hairs, and glabrous with age. Twigs often zig-zag, often with single or paired thorns, especially the larger thorns often with 1 (2 or 3) node(s) bearing flowers or a small leaf, the shorter twigs rigid and thorn tipped. Leaves gradually drought deciduous, mostly 2.5–7 × 0.5–2.5 (3.5) cm; blades more or less ovate to broadly elliptic, scabrous (dotted with hard, minute glands often sprouting a stiff, forward-pointing hair, which feel like sandpaper), the base mostly or nearly symmetrical, the margins with forward-pointing teeth to sometimes entire.

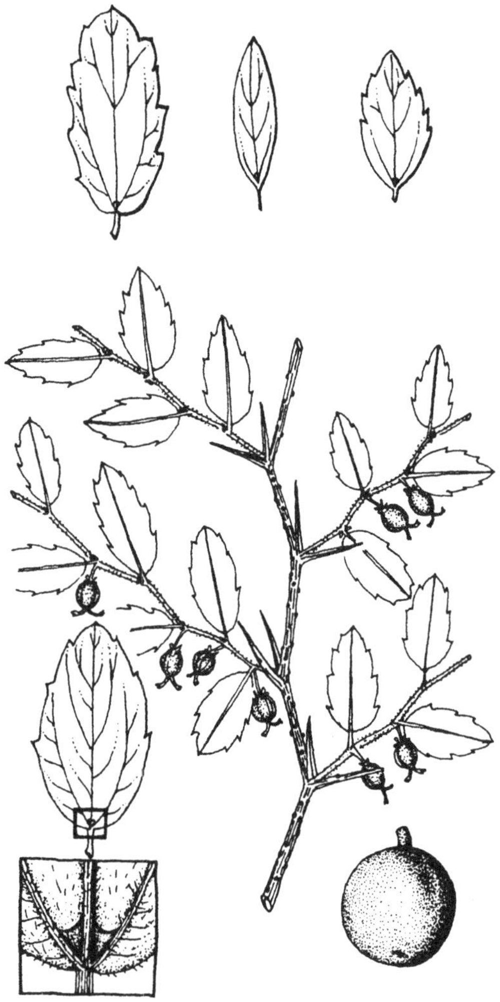

Figure 264. *Celtis pallida* Drawn by F. Bond (courtesy James Henrickson).

Fruits 8–10 mm, glabrous, bright orange; pericarp rather thin, fleshy, not hard-walled, edible, and moderately sweet and tart, the seed hard; often with an abundant crop of fruit in early fall. Flowering spring and other seasons.

Widespread: washes and drainageways, arroyos and canyons, bajadas, plains, and less commonly on rocky slopes. Sonoran desertscrub, Chihuahuan desertscrub, thornscrub, tropical deciduous forest; nearly throughout Sonora except higher elevations and extreme northwestern Sonora; near sea level–1250 m (4100 ft). Arizona to southern Texas and Oaxaca, both states of Baja California, southern Florida, Central America, Paraguay, and Argentina. An additional subspecies occurs in South America.

The Yaqui, Seri and others prized the strong, flexible wood for making bows.

***Celtis reticulata*** [*C. laevigata* var. *reticulata*]. Netleaf hackberry, canyon hackberry; *machaquí* (Guarijío); *cúmaro*; *cumbro* (Mayo) (figure 265)

Trees often 8–15 (20) m with a well-developed trunk. Northern Sonora populations with large, irregular, knobby, wartlike ridges and protuberances on trunks and major limbs, those in southern Sonora (vicinity of Sierra el Aguaje and Guaymas southward) generally with smooth, gray bark; twigs with reddish bark. Herbage pubescent. Winter bud scales several, intergrading with the stipules. Stipules 4.5–10 × 1.5–2.0 mm, paired, lanceolate-elliptic to linear-oblong, very thin and nearly transparent. Foliage winter deciduous at least in northern Sonora or in southern Sonora leaves falling as new ones appear in January. Leaves 3–5.5 (11.5) × 1.8–4 (10) cm; blades lanceolate to mostly ovate, often asymmetric, the apex acuminate, the surfaces dry and rough to the touch due to scabrous hairs, the margins mostly entire but at least some leaves often with teeth on the apical portion of the blade.

Flowers 3–5 mm wide. Fruits yellow-orange mostly in southern Sonora to mostly red-brown in the north, thin but hard-walled, globose, 7–8 mm in diameter, the slightly fleshy mesocarp marginally edible, the fruits soon becoming dry. Flowering March–June; fruiting summer–fall.

Riparian habitats, usually along streams and rivers; Sonoran desertscrub, Chihuahuan desertscrub, thornscrub, tropical deciduous forest, grassland, and oak woodland. Nearly statewide except lowland regions of northwestern and southwestern Sonora, 60–1830 m (200–6000 ft). The southernmost record is from near El Zapote, about 30 km southeast of Alamos near the Sinaloa border. Western United States as far north as Washington and Kansas and northern Mexico including Baja California Sur, Sonora, Chihuahua, and Coahuila.

The tallest and largest cúmaro known to us in Sonora is nearly 20 m tall. It was found by William Peachey (personal communication 1996) at Cueva del Cúmaros, north-northwest of Tepache (southeast of Moctezuma) at 710 m in thornscrub on quaternary basalt rock. This species is sometimes planted as a landscape subject in the southwestern United States.

## *Ulmus*—Elm

Trees or rarely shrubs, usually winter deciduous. Concentrated in temperate forests of the Northern Hemisphere, mostly Eurasia; 40 species, including 4 in Mexico, 3 of them native. Reference: Todzia and Panero (1998).

*Ulmus mexicana*, apparently the tallest tree species in Mexico, reaches 87 m with a trunk to 6 m in diameter (Pennington and Sarukhán 1968); it occurs in tropical forests from southern Mexico to Panama.

Figure 265. *Celtis reticulata* (Benson and Darrow 1945). Drawn by L. B. Hamilton.

***Ulmus pumila*.** Siberian elm (figure 266)

Trees often 5–8 m with a well-developed trunk; wood weak and brittle. Bark gray, deeply fissured on trunks and larger limbs. Herbage glabrate. Stipules 3–6 mm. Leaves winter deciduous, 4–8.5+ cm; petioles 0.5–1 cm; blades elliptic to lanceolate with a pointed tip, dull green, symmetric to asymmetric at the base, veins alternately pinnate, conspicuous on the lower surfaces, with 10–14 pairs of lateral veins, margins singly serrated. Flowers in small clusters, appearing in spring before the leaves; buds globose, forming in late fall and early winter, the scales purple-brown with long hairs on their margins. Flowers bisexual, approximately 4.5 mm including the exserted stamens. Fruit a flat, broad-winged samara 11.5–13 mm in diameter, notched at tip.

Roadsides; Chihuahuan desertscrub, grassland, and oak woodland; northeastern Sonora, 1250–1800 m (4100–5900 ft). Scattered trees and small colonies established and spreading along Highway 2 from about 50 km west of Cananea eastward to the Chihuahua border. Also fairly common in similar habitats in adjacent Cochise County, Arizona.

This tree apparently propagates by root sprouts and produces large quantities of viable seeds. This elm was widely cultivated in southeastern Arizona and northeastern and north-central Sonora but has fallen from popularity because of its susceptibility to pathogens and insects and because of the readily breaking branches. It is widely planted for quick-growing windbreaks and is weedy across much of the temperate United States. It is native to central Asia.

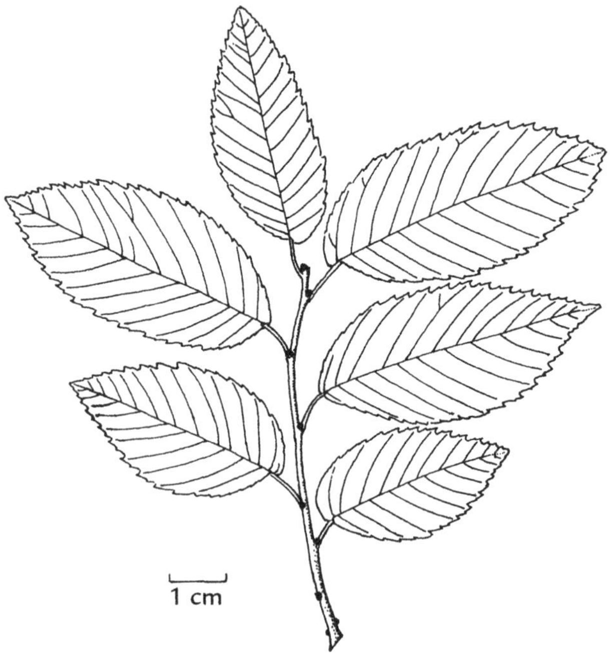


Figure 266. *Ulmus pumila*. Drawn by M. B. Johnson.

## Urticaceae—Nettle Family

Annual to perennial herbs, rarely shrubs or small trees; often with stinging hairs. Several herbaceous genera in Sonora. Mostly tropical, some in temperate regions; 45 genera, 800 species. Reference: Boufford (1997).

### *Urera*

Trees and shrubs with stinging hairs; usually dioecious. Tropical America, Africa, and Madagascar; 35 species.

***Urera*** cf. ***caracasana*.** *Ortiguilla* (figure 267)

In Sonora usually large, spindly shrubs, occasionally irregular, sprawling trees 7–10 m; trunk usually short; stems semiwoody, rather thick and hollow, sometimes with large, internodal prickles, and a dense, leafy crown. Herbage sparsely pubescent with slender, stinging hairs, which upon contact with the skin inflict mild, short-lasting (perhaps 5 minutes) pain. Drought deciduous or perhaps evergreen in the wetter habitats. Leaves alternate, simple; petioles 5–24 cm; blades 15–40 × 15–35 cm, broadly ovate to orbicular-ovate, thin, pinnately veined, with numerous minute cistoliths (small, white, smooth, calcium carbonate bodies), the margins coarsely and somewhat irregularly toothed; stipules deciduous even before the leaves expand.

Dioecious. Inflorescences dense, dichotomously several-times branched panicles, the branches often interlaced, along the bare stems below the leaves, and shorter than the petioles, or flowering when leafless; inflorescence branches slender with spinescent hairs. Flowers minute and inconspicuous, whitish green; calyx with 4 segments; petals absent; wind pollinated. Male flowers with 4 stamens, the dry powdery pollen catapulted by the elastic movement of the stamens at anthesis. Female flowers approximately 1.5 mm; sepals thin and bractlike; stigma reddish and brushlike. Fruits approximately 3 mm in diameter, rounded, red, berrylike achenes surrounded by a fleshy calyx; seeds somewhat disklike, moderately compressed. Ripe fruit September–October.

Understory shrubs and trees in riparian canyon-bottom habitats with tropical evergreen forest surrounded by tropical deciduous forest and lower oak woodland; southeastern Sonora. Tree-sized plants known from Sierra de Alamos, Arroyo Verde in the upper Río Cuchujaqui at about 800 m (2625 ft), and less common in Arroyo Tepopa at 1040 m (3410 ft). Also southwestern Chihuahua. *Urera caracasana* from Sinaloa southward in tropical America to

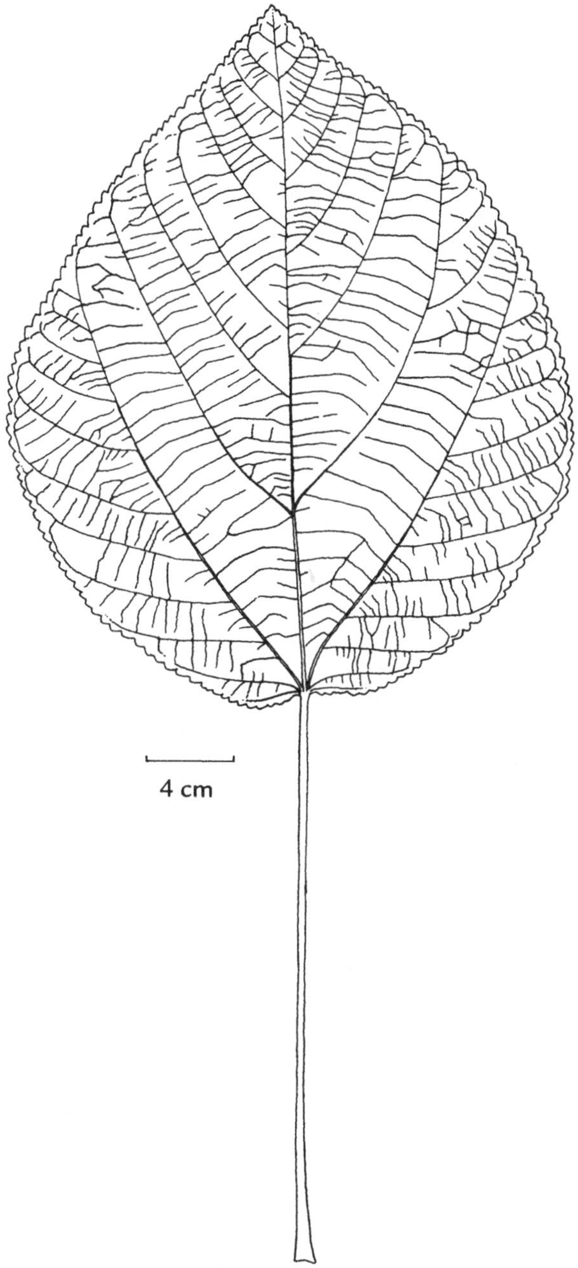


Figure 267. *Urera* cf. *caracasana*. Drawn by M. B. Johnson.

Argentina and the West Indies. In Sinaloa it becomes a spreading tree to 10 m.

The tree form in Sonora might not be the same species as the more widespread shrub form. These shrubs, often 2.5 m or more, are generally spindly and few-branched, and differ from the Arroyo Verde trees in lacking roselike stem prickles, having generally more slender stems, denser pubescence especially on the lower leaf surfaces, and more finely and evenly serrated leaf margins. These shrubs may have leaves nearly as large as the Arroyo Verde trees. The shrub form occurs in moist, riparian habitats in lower oak woodland and tropical deciduous forest as low as 390 m.

## Verbenaceae—Verbena Family

Plants of diverse habits including herbs (mostly perennials), vines, shrubs, or trees. Stems and branchlets mostly square in cross-section. Leaves and branching mostly opposite and decussate (each pair at a right angle to the pair above and below). Stipules absent. Leaves simple or compound.

Calyx mostly 4- or 5-lobed or -toothed. Corollas mostly 4- or 5-lobed, usually bilateral. Stamens 4 (in Sonora), the filaments attached to the corolla tube. Ovary superior; style 1, the stigma often 2-lobed. Fruits mostly enclosed by the calyx, separating into 1-seeded nutlets or drupes.

Worldwide, largely tropical but extending into deserts and temperate regions; 100 genera, 2600 species.

*Lantana* and *Verbena* are widely cultivated garden ornamentals. Many herbaceous and shrub species occur in Sonora.

1' Leaves compound. ________ ***Vitex***
1. Leaves simple.
   2. Leaves mostly 5–11 cm, relatively firm, not thin and not quickly wilting; flowers white; fruits fleshy. ***Citharexylum***
   2' Leaves mostly 10–30 cm, relatively thin and quickly wilting; flowers yellow to orange; fruits dry. ***Lippia***

### *Citharexylum*

Trees and shrubs. Tropical America, northward to Florida, Texas, and Sonora; 70 species. One other species in Sonora: *C. flabellifolium*, a woody shrub with thorn-tipped, rigid twigs and showy lavender flowers; coastal areas from the vicinity of San Carlos southward. Reference: Moldenke (1958).

***Citharexylum scabrum*** [*C. incanum*] *Chaparacoba* (Mayo)

Shrubs to 4 m and rarely small trees 6–7 m. Twigs striate, inconspicuously quadrangular in cross-section; leaf scars raised, slightly corky. Herbage with short, rather thick, spreading hairs. Leaves opposite and sometimes alternate near tips of twigs, 4.5–11 cm, broadly ovate to oval, the tips broadly obtuse, the margins entire (on Sonora specimens, but Sinaloa specimens have both entire and coarsely serrated margins); blades decurrent onto petioles; petioles 8–15 mm.

Flowers crowded on slender racemes 2–14 cm at the ends of small, leafy twigs, on pedicels 1 mm and subtended by bractlets 1 mm. Flowers white, sweet-scented, 5–6 mm. Calyx 3.0–3.5 mm, shallowly toothed, not spreading. Corollas short-pubescent outside and densely pubescent with longer hairs inside surrounding the anthers and nearly occluding the entrance to the tube; tube green; limb cream white, 5 mm wide; lobes 1 mm, erect to spreading and broadly obtuse. (This species is reported to have a fifth sterile stamen, or staminode, but none were seen among the few Sonora specimens.) Fruits 5–6 mm wide, with a fleshy, salmon to orange pericarp surrounding 2 wedge-shaped nutlets; calyx persistent in fruit, 5–6 mm wide. Flowering September.

Shrubs to about 4 m in localized populations in coastal thornscrub in southwestern Sonora. Tree-sized plants in Sonora known only on the western edge of tropical deciduous forest in the vicinity of Masiaca, at 65 m (215 ft), where it grows in fencerows. Southern Sonora, Sinaloa, and Baja California Sur.

The Sonoran specimens appear closely related to *C. berlandieri*, which is reported for southern Texas and northeastern Mexico to Veracruz and Sinaloa. Perhaps they are not distinct species; *C. scabrum* is the older name.

### *Lippia*

Mostly shrubs to trees, mainly tropical. Africa and the Americas; 200 species.

***Lippia umbellata*** [*L. pringlei*]. *Bacatón; talacáo* (Guarijío) (figure 268)

Mostly large spreading shrubs, less often small trees and rarely reaching 15 m; bark smooth, light tan or gray; smaller trees branching from near the base, the larger ones with a well-developed trunk and a high, spreading crown. Twigs often relatively thick and brittle. Herbage and inflorescences densely hairy. Foliage with a strong *Lantana*-like odor. Leaves 9.5–30+ × 2.5–12 cm; petioles 1.5–5 cm; blades lanceolate to broadly ovate, upper surface darker, scabrous like sandpaper, the lower surface with short, dense, woolly hairs between the veins, the apex mostly acute, the base short-winged onto the upper petiole, the margins with small, rounded teeth.

Inflorescenses of compound umbellate clusters from upper leaf axils, the flowers crowded in conelike, short spikes or heads approximately 1 cm wide, these on elongated peduncles with broad, acuminate bracts. Flowers fragrant. Corollas yellow or yellow-orange, 3 mm, the lobes hairy on the outer surfaces. Fruits of glabrous capsules 1.2 mm, separable into 2 nutlets, enclosed in the hairy calyx. Flowering March–April and July–December.

Canyons and ridges; tropical deciduous forest and oak woodland, often riparian: southeastern Sonora, 455–1350 m (1500–4430 ft). Sonora and adjacent southwestern Chihuahua to Central America.

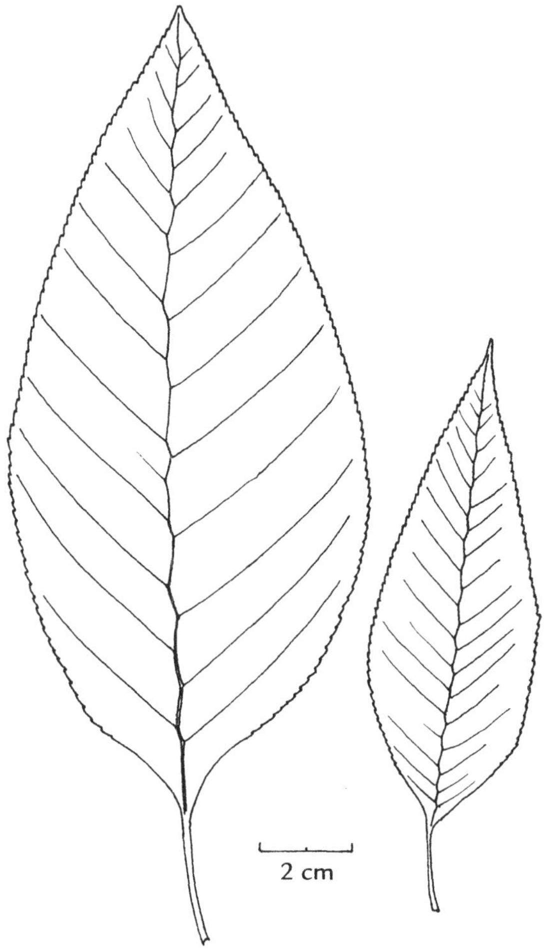


Figure 268. *Lippia umbellata*. Drawn by M. B. Johnson.

The Guarijío regard its medicinal qualities as particularly efficacious (Gentry 1942); the sap is used for toothache, and the leaves, steeped in hot water, are coated with mentholatum or another ointment or grease and applied to bruises, sores, or the forehead for headaches.

### *Vitex*

Leaves digitately compound (or occasionally reduced to one leaflet); leaflets short to long stalked, usually with entire margins, the veins pinnate and more conspicuous below. Flowers in axillary cymes. Calyx radial or nearly so, with 5 short teeth; corollas conspicuously bilateral, 5-lobed with a large lip. Fruit a fleshy drupe. Mostly tropical and subtropical; 250 species.

The monk's pepper or chaste tree, *V. agnus-castus*, native to Europe, is cultivated as an ornamental flowering tree in Sonora and Arizona. It has aromatic foliage with five to seven narrow and thin leaflets and purple, lavender, pink, or white flowers. The shrub *V. trifolia*, in Sonora called *carnavalito* or *hazme como quieres* ("do what you want with me"), is commonly grown as a hedge and ornamental shrub and can become a small tree. The leaves have three pale and strongly bicolored, broad leaflets.

1. Leaves usually with 3 leaflets; corollas densely hairy outside; canyon bottoms and streams. ___ ***V. mollis***

1' Leaves mostly with 5 leaflets; corollas sparsely hairy or glabrous; slopes and cliffs. ________ ***V. pyramidata***

***Vitex mollis.*** *Igualama, uvalama; cajúguari, juguari* (Guarijío); *júbare* (Mayo); *jupari* (Yaqui) (figure 269A)

Shrubs to trees 4–20 m. Bark on trunk and larger limbs light gray-brown, with age dark brown, gradually peeling in narrow, vertically elongated checks or fissures. (The largest trees are in the lower Guarijío region along canyons of the Río Mayo above Mesa Colorado and the nearby Río Guajaray. Here one often finds *uvalama* trees 10–20 m with massive, gnarled, and irregular trunks to 1 m in diameter.) Herbage, especially when young, and inflorescences and calyx densely pubescent with golden-brown hairs, the pubescence of summer rainy season leaves whitish and sparser; outer corolla surfaces also densely hairy. Mostly evergreen; leaves mostly with 3 leaflets, and sometimes with some unifoliolate leaves and vigorous long shoots occasionally with 5 leaflets; leaflets lighter green below, oblong-oval to obovate or elliptic, the tips rounded to notched or broadly short-pointed, the terminal leaflet blades 7.5–20 (23) cm, the lateral leaflets smaller with blades often 4–9 cm, sometimes much reduced or not developing; petioles 2.8–6.3 cm; young plants or watershoots sometimes with the margins serrated toward the tip.

Inflorescences shorter than the leaves, sparsely flowered. Flowers on relatively thick pedicels and with far-reaching, pungent fragrance. Calyx 2.5–4 mm; corollas bilaterally symmetrical, 17–20 mm, with a large lavender-blue lip and smaller lobes white, the tube pale bluish white with lavender to dark blue-purple guidelines. Stigma, style, and filaments white; anthers lavender. Fruits rounded, 1.5–2 cm in diameter; pericarp blackish when ripe, resembling and tasting like cooked prunes. Flowering (February) March–May (August); fruiting June–November.

Canyon bottoms and streams; thornscrub, tropical deciduous forest, lower elevations in oak woodland, and margins of Sonoran desertscrub; east-central and southern Sonora, 35–1200 m (115–3935 ft). Also Baja California Sur, and southwestern Chihuahua to Morelos and Oaxaca.

The fruits are eaten fresh and also cooked with *panocha* (brown sugar) and are sometimes sold in local markets. The ripe fruits have a strong, almost unpleasant floral scent. The fleshy, black pulp, which may vary from sweet to bitter, leaves a black stain on teeth and tongue. The seeds are not eaten. A decoction of the fruit and leaves has been used as a remedy for diarrhea. *Vitex mollis* is one of the preferred substrates for the common yellow-flowered epiphytic orchid *Oncidium cebolleta*.

***Vitex pyramidata.*** *Negrito; cajuari* (Guarijío); *jupa'are* (Mayo) (figure 269B)

Large shrubs or trees to 8 m; bark dark gray, rough, and flaking. Branchlets dichotomous. Young twigs, petioles, and leaflet stalks densely tomentose-hairy. Leaves mostly with (3) 5 leaflets; leaflets lanceolate to ovate and sometimes elliptic, the upper surfaces glabrous or sparsely hairy, the lower surfaces conspicuously net-veined, with dense, short, nonglandular and glandular hairs, the middle leaflet longest, often 7.5–15 cm; petioles 2.8–7 cm.

Inflorescence branches slender, relatively crowded and many flowered. Flowers on slender pedicels; calyx 1.5–2.0 mm; corollas 6–8 mm, bright blue with a white throat. Fruits rounded, 1–1.3 cm in diameter; pericarp blackish when ripe, semifleshy, edible fresh. Flowering April–August.

Rocky slopes, ridges, and cliffs; tropical deciduous forest and oak woodland; southeastern Sonora, 365–1065 m (1200–3500 ft). Sonora to Durango and Jalisco.

The fruits are similar to those of *V. mollis* and are occasionally sold in local markets.

## Zygophyllaceae—Caltrop Family

Herbs, shrubs, and some trees. Worldwide in tropical and mostly semiarid and desert climates, and some in temperate regions; 30 genera, 250 species.

Zygophylls are especially prominent elements of desert vegetation worldwide. Prominent genera in Sonora include three species of summer poppy (*Kallstroemia*) and creosotebush (*Larrea divaricata*).

### *Guajacum*

Tropical and subtropical trees and shrubs of the New World; 6 species.

*Guajacum officinale* and *G. sanctum*, closely related species from farther south in the Americas, are the source of lignum vitae, the hardest commercial wood. Exported to Europe as a supposed remedy for syphilis, *G. officinale* was almost exterminated due to overharvesting during colonial times. The genus name is often spelled *Guaiacum*. However, *Guajacum* is Linnaeus' original spelling in Species Plantarum of 1753.

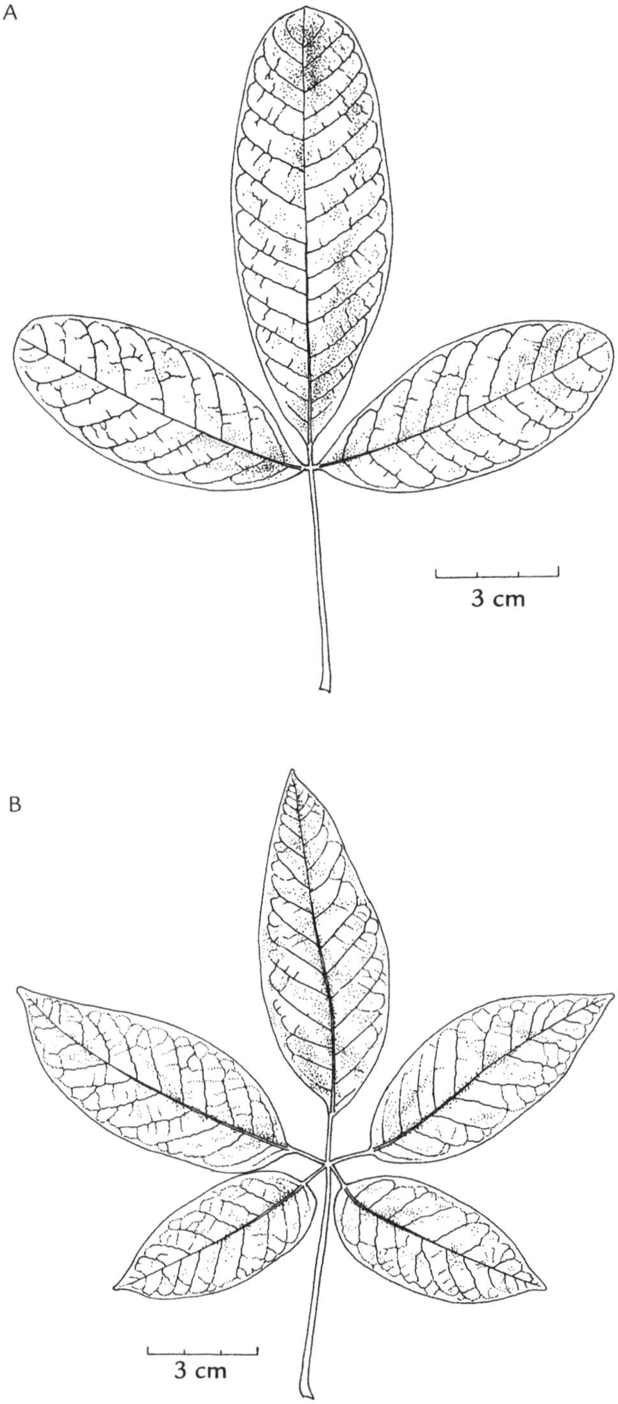


Figure 269. *Vitex*. (A) *V. mollis*, (B) *V. pyramidata*. Drawn by M. B. Johnson.

***Guajacum coulteri*** [*G. coulteri* var. *palmeri*]. *Guayacán*; *júyaguo* (Mayo); *mocni* (Seri); *huya'awo, huyawo* (Yaqui) (figure 270)

Rigidly branched large shrubs to small trees; wood extremely hard (it does not float in water); branches often thick; bark smooth, gray. Twigs and leaf stalks with short, whitish hairs; leaflets glabrous or sometimes sparsely to moderately pubescent, especially when young. Leaves opposite, compound, often crowded on short shoots, 1.5–6 cm, dark green, evergreen or tardily deciduous during extreme drought, even pinnate with 3–5 pairs of leaflets; leaflets 9–32 mm, asymmetric especially at the base, the margins entire; stipules more or less triangular, brown, 1–2 mm on short shoots, to 3 mm on long shoots.

Flowers crowded in small, axillary clusters, each lasting one day, on slender pedicels. Sepals 5, separate, green at the base, with brown- to purple-blue tips and margins. Petals 5, 10–15 mm, separate, each twisted at the base like a propeller blade, brilliant indigo blue, fading after anthesis of anthers. Stamens 10; filaments blue; anthers lemon yellow before anthesis. Ovary superior, green, short stalked, the ovary and style with white hairs. Fruit a semifleshy capsule, obovate, with 2–5 wings, yellow to yellow-orange when ripe, usually only 1 or 2 seeds developing; seeds 1 cm, enveloped in a thin, bright red aril, and conspicuously protruding from the carpels. Flowering (April) May–July.

Slopes, plains, and arroyos; typically a shrub in Sonoran desertscrub and thornscrub, and a small tree in tropical deciduous forest; central and southern Sonora, near sea level–610 m (2000 ft). Sonora to Puebla and Oaxaca.

Spectacular masses of indigo blue flowers are produced during the height of the premonsoon drought. It is suitable in nearly frost-free areas as an ornamental landscape subject, probably as a hedge or specimen tree. The sap is one of several ingredients used by the Seri to make a highly prized and unique blue dye (Felger and Moser 1985).

## MONOCOTS

### Agavaceae—Agave Family

Warm and mostly arid or semiarid regions of the New World; 18 genera, 410 species. Reference: Gentry (1972).

*Agave* is the largest genus. These striking rosette-leaved plants are conspicuous elements in the Sonoran landscape. More than two dozen species of *Agave* are native to Sonora, and several others are often cultivated. Two other genera, *Hesperaloe* and *Manfreda*, occur in Sonora.

#### *Yucca*

Shrubs to tree-sized plants; trunks thick and fibrous (in Sonora). Leaves thick and semifleshy, tough-fibrous, firm,

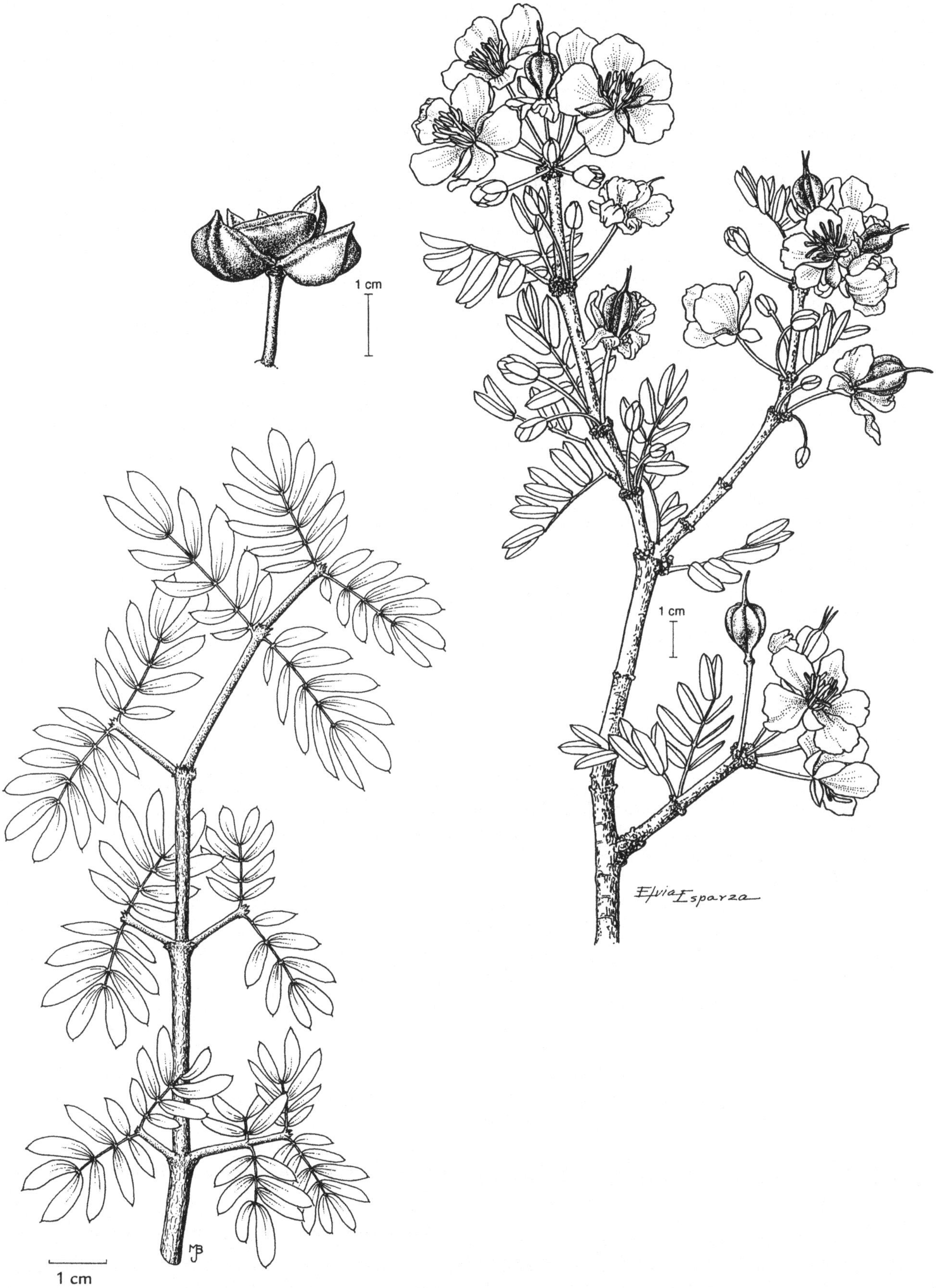


Figure 270. *Guajacum coulteri*. (Pennington and Sarukhán 1998). Drawn by E. Esparza and M. B. Johnson.

and spine-tipped, the margins entire, or minutely serrulate, and often with shredding fibers. Inflorescences large and terminal; flowers white, large, and showy, with 6 rather thick, waxy tepals. Fruits fleshy and indehiscent, or dry, dehiscent capsules. Seeds flattened, D-shaped or sometimes rounded (see *Y. grandiflora*), blackish, and numerous. North America, mostly in dry regions of Mexico and southwestern United States; 45 species. References: Clary and Simpson (1995), Matuda and Piña Lujan (1980).

Yuccas in Sonora mostly occur at elevations above the desert, thornscrub, and tropical deciduous forest. Only *Y. arizonica* occurs within the Sonora portion of the Sonoran Desert. In contrast, yuccas are widespread in the deserts of the Baja California peninsula. *Yucca whipplei*, which seems best treated as *Hesperoyucca whipplei*, occurs in extreme northwestern Sonora.

The roots of some species have been used for soap. Fruits of the fleshy-fruited species, known as banana yuccas, are edible, as are yucca flowers and seeds. *Yucca elata* has dry, dehiscent fruits; the other Sonoran yuccas have fleshy, indehiscent fruits.

Two nonarborescent yuccas occur in southeastern Sonora: *Y. madrensis*, in pine-oak forest forms a thick stem rarely to 2 m, and Gentry (1972) mentions an unidentified and probably unnamed species on the Sonora-Chihuahua border.

1. Leaves mostly 1–1.5 cm wide, relatively flexible; fruits dehiscent, dry at maturity. ______ ***Y. elata***
1' Leaves 2–6 cm wide, relatively rigid; fruits indehiscent, fleshy at maturity.
  2. Margins of young leaves with threadlike fibers. ***Y. arizonica***
  2' Leaf margins entire, not exfoliating into threads or with occasional threads.
    3. Inflorescences declined with glabrous branches. ***Y. declinata***
    3' Inflorescences erect with pubescent branches.
      4. Flowers subsessile, erect to spreading; flowering ovary 4.5–6 cm; fruits 13–25+ cm; east-central and southeastern Sonora. ***Y. grandiflora***
      4' Flowers pedicelled, nodding; flowering ovary 2.5 cm; fruits 4–12 cm; northeastern and north-central Sonora. ______ ***Y. schottii***

***Yucca arizonica***. Arizona yucca; *dátil; howij* (O'odham); *hamat* (Seri) (figures 271 and 272A)

Often forming colonies with thick trunks 1–1.5 (5+) m; branching from the base and also near the crown. Leaves 60–80 × 3–4 cm, stiff and thick, light green to often yellow-green, the margins with coarse, often brown, fibrous threads.

Inflorescences slightly shorter to about one-third longer than the leaves, and with glabrous branches. Flowers pedicelled, nodding, not opening wide, the tepals 6–7 cm; flowering ovary slender, 4–4.5 cm. Fruits green and fleshy, the flesh sweet and edible when ripe. Flowering April–May.

Hills, slopes, plains, arroyos, and canyons; Sonoran and Chihuahuan desertscrub, grassland, and lower elevations in oak woodland; northern Sonora, 305–1380 m (1000–4525 ft). Also southern Arizona. Widely scattered and often extensive populations in north-central and northeastern Sonora; mostly at the upper edge of the desert and above.

An isolated and notable population of arborescent plants sometimes reaching 5 m in height occurs in Sonoran desertscrub on granitic soils in the vicinity of Rancho el Dátil southwest of Caborca.

*Yucca arizonica* is readily distinguished from *Y. schottii* by its multiple stem habit, yellow-green rather than blue-green leaves, spring-flowering, and its lower elevation habitat.

***Yucca declinata***. Dátil.

Robust yuccas 4–6 m, resembling *Y. grandiflora* in habit. Trunk upright, thick, branching in the crown and suckering from the base. Leaves often 50–135+ × 4.5–6 cm at midpoint. Panicles 1–1.3 m, usually declined, the branches glabrous. Pedicels 10–15 mm. Tepals 4–5 cm. Fruits often 15–20 cm and fleshy.

Rocky, volcanic or limestone hills and small valleys often with clay soils. Open oak woodland with *Juniperus coahuilensis*. Central Sonora in the Bacadéhuachi drainage at Ciénaga el Palmar and the vicinity of Bacanora, 700–1250 m (2300–4100 ft).

Gentry (1972) noted that the habit of this yucca resembles that of *Y. arizonica*, *Y. grandiflora*, and *Y. schottii*, but differs from all of them in the declined inflorescence. The elongated ovary and glabrous inflorescence distinguish it from *Y. schottii* and aligns it with *Y. arizonica* and *Y. grandiflora*, but the much shorter tepals appear to separate it from those species. It differs further from *Y. grandiflora* in the long pedicels and glabrous rachis and from *Y. arizonica* in the more open inflorescence, shorter stamens, larger leaves and fruits, and taller habit. This is one of the few tree species wholly endemic to the state of Sonora. It remains unstudied since Gentry's (1972) work.

***Yucca elata***. Soap-tree yucca; *palmilla* (figure 272B)

Thick, solitary or few-branched stems 1–2 (3+) m, seldom arborescent in Sonora, clothed in persistent, dry leaves; thick taproot known to extend several meters in depth. Leaves often 40–70 × 1–1.5 cm, numerous, flexible, nar-

Figure 271. *Yucca arizonica*. Near La Unión, southwest of Caborca (Turner et al. 1995). Photograph by H. L. Shantz.

rowly linear, the margins white with very slender, threadlike, white fibers.

Flowering stalks often 2–3 m; flowers on the upper one-third, many but not crowded. Flowers pedicelled, opening wide during the night, partially closing the next day; tepals 4–7 × 2–4.5 cm; flowering pistil 2–3.2 cm; ovary 1.5–2 cm plus the style 8–12 mm. Fruits dehiscent, drying as persistent capsules. Flowers May–July; the capsules mature the same season.

Valley plains, bajadas, and arroyos; Chihuahuan desertscrub and grassland; not seen on hills or mountain slopes; northeastern Sonora, 1150–1550 m (3770–5085 ft). Also Coahuila, Chihuahua, Texas, New Mexico, and Arizona. Extending into Sonora near the Arizona border from the vicinity of Naco (approximately 10 km west of Naco junction on Mexico Highway 2) eastward, and southward to the vicinity of Fronteras and the east side of Sierra de los Ajos to about 65 km south of Agua Prieta. Characteristically grows on deep, often sandy soils.

In Arizona there is often considerable variation in flower size from plant to plant within a single population. This variation appears to be influenced by soil moisture, size and vigor of the plant, and exposure (shading). It is a desirable landscape subject that is often planted in southern Arizona and is suitable for landscaping in northern Sonora.

***Yucca grandiflora***. *Dátil*; *sahuiliqui* (Guarijío) (figure 272C)

Robust yuccas 4–6 m with a well-formed, upright trunk and branching in the crown. Leaves 70–150 × 4–7 cm, dark green, moderately thick, the margins entire to filiferous, the dry, dead leaves persisting for many years.

Inflorescences overtopping the leaves, 0.7–1 m, the lateral branches white-woolly. Flowers subsessile to short pedicelled, erect or spreading; tepals 6–9 cm, spreading open; flowering ovary elongated, 4.5–6 cm. Fruits large and fleshy, at least 13–25 cm. Seeds flattened or rounded. Flowering February–May.

Slopes and ridges; lower oaks with elements of upper desertscrub and thornscrub; east-central and southeastern Sonora, 760–1370 m (2500–4500 ft). Widely scattered pockets in mountains in Sonora from the vicinity of Nacozari southward and adjacent southwestern Chihuahua. Generally at lower elevations and ranging much farther south than *Y. schottii*.

Gentry (1972: 164) reports that the fruits are

> sweet like a banana. . . . Some of the rancheros about Bacanora expressed a liking for the sweet fruits, saying they were even better when roasted than raw. I ate

> them when hungry, but was soon satiated, as with a bar of candy. Some of these fruits reach very large size, 8 inches long and 2 inches in diameter and estimated to weigh about 2 pounds. They are quite bitter until fully ripe, when the thick flesh all about the seed core becomes soft. The seeds fill the cells irregularly, many of them undeveloped, apparently for lack of the pollen gamete. When seeds are remote among the cells, they grow very thick, being about round. Where they grow closely packed, they are strongly flattened as is typical of seeds of the Agavaceae, but large and thick like all seeds of the fleshy-fruited yuccas.

This elegant, large-flowered yucca is being cultivated in southern Arizona and is frost hardy in Tucson.

***Yucca schottii.*** Mountain yucca, Spanish bayonet (figures 272D and 273)

Trunks short, commonly solitary, or sometimes reaching 3–4.5 m and with few branches, the old, dry leaves persistent. Leaves often 50–85 × 3.5–5.5 cm, bluish-green glaucous to dull green, and relatively thin, the margins entire or with a few fibers.

Inflorescences about as long as to one-third longer than the leaves, densely flowered to near base; inflorescence branches densely pubescent with short hairs. Flowers pedicelled, nodding; tepals 3.5–3.8 cm, usually not opening wide; flowering ovary 2–3 cm. Fruits 6–12 cm, green and fleshy; pulp edible, tasting somewhat like a bland apple. Flowering June–July.

Hills, mountain slopes, and canyons; oak woodland, pine-oak woodland, pine forest, and occasional in grassland; northeastern Sonora, 1220–2135 m (4000–7000 ft), and north-central Sonora in the Sierra Aconchi. Often common and dominating the landscape, such as on rocky, north-facing, gentle, grassy slopes just below the oak zone southeast of Agua Prieta. Also southeastern Arizona, southwestern New Mexico, and western Chihuahua. It is suitable for cultivation at intermediate and higher elevations, undoubtedly tolerating considerable cold.

### Arecaceae (Palmae)—Palm Family

Worldwide, mostly in the tropics, many in subtropics, and a few at the edge of temperate zones; 191 genera, 2500 species. References: Felger and Joyal (1999), Henderson et al. (1995), Quero Rico (1989), Uhl and Dransfield (1987), Zona (2000).

The few palms in deserts around the world are usually restricted to riparian habitats. Economically important members include the coconut, date palm, oil palms, and many others.

Writing from Europe in the late eighteenth century, more than a decade after suffering through the Jesuit expulsion, Ignaz Pfefferkorn (1949: 74) stated: "Many palm trees grow in the region near the sea, especially in the vicinity of Guaymas. Since these palms neither yield fruit nor attain the height which is reached by genuine palms, they seem to be either degenerate or false members of their family. Hence, they do not merit the space necessary to describe them." We disagree and devote this space to describing the magnificent palms of Sonora.

The northernmost palms in western North America are single-trunked, fan-leaved palms. These palms have thicker trunks and tougher, thicker, and usually grayer or duller colored leaves than their southern and more tropical relatives. These xeromorphic characters seem to be adaptations to aridity and winter freezing, which seems to determine the northern limits for these plants.

The leaves of most Sonoran palms tend to be semi-persistent, often forming a protective skirt of dry leaves which ultimately falls away. The dry leaves are often purposely burned because they harbor wasp nests; the palms generally recover unless there is too much fuel buildup. Other characteristics of the native Sonoran palms include one large inflorescence in each leaf axil of usually most or all of the leaves making up one year's crown on mature trees. Each inflorescence produces large numbers of flowers and fruits, so that flowering trees are attractive, followed by abundant fruits. The flowers are bisexual and small; there are three sepals and three petals, and six stamens. The fruits are one seeded; the seeds are rounded to oval and not grooved.

The habitats where many palm groves occur are often heavily exploited for cattle grazing, settlements, and recreation, and their future is imperiled. The more accessible, magnificent palm oases in the vicinity of Bahía San Carlos have been decimated by development, clearing of undergrowth, and removal of palms for landscaping. Magnificent palm oases are thus quickly converted to the "living dead" with only old, soon to be senescent palms. In contrast, the pristine Cañón las Barajitas and other inaccessible places such as the top of Sierra Libre and the Magdalena palm canyon support groves of hundreds of thousands if not millions of native palms.

Sonoran palms have been used for thatching, weaving, building materials, and food (e.g., Felger and Joyal 1999, Felger and Moser 1985, Joyal 1996a, 1996b). *Sabal uresana*, the most widespread Sonoran palm, is an important nontimber resource and the most economically valuable of the Sonoran palms. Although other palm species and a few palmlike monocots are mentioned in the literature, *Sabal* is the most commonly and consistently cited Sonoran palm in more than three centuries of historical references (Joyal 1996a).

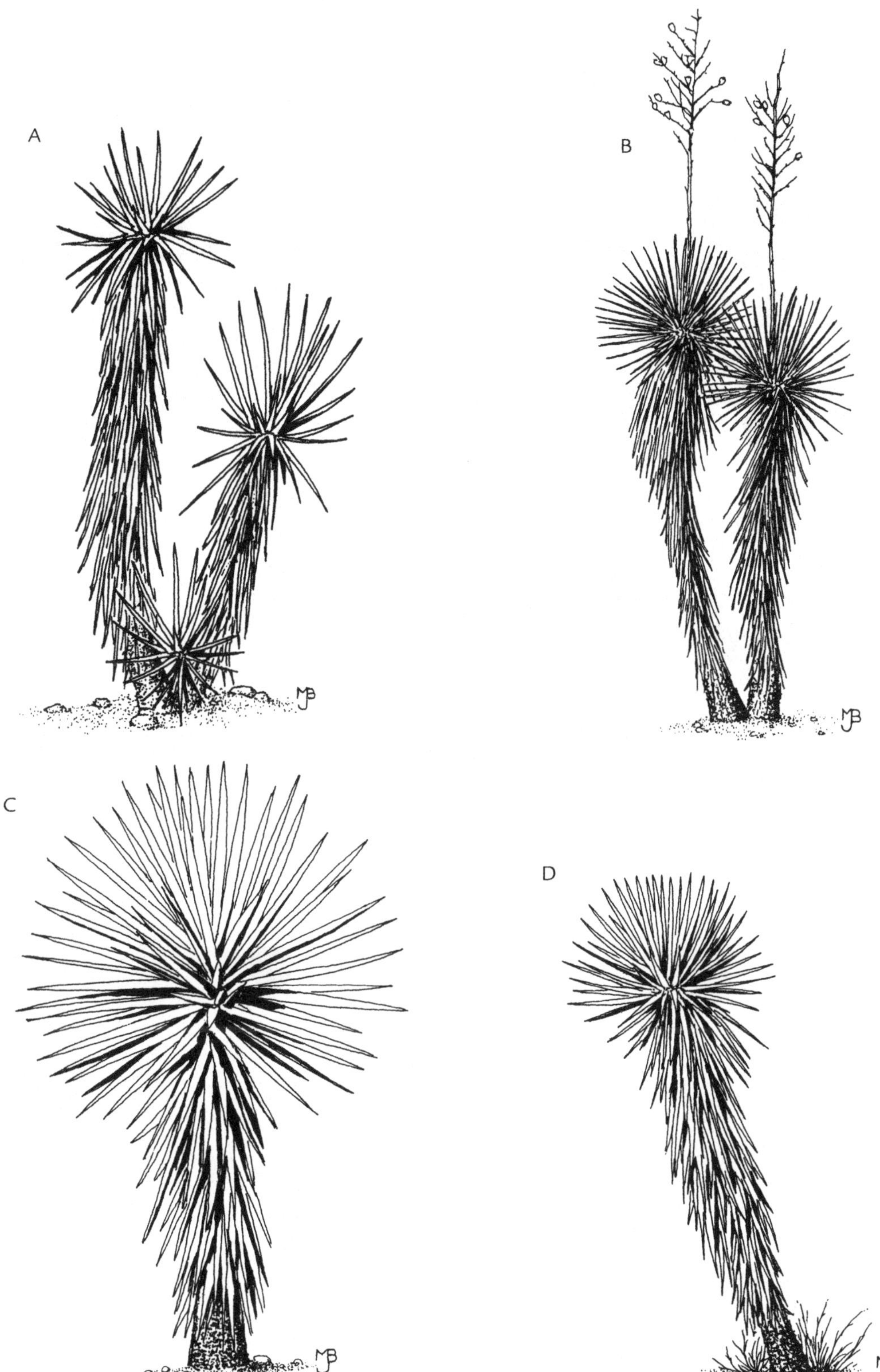


Figure 272. *Yucca*. (A) *Y. arizonica*, (B) *Y. elata*, (C) *Y. grandiflora*, (D) *Y. schottii*. Drawn by M. B. Johnson.

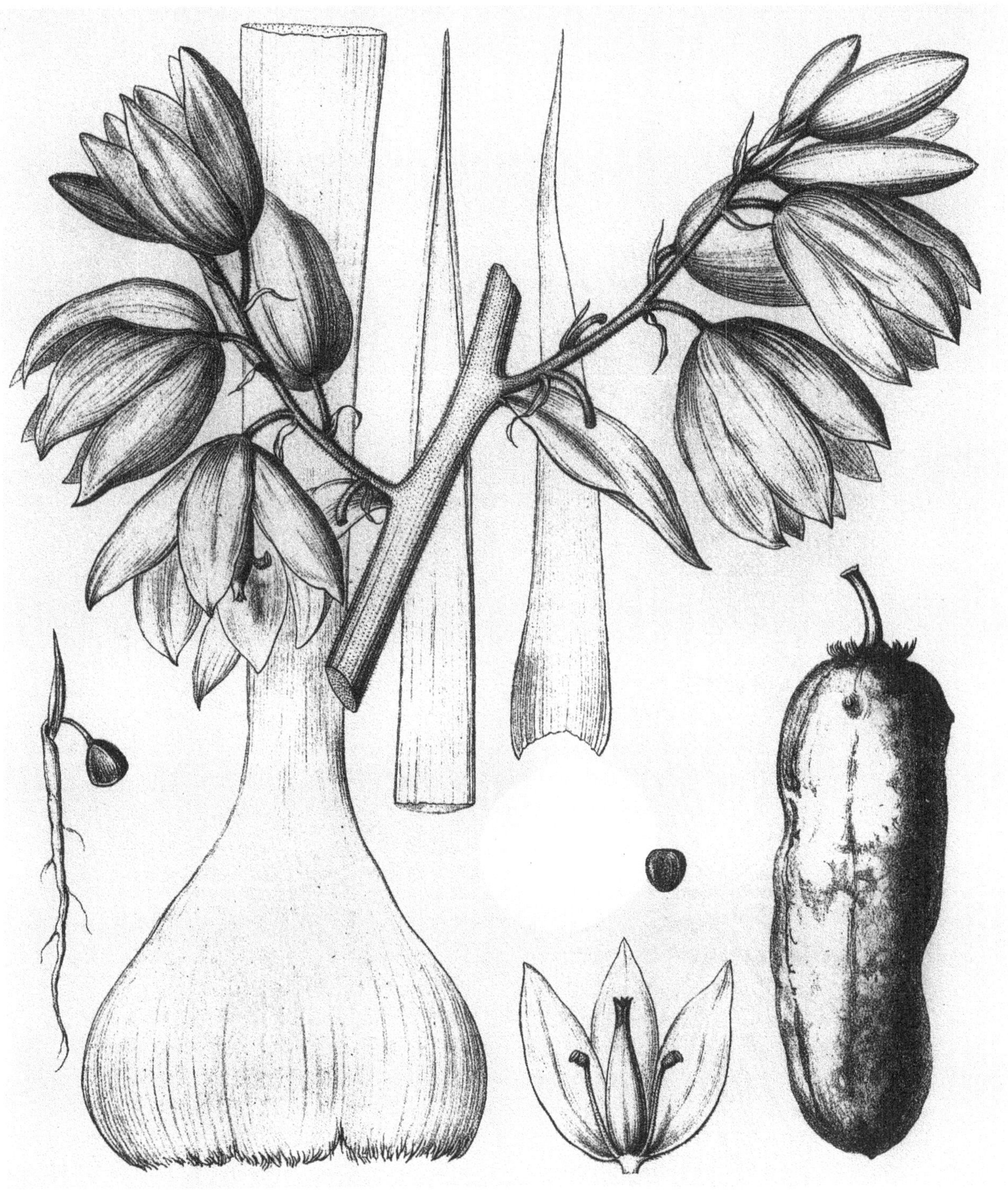

Figure 273. *Yucca schottii* (Sargent 1890). Drawn by C. E. Faxon.

The newly emerging, unexpanded palm leaf is called a *cojoyo* (this term is also used as a vulgar term for penis). *Cojoyos*, primarily of *Sabal* and sometimes of *Brahea aculeata*, are highly valued for weaving twilled and coiled baskets, mats (*petates*), and other items, and people travel considerable distances to harvest them. *Sabal* is the only Sonoran palm used in hat making. The fully expanded leaves called *pencas*, especially those of *Sabal*, are widely used for broom making. The leaves of *Sabal* and *Brahea aculeata* are extensively used for thatching. *Brahea aculeata* leaves are sometimes fashioned into saddle blankets. Strips of leaf blades, especially of *Brahea aculeata* and to a lesser degree those of *B. nitida*, are woven into a coarse cordage or rope and used for such purposes as tethering animals (Felger and Joyal 1999; Joyal 1996a).

In the desert region, especially near Ures, the trunks of *Sabal* have been used to build corrals and as crossbeams (*vigas*) for houses. Traditionally only the trunks of dead trees were used to make corrals, but in recent years living trees have been cut, contrary to the policy of local forestry offices. Farther east and southward in the mountains, palm logs are occasionally used in construction and as benches and are also hollowed out to make planters. During the nineteenth century the trunks of *Washingtonia robusta* were used in Guaymas "as rafters for houses. It is said to be a durable wood" (Rose 1899: 255).

The blackish, sugary mesocarp of the ripe fruit of *Sabal*, locally called *taco*, is eaten fresh in northeastern and east-central Sonora but is considered unpalatable in the desert area near Ures. The fruits of *Washingtonia* are sweet and edible and apparently were used for food by the Seri (Felger and Moser 1985). The fruits of certain *Brahea* were used for food in Sonora, but the information is fragmentary. For example, the fruits of *B. aculeata* are eaten by the Guarijío (Felger and Joyal 1999), and those of *B. brandegeei* are eaten in the Ures region (Joyal 1996a).

Writing in the eighteenth century, Nentvig (1977) reported the use of palm hearts (terminal buds) for food in northeastern Sonora. Gentry (1942, 1963) likewise reported that the Guarijío ate palm heart (*corazón*), apparently of *Sabal* and perhaps *Brahea aculeata*, a practice that kills the plant. Palm hearts are apparently no longer harvested in Sonora.

The Sonoran palms are well suited for horticulture in most parts of the state and other regions with similar climate. They are readily grown from fresh seeds. In addition, many other palm species from various regions of the world are cultivated as ornamentals and shade trees in Sonora. Many other palm species can be cultivated in the state if given sufficient water, especially in the nearly frost-free regions. However, yellowed, discolored leaves can result from highly alkaline or salinized irrigation water. The more commonly cultivated palms in Sonora include:

*Chamaedorea seifrizii*. Bamboo palm. The narrow-leaved form; multiple, bamboolike stems. Common landscape subject from San Luis Río Colorado and Hermosillo southward, both as a container plant and planted in the ground. Grows best in light shade or protection from the hot, afternoon sun, but also tolerates full sun. It is unusual in the genus in tolerating full desert sun.

*Chamaerops humilis*. European fan palm. Hardy state-wide.

*Cocos nucifera*. *Coco* or coconut. Successfully grown as a landscape subject along the coast from the vicinity of Guaymas and San Carlos southward and occasionally in Hermosillo. Sometimes planted on the coast as far north as Bahía Kino, but these plants often suffer from cold weather.

*Phoenix*, the date palms, including *P. canariensis* and *P. dactylifera*. Nearly statewide; also the dwarf *P. roebelini*.

*Roystonea regia*. *Palma real* or royal palm. Hermosillo southward.

*Syagrus romanzoffiana*. Queen palm. Lowlands statewide.

*Trachycarpus fortunei*. Windmill palm. Most successful in the northern part of the state, often at higher elevations.

*Washingtonia*. Both species widely cultivated nearly statewide.

1. Plants often branching from the base (suckering); leaves pinnate; dioecious; seeds more than twice as long as wide, conspicuously grooved; date palm, cultivated, rarely surviving at abandoned settlements. _________ ***Phoenix***

1' Plants solitary (not branching from the base); leaves palmate (fan palms); flowers bisexual; seeds less than twice as long as wide, not grooved.

   2. Petioles always entire; leaves prominently costapalmate (petiole extending well into blade on the lower surface); blade not flat; inflorescences shorter than or sometimes as long as the leaves; seeds broader than long, depressed globose. ___ ***Sabal***

   2' Petioles entire or armed; leaves not or only moderately costapalmate; inflorescences often longer than the leaves; leaf blades flat or nearly so; seeds rounded to longer than broad.

      3' Leaf blades green above and silvery-blue glaucous beneath, or tough and dull green or gray-green on both surfaces; sepals separate, entire; widespread. ____________________ ***Brahea***

      3. Leaf blades shiny green on both surfaces; sepals united below, the lobes with ragged margins; in Sonora native only in the Sierra el Aguaje (north of Bahía San Carlos). ___ ***Washingtonia***

## *Brahea*—Hesper Palm

Upper leaves usually persisting to form a short skirt of dead leaves, the lower leaves ultimately falling away, resulting in a bare trunk. Flowers cream colored, sessile; sepals separate, with entire margins. Northern Mexico, including both states of Baja California, to southern Mexico and two extending to Central America; 12 species. References: Bailey (1937a, 1937b, 1943).

The Baja California blue palm (*B. armata*) is sometimes cultivated in the Sonoran Desert region.

1. Petioles spiny; fruits globose, 1.5–3.0 cm in diameter, when ripe with a thin, edible, datelike, sweet, fleshy pericarp.
   2. Small to medium-sized palms; fruits 2.5–3 cm in diameter; southeastern Sonora. ____ ***B. aculeata***
   2' Medium-sized to large palms; fruits 1.5–2 cm in diameter; central and northeastern Sonora. ***B. brandegeei***

1' Petioles unarmed or with small spines near base; fruits ovoid, 8.5-10 mm, the pericarp not edible.
   3. Hastula with a woolly or feltlike patch; flowering-bearing branchlets (rachillae) densely hairy, > 2.2–3 mm in diameter. ____ ***B. dulcis***
   3' Hastula glabrous; flowering branchlets glabrous, < 1 mm in diameter. ____ ***B. nitida***

**_Brahea aculeata_** [*Erythea aculeata*]. Sonoran hesper palm; *cohuehui, palma; tajcú* (Guarijío) (figure 274)

Trunks often 2–9+ m, 12.5–25 cm in diameter. Leaves often 10–30, 95–180 cm. Petioles often about as long as the leaf blade, 40–116 × 0.7–1.9 cm just below the hastula, armed with mostly stout, often yellow, single or sometimes double marginal teeth 1–6 mm, reduced apically (toward the hastula), or sometimes the young plants and smaller-leaved adult plants with the teeth absent or greatly reduced. Leaf blades 55–90 × 60–110 cm, glaucous bluish-green, often discoloring with darker blotches upon drying.

Inflorescences about equal to or longer than the leaves, often reaching 1.7 m, arching, then drooping with weight of the fruits. Inflorescence branches and rachillae sparsely to densely tomentose; rachillae (3) 5–11 cm. Flowers white, very fragrant (at least in the late afternoon), the buds 1, 2 or 3 together, usually only 1 develops into a flower; sepals 1 mm, broadly triangular, greenish white; petals 1 mm; filaments united below into a staminal ring, the free portion short and slender. Fresh fruits 2.5–3 cm maximum diameter, nearly round, with a thin, fleshy, sweet, datelike pericarp, the surfaces smooth, yellow-brown and somewhat mottled. Seeds 1.5–2.0 cm maximum diameter, rounded or nearly so. Flowering March–May; fruits ripening late August–early September.

Valleys, hills, mountains, and canyon bottoms and slopes, the distribution decidedly patchy but the plants often locally abundant. Tropical deciduous forest and lower oak woodland and occasionally into higher elevation zones in oak woodland and lower pine-oak woodland; southeastern Sonora in the Municipio de Alamos, 320–1500 m (1050–4920 ft). Also southwestern Chihuahua, Durango, and Sinaloa.

The northern limit for this palm and its relationship with *B. brandegeei* remains unresolved. *Brahea brandegeei* (from Sonora) is a larger palm but with smaller fruits. Grown side by side in Tucson, these palms demonstrate clear differences in size and growth rates (figure 274); also see Felger and Joyal (1999).

**_Brahea brandegeei_** [*Erythea brandegeei; Brahea elegans; Erythea elegans; Brahea clara; Erythea clara*]. *Babiso* (north-central Sonora); *coguegue* (central Sonora); *palmilla* (central Sonora); *zamij cmaam* (Seri) (figures 274, 275A, and 276)

Robust palms; trunk reaching 6–16 (20) m, 17–35 cm in diameter. Leaves about 40 or fewer per tree, 1.2–2.3 m. Petioles stout, often 80–135 cm, with usually strong, straight to curved, solitary and double spines reaching 7.5–8.5 mm. Leaf blades often strongly bluish- or silvery-glaucous to dull green on both surfaces; often with much variation in leaf color (especially in the San Carlos-Ensenada Grande populations, where different trees even in the same grove may have different colored leaves).

Inflorescences often 10–12 per tree, longer than the leaves, often 2.0–3.4 m. Flowers white, in clusters of (1) 2 or 3, subtended by bractlets 1.0–1.5 mm; sepals 1 mm, cupped and broad, thin, and scarious; corollas thick, the lobes 1.6 mm, broadly deltate; filaments united below into a staminal ring, the free portions short and slender. Fruits rounded, (15.8) 18.0–19.5 mm maximum diameter. Seeds 14.5–19.0 mm maximum diameter, nearly rounded or with the maximum diameter 10–20% larger than the minimum diameter. Flowering mostly late spring; fruits generally ripening the following winter or spring.

Canyons, mountain slopes, ridges, cliffs including sea cliffs, and less common on coastal dunes and valley bottoms. Sonoran desertscrub, thornscrub, tropical deciduous forest, and lower oak zones; west-central to east-central and northeastern Sonora, near sea level–1160 m (3800 ft). Also Baja California Sur in the Sierra San Francisco, Sierra de la Giganta, and Sierra de la Laguna.

In Sonora these palms can be divided into a western and an eastern population. The western population occurs at or near the coast in three locales: (1) Large groves are found in canyons and on cliffs and steep rock slopes between Bahía San Carlos and the vicinity of Bahía San Pedro, including Cañón de Nacapule and Cañón las Barajitas, ex-

Figure 274. *Brahea aculeata* from Alamos (front), and *B. brandegeei* from Cañón de Nacapule. Twenty-eight year-old palms in Tucson. Photograph by R. Felger 1998.

tending to peak elevation of about 800 m; rarely on low beach dunes at Bahía Algodones. These palms often grow from crevices in otherwise solid rock faces including sea cliffs. (2) Another enormous population occurs in similar, yet inland habitats in the Sierra Libre, where it is especially abundant at higher elevations to the summit at about 1160 m. (3) An isolated group of apparently fewer than one dozen palms grows below north-facing cliffs on the north side of Sierra Seri below Pico Johnson. This is the most isolated and smallest palm population in Sonora. There are juvenile and adult plants. Known to us only from photographs, they look like the Sierra el Aguaje hesper palms about 185 km to the south.

The eastern population extends from the Río Yaqui drainage in the vicinity of Highway 16 northward to near Nacori Chico and Arizpe. Their distribution is patchy. The northernmost of these palms, such as those in valley bottoms near Arizpe, become strikingly large and robust and have comparatively green leaves. From a distance they resemble *Washingtonia robusta*.

Henderson et al. (1995) list *B. elegans* as a synonym of *B. brandegeei* but offer no explanation for their decision. Felger and Joyal (1999) treated them as separate species, but after examining additional specimens and field work in Baja California Sur we find that we are unable to separate them. In both Sonora and the Baja California peninsula there seem to be greater latidudinal (and in Sonora, longitidinal) differences than between the peninsula and the mainland. Hesper palms in western Sonora have been called *B. roezlii* Linden (e.g., Wiggins 1964), based on a poorly documented cultivated specimen. Bailey (1943) showed *B. roezlii* to be a synonym of *B. armata* of Baja California.

Although relatively slow growing, these palms thrive under cultivation in the Sonoran Desert and those from the Sierra el Aguaje tolerate freezing weather in Tucson without damage. It is worthy of wide cultivation in very hot climates.

The relationships with the seemingly closely related and geographically neighboring hesper palms, *B. aculeata* and *B. armata*, remain unresolved and in-depth studies are much needed. The hesper palm of western Sonora is distinguished from *B. armata* by its more slender trunks, generally greener (not as "white" or "silvery") leaves that are more variable in color, and smaller inflorescences. However, some cultivated hesper palms at the Huntington Botanical Garden in southern California, grown from seed obtained in the vicinity of Bahía San Carlos and planted in full sun in an arid microhabitat, are exceptionally robust and resemble *B. armata* in gross appearance. *Brahea armata* is endemic to Baja California and Isla Angel de la Guarda in the Gulf of California.

***Brahea dulcis.*** *Palma ceniza* (figure 277)

Small palms resembling *B. nitida*, reaching 5+ m; trunk often 10+ cm in diameter; the older leaves falling away. Leaves reaching 1.2 m. Petioles 52–80+ cm, entire or often with some broad teeth 1–1.5 mm near the base, glaucous, 14–25 mm wide at base, tapering to 8–15 mm wide just below the hastula. Leaf blades 65–85 cm, about as wide, shiny green above, glaucous-blue below; bearing deciduous, sparse, brownish woolly pubescence on segment midribs near the base of the blade on both surfaces, this tomentum and the hastula usually wearing away with age. Hastula to ± 1.5 cm, thin, brown, and scarious, with a woolly patch at its base, fraying with age.

Inflorescences arching, usually longer than the leaves; flowering branches and rachillas purple-brown, often becoming cream colored, densely woolly-tomentose; branches crowded, the branching angles acute, the rachillas catkin-like, floriferous to their bases, 4–10 cm long, 2.2–3.0 mm in diameter. Flowers greenish white, spirally arranged, not paired or clustered (more or less equally spaced from one another), their bases submerged in the woolly tomentum,

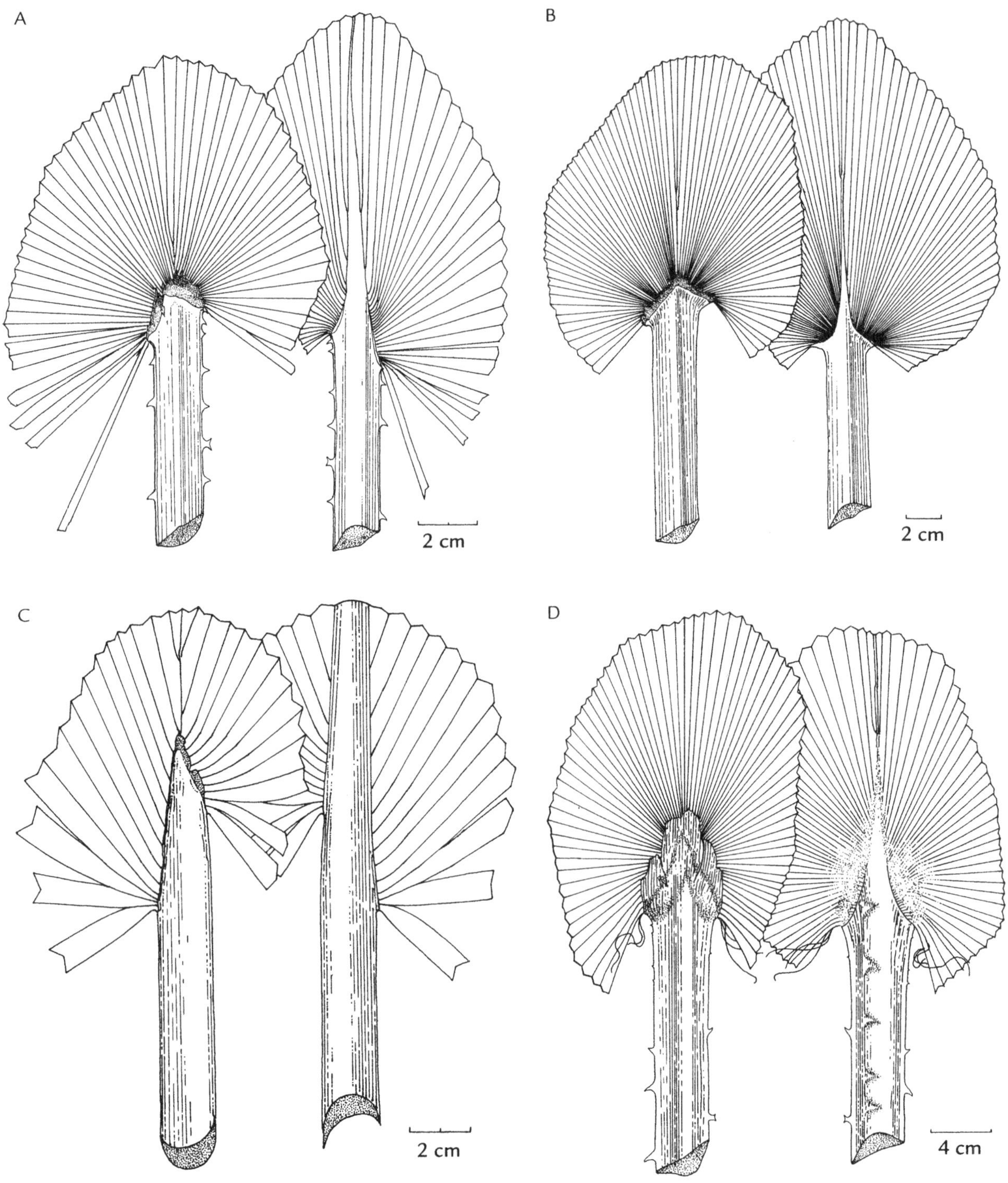


Figure 275. Arecaceae. (A) *Brahea brandegeei*, (B) *B. nitida*, (C) *Sabal uresana*, (D) *Washingtonia robusta*. Drawn by M. B. Johnson.

Figure 276. *Brahea brandegeei* and *Sabal uresana*. Cañón Las Barajitas, west-central Sonora. The tallest palm is *Sabal*, the others are *Brahea* (Felger and Joyal 1999). Photograph by R. S. Felger.

the buds longer than wide; sepals ± 1 mm, overlapping, short woolly (tomentose-canescent) on exposed surfaces, the margins scarious; petals 2.3–2.4 mm, glabrous except sometimes with few small hairs (canescent) at base, striate (longitudinal lines formed by prominent veins); filaments slender from broadly expanded, thick, deltoid bases. Young fruits densely pubescent with short hairs; mature fruits with a fleshy pericarp, smooth, green turning purple-black when ripe, rounded, 10.5–12.5 mm in diameter, glabrous or sparsely pubescent at base, wrinkling on drying, 8.5–11.5 mm in diameter. Seeds brown, 6.5–9.0 mm, broadly ovoid. Flowering (March) June–July; fruits maturing January–February.

Moist slopes, cliff tops, and riparian canyons; upper elevations of tropical deciduous forest, oak woodland, and lower pine-oak woodland; southeastern Sonora, 700–1400 m (2300–4590 ft). Southward from Sonora and northeastern Mexico to Chiapas and Central America; probably also in southwestern Chihuahua.

In Sonora known for certain from above 1100 m on Sierra de Alamos, and two remote areas farther east along the Río Cuchujaqui drainage: near Choquincahui and at Rancho Santa Bárbara. This small, stately palm often grows out of rock crevices on precipitous cliffs and canyon walls. The substrates do not seem to be limestone, unlike those of populations of this species elsewhere in Mexico. *Brahea dulcis* is quite variable across its entire range with a number of poorly known segregates, but these probably represent only local morphological variation (Henderson et al. 1995). In Sonora the stems are solitary but elsewhere in Mexico the stems are often clustered.

***Brahea nitida*** [*B. prominens*]. *Babiso; jabehui* (Río Satachic); *metajcú* (Guarijío) (figures 275B, 278, and 279)

A graceful palm with a relatively slender trunk to 10–12 m, usually much shorter, 15.5–30+ cm in diameter. Leaf blades of larger trees often 80–100 cm and often

Figure 277. *Brahea dulcis*. Rancho Santa Bárbara, Río Cuchujaqui drainage (Felger and Joyal 1999). Photograph by E. Joyal.

slightly wider than long, flat, shiny green above, bluish glaucous below, segments divided to about half way, with a slender, brown, and curling fiber in the sinus of each leaf segment; petioles entire (unarmed) or sometimes with few minute teeth near the base, bluish glaucous, about as long the blades, 1.6–4 cm wide near base, tapering to 1.1–2 cm wide near the blade.

Inflorescences arching, often reaching 2.2–2.3 m, branched 4-times; inflorescence branches and branchlets (includes rachillas) with fine, whitish hairs; rachillae floriferous nearly to the base, 2.0–5.6 cm, 0.7–0.9 mm in diameter. Flowers spirally arranged, not paired or clustered, cream white, sessile, subtended by minute, broadly deltoid bractlets with acute tips. Calyx 1.2–1.5 mm, cuplike, the sepals overlapping, separate, broadly triangular, thick toward the base, puberulent with short, thickish, white hairs. Petals 2.2–2.8 mm, with sparse hairs below like those of the calyx, closed (valvate) in bud, slightly cupped near the apex, faintly indented (grooved) to accommodate the anthers. Filaments only united basally, broadly triangular with the upper portion slender and free. Young fruits pubescent; mature fruits 11.0–13.0 mm, 8.5–9.0 mm maximum width, broadly ellipsoid, smooth and yellow-brown to purplish when fresh with a thin, fleshy pericarp, wrinkled when dry, and often with short, sparse hairs near the base, otherwise glabrous. Seeds 8.3–11.0 mm, 6.4–7.2 mm maximum width, ovoid-ellipsoid, light brown, smooth. Flowering primarily June–July; fruits maturing late July of the same season, at least among northern Sonora populations.

Sonoran desertscrub-oak woodland ecotone, oak woodland, probably in pine-oak woodland in southeastern Sonora, and tropical deciduous forest. Eastern Sonora, 860–1850 m (2720–6070 ft). Mountain regions, often on limestone or rhyolite, on steep slopes, sheer cliffs, canyon walls, and along canyon bottoms. About 25 km southeast of Magdalena de Kino and the vicinity of Nacori Chico, and widely scattered southward in mountains east of the

Figure 278. *Brahea nitida*. El Baviso canyon, southeast of Magdalena, north-central Sonora (Felger and Joyal 1999). Photograph by G. M. Ferguson.

Figure 279. *Brahea nitida*. Sierra Toro Muerto, northeastern Sonora (Felger and Joyal 1999). Photograph by E. Joyal.

Sonoran Desert to Sinaloa and southwestern Chihuahua. This species, as defined by Quero Rico (1989), ranges southward in western Mexico to Guatemala.

El Baviso, or Palm Canyon, southeast of Magdalena de Kino, on the southwest side of a limestone and rhyolite mountain, supports the northernmost, and largest and densest population of this species in Sonora. Thousands of palms are crowded in the rocky canyon bottom and below north-facing cliffs of this deep, rugged, and steep bowl-shaped box canyon. Cattle are excluded from the canyon by huge boulders and rock faces at its lower end. Smaller populations occur in other nearby canyons and mountains that are likewise inaccessible to cattle.

In riparian canyons along the Río Sátachi, near Nácori Chico, in northeastern Sonora, these palms grow alongside maples (*Acer grandidentatum*), oaks (*Quercus* spp.), and organpipe cacti (*Stenocereus thurberi*). In southeastern Sonora this palm is generally localized, often on sheer rock walls of narrow, riparian canyons in tropical deciduous forest of the Río Mayo region such as Arroyo Gochico east of San Bernardo.

This magnificent palm is worthy of extensive cultivation in hot, arid to semiarid climates. It is not as frost hardy as members of the *B. aculeata* complex, and in cultivation in Tucson it is often severely damaged by freezing weather.

We are following Quero Rico (1989) in treating *B. prominens* as a synonym of *B. nitida*. There are large distributional gaps between the Sonoran populations and those of southern Mexico. Some of these gaps, however, could be an artifact of the collection records since this palm tends to grow on cliffs in remote mountain region, often with dangerous access due to illicit agriculture.

### *Phoenix*

Native to the Old World; 12 species. Reference: Barrow (1998).

The Canary Island date palm (*P. canariensis*) is extensively cultivated in all but the higher elevations of Sonora. It is a large palm with a massive, solitary trunk, and green leaves. Several other species are sometimes cultivated in the state.

***Phoenix dactylifera.** Date palm; *dátil* (figures 5 and 281)

Large palms, branching (suckering) from the base. Leaf bases persistent on the trunk. Leaves gray-green (glaucous), pinnately compound, several meters long; leaflets induplicate (V-shaped in cross-section, a unique character among genera of pinnate-leaved palms), the basal leaflets modified as large, sharp spines.

Dioecious. Inflorescence branches orange, much shorter than the leaves. Flowers white; male flowers with short filaments. Seeds 1.5–3 cm, more than twice as long as wide, conspicuously grooved. Flowering in spring.

Sonoran desertscrub and thornscrub. Central and southwestern Sonora, near sea level to 230+ m (750+ ft).

A few date palms are weakly naturalized, or perhaps only persistent, on Isla Lobos, a coastal, sandbar island of southwestern Sonora. Elsewhere date palms occasionally persist from cultivation. In contrast, it is extensively naturalized in scattered riparian or semiriparian habitats in Baja California Sur. It is widely cultivated, sometimes commercially, in the lowlands in various parts of Sonora. The date palm is native to the Nile region and grown in many hot, dry regions of the world.

Date palms had been introduced into Baja California by the eighteenth century and were producing by the end of the century (Aschmann 1957). Propagation was by seed. The named horticultural varieties are propagated vegetatively by large offshoots, and were not brought to southwestern North America until the late nineteenth or early twentieth centuries.

### *Sabal*—Palmetto

Sonora and Chihuahua and southeastern United States to northern South America, Bermuda and the Caribbean; 15 species. References: Joyal (1996a, 1996b), Zona (1990).

***Sabal uresana.*** Sonoran palmetto; *babiso, palma real, palma verde, coguegue* (figures 275C, 276, and 280)

Large palms reaching 5–15+ m; trunks often 26–36 cm in diameter and fire scarred. Leaves very large; markedly costapalmate (the leafstalk extends well into the leaf blade on the lower or adaxial surface); petioles entire (unarmed), (0.6) 0.8–1.7 (2) m × 2.5–5 cm just below the hastula, split at base; blades often 1–1.5 m, glaucous bluish-green, paler and "windowed" along costapalmate leafstalk, and often with coarse intersegmental fibers.

Figure 280. *Sabal uresana*. Rancho Napopa, near Nacori Chico (Felger and Joyal 1999). Photograph by E. Joyal.

Inflorescences often about 2 m, spreading to arching (not drooping), much branched, shorter than to sometimes nearly as long as the leaves. Flowers white, sessile; sepals united below, the lobes entire; petals 4.4–5.3 mm. Fruits 10–23 mm in diameter, globose, brown to blackish; seeds 8–16 mm in diameter, depressed-globose, dark red-brown to blackish, glossy. Flowering May–June; fruits ripening the following spring.

Sonoran desertscrub, thornscrub, tropical deciduous forest, and lower elevations in oak woodland; west-central and eastern Sonora, near sea level–1220 m (4000 ft). Coastal dunes, low-lying habitats, and riparian canyon bottoms in the San Carlos-Barajitas region in west-central Sonora. Sandy plains, floodplains, canyon bottoms, low hills, and sometimes at springs on lower mountain slopes from Opodepe southward in central and eastern Sonora. Also southwestern Chihuahua and northernmost Sinaloa.

*Sabal uresana* is best developed in deep, sandy soils and does not grow on rocky slopes. This palm is often left standing in fields and orchards. Southward in Sinaloa it is replaced by *S. rosei*, distinguished by its more slender trunks and greener leaves. The latter is replaced southward by *S. pumos* in the Río Balsas Basin region. These three sabals constitute the phylogenetic arm of the genus in western Mexico.

Plants of the northern populations are larger than those of southeastern Sonora, and are among the largest of any species of *Sabal*. There seems to be north-south clinal variation in overall plant and trunk size as well as fruit size, with the larger-fruited populations in the north and the smallest-fruited plants in the south and along the coast in the Guaymas region. Orioles attach their nests, woven from sabal leaf fibers, to the leaf blades. In cultivation the plants are not damaged by freezing temperatures at least as far north as Tucson. This palm is occasionally cultivated in Sonora and Arizona and is highly recommended as a landscape tree for large areas. The young plants need to be planted in the ground as they do poorly in containers, probably because of their deep roots.

The fruits are eaten by many animals including coyotes, foxes, and ringtail cats. Bruchid beetle larvae in the genus *Caryobruchus* (Bruchidae: Pachymerinae) feed on *Sabal* seeds, and the adult beetles feed on the floral nectar (Zona 1990). Bruchid seed predation in *S. uresana* is as high as 50%.

## *Washingtonia*—Fan Palm

Two closely related species. *Washingtonia filifera*, native to waterholes and riparian canyons in northern Baja California, southeastern California, and southwestern Arizona, is distinguished by its much thicker trunk and dull green leaves. Both species are extensively cultivated worldwide in Mediterranean and dry subtropical regions; they thrive with irrigation in hot, arid climates. Reference: Bailey 1936.

***Washingtonia robusta*** [*W. sonorae*] Mexican fan palm; *abanico* (figures 275D and 281)

Trunks reaching 15–20 m, frequently fire-scarred. Petioles often reaching 1.4 m, split at base; juvenile to half-grown trees have petioles armed with strong, wicked, single or sometimes double, orange-brown teeth 10–15 mm along the full length; mature or taller trees have unarmed petioles reaching 1.4 m, or with few, small blunt spines (3.0–3.5 mm) at the petiole base. Hastula large, papery, more or less triangular, the margins irregular, fraying on both surfaces. Leaf blades uniformly shiny green on both surfaces, nearly flat, the free leaf segments usually drooping.

Flowering branches often reaching 2.6 m, longer than the leaves, arching, then drooping with weight of the fruits. Flowers white, short-pedicelled, 6–8 mm; sepals somewhat papery, united below into a tube, the lobes overlapping with ragged margins; petals longer than the sepals, untied below, the lobes slender, pointed, and reflexed; stamens separate, borne on the corolla tube. Fruits 7.5–9.0 mm, ellipsoid to rounded, blackish when ripe; the thin, fleshy pericarp edible and sweet like a date; fruits mostly falling together with the dry, persistent calyx and short pedicel. Seeds 4.7–6.5 × 4.5–4.9 mm, ellipsoid to globose, dark red-brown, smooth or nearly so on the dorsal side, the lower surface (hilum side) wrinkled. Flowering mostly May–June; fruits ripening in fall to early spring.

Along permanent or nearly permanent small streams in rhyolite canyons; Sonoran desertscrub. West-central Sonora, (10) 30–135 m (100–440 ft). In Sonora native only to various canyons in the Sierra el Aguaje north of Bahía San Carlos. Locally dense populations occur in the upper part of Cañón de Nacapule, several very steep, seaward-facing canyons in mountains along the coast northwest of Bahía San Carlos (e.g., La Huerta and other nearby palm canyons such as Los Anegados, and Cañón las Barajitas). Limited to immediate margins of streams or at springs with permanent water or moist soil, where it locally outnumbers *Brahea* and *Sabal*. Otherwise native only in Baja California Sur.

Its extremely limited distribution makes wild populations of this palm vulnerable to local extirpation. The fruits are eagerly eaten by many birds and various mammals including badgers and coyotes. Seeds are often found germinating in the decaying droppings of raccoons. Bailey (1936) pointed out that petioles of mature trees have few or no marginal spines. Cornett's (1986) note on *W. filifera* confirms Bailey's earlier observations and speculates that there might have been a relationship with Pleistocene mastodons which had a vertical reach of about 6 m. However, *Sabal* has entire petioles throughout its life history, and certain

Figure 281. *Washingtonia robusta*. Cultivated in plaza at Alamos; the palm with the thicker trunk is *W. filifera*, and the pinnate-leaved palm at left is *Phoenix dactylifera*. (Photographer unknown, ca. 1960).

*Brahea* species retain stout petiole spines throughout their lifespan.

This tall, stately palm is one of the most commonly cultivated palms in southern Arizona, southern California, and northwestern Mexico. It is the fastest-growing of the Sonoran palms. This species is raised in containers or field-grown and commonly transplanted even when large. Together with the date palm it is one of the most widely cultivated extra-tropical palms in the world.

## Nolinaceae—Beargrass Family

Warm and mostly semiarid or arid regions; Central America to southern United States, mostly in Mexico; 4 genera, 55 species. Perennial herbs to small trees. References: García-Mendoza and Galván (1995), Gentry (1972).

Several species of *Dasylirion* occur in Sonora; the other two genera are restricted to Mexico. These genera are allied to or sometimes included in the Dracaenaceae or both placed in the Convallariaceae (includes tropical to temperate perennial herbs). Nolinaceae have dry fruits, whereas Dracaenaceae have fleshy fruits and are widely distributed in tropical to subtropical regions worldwide.

### *Nolina*—Beargrass

Trunks thickened at the base; bark becoming corky and fissured with age. Leaves numerous, long, slender, flat, flexible, and thickish but not succulent, unarmed except for minutely serrated margins with tiny, forward-facing spines (running your fingers backward along the leaves results in razorlike cuts), the leaf tip brushlike with frayed fibers.

Dioecious (in Sonora). Inflorescences large, many flowered, compound racemose panicles, appearing in the center of the rosette and overtopping the leaves. Pedicels slender, jointed; flowers small; tepals 6. Fruit a small, 3-winged capsule with 1–3 small, rounded seeds.

Mexico and southern United States; 24 species.

*Sacahuista*, also known as *palmilla* or beargrass (*Nolina microcarpa*), is abundant at mid-elevations in north-central and northeastern Sonora; it has a thick, mostly subterranean trunk. One other species (*N. texana*) occurs in Sonora; it is similar to *N. microcarpa* but has inflorescences shorter than the leaves. The two arborescent *Nolina* species in Sonora are widely disjunct.

1. Leaves light green; capsules 8–10 mm long and about as wide; extreme northwestern Sonora. ____***N. bigelovii***

1' Leaves green to glaucous blue-green; capsules 5–6 mm and wider than long; east-central and southeastern Sonora. ____***N. matapensis***

***Nolina bigelovii*.** Desert tree-beargrass; *moho* (O'odham) (figure 282A)

Trunk 0.5–3 m, thick, and unbranched or sometimes developing a few short branches, the base thick with checkered bark. (Plants with trunks 4–4.5 m tall occur in an isolated canyon in the Tinajas Altas Mountains in southwestern Arizona within 1 km of the Sonoran border, and a dead fallen plant had a trunk slightly more than 6 m long.) Leaves usually more than 1 m × 1.5–3 cm, linear, flat, and easily bent, light green; leaf margins at first minutely serrated, the margins soon peeling away in persistent fibrous strips; old, dry leaves persistent on the trunk for a number of years, forming a palmlike, shaggy "skirt."

Flowering stalks and inflorescences 1–3 m, stout, near terminal. Flowers white; tepals 2.5–3.0 mm. Capsules 8–10 mm and about as wide, papery; seeds ovoid or oblong, 3.5–4 mm. Usually flowering May–June; fruits June.

Sonoran desertscrub in extreme northwestern Sonora, 250–700 m (820–2300 ft). Granitic ranges west of the Pinacate volcanic complex (i.e., Sierra del Rosario, Sierra del Viejo, Sierra Lechuguilla, and Cerro Pinto). Rock slopes, cliffs, and canyons on steep slopes, often growing from crevices in cliff faces, and along dry canyon bottoms. The densest stands are at higher elevations on north- and east-facing slopes, generally along the sides of very steep canyons. Also western Arizona, southeastern California, and northeastern Baja California.

The inflorescences are massive white plumes that swarm with honeybees and large spider wasps (*Hemipepsis*). The

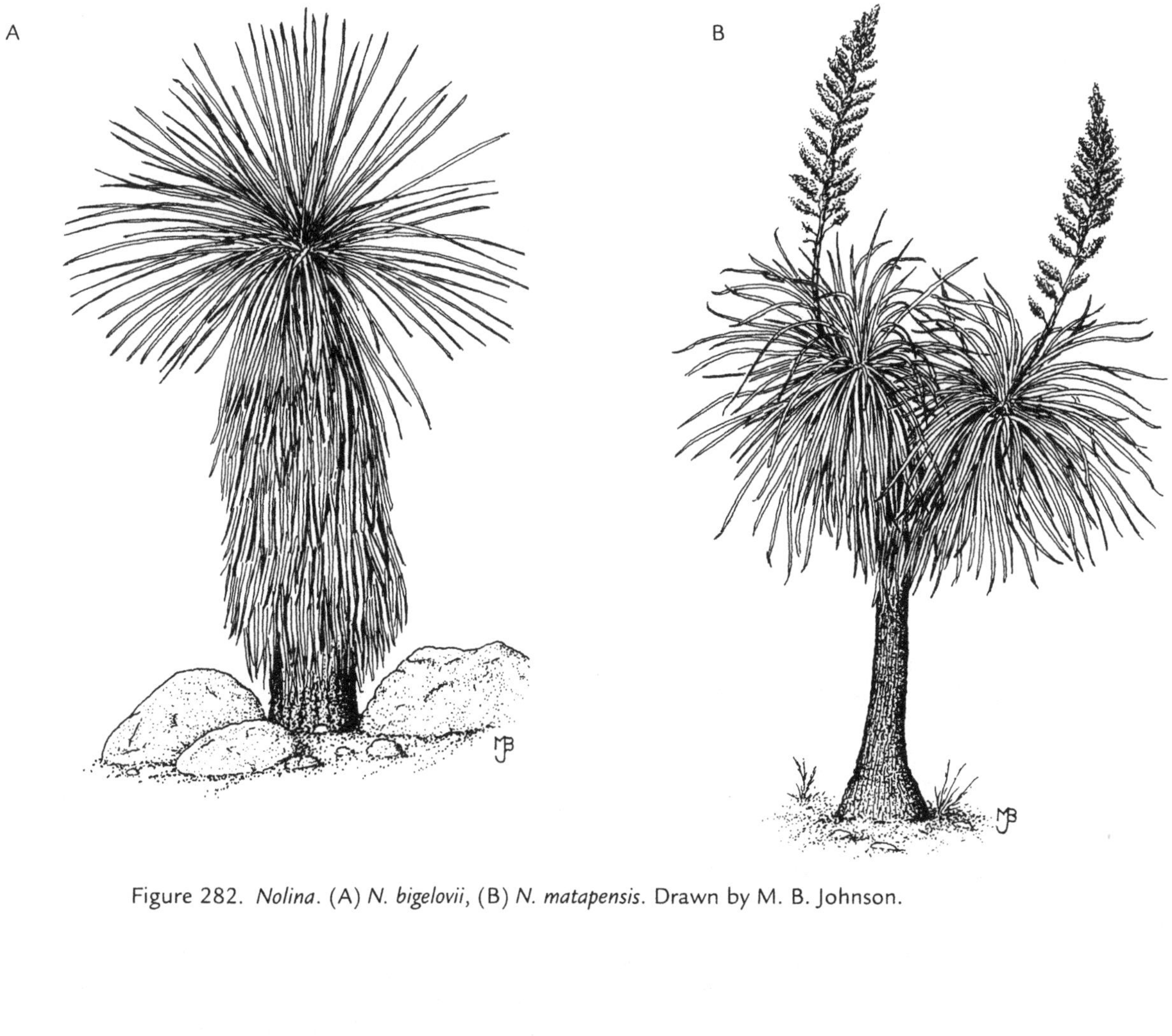


Figure 282. *Nolina*. (A) *N. bigelovii*, (B) *N. matapensis*. Drawn by M. B. Johnson.

bees and wasps cling and crawl all about the floriferous branches, even while buffeted by the hot, dry winds of early summer.

The plants are amazingly drought tolerant. In the Sierra del Rosario they continued to look healthy in April 1972, following two-and-a-half years of no measurable rainfall (Felger 1980). Flowering ceased until after the rains resumed. This nolina is readily grown from seeds, although it is slow growing. It makes a handsome drought-tolerant landscape subject. The plants thrive with additional water during the hot seasons and are frost hardy in southern and western Arizona.

***Nolina matapensis.*** Sonoran tree-beargrass; *palmilla, palmita; tueya* (Guarijío) (figure 282B)

Yuccalike plants to 7 m, often forming several branches at a height of 2–3 m; branches often becoming contorted; trunk expanded at base; bark on the lower trunk dark gray, longitudinally and deeply furrowed and checkered. Dead leaves forming a short thatch, the trunks of larger plants leafless below. Leaves 70–170 × 1–2 cm wide near base, green to glaucous green, gradually tapering to stringlike dry, dead tips, the margins minutely serrated.

Flowering stalks and inflorescences 1.5–2.8 m. Flowers dull white; tepals 2 mm. Capsules 5–6 × 8–9 mm. Seeds rounded to longer than wide, 2.5–3.8 mm. Flowering April–May; fruits September–October.

Hills, slopes, ridges, and arroyos; oak woodland, pine-oak woodland, and occasionally at the upper limits of tropical deciduous forest; eastern Sonora, 760–1830 m (2500–6000 ft). In widely scattered ranges, from near Matapé and the Sierra el Tigre to the Sierra Orejón on the Chihuahua border in the upper Río Mayo drainage. Also southwestern Chihuahua and apparently northern Sinaloa.

This graceful, slender-trunked plant is cultivated in southern Arizona. It has withstood freezing weather of approximately −12°C (10°F) near Catalina, Arizona. Although slow growing, it grows faster than *N. bigelovii*. The Guarijío made baskets from leaves of the terminal conical leaf bud. *Nolina matapensis* resembles *N. beldingii* of Baja California Sur mountains, and they are perhaps conspecific. *Nolina beldingii* is the older name.

# Literature Cited


Adams, R. P. 1993. *Juniperus*. Flora of North America 2: 412–420.

Adams, R. P. 2000. Systematics of smooth leaf margin *Juniperus* of the western hemisphere based on leaf essential oils and RAPD DNA fingerprinting. Biochemical Systematics and Ecology 28: 149–162.

Adams, R. P. and T. A. Zanoni. 1979. The distribution, synonymy, and taxonomy of three junipers of Southwestern United States and northern Mexico. Southwestern Naturalist 24: 323–329.

Allen, C. A. 1945. Studies in the Lauraceae VI. A preliminary overview of the Mexican and Central American species. Journal of the Arnold Arboretum 26: 280–434.

Argus, G. W. 1986. The genus *Salix* (Salicaceae) in the Southeastern United States. Systematic Botany Monographs 9.

Argus, G. W. 1995. Salicaceae part I. *Salix*. Journal of the Arizona Academy of Sciences 29: 39–62.

Argus, G. W. and C. L. McJannet. 1992. A taxonomic reconsideration of *Salix taxifolia* sensu lato (Salicaceae). Brittonia 44: 461–474.

Aschmann, H. 1957. The introduction of date palms into Baja California. Economic Botany 11: 174–177.

Austin, D. A. and Z. Huáman. 1996. A synopsis of *Ipomoea* (Convolvulaceae) in the Americas. Taxon 45: 3–38.

Austin, D. A. and R. Simao Bianchini. 1998. Additions and corrections in American *Ipomoea* (Convolvulaceae). Taxon 47: 833–838.

Bailey, D. K. and F. G. Hawksworth. 1979. Pinyons of the Chihuahuan Desert region. Phytologia 44: 129–133.

Bailey, L. H. 1936. *Washingtonia*. Gentes Herbarum 4: 53–82.

Bailey, L. H. 1937a. *Erythea*—the hesper palms. Gentes Herbarum 4: 85–118.

Bailey, L. H. 1937b. Notes on *Brahea*. Gentes Herbarum 4: 119–125.

Bailey, L. H. 1943. *Brahea*, and an *Erythea*. Gentes Herbarum 4: 177–197.

Bailey, V. L. 1962. Revision of the genus *Ptelea* (Rutaceae). Brittonia 14: 1–45.

Barneby, R. C. 1977. *Dalea* imagines. Memoirs of the New York Botanical Garden 27.

Barneby, R. C. 1989. Fabales. Intermountain flora, vol. 3, part B. New York Botanical Garden, Bronx, NY.

Barneby, R. C. 1991. Sensitivae Censitae. A description of the genus *Mimosa* Linnaeus (Mimosaceae) in the New World. Memoirs of the New York Botanical Garden 65.

Barneby, R. C. and J. W. Grimes. 1996. Silk tree, Guanacaste, monkey's earring: *Abarema*, *Albizia*, and allies. Memoirs of the New York Botanical Garden 74, part 1.

Barneby, R. C. and J. W. Grimes. 1997. Silk tree, Guanacaste, monkey's earring: *Pithecellobium*, *Cojoba*, and *Zygia*. Memoirs of the New York Botanical Garden 74, part 2.

Barrow, S. 1998. A monograph of *Phoenix* L. (Palmae: Coryphoideae). Kew Bulletin 53: 513–575.

Bartel, J. A. 1993. Cupressaceae, cypress family. Journal of the Arizona-Nevada Academy of Science 27: 195–200.

Bates, J. 1987. Winter ecology of the Gray Vireo (*Vireo vicinior*) in Sonora, Mexico. Masters thesis, The University of Arizona, Tucson.

Bean, L. J. and K. S. Saubel. 1972. Temalpakh. Malki Museum Press, Banning, CA.

Becerra, J. X. 1997. Insects on plants: macroevolutionary chemical trends in host use. Science 276: 253–256.

Becerra, J. X. and D. L. Venable. 1990. Rapid-terpene-bath and "squirt-gun" defense in *Bursera schlechtendalii* and the counterploy of chrysomelid beetles. Biotropica 22: 320–323.

Becerra, J. X. and D. L. Venable. 1999. Nuclear ribosomal DNA phylogeny and its implications for evolutionary trends in Mexican *Bursera* (Burseraceae). American Journal of Botany 86: 1047–1057.

Bell, W. H. and E. F. Castetter. 1937. The utilization of mesquite and screwbean by the aborigines in the American southwest. University of New Mexico Biological Series 5: 1–55.

Benítez de Rojas, C. and W. G. D'Arcy. 1998. The genera *Cestrum* and *Sessea* (Solanaceae: Cestreae) in Venezuela. Annals of the Missouri Botanical Garden 85: 273–351.

Benson, L. 1941. The mesquites and screw-beans of the United States. American Journal of Botany 28: 748–754.

Benson, L. 1969. The cacti of Arizona, 3rd ed. The University of Arizona Press, Tucson.

Benson, L. 1982. The cacti of the United States and Canada. Stanford University Press, Stanford, CA.

Benson, L. and R. A. Darrow. 1944. A manual of southwestern desert trees and shrubs. University of Arizona Bulletin Biological Sciences no. 6. Tucson, AZ.

Benson, L. and R. A. Darrow. 1981. Trees and shrubs of the Southwestern deserts, 3rd ed. The University of Arizona Press, Tucson.

Berg, C. C. 1986. The delimitation and subdivision of the genus *Maclura* (Moraceae). Proceedings of the Koninklijke Nederlandse Akademie van Wetenschappen, series C 89: 241–247.

Berg, C. C. 1988. The genera *Trophis* and *Streblus* (Moraceae) remodelled. Proceedings of the Koninklijke Nederlandse Akademie van Wetenschappen, series C 91: 345–362.

Bolli, R. 1994. Revision of the genus *Sambucus*. Dissertationes Botanicae 22. J. Cramer, Berlin.

Borhidi, A., E. Gondár, and Z. Orosz-Kovács. 1988. The reconsideration of the genus *Cordia* L. Acta Botanica Academiae Scientiarum Hungaricae 34: 375–423.

Boufford, D. E. 1997. Urticaceae. Flora of North America 3: 400–413.

Bowden, C., J. W. Dykinga, and P. S. Martin. 1993. The secret forest. University of New Mexico Press, Albuquerque.

Bowers, J. E. 1980. Catastrophic freezes in the Sonoran Desert. Desert Plants 2: 232–236.

Bravo-Hollis, H. 1978. Las cactáceas de México, 2nd ed., vol. I. Universidad Nacional Autónoma de México, México, D.F.

Britton, N. L. and J. N. Rose. 1919–1923. The Cactaceae, 4 vols. Carnegie Institute of Washington Publication 248. Washington, D.C.

Brizicky, G. K. 1962. The genera of Rutaceae in the southeastern United States. Journal of the Arnold Arboretum 43: 1–22.

Brizicky, G. K. 1963. The genera of Sapindales in the southeastern United States. Journal of the Arnold Arboretum 44: 462–501.

Brizicky, G. K. 1964a. The genera of Celastrales in the southeastern United States. Journal of the Arnold Arboretum 45: 227–234.

Brizicky, G. K. 1964b. The genera of Rhamnaceae in the southeastern United States. Journal of the Arnold Arboretum 45: 439–463.

Brown, D. E. (ed). 1982. Biotic communities of the American Southwest—United States and Mexico. Desert Plants 4: 3–341.

Brummitt, R. K. and C. E. Powell. 1992. Authors of plants names. Royal Botanic Gardens, Kew, UK.

Bruneau, A. 1996. Phylogenetic and biogeographical patterns in *Erythrina* (Leguminosae: Phaseoleae) as inferred from morphological and chloroplast DNA characters. Systematic Botany 21: 587–605.

Buchmann, S. L. 1987. The ecology of oil flowers and their bees. Annual Review of Ecology and Systematics 18: 343–369.

Bullock, A. A. 1936. Notes on the Mexican species of the genus *Bursera*. Kew Bulletin 6: 346–387.

Burkart, A. 1976. A monograph of the genus *Prosopis* (Leguminosae subfam. Mimosoideae). Journal Arnold Arboretum 57: 219–249, 450–525.

Búrquez, A., A. Martínez-Yrízar, R.S. Felger, and D. Yetman. 1999. Vegetation and habitat diversity at the southern edge of the Sonoran Desert. Pp. 36–67 in: R. H. Robichaux (ed). Ecology of Sonoran Desert plants and plant communities. The University of Arizona Press, Tucson.

Búrquez, A., A. Martínez-Yrízar, and P. S. Martin. 1992. From the high Sierra Madre to the coast: changes in vegetation along highway 16, Maycoba-Hermosillo. Pp. 239–252 in: K. F. Clark, J. Roldán-Quintana, and R. Schmidt (eds). Geology and mineral resources of northern Sierra Madre Occidental, México. El Paso Geological Society, El Paso, TX.

Butanda-Cervera, A., C. Vásquez-Yanes, and L. Trejos. 1978. La polinización Quiropterófila: Una revisión bibliográfica. Biotica 3: 29–35.

Carter, A. M. 1974. The genus *Cercidium* (Leguminosae: Caesalpinioideae) in the Sonoran Desert of Mexico and the United States. Proceedings of the California Academy of Sciences, 4th series 40: 17–57.

Carvajal, S. and R. McVaugh. 1992. *Pinus*. Flora Novo-Galiciana 17: 32–100.

Castetter, E. F. and W. H. Bell. 1937. The aboriginal utilization of the tall cacti in the American Southwest. University of New Mexico Bulletin 307.

Clarke, H. D. and D. S. Seigler. 1989. *Acacia farnesiana* (Fabaceae:Mimosoideae) and related species from Mexico, the southwestern U.S., and the Caribbean. Systematic Botany 14: 549–564.

Clary, K. H. and B. B. Simpson. 1995. Systematics and character evolution of the genus *Yucca* L. (Agavaceae): Evidence from morphology and molecular analyses. Boletín de la Sociedad Botánica de México 56: 77–88.

Cocucci, A. E. 1961. Revisión del género *Ruprechtia* (Polygonaceae). Kurtziana 1: 217–269.

Cocucci, A. E. 1997. *Ruprechtia* C.A. Meyer. Pp. 39–41 in: Flora Fanerogámica Argentina 43, Polygonaceae, part 1. PROFLORA, Conicet, Córdoba, Argentina.

Condit, I. J. 1947. The fig. Chronica Botanica, Waltham, MA.

Conn, J. S. and E. K. Snyder-Conn. 1981. The relationship of the rock outcrop microhabitat to germination, water relationships, and phenology of *Erythrina flabelliformis* (Fabaceae) in southern Arizona. Southwestern Naturalist 25: 443–451.

Cozzo, D. 1946. Relación anatómica entre la estructura del leño de las especias argentinas de *Capparis* y *Atamisquea*. Lilloa 12: 29–37.

Crisp, M. D. and J. J. Doyle (eds). 1995. Advances in legume systematics, part 7. Royal Botanic Garden, Kew, UK.

Critchfield, W. B. and E. L. Little, Jr. 1966. Geographic distribution of the pines of the world. USDA Forest Service Miscellaneous Publication 991. Government Printing Office, Washington, D.C.

Cronquist, A. 1944. Studies in the Simaroubaceae-IV. Resume of the American genera. Brittonia 5: 128–147.

Cronquist, A. 1981. A integrated system of classification of flowering plants. Columbia University Press, New York.

Cronquist, A., N. H. Holmgren, P. K. Holmgren. 1997. Subclass Rosidae (except Fabales.) Intermountain Flora, vol. 3, part A. New York Botanical Garden, Bronx, NY.

Cronquist, A., A. H. Holmgren, N. H. Holmgren, and J. L. Reveal. 1972. Vascular plants of the Intermountain West, U.S.A. Intermountain Flora, vol. 1. New York Botanical Garden, Bronx, NY.

D'Antoni, H. L. and O. T. Solbrig. 1977. Algarrobos in South American cultures: Past and present. Pp. 189–199 in: B. B. Simpson (ed). Mesquite. Dowden, Hutchinson & Ross, Stroudsburg, PA.

Debreczy, Z. and I. Rácz. 1995. New species and varieties of conifers from Mexico. Phytologia 78: 217–243.

Dimmitt, M. A. 1987. The hybrid palo verde 'Desert Museum': a new, superior tree for desert landscapes. Desert Plants 8: 99–103.

Dorado, O. 1992. A systematic and evoultionary study of the genus *Brongniartia* (Fabaceae). Ph.D. dissertation, Claremont Graduate School, Claremont, CA.

Dorado, O., G. Avila, D. M. Arias, R. Ramirez, D. Salinas, and G. Valladares. 1996. The Arbol del Tule (*Taxodium mucronatum* Ten.) is a single genetic individual. Madroño 43: 445–452.

Dorn, R. D. 1976. A synopsis of American *Salix*. Canadian Journal of Botany 54: 2769–2789.

Dorn, R. D. 1977. Willows of the Rocky Mountain States. Rhodora 79: 390–429.

Dorn, R. D. 1998. A taxonomic study of *Salix* section *Longifoliae* (Salicaceae). Brittonia 50: 193–210.

Dugand, A. 1943. Revalidación de *Bombax ceiba* L. como especie tipica del género *Bombax* L. y descripción de *Pseudobombax* gen. nov. Caldasia 6: 47–68.

Eckenwalder, J. E. 1977. North American cottonwoods (*Populus*, Salicaceae) of sections *Abaso* and *Aigeiros*. Journal of the Arnold Arboretum 58: 193–208.

Elias, T. S. 1970. The genera of Ulmaceae in the Southeastern United States. Journal of the Arnold Arboretum 51: 18–40.

Elias, T. 1976. A monograph of the genus *Hamelia* (Rubiaceae). Memoirs of the New York Botanical Garden 26: 81–144.

Ezcurra, E., M. Equihua, and J. López-Portillo. 1987. The desert vegetation of El Pinacate, Sonora, Mexico. Vegetatio 71: 49–60.

Ezcurra, E., R. S. Felger, A. Russell, and M. Equihua. 1988. Fresh water islands in a desert sand sea: the hydrology, flora, and phytogeography of the Gran Desierto oases of northwestern Mexico. Desert Plants 9: 35–44, 55–63.

Farjon, A. 1984. Pines: drawings and descriptions of the genus *Pinus*. E. J. Brill, Leiden, the Netherlands.

Farjon, A. 1993. Nomenclature of the Mexican cypress or "cedar of Goa", *Cupressus lusitanica* Mill. (Cupressaceae). Taxon 42: 81–84.

Farjon, A. and B. T. Styles. 1997. *Pinus* (Pinaceae). Flora Neotropica Monograph 75.

Felger, R. S. 1971. *Magnolia* in northwestern Mexico. Journal of the Arizona Academy of Science 6: 251–253.

Felger, R. S. 1977. Mesquite in Indian cultures of southwestern North America. Pp. 150–176 in: B. B. Simpson (ed). Mesquite. Dowden, Hutchinson & Ross, Stroudsburg, PA.

Felger, R. S. 1980. Vegetation and flora of the Gran Desierto, Sonora, Mexico. Desert Plants 2: 87–114.

Felger, R. S. 1992. Synopsis of the vascular plants of Northwestern Sonora, Mexico. Ecologica 2: 11–44 (Hermosillo).

Felger, R. S. 1998. Cupressaceae—Cypress Family, Pp. 191–192 in: P. S. Martin et al. Gentry's Rio Mayo Plants. The University of Arizona Press.

Felger, R. S. 1999. Flora of Cañón de Nacapule: a desert-bounded tropical canyon near Guaymas, Sonora, Mexico. Proceedings of the San Diego Society of Natural History no. 35.

Felger, R. S. 2000. Flora of the Gran Desierto and Río Colorado of northwestern Mexico. The University of Arizona Press, Tucson.

Felger, R. S., B. Broyles, M. Wilson, and G. P. Nabhan. 1997. The binational Sonoran Desert Biosphere Network and it plant life. Journal of the Southwest 39: 411–560.

Felger, R. S. and M. A. Dimmitt. 1998a. Bromeliaceae. Pp. 475–477 in: P. S. Martin et al. (eds). Gentry's Río Mayo Plants. The University of Arizona Press, Tucson.

Felger, R. S. and M. A. Dimmitt. 1998b. Orchidaceae. Pp. 492–498 in: P. S. Martin et al. (eds). Gentry's Río Mayo Plants. The University of Arizona Press, Tucson.

Felger, R. S. and M. B. Johnson. 1995. Trees of the Northern Sierra Madre Occidental and Sky Islands of southwestern North America. Pp. 71–83 in: L. DeBano et al. (eds). Biodiversity and management of the madrean archipelago: the Sky Islands of Southwestern United States and northern Mexico. Gen. Tech. Rep. RM-GTR-264. USDA, Forest Service, Rocky Mountain Forest Range Experiment Station, Fort Collins, CO.

Felger, R. S. and E. Joyal. 1999. The Palms (Arecaceae) of Sonora, Mexico. Aliso 18: 1–18.

Felger, R. S. and C. H. Lowe. 1967. Clinal variation in the surface-volume relationships of the columnar cactus *Lophocereus schottii* in northwestern Mexico. Ecology 48: 530–536.

Felger, R. S. and C. H. Lowe. 1976. The island and coastal vegetation and flora of the Gulf of California, Mexico. Natural History Museum of Los Angeles County, Contributions in Science 285.

Felger, R. S. and M. B. Moser. 1985. People of the desert and sea: Ethnobotany of the Seri Indians. The University of Arizona Press, Tucson.

Felger, R. S., P. L. Warren, S. A. Anderson, and G. P. Nabhan. 1992. Vascular plants of a desert oasis: Flora and ethnobotany of Quitobaquito, Organ Pipe Cactus National Monument, Arizona. Proceedings of the San Diego Society of Natural History no. 8.
Felger, R. S. and M. F. Wilson (eds). 1995. Northern Sierra Madre Occidental and its Apachian outliers: A neglected center of biodiversity. Pp. 36–59 in: L. DeBano et al. (eds). Biodiversity and management of the madrean archipelago: The Sky Islands of Southwestern United States and northern Mexico. Gen. Tech. Rep. RM-GTR-264. U.S. Department of Agriculture, Forest Service, Rocky Mountain Forest Range Experiment Station, Fort Collins, CO.
Ferguson, G. M. 1998. Pinaceae. Pp. 192–195 in: P. S. Martin et al. (eds). Gentry's Río Mayo Plants. The University of Arizona Press, Tucson.
Fernández de Oviedo y Valdés, G. 1959. Historia general y natural de las Indias. Ediciones Atlas, Madrid.
Fernández Nava, R. 1996. Flora del Bajio y de Regiones adyacentes 43. Rhamnaceae. Insitituto de Ecología, Pátzcuaro, Mexico.
Fleming, T. H. and J. L. Hamrick. 1998. Geographic variation in the breeding system and the evolutionary stability of trioecy in *Pachycereus pringlei* (Cactaceae). Evolutionary Ecology 12: 279–289.
Fleming, T. H. and J. N. Holland. 1998. The evolution of obligate pollination mutalism: Senita cactus and senita moth. Oecologia 114: 368–375.
Fleming, T. H., S. Maurice, S. L. Buchmann, and M. D. Tuttle. 1994. Reproductive biology and relative male and female fitness in trioecious cactus, *Pachycereus pringlei* (Cactaceae). American Journal of Botany 8: 858–867.
Fleming, T. H., M. D. Tuttle, and M. A. Horner. 1996. Pollination biology and relative importance of nocturnal and diurnal pollinators in three species of Sonoran Desert columnar cacti. Southwestern Naturalist 41: 257–269.
Flora of North America Editorial Committee. 1993. Flora of North America, vol. 2. Oxford University Press, New York.
Flora of North America Editorial Committee. 1997. Flora of North America, vol. 3. Oxford University Press, New York.
Friedman, S. L. 1996. Vegetation and flora of the coastal plains of the Río Mayo region, southern Sonora, México. Masters thesis, Arizona State University, Tempe.
Freytag, G. F. 1951. A revision of the genus *Guazuma*. Ceiba 1: 193–225.
Funk, V. A. 1982. The systematics of *Montanoa* (Asteraceae, Heliantheae). Memoirs of the New York Botanical Garden 36.
Furlow, J. J. 1979. The systematics of the American species of *Alnus* (Betulaceae). Rhodora 81: 1–121, 151–248.
Furlow, J. J. 1997. Betulaceae. Flora of North America 3: 507–538.
García-Mendoza, A. and R. Galván V. 1995. Riqueza de las familias Agavaceae y Nolinaceae en México. Boletín de la Sociedad Botánica de México 56: 7–24.
Gentry, A. H. 1980. Bignoniaceae, part I (Crescentieae and Tourrettieae). Flora Neotropica Monograph 25.
Gentry, A. H. 1992a. Bignoniaceae, part II (tribe Tecomeae). Flora Neotropica Monograph 25.
Gentry, A. H. 1992b. A synopsis of Bignoniaceae ethnobotany and economic botany. Annals of the Missouri Botanical Garden 79: 53–64.
Gentry, H. S. 1942. Rio Mayo plants—a study of the flora and vegetation of the valley of the Rio Mayo, Sonora. Carnegie Institution of Washington, Publication 527. Washington, D. C.
Gentry, H. S. 1946a. Sierra Tacuichamona—a Sinaloa plant locale. Bulletin of the Torrey Botanical Club 73: 356–362.
Gentry, H. S. 1946b. Notes on the vegetation of Sierra Surotato in northern Sinaloa. Bulletin of the Torrey Botanical Club 73: 451–462.
Gentry, H. S. 1948. Additions to the flora of Sinaloa and Nuevo León. Brittonia 6: 309–331.
Gentry, H. S. 1949. Land plants collected by the Vallero III, Allan Hancock Pacific Expeditions 1937–1951. Allan Hancock Pacific Expeditions 13. University of Southern California Press, Los Angeles.
Gentry, H. S. 1963. The Warihio Indians of Sonora-Chihuahua: An ethnographic survey. Anthropological Papers 65, Bureau of American Ethnology Bulletin 186. Smithsonian Institution, Washington, D.C.
Gentry, H. S. 1972. The agave family in Sonora. USDA Handbook 399. Government Printing Office, Washington, D.C.
Gentry, H. S. 1982. Sinaloan deciduous forest. Desert Plants 4: 73–77.
Gibson, A. C. 1990. The systematics and evolution of subtribe Stenocereinae, 8. Organ pipe cactus and its closest relatives. Cactus and Succulent Journal (U.S.) 6: 13–24.
Gibson, A. C. and K. E. Horak. 1978. Systematic anatomy and phylogeny of Mexican columnar cacti. Annals of the Missouri Botanical Garden 65: 999–1057.
Gibson, A. C. and P. S. Nobel. 1986. The cactus primer. Harvard University Press, Cambridge, MA.
Goldberg, D. E. 1982. The distribution of evergreen and deciduous trees relative to soil type: An example from the Sierra Madre, Mexico, and a general model. Ecology 63: 942–951.
González Villarreal, L. M. 1986. Contribución al conocimiento del genero *Quercus* (Fagaceae) en el estado de Jalisco. Colección Flora de Jalisco. Instituto de Botánica, Universidad de Guadalajara.
Graham, S. A. 1966. The genera of Araliaceae in the southeastern United States. Journal of the Arnold Arboretum 47: 126–136.
Hardin, J. W. 1976. Terminology and classification of *Quercus* trichomes. Journal of the Elisha Mitchell Society 92: 151–161.
Hardin, J. W. 1990. Variation patterns and recognition of varieties of *Tilia americana* s.l. Systematic Botany 15: 33–48.

Hastings, J. R., R. M. Turner, and D. K. Warren. 1972. An atlas of some plant distributions in the Sonoran Desert. Technical reports on the meteorology and climatology of arid regions. University of Arizona, Institute of Atmospheric Physics, Tucson.

Hawkins, J. A. 1996. Systematics of *Parkinsonia* L. and *Cercidium* Tul. (Leguminosae, Caesalpinoideae). Ph.D. thesis, University of Oxford.

Hawkins, J. A., L. White Olascoaga, C. E. Hughes, J. R. Contreras Jiménez, and P. Mercado Ruaro. 1999. Investigation and documentation of hybridization between *Parkinsonia aculeata* and *Cercidium praecox* (Leguminosae: Caesalpinoideae). Plant Systematics and Evolution 216: 49–68.

Henderson, A., G. Galeano, and R. Bernal. 1995. Field guide to the palms of the Americas. Princeton University Press, Princeton, NJ.

Henrickson, J. 1972. A taxonomic revision of the Fouquieriaceae. Aliso 7: 439–537.

Henrickson, J. 1985. A taxonomic revision of *Chilopsis* (Bignoniaceae). Aliso 11: 179–197.

Hess, W. J. and J. Henrickson. 1987. A taxonomic revison of *Vauquelinia* (Rosaceae). Sida 12: 101–163.

Hickman, J. C. (ed). 1993. The Jepson manual. University of California Press, Berkeley.

Howard, R. A. 1959. Studies in the genus *Coccoloba*, VII. Journal of the Arnold Arboretum 40: 176–220.

Hughes, C. 1998. Monograph of *Leucaena* (Leguminosae-Mimosoideae). Systematic Botany Monographs 55.

Hughes, C. E. and B. T. Styles. 1984. Exploration and seed collection of multiple-purpose dry-zone trees in Central America. International Tree Crops Journal 3: 1–32.

Humphrey, R. R. 1960. Forage production on Arizona ranges. University of Arizona Agricultural Station Bulletin no. 302.

Humphrey, R. R. 1974. The boojum and its home. The University of Arizona Press, Tucson.

Hunziker, A. T. 1984. Capparaceae. Pp. 94–95 in: L. Hauman (ed), Los géneros de fanerogamas de Argentina, Boletín de la Sociedad Argentina de Botanica, vol. 23. Sociedad Argentina de Botanica, Córdoba.

Irwin, H. S. and R. C. Barneby. 1982. The American *Cassiinae*. Memoirs of the New York Botanical Garden 35.

Isely, D. 1969. Legumes of the United States, part 1: Native *Acacia*. Sida 3: 365–386.

Isely, D. 1998. Native and naturalized Leguminosae (Fabaceae) of the United States. Monte L. Bean Life Science Museum, Brigham Young University, Provo, UT.

Isely, D. and F. J. Peabody. 1984. *Robinia* (Leguminosae: Papilionoideae). Castanea 49: 187–202.

Janzen, D. H. (ed) 1983. Costa Rican Natural History. University of Chicago Press, Chicago.

Johnson, M. B. 1988. Horticultural characteristics of seven Sonoran Desert woody legumes which show potential for southwestern landscaping. Masters thesis, The University of Arizona, Tucson.

Johnson, M. B. 1992. The genus *Bursera* (Burseraceae) in Sonora, Mexico and Arizona, U.S.A. Desert Plants 10: 126–144.

Johnston, I. M. 1924. Expedition of the California Academy of Sciences to the Gulf of California in 1921: The botany (vascular plants). Proceedings of the California Academy of Sciences, series 4, 12: 951–1218.

Johnston, M. C. 1962a. The North American mesquites *Prosopis* sect. *Algarobia* (Leguminosae). Brittonia 14: 72–90.

Johnston, M. C. 1962b. Revision of *Condalia* including *Microrhamnus* (Rhamnaceae). Brittonia 14: 332–368.

Johnston, M. C. 1963. The species of *Ziziphus* indigenous to the United States and Mexico. American Journal of Botany 50: 1020–1027.

Johnston, M. C. 1966. Systematic studies of the plant genus *Karwinskia* in Mexico and Central America. Pp. 331–357 in: Yearbook of the American Philosophical Society.

Johnston, M. C. 1971. Revision of *Colubrina* (Rhamnaceae). Brittonia 23: 2–53.

Johnston, M. C. 1990. The vascular plants of Texas, a list up-dating the manual of the vascular plants of Texas, 2nd ed. Marshall C. Johnston, Austin, TX.

Jones, C. E., L. J. Colin, T. R. Ericson, and D. K. Dorsett. 1998. Hybridization between *Cercidium floridum* and *C. microphyllum* (Fabaceae) in California. Madroño 45: 110–118.

Joyal, E. 1996a. The use of *Sabal uresana* (Arecaceae) and other palms in Sonora, Mexico. Economic Botany 50: 429–445.

Joyal, E. 1996b. The palm has its time: an ethnoecology of *Sabal uresana* in Sonora, Mexico. Economic Botany 50: 446–462.

Kaastra, R. C. 1973. Revision of *Chlorophora* (Moraceae) in America. Acta Botanica Neerlandica 21: 657–670.

Kaul, R. B. 1997. Platanaceae. Flora of North America 3: 358–361.

Kearney, T. H. and R. H. Peebles. 1960. Arizona flora, 2nd ed. with supplement by J. T. Howell and E. McClintock. University of California Press, Berkeley.

Kevan, P. G. and R. A. Bye. 1991. The natural history, sociobiology, and ethnobiology of *Eucheira socialis* Westwood (Lepidoptera: Pieridae), a unique and little-known butterfly from Mexico. The Entomologist 110: 146–165.

Klaus, W. 1987. Mediterranean pines and their history. Plant Systematics and Evolution 162: 133–163.

Kohlmann, B. and S. Sánchez-Colón. 1984. Estudio areográfico del género *Bursera* en Mexico. Cactáceas y Succulentas Mexicanas 29: 27–32.

Kopp, L. 1966. A taxonomic revision of the genus *Persea* in the western Hemisphere. Memoirs of the New York Botanical Garden 14: 1–120.

Kostermans, A. J. H. 1961. New World species of *Cinnamomum* Trew (Lauraceae). Reinwardtia 6: 17–24.

Krizman, R. D. 1972. Environment and season in a tropical deciduous forest in Northwestern Mexico. Ph.D. dissertation, The University of Arizona, Tucson.

Laferrière, J. E. 1992. Berberidaceae, barberry family [vascular plants of Arizona]. Journal of the Arizona-Nevada Academy of Science 26: 2–4.

Laferrière, J. E. 1994. Vegetation and flora of the Mountain Pima village of Nabogame, Chihuahua, Mexico. Phytologia 77: 102–140.

Landrum, L. R. 1992. Sapotaceae, sapodilla family [vascular plants of Arizona]. Journal of the Arizona-Nevada Academy of Science 26: 34–35.

Landrum, L. R. 1994. Fagaceae, oak family [vascular plants of Arizona]. Journal of the Arizona-Nevada Academy of Science 27: 203–214.

Landrum, L. R. 1995. Aceraceae, maple family [vascular plants of Arizona]. Journal of the Arizona-Nevada Academy of Science 29: 2–5.

Lang, J. M. and D. Isely. 1982. *Eysenhardtia* (Leguminosae: Papilionoideae). Iowa State Journal of Research 56: 394–417.

Lavin, M. 1988. Systematics of *Coursetia* (Leguminosae-Papilionoideae). Systematic Botany Monographs 21.

Lavin, M. and M. Sousa S. 1995. Phylogenetic systematics and biogeography of the tribe Robinieae (Leguminosae). Systematic Botany Monographs 45.

Lay, K. K. 1949. A revision of the genus *Heliocarpus*. Annals of the Missouri Botanic Gardens 36: 507–541.

Lee, Y. S., D. S. Seigler, and J. E. Ebinger. 1989. *Acacia rigidula* (Fabaceae) and related species in Mexico and Texas. Systematic Botany 14: 91–100.

Leeuwenberg, A. J. M. 1994. A revision of *Tabernaemontana*, The New World species and *Stemmadenia*. Royal Botanic Gardens, Kew, UK.

Lewis, G. P. 1998. *Caesalpinia*. A revision of the *Poinci-anella—Erythrostemon* group. Royal Botanic Gardens, Kew, UK.

Lindsay, G. 1963. The genus *Lophocereus*. Cactus and Succulent Journal (U.S.) 35: 176–192.

Little, E. L. 1950. Southwestern trees—a guide to the native species of New Mexico and Arizona. USDA Agricultural Handbook no. 9. Government Printing Office, Washington, D.C.

Little, E. L., Jr. 1952. Notes on *Fraxinus* (ash) in the United States. Journal of the Washington Academy of Sciences 42: 369–380.

Little, E. L., Jr. 1953. Check list of native and naturalized trees of the United States (including Alaska). USDA Handbook no. 41. Government Printing Office, Washington, D.C.

Little, E. L., Jr. 1970. Names of New World cypresses (*Cupressus*). Phytologia 20: 429–445.

Little, E. L., Jr. 1976. Atlas of United States trees, vol. 3. Minor western hardwoods. USDA Forest Service Miscellaneous Publication 1134. Government Printing Office, Washington, D.C.

Little, E. L., Jr. 1979. Checklist of United States Trees. USDA Agricultural Handbook no. 541. Government Printing Office, Washington, D.C.

Little, E. L. and W. B. Critchfield. 1969. Subdivisions of the genus *Pinus* (Pines). USDA Forest Service Miscellaneous Publication 1144. Government Printing Office, Washington, D.C.

Liu, T.-S. 1971. A monograph of the genus *Abies*. Department of Forestry, National Taiwan University, Taipei.

Loock, E. E. M. 1950. The pines of Mexico and British Honduras. Union of South Africa Department of Forestry Bulletin 35. Johannesburg.

Lumholtz, C. S. 1912. New trails in Mexico. Scribner, New York.

Lundell, C. L. 1939. Revision of the American Celastraceae I. *Wimmeria*, *Microtropis*, and *Zinowiewia*. Contributions from the University of Michigan Herbarium 3: 1–48.

Lundell, C. L. 1971. Myrsinaceae, in Flora of Panama. Annals of the Missouri Botanical Garden 58: 285–353.

Malusa, J. 1992. Phylogeny and biogeography of the pinyon pines (*Pinus* subsect. *Cembroides*). Systematic Botany 17: 42–66.

Marroquín, J. S. 1972. A monographic study of the genus *Berberis* in Mexico. Ph.D. dissertation, Northeastern University, Boston, MA.

Marshall, J. T., Jr. 1957. Birds of pine-oak woodland in southern Arizona and adjacent Mexico. Pacific Coast Avifauna no. 22.

Martin, P. S., D. Yetman, M. Fishbein, P. Jenkins, T. R. Van Devender, and R. K. Wilson. 1998. Gentry's Río Mayo Plants: The tropical deciduous forest & environs of northwest Mexico. The University of Arizona Press, Tucson.

Martin, S. C. 1948. Mesquite seeds remain viable after 44 years. Ecology 29: 393.

Martínez, M. 1946. Los *Juniperus* mexicanos. Anales, Instituto de Biología, México 17: 3–128.

Martínez, M. 1947. Los *Cupressus* de México. Anales, Instituto de Biología, México 18: 71–149.

Martínez, M. 1948a. Los *Abies* mexicanos. Anales, Instituto de Biología, México 19: 11–104.

Martínez, M. 1948b. Los pinos mexicanos. Ediciones Botas, México, D.F.

Martínez, M. 1963. Las pináceas Mexicanas, 3rd ed. Universidad Nacional Autónoma de México, México, D.F.

Martínez, M. 1969. Las plantas medicinales de México. Ediciones Botas, México, D.F.

Martínez-Yrizar, A. 1995. Biomass distribution and primary productivity of tropical dry forests. Pp. 326–345 in: S. H. Bullock, H. A. Mooney, and E. Medina (eds). Seasonally dry tropical forests. Cambridge University Press, Cambridge, UK.

Martínez-Yrízar, A., A. Búrquez, and M. Maass. 2000. Structure and functioning of tropical dedicíduous forests in western Mexico. Pp. 19–35 in: R. H. Robichaux and D. A. Yetman (eds). The tropical deciduous forest of Alamos: Biodiversity of a threatened ecosystem in Mexico. The University of Arizona Press, Tucson.

Matuda, E. and I. Piña Lujan. 1980. Las plantas mexicanas del género *Yucca*. FONAPAS, Gobierno del Estado de México, Toluca.

May, L. A. 1973. Resource reconnaissance of the Gran Desierto region, northwestern Sonora, Mexico. Master's thesis, The University of Arizona, Tucson.

McClaran, M. P. and T. R. Van Devender (eds). 1995. The desert grassland. The University of Arizona Press, Tucson.

McPherson, G. 1981. Studies in *Ipomoea* (Convolvulaceae) 1. The Arborescens group. Annals of the Missouri Botanical Garden 68: 527–545.

McVaugh, R. 1951. A revision of the North American black cherries (*Prunus serotina* Ehrh., and relatives). Brittonia 7: 279–315.

McVaugh, R. 1961. Euphorbiaceae Novae Novo-Galicianae. Brittonia 13: 145–205.

McVaugh, R. 1963. Tropical American Myrtaceae, II. Fieldiana: Botany 29: 393–532.

McVaugh, R. 1974. Flora Novo-Galiciana: Fagaceae, beech family. Contributions from the University of Michigan Herbarium 12, I (3): 1–93.

McVaugh, R. 1987. Leguminosae. Flora Novo-Galiciana 5.

McVaugh, R. 1992. Gymnosperms and pteridophytes. Flora Novo-Galiciana 17.

McVaugh, R. and J. Rzedowski. 1965. Synopsis of the genus *Bursera* L. in western Mexico, with notes on the material of *Bursera* collected by Sesse and Mociño. Kew Bulletin 18: 317–382.

Mearns, E. A. 1907. Mammals of the Mexican boundary of the United States. United States National Museum Bulletin 56.

Meyer, F. G. 1997. Magnoliaceae. Flora of North America 3: 3–10.

Miller, G. N. 1955. The genus *Fraxinus*, the ashes, in North America, north of Mexico. Cornell University Agricultural Experiment Station Memoir 335. Ithaca, NY.

Miller, J. S. 1985. Systematics of the genus *Cordia* (Boraginaceae) in Mexico and Central America. Ph.D. dissertation, St. Louis University, Missouri.

Mirov, N. T. 1967. The genus *Pinus*. The Ronald Press, New York.

Moldenke, H. N. 1958–1959. Materials toward a monograph of the genus *Citharexylum*. Phytologia 6: 242–256; 6: 262–320; 6: 332–368; 7: 7–48.

Mooney, H. A. and W. A. Emboden. 1968. The relationship of terpene composition, morphology, and distribution of populations of *Bursera microphylla* (Burseraceae). Brittonia 20: 44–51.

Mooney, H. A. and B. R. Stain. 1964. Bark photosynthesis in ocotillo. Madroño 17: 230–233.

Moran, R. 1962. Visitors to the flowers of *Pachycereus pringlei*. National Cactus and Succulent Journal 17: 21.

Moran, R. and R. S. Felger. 1968. *Castela polyandra*, a new species in a new section; union of *Holacantha* with *Castela* (Simaroubaceae). Transactions of the San Diego Society of Natural History 15: 31–40.

Muller, C. H. 1942. Notes on the American flora, chiefly Mexican. American Midland Naturalist 27: 470–490.

Munz, P. A. 1968. Supplement to a California flora. University of California Press, Berkeley.

Munz, P. A. 1974. A flora of southern California. University of California Press, Berkeley.

Nabhan, G. P. and J. L. Carr. 1994. Ironwood: An ecological and cultural keystone of the Sonoran Desert. Conservation International Occasional Paper no. 1. Washington, D.C.

Nabhan, G. P. and T. E. Sheridan. 1977. Living fencerows of the Rio San Miguel, Sonora, Mexico: traditional technology for floodplain management. Human Ecology 5: 97–111.

National Register of Big Trees. 1996. American forests. Washington, D.C.

Nentvig, J. 1977. El rudo ensayo. SEP Instituto Nacional de Antropología e Historia. Proyectos Especiales 58. Colección Científica, Etnología, México, D.F.

Nixon, K. C. 1993. The genus *Quercus* in Mexico. Pp. 447–458 in: T. P. Ramamoorthy, R. Bye, and A. Lot (eds). Biological diversity of México: Origins and distribution. Oxford University Press, New York.

Nixon, K. C. 1997. Fagaceae. Flora of North America 3: 436–506.

Noad, T. and A. Birnie. 1989. Trees of Kenya. T. C. Noad and A. Birnie, Nairobi.

Nobel, P. 1982. Low-temperature tolerance and cold hardening of cacti. Ecology 63: 1650–1656.

Norman, E. M. 1967. The genus *Buddleja* in North America. Gentes Herbarium 10: 47–114.

Norman, E. M. 1992. Buddlejaceae, butterfly bush family [vascular plants of Arizona]. Journal of the Arizona-Nevada Academy of Science 26: 5–6.

Palgrave, K. C. 1977. Trees of Southern Africa. C. Struik Publishers, Capetown.

Paredes Aguilar, R., T. R. Van Devender, and R. S. Felger. 2000. Cactaceas de Sonora, México: su diversidad, uso y conservación. Arizona-Sonora Desert Museum Press, Tucson.

Parker, K. F. 1958. Arizona ranch, farm, and garden weeds. Agricultural Extension Service circular 265. The University of Arizona, Tucson.

Parker, K. F. 1972. An illustrated guide to Arizona weeds. The University of Arizona Press, Tucson.

Pascarella, J. B. 1997. Breeding systems of *Ardisia* Sw. (Myrsinaceae). Brittonia 49: 45–53.

Passini, M.-F. 1994. Synonymie entre *Pinus discolor* Bailey & Hawksworth et *Pinus johannis* M.-F. Robert. Acta Botanica Gallica 141: 387–388.

Pennington, C. W. 1980. The Pima Bajo of Central Sonora, Mexico. University of Utah, Salt Lake City.

Pennington, T. D. 1981. Meliaceae. Flora Neotropica Monograph 28.

Pennington, T. D. 1990. Sapotaceae. Flora Neotropica Monograph 52.

Pennington, T. D. and J. Sarukhán. 1968. Arboles tropicales de México. Instituto Nacional de Investigaciones Forestales, Secretaría de Agricultura y Ganadería and FAO, México, D.F.

Pennington, T. D. and J. Sarukhán. 1998. Arboles tropicales de México, 2nd ed. Universidad Nacional Autónoma de México, Fondo de Cultura Económica, México, D.F.

Perry, Jr., J. P. 1991. The pines of Mexico and Central America. Timber Press, Portland, OR.

Pesman, W. M. 1962. Meet flora Mexicana. Dale S. King, Globe, AZ.

Pinkava, D. J. 1995. Cactaceae part I. The cereoid cacti. Journal of the Arizona-Nevada Academy of Science 29: 6–12.

Pinkava, D. J., L. A. McGill, T. Reeves, and M. G. McLeod. 1977. Chromosome numbers in some cacti of western

North America, part 3. Bulletin of the Torrey Botanical Club 104: 105–110.
Pinkava, D. J., M. G. McLeod, L. A. McGill, and R. C. Brown. 1973. Chromosome numbers in some cacti of western North America, part 2. Brittonia 25: 2–9.
Polhill, R. M. and P. H. Raven (eds). 1981. Advances in legume systematics, parts I and II. Royal Botanic Gardens, Kew, UK.
Poppendieck, H. H. 1981. Cochlospermaceae. Flora Neotropica Monograph 27.
Porter, D. M. 1974. The Burseraceae in North America north of Mexico. Madroño 22: 273–276.
Powell, A. M. 1988. Trees and shrubs of Trans-Pecos Texas. Big Bend Natural History Association. Big Bend National Park, TX.
Price, R. A., A. Liston, and S. H. Stauss. 1998. Phylogeny and systematics of *Pinus*. Pp. 49–68 in: D. M. Richardson (ed). Ecology and biogeography of *Pinus*. Cambridge University Press, Cambridge, UK.
Quero Rico, H. J. 1989. Flora genérica de arecaceas de México. Ph.D. dissertation, Universidad Nacional Autónoma de México, México, D.F.
Ramirez, B. W. 1969. Fig wasps: Mechanism of pollen transfer. Science 163: 580–581.
Ramirez, B. W. 1970. Host specificity of fig wasps (Agaonidae). Evolution 24: 680–691.
Ramirez, B. W. 1986. Artificial hybridization and self-fertilization in *Ficus* (Moraceae). Brenesia 25–26: 265–272.
Record, S. J. and R. W. Hess. 1943. Timbers of the New World. Yale University Press, New Haven, CT.
Rea, A. 1997. At the desert's green edge: An ethnobotany of the Gila River Pima. The University of Arizona Press, Tucson.
Reina Guerrero, A. L. 1993. Contribución a la introdución de nuevos cultivos en Sonora: Las plantas medicinales de los pimas bajos del Municipio de Yécora, Sonora. Tesis de Ingeniero, Universidad de Sonora, Hermosillo.
Richardson, D. M. (ed). 1998. Ecology and biogeography of *Pinus*. Cambridge University Press, Cambridge, UK.
Ricketson, J. M. and J. J. Pipoly III. 1997. Nomenclatural notes and synopsis of the genus *Myrsine* (Myrsinaceae) in Mesoamerica. Sida 17: 579–589.
Ridsdale, C. E. 1976. A revision of the tribe Cephalantheae (Rubiaceae). Blumea 23: 177–188.
Robertson, K. R. 1972. The Malpighiaceae in the southeastern United States. Journal of the Arnold Arboretum 53: 101–112.
Robichaux, R. H. (ed). 1999. Ecology of Sonoran Desert plants and plant communities. The University of Arizona Press, Tucson.
Robichaux, R. H. and D. A. Yetman (eds). 2000. The tropical deciduous forest of Alamos: Biodiversity of a threatened ecosystem in Mexico. The University of Arizona Press, Tucson.
Rogers, D. J. and S. G. Appan. 1973. *Manihot, Manihotoides* (Euphorbiaceae). Flora Neotropica Monograph 13.
Rollins, R. 1950. The guayule rubber plant and its relatives. Contributions Gray Herbarium 172: 1–72.
Rosatti, T. J. 1989. The genera of suborder Apocynineae (Apocynaceae and Asclepiadaceae) in the southeastern United States. Journal of the Arnold Arboretum 70: 307–401.
Rose, J. N. 1899. Notes on useful plants of Mexico. Contributions from the United States National Herbarium 5: 209–259.
Rudd, V. E. 1969. A synopsis of the genus *Piscidia* (Leguminosae). Phytologia 18: 473–499.
Rudd, V. E. and A. M. Carter. 1983. *Acacia pacensis* (Leguminosae: Mimosoideae), a new species from Baja California Sur, Mexico. Madroño 30: 176–180.
Rzedowski, J. 1978. Vegetación de México. Editorial Limusa, México, D. F.
Rzedowski, J. 1985. Análisis de la distribución geográfica de las especias Mexicanas del género *Populus* (Salicaceae). Comptes Rendus de la Société de Biogeographie 60: 141–150.
Rzedowski, J. and G. Calderón de Rzedowski. 1988. Análisis de la distribución geográfica del complejo *Prosopis* (Leguminosae, Mimosoideae) en Norteamérica. Acta Botanica Mexicana 3: 7–19.
Rzedowski, J. and H. Kruse. 1979. Algunas tendencias evolutivas en *Bursera* (Burseraceae). Taxon 28: 103–116.
Sargent, C. S. 1890–1902. The Silva of North America. Houghton, Mifflin, Boston, MA.
Schmutz, E. M, B. N. Freeman, and R. E. Reed. 1968. Livestock poisoning plants of Arizona. The University of Arizona Press, Tucson.
Schmutz, E. M. and L. B. Hamilton. 1979. Plants that poison. Northland Press, Flagstaff, AZ.
SEDESOL (Secretaría de Desarollo Social). 1994. Norma oficial Mexicana NOM-059–ECOL-1994, que determina las especies y subespecies de flora y fauna silvestres terrestres y acuáticas en peligro de extinción, amenazadas, raras y las sujetas a protección especial, y que establece especificaciones para su protección. Diario oficial de la federación TOMO CDLXXXVIII 10: 22, México, D.F.
Seigler, D. S. and J. E. Ebinger. 1988. *Acacia macracantha, A. pennatula*, and *A. cochliacantha* (Fabaceae: Mimosoideae) species complexes in Mexico. Systematic Botany 13: 7–15.
Shaw, G. R. 1909. The pines of Mexico. Publications of the Arnold Arboretum 1.
Sherbrooke, W. C. and J. C. Scheerens. 1979. Ant-visited extrafloral (calyx and foliar) nectaries and nectar sugars of *Erythrina flabelliformis* Kearney in Arizona. Annals of the Missouri Botanical Garden 66: 472–481.
Sherman-Broyles, S. L., W. T. Barker, and L. M. Schulz. 1997. Ulmaceae. Flora of North America 3: 368–369.
Shreve, F. 1951. Vegetation of the Sonoran Desert. Carnegie Institution of Washington Publication 591.
Shreve, F. and I. L. Wiggins. 1964. Flora and vegetation of the Sonoran Desert, 2 vols. Stanford University Press, Stanford, CA.
Simpson, B. B. (ed). 1977. Mesquite. Dowden, Hutchinson & Ross, Stroudsburg, PA.

Sleumer, H. 1967. Monographia Clethracearum. Botanische Jahrbücher für Systematik 87: 36–175.

Sleumer, H. O. 1980. Flacourtiaceae. Flora Neotropica 22.

Sleumer, H. O. 1984. Olacaceae. Flora Neotropica Monograph 38.

Smith, A. C. 1944. Araliaceae. North American Flora 28B: 3–41.

Sobarzo, H. 1966. Vocabulario Sonorense. Editorial Porrua, México, D.F.

Solbrig, O. T. and P. D. Cantino. 1975. Reproductive adaptations in *Prosopis* (Leguminosae: Mimosoideae). Journal of the Arnold Arboretum 56: 185–210.

Sorenson, P. D. 1995. *Arbutus*. Pp. 194–221 in: S. L. Luteyn et al. (eds). Ericaceae, part II. Flora Neotropica Monograph 66.

Spellenberg, R. 1992. A new species of black oak (*Quercus*, subg. *Erythrobalanus*, Fagaceae) from the Sierra Madre Occidental, Mexico. American Journal of Botany 79: 1200–1206.

Spellenberg, R. W. in press. Oaks of La Frontera. Appendix, in: G. L. Webster and C. J. Bahre, eds. Changing plant life of La Frontera: Observations on vegetation in the United States/Mexico Borderlands. University of New Mexico Press, Albuquerque.

Spellenberg, R. and J. R. Bacon. 1996. Taxonomy and distribution of a natural group of black oaks of Mexico (*Quercus*, section *Lobatae*, subsection *Racemiflorae*). Systematic Botany 21: 85–99.

Spellenberg, R., J. R. Bacon, and D. E. Breedlove. 1995. A new species of *Quercus* (Fagaceae, Sect. *Lobatae*, group *Racemiflorae*) from the Sierra Madre Occidental, Mexico. Madroño 42: 26–33.

Spellenberg, R., T. Lebgue, and R. Corral-Díaz. 1996. A specimen-based, annotated checklist of the vascular plants of Parque Nacional "Cascada de Basaseachi" and adjacent areas, Chihuahua, México. Listados Florísticos de México XIII. Instituto de Biología, UNAM, México, D.F.

Stahl, B. 1989. A synopsis of Central American Theophrastaceae. Nordic Journal of Botany 9: 15–30.

Standley, P. C. 1920. The North American species of *Agonandra*. Journal of the Washington Academy of Science 10: 505–508.

Standley, P. C. 1920–1926. Trees and shrubs of Mexico. Contributions from the United States National Herbarium 23: 1–1721.

Standley, P. C. and L. O. Williams. 1966. Clethraceae. Flora of Guatemala. Fieldiana: Botany 24, part 8, no. 2: 74–81.

Standley, P. C. and L. O. Williams. 1974. Bignoniaceae. Flora of Guatemala, part 10, no. 3. Fieldiana: Botany 24: 153–232.

Stead, J. W. and B. T. Styles. 1984. Studies of Central American pines: a revision of the 'pseudostrobus' group (Pinaceae). Botanical Journal of the Linnean Society 89: 249–275.

Steinmann, V. W. and R. S. Felger. 1997. The Euphorbiaceae of Sonora, Mexico. Aliso 16: 1–71.

Sykes, G. G. 1982. The naming of the boojum. Journal of Arizona History, winter 1982: 351–356.

Todzia, C. A. and J. L. Panero. 1998. A new species of *Ulmus* (Ulmaceae) from southern Mexico and a synopsis of the species in Mexico. Brittonia 50: 343–347.

Toledo, V. M. 1974. Observations on the relationships between hummingbirds and *Erythrina* species. Lloydia 37: 482–487.

Tomlinson, P. B. 1980. The biology of trees native to tropical Florida. Harvard University Printing Office, Allston, MA.

Tomlinson, P. B. 1986. The botany of mangroves. Cambridge University Press, Cambridge, UK.

Tucker, J. M. and H. S. Haskell. 1960. *Quercus dunnii* and *Q. chrysolepis*. Brittonia 12: 196–219.

Turner, R. M., T. L. Burgess, and J. E. Bowers. 1995. Sonoran Desert plants, an ecological atlas. The University of Arizona Press, Tucson.

Turner, R. M. and C. L. Busman. 1984. Vegetative key for identification of the woody legumes of the Sonoran Desert region. Desert Plants 6: 189–202.

Uhl, N. W. and J. Dransfield. 1987. Genera palmarum. A classification of palms based on the work of Harold E. Moore, Jr. Allen Press, Lawrence, KS.

Uphof, J. C. T. 1968. Dictionary of economic plants, 2nd ed., revised. J. Cramer, Lehre, Germany.

van der Werff, H. 1991. Key to New World Lauraceae. Annals of the Missouri Botanical Garden 78: 377–387.

van der Werff, H. 1997. Lauraceae. Flora of North America 3: 26–36.

Van Devender, T. R., A. C. Sanders, R. K. Wilson, and S. A. Meyer. 2000. Vegetation, flora, and seasons of the Río Cuchujaqui, a tropical deciduous forest near Alamos, Sonora. Pp. 36–101 in: R. H. Robichaux and D. A. Yetman (eds). The tropical deciduous forest of Alamos: Biodiversity of a threatened ecosystem in Mexico. The University of Arizona Press, Tucson.

Vázquez-G., J. A. 1994. *Magnolia* (Magnoliaceae) in Mexico and Central America: A synopsis. Brittonia 46: 1–23.

Watson, E. E. 1923. The genus *Heliocarpus*. Bulletin of the Torrey Botanical Club 50: 109–128.

Watson, F. D. 1993. *Taxodium*. Flora of North America 2: 403–404.

Webster, G. L. 1967. The genera of Euphorbiaceae in the southeastern United States. Journal of the Arnold Arboretum 48: 363–430.

Webster, G. L. 1975. Conspectus of a new classification of the Euphorbiaceae. Taxon 24: 593–601.

Webster, G. L. 1993. A provisional synopsis of the sections of the genus *Croton* (Euphorbiaceae). Taxon 42: 793–823.

Webster, G. L. 1994. Synopsis of the genera and suprageneric taxa of Euphorbiaceae. Annals of the Missouri Botanical Gardens 81: 33–144.

West, J. G. 1984. A revision of *Dodonaea* Miller (Sapindaceae) in Australia. Brunonia 7: 1–194.

White, S. S. 1948. The vegetation and flora of the region of the Rio de Bavispe in northeastern Sonora, Mexico. Lloydia 11: 229–302.

Whittemore, A. T. 1997. *Berberis*. Flora of North America 3: 276–286.

Whittemore, A. T. and D. E. Stone. 1997. *Juglans*. Flora of North America 3: 425–428.

Wiebes, J. T. 1986. Agaonidae (Hymenoptera Chalcidoidea) and *Ficus* (Moraceae): Fig wasps and their figs. Proceedings of the Koninklijke Nederlandse Akademie van Wetenschappen 89: 335–355.

Wiggins, I. L. 1964. Flora of the Sonoran Desert. Pp. 189–1740 in: F. Shreve and I. L. Wiggins (eds). Flora and vegetation of the Sonoran Desert, 2 vols. Stanford University Press, Stanford, CA.

Wiggins, I. L. 1980. Flora of Baja California. Stanford University Press, Stanford, CA.

Williams, J. K. 1996. The Mexican genera of the Apocynaceae (sensu A. DC.), with key and additional taxonomic notes. Sida 17: 197–213.

Wilson, K. A. 1960. The genera of Myrtaceae in the southeastern United States. Journal of the Arnold Arboretum 41: 270–278.

Wolf, C. B. and W. W. Wagener. 1948. The New World cypresses. Aliso 1: 1–444.

Woodson, R. E., Jr. 1928. Studies in the Apocynaceae, II: A revision of the genus *Stemmadenia*. Annals of the Missouri Botanical Garden 15: 341–378.

Woodson, R. E., Jr. 1938. Apocynaceae. North American Flora 29: 103–192.

Wunderlin, R. P. 1983. Revision of the arborescent Bauhinias (Fabaceae: Caesalpiniodeae: Cercideae) native to Middle America. Annals of the Missouri Botanical Garden 70: 95–127.

Wunderlin, R. P. 1997. Moraceae. Flora of North America 3: 388–399.

Yetman, D. 1998. Scattered round stones, a Mayo village in Sonora, Mexico. University of New Mexico Press, Albuquerque.

Yetman, D. in press. The Guarijios of Northwest Mexico and their ethnobotany. University of New Mexico Press, Albuquerque.

Yetman, D. and T. R. Van Devender. in press. The Mayos of northwest Mexico, history, culture, and ethnobotany. University of California Press, Berkeley.

Zanoni, T. A. 1980. Notes on *Cupressus* in Mexico. Boletín de la Sociedad Botanica de Mexico 39: 128–133.

Zanoni, T. A. and R. P. Adams. 1975. The genus *Juniperus* in Mexico and Guatemala: numerical and morphological analysis. Boletín de la Sociedad Botanica de México 35: 69–92.

Zanoni, T. A. and R. P. Adams. 1979. The genus *Juniperus* in Mexico and Guatemala: synonymy, key, and distributions of the taxa. Boletín de la Sociedad Botanica de México 38: 83–121.

Zárate P., S. 1994. Revisión del género *Leucaena* Benth. en México. Anales Instituto Biología, Universidad Nacional Autónoma de México, Botanica 51: 238–250.

Zona, S. 1990. A monograph on the genus *Sabal* (Arecaceae: Coryphoideae). Aliso 12: 583–666.

Zona, S. 2000. Arecaceae. Flora of North America 22: 95–123.

# Appendix: Habitat Distributions of the Trees of Sonora, Mexico

| | MNG | TDF | THS | SDS | CDS | GRS | OWD | POW | MCF |
|---|---|---|---|---|---|---|---|---|---|
| **CONIFERS** | | | | | | | | | |
| Cupressaceae | | | | | | | | | |
| *Cupressus arizonica* | | | | | | | OWD | POW | |
| *C. lusitanica* | | | | | | | OWD | POW | |
| *Juniperus coahuilensis* | | | | (SDS) | (CDS) | GRS | OWD | POW | |
| *J. deppeana* | | | | | | GRS | OWD | POW | |
| *J. durangensis* | | | | | | | | POW | |
| *J. scopulorum* | | | | | | | | POW | |
| *Taxodium distichum* | | TDF | | | | | | | |
| Pinaceae | | | | | | | | | |
| *Abies concolor* | | | | | | | | | MCF |
| *A. durangensis* | | | | | | | | POW | MCF |
| *Pinus arizonica* | | | | | | | | POW | MCF |
| *P. discolor* | | | | | | | OWD | POW | |
| *P. engelmannii* | | | | | | | OWD | POW | |
| *P. herrerae* | | | | | | | | POW | |
| *P. leiophylla* | | | | | | | OWD | POW | |
| *P. lumholtzii* | | | | | | | | POW | |
| *P. maximinoi* | | | | | | | | POW | |
| *P. oocarpa* | | | | | | | | POW | |
| *P. ponderosa* | | | | | | | | | MCF |
| *P. strobiformis* | | | | | | | | POW | MCF |
| *P. yecorensis* | | | | | | | OWD | POW | |
| *Pseudotsuga menziesii* | | | | | | | | POW | MCF |

Mangroves, MNG; tropical deciduous forest, TDF; thornscrub, THS; Sonoran desertscrub, SDS; Chihuahuan desertscrub, CDS; grassland, GRS; oak woodland, OWD; pine-oak woodland/pine forest, POW; mixed conifer forest, MCF. *Non-native species.

Habitats in which species infrequently occur are indicated by ( ). Totals do not include species of infrequent occurrence.

*(continued)*

| | MNG | TDF | THS | SDS | CDS | GRS | OWD | POW | MCF |
|---|---|---|---|---|---|---|---|---|---|
| **DICOTS** | | | | | | | | | |
| Aceraceae | | | | | | | | | |
| *Acer grandidentatum* | | | | | | | OWD | POW | MCF |
| *A. negundo* | | | | | | | | POW | (MCF) |
| Anacardiaceae | | | | | | | | | |
| **Spondias purpurea* | | TDF | | | | | | | |
| Apocynaceae | | | | | | | | | |
| *Plumeria rubra* | | TDF | THS | SDS | | | (OWD) | | |
| *Stemmadenia tomentosa* | | TDF | | | | | | | |
| *Vallesia glabra* | | TDF | THS | SDS | | | | | |
| *V. laciniata* | | | THS | SDS | | | | | |
| Aquifoliaceae | | | | | | | | | |
| *Ilex rubra* | | | | | | | OWD | POW | |
| *I. tolucana* | | | | | | | | POW | |
| Araliaceae | | | | | | | | | |
| *Aralia humilis* | | (TDF) | (THS) | | | | OWD | POW | |
| *Oreopanax peltatum* | | (TDF) | | | | | OWD | POW | |
| Asteraceae | | | | | | | | | |
| *Montanoa tomentosa* | | TDF | | | | | | | |
| *Parthenium tomentosum* | | TDF | THS | | | | OWD | | |
| Avicenniaceae | | | | | | | | | |
| *Avicennia germinans* | MNG | | | | | | | | |
| Berberidaceae | | | | | | | | | |
| *Berberis longipes* | | TDF | | | | | OWD | | |
| Betulaceae | | | | | | | | | |
| *Alnus oblongifolia* | | | | | | | OWD | POW | MCF |
| *Ostrya virginiana* | | | | | | | OWD | POW | |
| Bignoniaceae | | | | | | | | | |
| *Chilopsis linearis* | | | | | CDS | GRS | OWD | | |
| **Crescentia alata* | | TDF | | | | | | | |
| *Tabebuia chrysantha* | | TDF | | | | | | | |
| *T. impetiginosa* | | TDF | THS | | | | | | |
| *Tecoma stans* | | TDF | THS | SDS | CDS | | OWD | | |
| Bombacaceae | | | | | | | | | |
| *Ceiba acuminata* | | TDF | THS | | | | (OWD) | | |
| *Pseudobombax palmeri* | | TDF | | | | | (OWD) | | |

Mangroves, MNG; tropical deciduous forest, TDF; thornscrub, THS; Sonoran desertscrub, SDS; Chihuahuan desertscrub, CDS; grassland, GRS; oak woodland, OWD; pine-oak woodland/pine forest, POW; mixed conifer forest, MCF. *Non-native species.

Habitats in which species infrequently occur are indicated by ( ). Totals do not include species of infrequent occurrence.

| | MNG | TDF | THS | SDS | CDS | GRS | OWD | POW | MCF |
|---|---|---|---|---|---|---|---|---|---|
| Boraginaceae | | | | | | | | | |
| *Cordia sonorae* | | TDF | THS | SDS | | | | | |
| Buddlejaceae | | | | | | | | | |
| *Buddleja cordata* | | | | | | | OWD | POW | |
| *B. parviflora* | | | | | | | OWD | POW | |
| *B. sessiliflora* | | TDF | (THS) | (SDS) | | | OWD | | |
| Burseraceae | | | | | | | | | |
| *Bursera fagaroides* | | TDF | THS | SDS | | | (OWD) | | |
| *B. grandifolia* | | TDF | | | | | | | |
| *B. hindsiana* | | | | SDS | | | | | |
| *B. lancifolia* | | TDF | THS | | | | (OWD) | | |
| *B. laxiflora* | | TDF | THS | SDS | | | | | |
| *B. microphylla* | | | | SDS | | | | | |
| *B. penicillata* | | TDF | (THS) | | | | (OWD) | | |
| *B. simaruba* | | TDF | | | | | | | |
| *B. stenophylla* | | TDF | | | | | (OWD) | | |
| Cactaceae | | | | | | | | | |
| *Carnegiea gigantea* | | | THS | SDS | | (GRS) | (OWD) | | |
| *Lophocereus schottii* | | | (THS) | SDS | | | | | |
| **Opuntia ficus-indica* | | | (THS) | (SDS) | | | | | |
| *O. fulgida* | | | | SDS | | | | | |
| *O. thurberi* | | TDF | THS | SDS | | | | | |
| *O. versicolor* | | | THS | SDS | | | | | |
| *O. wilcoxii* | | TDF | THS | | | | | | |
| *Pachycereus pecten-aboriginum* | | TDF | THS | (SDS) | | | | | |
| *P. pringlei* | | | | SDS | | | | | |
| *Pilosocereus alensis* | | TDF | | | | | | | |
| *Stenocereus montanus* | | TDF | | | | | | | |
| *S. thurberi* | | TDF | THS | SDS | | (GRS) | (OWD) | | |
| Capparaceae | | | | | | | | | |
| *Capparis atamisquea* | | | | SDS | | | | | |
| *C. flexuosa* | | | THS | | | | | | |
| *Crateva palmeri* | | TDF | THS | | | | | | |
| *Forchhammeria watsonii* | | | THS | SDS | | | | | |
| Caprifoliaceae | | | | | | | | | |
| *Sambucus nigra* | | | | (SDS) | | GRS | OWD | POW | |
| Celastraceae | | | | | | | | | |
| *Canotia holacantha* | | | | SDS | | | | | |
| *Maytenus phyllanthoides* | MNG | | THS | SDS | | | | | |
| *Wimmeria mexicana* | | TDF | | | | | (OWD) | | |

Mangroves, MNG; tropical deciduous forest, TDF; thornscrub, THS; Sonoran desertscrub, SDS; Chihuahuan desertscrub, CDS; grassland, GRS; oak woodland, OWD; pine-oak woodland/pine forest, POW; mixed conifer forest, MCF. *Non-native species.

Habitats in which species infrequently occur are indicated by ( ). Totals do not include species of infrequent occurrence.

*(continued)*

| | MNG | TDF | THS | SDS | CDS | GRS | OWD | POW | MCF |
|---|---|---|---|---|---|---|---|---|---|
| Clethraceae | | | | | | | | | |
| *Clethra mexicana* | | | | | | | OWD | POW | |
| Cochlospermaceae | | | | | | | | | |
| *Cochlospermum vitifolium* | | TDF | | | | | | | |
| Combretaceae | | | | | | | | | |
| *Laguncularia racemosa* | MNG | | | | | | | | |
| Convolvulaceae | | | | | | | | | |
| *Ipomoea arborescens* | | TDF | THS | SDS | | | (OWD) | | |
| Cornaceae | | | | | | | | | |
| *Cornus disciflora* | | | | | | | OWD | POW | |
| Ebenaceae | | | | | | | | | |
| *Diospyros sonorae* | | TDF | THS | | | | | | |
| Ericaceae | | | | | | | | | |
| *Arbutus arizonica* | | | | | | | OWD | POW | |
| *A. xalapensis* | | | | | | | | POW | |
| Erythroxylaceae | | | | | | | | | |
| *Erythroxylum mexicanum* | | TDF | | | | | | | |
| Euphorbiaceae | | | | | | | | | |
| *Adelia virgata* | | | THS | SDS | | | | | |
| *Croton fantzianus* | | TDF | | | | | | | |
| *Drypetes gentryi* | | TDF | | | | | | | |
| **Hura polyandra* | | | THS | | | | | | |
| *Jatropha cordata* | | TDF | THS | SDS | | | (OWD) | | |
| *J. malacophylla* | | TDF | | | | | | | |
| *Manihot aesculifolia* | | TDF | | | | | | | |
| *Manihot* sp. | | TDF | | | | | | | |
| **Ricinus communis* | | TDF | THS | SDS | | | OWD | | |
| *Sebastiania cornuta* | | TDF | | | | | | | |
| *S. pavoniana* | | TDF | | | | | | | |
| Fabaceae | | | | | | | | | |
| Caesalpinioideae: | | | | | | | | | |
| *Bauhinia pringlei* | | TDF | | | | | | | |
| *Caesalpinia caladenia* | | TDF | THS | | | | | | |
| *C. palmeri* | | (TDF) | THS | SDS | | | | | |
| *C. platyloba* | | TDF | THS | | | | | | |
| *C. sclerocarpa* | | | THS | | | | | | |
| *Conzattia multiflora* | | TDF | | | | | | | |
| *Haematoxylum brasiletto* | | TDF | THS | (SDS) | | | | | |
| **Parkinsonia aculeata* | | TDF | THS | SDS | | GRS | (OWD) | | |
| *P. florida* | | | THS | SDS | | (GRS) | | | |

Mangroves, MNG; tropical deciduous forest, TDF; thornscrub, THS; Sonoran desertscrub, SDS; Chihuahuan desertscrub, CDS; grassland, GRS; oak woodland, OWD; pine-oak woodland/pine forest, POW; mixed conifer forest, MCF. *Non-native species.

Habitats in which species infrequently occur are indicated by ( ). Totals do not include species of infrequent occurrence.

| | MNG | TDF | THS | SDS | CDS | GRS | OWD | POW | MCF |
|---|---|---|---|---|---|---|---|---|---|
| *P. microphylla* | | | | SDS | | | | | |
| *P. praecox* | | TDF | THS | SDS | | (GRS) | | | |
| *P.* × *sonorae* | | | | SDS | | | | | |
| *Senna atomaria* | | TDF | THS | | | | | | |
| Mimosoideae: | | | | | | | | | |
| *Acacia cochliacantha* | | TDF | THS | SDS | | (GRS) | | | |
| *A. constricta* | | | THS | SDS | CDS | GRS | | | |
| *A. coulteri* | | TDF | THS | | | | | | |
| *A. farnesiana* | | TDF | THS | SDS | | GRS | (OWD) | (POW) | |
| *A. greggii* | | | | SDS | CDS | GRS | | | |
| *A. occidentalis* | | TDF | THS | SDS | | GRS | | | |
| *A.* aff. *pacensis* | | | THS | | | | | | |
| *A. pennatula* | | (TDF) | (THS) | | | | OWD | (POW) | |
| *A. pringlei* | | | THS | SDS | | | | | |
| *A. willardiana* | | | THS | SDS | | | | | |
| *Albizia sinaloënsis* | | TDF | THS | | | | | | |
| *Chloroleucon mangense* | | TDF | (THS) | | | | | | |
| *Havardia mexicana* | | TDF | THS | SDS | | | | | |
| *H. sonorae* | | (TDF) | THS | SDS | | | | | |
| *Leucaena involucrata* | | | THS | | | | | | |
| *L. lanceolata* | | TDF | THS | | | | | | |
| **L. leucocephala* | | TDF | THS | SDS | | | | | |
| *Lysiloma candidum* | | | | SDS | | | | | |
| *L. divaricatum* | | TDF | THS | SDS | | | | | |
| *L. watsonii* | | TDF | THS | (SDS) | | (GRS) | OWD | | |
| *Mimosa palmeri* | | TDF | THS | | | | | | |
| **Pithecellobium dulce* | | TDF | THS | | | | | | |
| *Prosopis articulata* | | | THS | SDS | | | | | |
| *P. glandulosa* | | TDF | THS | SDS | CDS | GRS | | | |
| *P. pubescens* | | | | SDS | | | | | |
| *P. velutina* | | | THS | SDS | | GRS | (OWD) | | |
| Papilionoideae: | | | | | | | | | |
| *Brongniartia alamosana* | | TDF | THS | | | | | | |
| *Coursetia glandulosa* | | TDF | THS | SDS | | (GRS) | | | |
| *Diphysa occidentalis* | | TDF | THS | SDS | | | | | |
| *D. suberosa* | | TDF | THS | (SDS) | | | (OWD) | | |
| *Erythrina flabelliformis* | | TDF | THS | SDS | | GRS | OWD | | |
| *Eysenhardtia orthocarpa* | | TDF | THS | SDS | | GRS | OWD | | |
| *Lonchocarpus hermannii* | | TDF | THS | | | | | | |
| *Olneya tesota* | | | THS | SDS | | | | | |
| *Piscidia mollis* | | TDF | THS | | | | | | |
| *Platymiscium trifoliolatum* | | TDF | | | | | | | |
| *Psorothamnus spinosus* | | | | SDS | | | | | |
| *Robinia neomexicana* | | | | | | | OWD | POW | |
| Fagaceae | | | | | | | | | |
| *Quercus albocincta* | | | | | | | OWD | POW | |
| *Q. arizonica* | | | | | | GRS | OWD | POW | |
| *Q. chihuahuensis* | | (TDF) | (THS) | (SDS) | | GRS | OWD | POW | |
| *Q. chrysolepis* | | | | | | | OWD | POW | |
| *Q. coccolobifolia* | | | | | | | | POW | |

Mangroves, MNG; tropical deciduous forest, TDF; thornscrub, THS; Sonoran desertscrub, SDS; Chihuahuan desertscrub, CDS; grassland, GRS; oak woodland, OWD; pine-oak woodland/pine forest, POW; mixed conifer forest, MCF. *Non-native species.

Habitats in which species infrequently occur are indicated by ( ). Totals do not include species of infrequent occurrence.

*(continued)*

| | MNG | TDF | THS | SDS | CDS | GRS | OWD | POW | MCF |
|---|---|---|---|---|---|---|---|---|---|
| *Q. durifolia* | | | | | | | | POW | |
| *Q. emoryi* | | | | | | GRS | OWD | POW | |
| *Q. gambelii* | | | | | | | | POW | MCF |
| *Q. grisea* | | | | | | GRS | (OWD) | | |
| *Q. hypoleucoides* | | | | | | | (OWD) | POW | |
| *Q. mcvaughii* | | | | | | | OWD | POW | |
| *Q. oblongifolia* | | | | | | GRS | OWD | | |
| *Q. perpallida* | | | | | | | OWD | POW | |
| *Q. rugosa* | | | | | | | | POW | |
| *Q. scytophylla* | | | | | | | OWD | POW | |
| *Q. sideroxyla* | | | | | | | | POW | |
| *Q. subspathulata* | | | | | | | | POW | |
| *Q. tarahumara* | | | | | | | | POW | |
| *Q. toumeyi* | | | | | | GRS | OWD | POW | |
| *Q. tuberculata* | | (TDF) | | | | GRS | OWD | | |
| *Q. viminea* | | | | | | | OWD | POW | |
| Flacourtiaceae | | | | | | | | | |
| *Xylosma flexuosum* | | (TDF) | | | | | OWD | POW | |
| *Xylosma* sp. | | TDF | THS | | | | | | |
| Fouquieriaceae | | | | | | | | | |
| *Fouquieria columnaris* | | | | SDS | | | | | |
| *F. diguetii* | | | | SDS | | | | | |
| *F. macdougalii* | | TDF | THS | SDS | | | | | |
| *F. splendens* | | | THS | SDS | CDS | GRS | OWD | (POW) | |
| Juglandaceae | | | | | | | | | |
| *Juglans major* | | (TDF) | (THS) | (SDS) | | GRS | OWD | POW | MCF |
| Koeberliniaceae | | | | | | | | | |
| *Koeberlinia spinosa* | | | (THS) | SDS | CDS | | | | |
| Lauraceae | | | | | | | | | |
| *Cinnamomum hartmanii* | | TDF | | | | | OWD | | |
| *Persea podadenia* | | | | | | | OWD | POW | |
| Magnoliaceae | | | | | | | | | |
| *Magnolia pacifica* | | | | | | | OWD | POW | |
| Malpighiaceae | | | | | | | | | |
| *Bunchosia sonorensis* | | TDF | THS | | | | | | |
| *Malpighia emarginata* | | TDF | THS | | | | | | |
| Meliaceae | | | | | | | | | |
| *Cedrela odorata* | | TDF | | | | | | | |
| **Melia azedarach* | | TDF | | | | | (OWD) | | |
| *Trichilia americana* | | TDF | | | | | | | |
| *T. hirta* | | TDF | | | | | OWD | | |

Mangroves, MNG; tropical deciduous forest, TDF; thornscrub, THS; Sonoran desertscrub, SDS; Chihuahuan desertscrub, CDS; grassland, GRS; oak woodland, OWD; pine-oak woodland/pine forest, POW; mixed conifer forest, MCF. *Non-native species.

Habitats in which species infrequently occur are indicated by ( ). Totals do not include species of infrequent occurrence.

| | MNG | TDF | THS | SDS | CDS | GRS | OWD | POW | MCF |
|---|---|---|---|---|---|---|---|---|---|
| Moraceae | | | | | | | | | |
| *Ficus cotinifolia* | | TDF | | | | | | | |
| *F. insipida* | | TDF | THS | (SDS) | | | | | |
| *F. maxima* | | TDF | | | | | | | |
| *F. pertusa* | | TDF | THS | (SDS) | | | | | |
| *F. petiolaris* subsp. *palmeri* | | | | SDS | | | | | |
| *F. petiolaris* subsp. *petiolaris* | | TDF | THS | SDS | | | (OWD) | (POW) | |
| *F. trigonata* | | TDF | | | | | | | |
| *Maclura tinctoria* | | TDF | | | | | | | |
| *Morus microphylla* | | (TDF) | | SDS | CDS | GRS | OWD | POW | |
| *Trophis racemosa* | | TDF | | | | | (OWD) | | |
| Myrsinaceae | | | | | | | | | |
| *Ardisia revoluta* | | TDF | | | | | | | |
| *Myrsine coriacea* | | | | | | | OWD | POW | |
| Myrtaceae | | | | | | | | | |
| **Psidium guajava* | | TDF | | | | | OWD | | |
| *Psidium sartorianum* | | TDF | | | | | | | |
| Nyctaginaceae | | | | | | | | | |
| *Pisonia capitata* | | TDF | THS | (SDS) | | | | | |
| Olacaceae | | | | | | | | | |
| *Schoepfia shreberi* | | TDF | | | | | | | |
| *S. shreveana* | | | THS | SDS | | | | | |
| Oleaceae | | | | | | | | | |
| *Fraxinus gooddingii* | | | | SDS | | GRS | OWD | | |
| *F. papillosa* | | | | | | GRS | OWD | POW | MCF |
| *F. velutina* | | | | SDS | CDS | GRS | OWD | POW | |
| Opiliaceae | | | | | | | | | |
| *Agonandra racemosa* | | TDF | THS | | | | | | |
| Platanaceae | | | | | | | | | |
| *Platanus wrightii* | | | | (SDS) | CDS | GRS | OWD | POW | |
| Polygonaceae | | | | | | | | | |
| *Coccoloba goldmannii* | | TDF | THS | SDS | | | | | |
| *Ruprechtia fusca* | | | THS | | | | | | |
| Rhamnaceae | | | | | | | | | |
| *Colubrina triflora* | | TDF | | | | | | | |
| *Condalia globosa* | | | THS | SDS | | | | | |
| *Karwinskia humboldtiana* | | TDF | THS | SDS | | | | | |
| *Rhamnus scopulorum* | | | | | | | OWD | POW | MCF |
| *Ziziphus amole* | | TDF | THS | SDS | | | | | |

Mangroves, MNG; tropical deciduous forest, TDF; thornscrub, THS; Sonoran desertscrub, SDS; Chihuahuan desertscrub, CDS; grassland, GRS; oak woodland, OWD; pine-oak woodland/pine forest, POW; mixed conifer forest, MCF. *Non-native species.

Habitats in which species infrequently occur are indicated by ( ). Totals do not include species of infrequent occurrence.

*(continued)*

| | MNG | TDF | THS | SDS | CDS | GRS | OWD | POW | MCF |
|---|---|---|---|---|---|---|---|---|---|
| Rhizophoraceae | | | | | | | | | |
| *Rhizophora mangle* | MNG | | | | | | | | |
| Rosaceae | | | | | | | | | |
| *Prunus gentryi* | | | | | | | | POW | |
| *P. serotina* subsp. *capuli* | | | | | | | OWD | POW | |
| *P. serotina* subsp. *virens* | | | | | | | OWD | POW | |
| *P. zinggii* | | TDF | | | | | OWD | | |
| *Vauquelinia californica* subsp. *pauciflora* | | | | (SDS) | | | (OWD) | | |
| *V. californica* subsp. *sonorensis* | | | | SDS | | | | | |
| Rubiaceae | | | | | | | | | |
| *Cephalanthus salicifolius* | | TDF | THS | | | | | | |
| *Hamelia xorullensis* | | TDF | | | | | | | |
| *Hintonia latiflora* | | TDF | THS | SDS | | | (OWD) | | |
| *Randia echinocarpa* | | TDF | THS | | | | | | |
| *R. laevigata* | | TDF | THS | | | | (OWD) | | |
| *R. obcordata* | | TDF | THS | SDS | | | | | |
| Rutaceae | | | | | | | | | |
| *Amyris balsamifera* | | | THS | SDS | | | | | |
| **Casimiroa edulis* | | TDF | | | | | | | |
| *Esenbeckia hartmanii* | | TDF | THS | SDS | | | | | |
| *Ptelea trifoliata* | | | | | | GRS | OWD | POW | |
| *Zanthoxylum fagara* | | TDF | THS | SDS | | | | | |
| Salicaceae | | | | | | | | | |
| *Populus angustifolia* | | | | | | | | POW | |
| *P. brandegeei* | | (TDF) | (THS) | (SDS) | | | OWD | | |
| *P. fremontii* | | | THS | SDS | | GRS | OWD | | |
| *P. mexicana* | | TDF | THS | | | | | | |
| *P. tremuloides* | | | | | | | | POW | MCF |
| *Salix bonplandiana* | | TDF | THS | SDS | | GRS | (OWD) | | |
| *S. exilifolia* | | | | | (CDS) | GRS | OWD | | |
| *S. gooddingii* | | TDF | THS | SDS | CDS | GRS | OWD | POW | |
| Sapindaceae | | | | | | | | | |
| *Dodonaea viscosa* | | TDF | THS | SDS | | | OWD | | |
| *Sapindus drummondii* | | | | SDS | CDS | GRS | OWD | | |
| *S. saponaria* | | TDF | THS | SDS | | | | | |
| *Thouinia acuminata* | | TDF | | | | | | | |
| *T. villosa* | | TDF | | | | | OWD | POW | |
| Sapotaceae | | | | | | | | | |
| *Sideroxylon lanuginosum* | | | | | | GRS | OWD | | |
| *S. occidentale* | | | THS | SDS | | GRS | | | |
| *S. persimile* | | TDF | | | | | OWD | | |
| *S. tepicense* | | TDF | | | | | | | |

Mangroves, MNG; tropical deciduous forest, TDF; thornscrub, THS; Sonoran desertscrub, SDS; Chihuahuan desertscrub, CDS; grassland, GRS; oak woodland, OWD; pine-oak woodland/pine forest, POW; mixed conifer forest, MCF. *Non-native species.

Habitats in which species infrequently occur are indicated by ( ). Totals do not include species of infrequent occurrence.

| | MNG | TDF | THS | SDS | CDS | GRS | OWD | POW | MCF |
|---|---|---|---|---|---|---|---|---|---|
| Simaroubaceae | | | | | | | | | |
| *Alvaradoa amorphoides* | | TDF | THS | | | GRS | | | |
| *Castela emoryi* | | | | SDS | | | | | |
| Solanaceae | | | | | | | | | |
| *Cestrum tomentosum* | | TDF | | | | | OWD | POW | |
| **Nicotiana glauca* | | TDF | THS | SDS | CDS | GRS | (OWD) | | |
| *Solanum erianthum* | | TDF | THS | | | | | | |
| Sterculiaceae | | | | | | | | | |
| *Guazuma ulmifolia* | | TDF | THS | SDS | | | | | |
| Tamaricaceae | | | | | | | | | |
| **Tamarix aphylla* | | | THS | SDS | | | | | |
| **T. ramosissima* | | | THS | SDS | (CDS) | GRS | | | |
| Theophrastaceae | | | | | | | | | |
| *Jacquinia macrocarpa* | | TDF | THS | SDS | | | | | |
| Tiliaceae | | | | | | | | | |
| *Heliocarpus attenuatus* | | TDF | THS | SDS | | | (OWD) | | |
| *H. palmeri* | | TDF | | | | | (OWD) | | |
| *Tilia americana* | | | | | | | | POW | |
| Ulmaceae | | | | | | | | | |
| *Aphananthe monoica* | | TDF | | | | | (OWD) | | |
| *Celtis iguanea* | | TDF | THS | | | | | | |
| *C. pallida* | | TDF | THS | SDS | CDS | | | | |
| *C. reticulata* | | TDF | THS | SDS | CDS | GRS | OWD | | |
| **Ulmus pumila* | | | | | CDS | GRS | OWD | | |
| Urticaceae | | | | | | | | | |
| *Urera* cf. *caracasana* | | TDF | | | | | OWD | | |
| Verbenaceae | | | | | | | | | |
| *Citharexylum scabrum* | | TDF | THS | | | | | | |
| *Lippia umbellata* | | TDF | | | | | OWD | | |
| *Vitex mollis* | | TDF | THS | (SDS) | | | (OWD) | | |
| *V. pyramidata* | | TDF | | | | | OWD | | |
| Zygophyllaceae | | | | | | | | | |
| *Guajacum coulteri* | | TDF | THS | SDS | | | | | |
| **MONOCOTS** | | | | | | | | | |
| Agavaceae | | | | | | | | | |
| *Yucca arizonica* | | | | SDS | CDS | GRS | OWD | | |
| *Y. declinata* | | | | | | | OWD | | |

Mangroves, MNG; tropical deciduous forest, TDF; thornscrub, THS; Sonoran desertscrub, SDS; Chihuahuan desertscrub, CDS; grassland, GRS; oak woodland, OWD; pine-oak woodland/pine forest, POW; mixed conifer forest, MCF. *Non-native species.

Habitats in which species infrequently occur are indicated by ( ). Totals do not include species of infrequent occurrence.

(*continued*)

| | MNG | TDF | THS | SDS | CDS | GRS | OWD | POW | MCF |
|---|---|---|---|---|---|---|---|---|---|
| *Y. elata* | | | | | CDS | GRS | | | |
| *Y. grandiflora* | | | | | | | OWD | | |
| *Y. schottii* | | | | | | (GRS) | OWD | POW | |
| | | | | | | | | | |
| Arecaceae | | | | | | | | | |
| *Brahea aculeata* | | TDF | | | | | OWD | | |
| *B. brandegeei* | | TDF | THS | SDS | | | OWD | | |
| *B. dulcis* | | TDF | | | | | OWD | POW | |
| *B. nitida* | | TDF | | SDS | | | OWD | | |
| **Phoenix dactylifera* | | | THS | SDS | | | | | |
| *Sabal uresana* | | TDF | THS | SDS | | | OWD | | |
| *Washingtonia robusta* | | | | SDS | | | | | |
| | | | | | | | | | |
| Nolinaceae | | | | | | | | | |
| *Nolina bigelovii* | | | | SDS | | | | | |
| *N. matapensis* | | (TDF) | | | | | OWD | POW | |
| Number of species | 4 | 144 | 114 | 97 | 18 | 42 | 87 | 75 | 13 |
| Percentage of tree flora | 1 | 51 | 40 | 34 | 6 | 15 | 31 | 26 | 5 |

Mangroves, MNG; tropical deciduous forest, TDF; thornscrub, THS; Sonoran desertscrub, SDS; Chihuahuan desertscrub, CDS; grassland, GRS; oak woodland, OWD; pine-oak woodland/pine forest, POW; mixed conifer forest, MCF. *Non-native species.

Habitats in which species infrequently occur are indicated by ( ). Totals do not include species of infrequent occurrence.

# Index

Scientific names of trees recognized and treated in this book are shown in **boldface**; other scientific names including synonyms are italicized. Authors of scientific names follow the recommendations of Brummitt & Powell (1992) but are not abbreviated. Page numbers in **bold** indicate treatment of taxa (e.g. family, genus, and species), those in *italics* indicate illustration.

The manufacturer's authorised representative in the EU for product
safety is Oxford University Press España S.A. of El Parque Empresarial
San Fernando de Henares, Avenida de Castilla, 2 - 28830 Madrid
(www.oup.es/en or product.safety@oup.com). OUP España S.A. also acts
as importer into Spain of products made by the manufacturer.
Printed and bound by CPI Group (UK) Ltd, Croydon, CR0 4YY
16/09/2026
02217860-0001